STABILITY DESIGN OF SEMI-RIGID FRAMES

STABILITY DESIGN OF SEMI-RIGID FRAMES

W. F. CHEN
School of Civil Engineering
Purdue University

YOSHIAKI GOTO
Department of Civil Engineering
Nagoya Institute of Technology

J. Y. RICHARD LIEW
Department of Civil Engineering
National University of Singapore

A Wiley-Interscience Publication
JOHN WILEY & SONS, INC.
New York • Chichester • Brisbane • Toronto • Singapore

This text is printed on acid-free paper.

REQUIREMENTS:
An IBM® PC/AT 486 or higher computer or compatible computer with 4MB RAM, 11 MB hard disk space, 3.5" high-density floppy disk drive, PC DOS or MS-DOS version 3.1 or later, Microsoft Windows 3.1 and a printer (optional)

Library of Congress Cataloging-in-Publication Data:

Chen, Wai-Fah, 1936–
Stability design of semi-rigid frames / W.F. Chen, Yoshiaki Goto, J.Y. Richard Liew.
p. cm.
Includes index.
ISBN 0-471-07670-8 (alk. paper)
1. Building, Iron and steel. 2. Structural frames. 3. Structural design. 4. Structural stability. I. Goto, Yoshiaki. II. Liew, J. Y. Richard. III. Title.
TA684.C437 1996
624.1'773—dc20 95-10783
CIP

Printed in the United States of America

10 9 8 7 6 5 4 3 2 1

PREFACE

The steel framework is one of the most commonly used structural systems in modern construction. The analysis of such structural systems is governed by the assumptions adopted in the modeling of their structural elements, especially those concerning the behavior of beam-to-beam column connections. Conventional methods of steel frame analysis use two idealized connection models: the rigid-joint model and the pinned-joint model. Because the actual behavior of frame connections always falls in between these two extremes, much attention has been focused in recent years toward a more accurate modeling of such connections. Recent research on structural connections has resulted in considerable progress and understanding of the subject that prompts changes in design provisions.

The 1986 AISC LRFD specification identifies Type PR (Partially Restrained) construction as one of the two basic types of construction. In the PR type of construction, the structural connections are assumed to offer some restraint to the members they connect. When the flexibility of connections is considered in the design of frames, the LRFD specification requires an explicit evaluation of the influence of the flexibility of connections on frame stability by rational analytical or empirical means. For the FR type of construction (Fully Restrained), the specification provides a simplified second-order analysis method with B_1 and B_2 amplification factors. In contrast to this, the design specification provides for Type PR construction only broad principles for analysis and design. It is left to the engineer to implement these principles in a quantitative manner.

The aim of this book is to provide practical methods for the stability design of semi-rigid frames. To this end, we present three types of analytical means for the evaluation of the influence of connection flexibility on frame stability in PR type of construction, ranging from a simplified member-by-member design approach to more sophisticated computer-based advanced analysis/ design approach.

The first one is a versatile computer-based analysis method in conjunction with the use of an extensive database of connections. This analysis method takes into account both the material nonlinearity of connections and the geometrical nonlinearity of frames and can be used accurately to evaluate the behavior of Type PR construction in a direct manner. In this method, the required flexural strength in the design interaction equations can be obtained directly without employing amplification factors such as B_1 and B_2 in the AISC LRFD specification. For the convenience of users, a computer program based on this method is developed and provided in the book.

The other method is a simplified version of the preceding procedure in which the philosophy of the present B_1 and B_2 factor method is again replaced by a direct second-order elastic-analysis program, but the behavior of beam-to-column connections is now replaced by simple analytical models.

The third method is the computer-based analysis method called advanced analysis. It is the most direct approach to structural design in which all significant nonlinear effects of material and geometry on system and member strength and stability are rigorously considered so that separate specification member-capacity checks are not required. There are two practical advanced analysis methods: one is called the refined plastic-hinge-based analysis, and the other is called the notational-load plastic-hinge analysis. They are suitable for direct semi-rigid frame design and have been developed, refined, and presented here for general use. A computer program based on these advanced analysis methods is also provided in the book.

The book is intended primarily for structural engineers familiar with the conventional processes of analysis and design in steel, but less familiar with direct second-order elastic or inelastic frame analysis with or without flexible joints. The main purposes of the book are to summarize in a unified manner the recent developments in stability design of semi-rigid frames and to provide some educational types of computer software for engineers to perform planar frame analysis for a better understanding of the behavior and more realistic prediction of a system's strength and stability. The information and tools can be used in the ultimate processes of practical semi-rigid frame design.

The authors are greatly indebted to Dr. N. Kishi for help in preparing Chapters 2 and 3.

W. F. Chen
Y. Goto
J. Y. R. Liew

January 1995

CONTENTS

COMPUTER REQUIREMENTS

The enclosed diskette requires an IBM-PC 486/66 MHz compatible computer with at least 4 MB RAM capable of running either Microsoft FORTRAN compiler Version 5.1 or a standard FORTRAN 77 compiler. The user should also have a hard disk with 11 MB of free space and 3.5" high-density floppy drive. The program can be used with both MS-DOS 3.1 or later and Microsoft Windows 3.1.

OVERVIEW

The disk accompanying this book is a computer-based analysis method called advanced analysis. The disk contains two practical advanced analysis methods: refined plastic hinge based analysis and notational load plastic hinge analysis, suitable for direct semi-rigid frame design. The main purpose of the computer program is to perform planar frame analysis for prediction of system's strength and stability.

Chapter 11 of the book is a tutorial which covers the computer programs contained on the diskette and referenced above. Throughout this chapter is an explanation of how to utilize the programs. PART 1 of the chapter covers the program PHINGE while PART 11 deals with the CON_ANA program. PHINGE is a FORTRAN based computer program for advanced inelastic analysis of two-dimensional steel frames while the CON_ANA program can be used to generate the moment-rotation characteristic of selected types of connections for use in advanced analysis of semi-rigid frames.

STEEL CONNECTION DATA BANK PROGRAM

The Steel Connection Data Bank Program that is referenced in Chapter 2 of the book is optional. The program is written in FORTRAN language and comes on a 3.5" floppy disk. To order the free disk, please call the following number, 1-800-225-5945.

USER ASSISTANCE AND INFORMATION

John Wiley and Sons, Inc. is pleased to answer questions regarding the program or a damaged disk. If you have any problems with a damaged disk please call our technical support number at (212) 850-6194, weekdays between 9 A.M. and 4 P.M. Eastern Standard Time.

To place additional orders or to request general information about orders or other Wiley products, please call Wiley customer service at (800) 879-4539.

1

INTRODUCTION AND HISTORICAL NOTE

1.1 INTRODUCTION

The main purpose of structural analysis is to ensure that a structure shall have a suitable safety factor against failure under its extreme design load. In addition, a structure shall meet its service requirements when subjected to its design working load.

There is, therefore, a fundamental two-stage process in the present design operation: firstly, the forces acting on the structural members and joints are determined by conducting a structural system analysis, and, secondly, the sizes of various structural members and the details of structural joints are selected by checking against specification member-capacity formulas and local stress criteria for joint detailing. Despite its popular use in the past and present, however, as a basis for design, the two-stage process has its limitations.

The first of these is that it does not give an accurate indication of the factor against failure, because it does not consider the interaction of strength and stability between the member and structural system in a direct manner. In current engineering design practice, this *interaction effect* is accounted for approximately by the use of the *effective length factor K-concept*. It is a well recognized fact that the actual failure mode of the structural system often does not have any resemblance whatsoever to the elastic buckling mode of the structural system that is the basis for the determination of the *K*-factor.

A second and perhaps more serious limitation is that for certain types of framed structures such as nonrectangular and irregular frames, it is difficult and complicated to find the *K*-factor by the elastic theory, with the consequence that designers have to use several empirical correction factors to make adjustments for the K-factor obtained by the usual alignment charts or formulas available in the open literature. This is particularly true for frames with leaned columns, whose resistance to sidesway must be provided by other restraint columns, the determination of the K-factor for this case may become so complex that its solution is often difficult to obtain by a direct use of computer because judgment is frequently needed in the computational process (Chen and Lui, 1987).

The most serious limitation is probably the rationale of the current two-stage design process in design: elastic analysis is used to determine the forces acting on each member of the structural system, whereas plastic analysis is used to determine the strength of each member treated as an isolated member. There is no verification of the *comparability* between the isolated member and the member as part of a frame. Basically, the difficulty lies in that member strength equations as specified by design specifications are unconcerned with system compatibility. There is no explicit guarantee that all members will sustain their design loads under the geometric configuration imposed by the framework.

There is an increasing awareness of the need of practical second-order analysis approaches that can model the behavior of structural members, connections, and other components in the determination of overall system response. Such approaches can model the true member performance explicitly in the analysis, and thus account for the compatibility between the member and system. The basic theory for *second-order inelastic analysis* is well established and documented in open literature (see, e.g., Chen and Toma, 1994) and the rapid increase in the power of desktop computers in recent years is making the use of nonlinear analysis in design more and more feasible. The real challenge is making this type of analysis work in engineering practice. This book attempts to take stock of where the structural engineering profession stands with regard to direct second-order elastic and inelastic analysis for semi-rigid frame design, and where it might be going. The last two of the following three design approaches will be described in detail in this book:

1. Present Approach: First-order elastic analysis with the K-factor and amplification factors B_1 and B_2.
2. Alternative Approach: Second-order elastic analysis with the K-factor.
3. Advanced Approach: Second-order inelastic analysis without the K-factor.

1.2 ELASTICITY IN FRAME DESIGN

The majority of building structures, in this and other countries, have been designed by the elastic theory by simply choosing suitable allowable stress values for the materials and by imposing limiting span/depth ratios for the service requirements. The analysis methods can be examined from the standpoint of deflection. All structures deflect under loading, but in general, the effect of this upon the overall geometry can be ignored. In the case of high-rise buildings, however, the lateral deflections may be such as to add a significant additional moment. This is known as the P–Δ moment or the P–Δ effect.

To cover this behavior rigorously, the governing equilibrium equations must be written with respect to the deformed geometry of the structure; the analysis is referred to as the *second-order analysis*. On the other hand, when the lateral deflections can be ignored and the equilibrium equations are written with respect to the undeformed (or original) geometry of the structure, the analysis is referred to as the *first-order analysis*. The load-deflection behaviors of a structure analyzed by first- and second-order elastic methods are illustrated in Fig. 1.1 by dashed curves. Second-order analysis is always necessary for the stability consideration of structures. The need for a direct second-order analysis of steel frames is increasing in view of the AISC LRFD specifications (AISC, 1993), which give explicit permission for the engineer to compute member forces from a direct second-order analysis. Although amplification factors B_1 and B_2 are provided by the LRFD specifications for computing second-order forces in a structure using a first-order analysis, it is rather cumbersome and tedious to calculate the amplification factors applied to first-order analysis. In view of this, it is more convenient and straightforward to use computer-based methods to calculate member forces directly as long as the method can be easily implemented in a computer. A practical second-order analysis method for elastic semi-rigid frame design is described in detail in Chapter 3.

The stability of individual members of the structure and stability of the structural system as a whole are considered independently in the current design process, so the interaction of the two is accounted for only approximately by the use of *effective length*, as shown in Fig. 1.2 for a rectangular rigid frame. For the braced column, the effective length must always be less than the actual story height (Fig. 1.2b), but that for the unbraced column, it will always be greater (Fig. 1.2c). When the relative stiffness of the beams and columns are known, the effective length can be computed directly by using *bifurcation analysis* (or *eigenvalue analysis*, as illustrated by the horizontal line in Fig. 1.1) for the ideal structure without imperfections (Fig. 1.2a). Once this effective-length factor K is determined, the second-order moments in each member of the structure can be obtained by the use of two amplification factors B_1 and B_2, as provided by the LRFD specifications. By multiplying the maximum first-order elastic moment by the B_1-factor, we can obtain the maximum moment in a member accounting for the member lateral-deflection effect known as the P–δ effect. While multiplying the first-order elastic sway moment by the B_2-factor, we account for the P–Δ effect due to the sidesway of the structure. The moment amplification factors are derived by using the

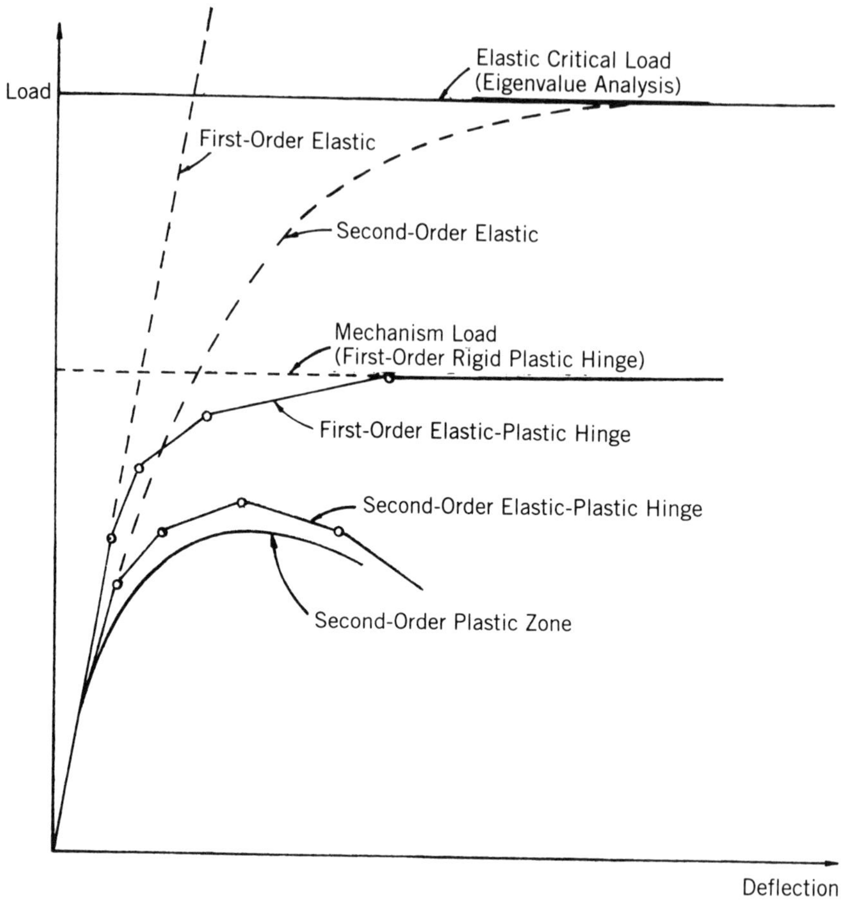

Fig. 1.1 Type of analysis methods.

second-order elastic analysis for an idealized substructural system (Chen and Lui, 1991). Because they are derived based on several simplified assumptions and applicable only to rectangular frames with rigid connections and small displacement, and are rather cumbersome and tedious to calculate, it is more convenient and straightforward to use a direct second-order analysis. In fact, many limit states code provisions prefer the direct second-order elastic analysis for frame design (Beedle, 1991), most notably the Canadian limit states design specifications (CSA, 1989).

Unlike a first-order analysis in which solutions can be obtained in a rather simple and direct manner, a second-order analysis often entails an iterative-type procedure to obtain solutions. This is because the deformed geometry of the structure is not known during the formulation of the equilibrium and

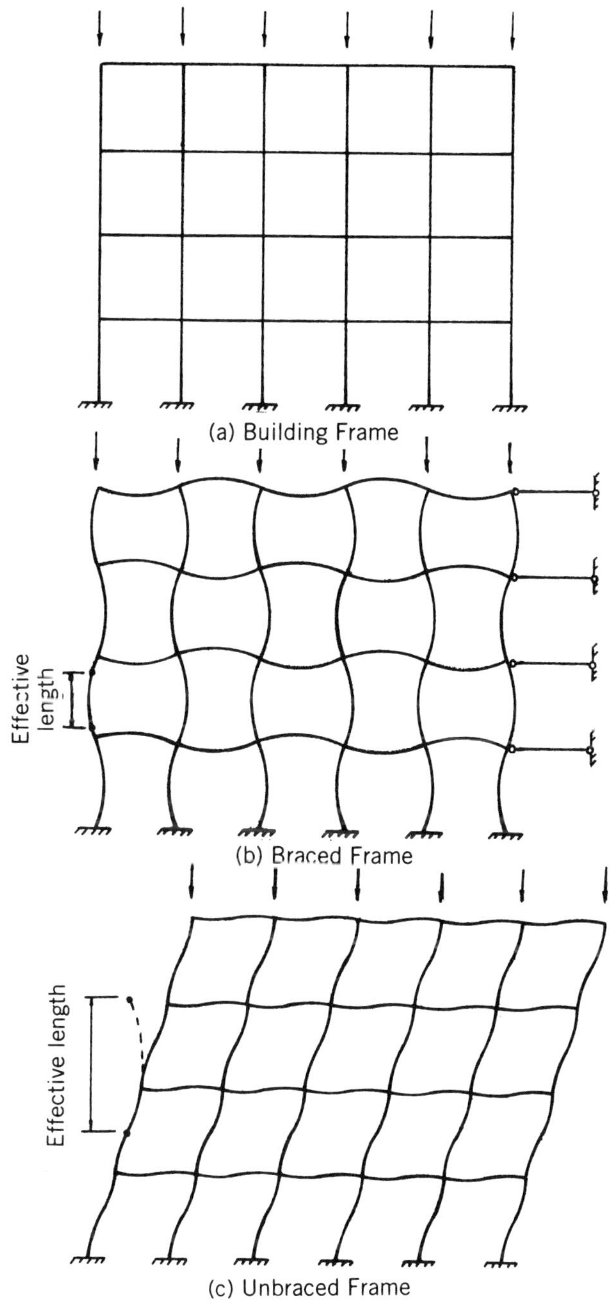

Fig. 1.2 Buckling modes for rectangular frames and effective-length concept.

kinematic relationships. Thus, the analysis usually proceeds in a step-by-step incremental manner. The deformed geometry of the structure obtained from a preceding cycle of calculation is used as the basis for formulating the equilibrium and kinematic relationships for the current cycle of calculations. There are various iterative schemes available for the solution of second-order problems in elastic as well as inelastic range. Details of the methods selected as the most useful and practical applicable means currently available for a particular application are described later in each relevant chapter when the procedures for computer implementation are presented.

To clarify the terms used in this book, a brief description of the three basic elastic methods of analysis for frame design is summarized here. A full description of each method is given later in each relevant chapter. Figure 1.1 contains a schematic representation of load-displacement diagrams obtained for a frame by each type of analysis methods (SSRC, 1988).

Bifurcation Analysis (or Eigenvalue Analysis). This type of analysis determines the load at which a bifurcation occurs from some original mode of deformation into a buckled-mode shape. It is the basis for determining the effective length factor K for design of members in frame structures.

First-Order Elastic Analysis. Linear elastic is, of course, the current staple of the profession. One of its distinguishing features is that equilibrium is formulated on the undeflected structure. This type of analysis allows a simple estimate of the distribution of forces in the structural system, but it does not provide any information on the strength or the stability of the frame.

Second-Order Elastic Analysis. Its distinguishing characteristic is that equilibrium is formulated on the deformed structural geometry. When derived on a consistent mechanics basis, it includes both P–Δ (frame sidesway) and P–δ (member curvature) effects, although few programs can model directly the P–δ effects (or the effects of geometric imperfections such as initial out-of-straightness of members), unless the members are subdivided into a number of elements. Second-order elastic analysis accounts for elastic stability effects, but it does not provide any direct information with regard to inelastic strength and stability. For typical frame-type structures, the second-order elastic load-deflection response tends to asymptotically approach the horizontal line corresponding to the elastic buckling load as the loads are increased.

1.3 SEMI-RIGID CONNECTIONS IN FRAME DESIGN

Conventional analysis and design of steel frameworks are usually carried out under the assumption that the connections joining the beams to the columns are either fully rigid or ideally pinned. The rigid-joint assumption implies that full slope continuity exists between the adjoining members and that the full (or a substantial percentage of) gravity moment is transferred from the beam to the column. On the other hand, the assumption of ideally pinned connections implies that the beams will behave as if they are simply supported and that the columns will carry no gravity moment from the beams. Although the assumption of fully rigid or ideally pinned connection behavior drastically simplifies the analysis and design procedures, the validity of these assumptions may be questionable for cases in which the rigidities of the connections, which here can be considered as the slope of the relation between the connection moment M and the connection relative rotation θ_r, are intermediate between the fully rigid and ideally pinned cases. In fact, as evident from experimental observations shown in Fig. 1.3, all connections used in current practice possess stiffnesses that fall between the two extreme cases of fully rigid and ideally pinned (Chen, 1987, 1988; Chen and Lui, 1986; Chen and Kishi, 1989; Beedle, 1993).

Realizing the potential implication of connections on frame design, the American Institute of Steel Construction (LRFD, 1993) has introduced provisions to allow designers to consider explicitly the behavior of connections in the design of structural steel frames.

The Allowable Stress Design (ASD) Specifications (AISC, 1989) list three types of constructions:

1. *Type 1, or "Rigid Framing":* This construction assumes that the beam-to-column connections have sufficient rigidity to maintain the original geometric angle between intersection members. Type 1 connections are assumed for elastic structural analysis.
2. *Type 2, or "Simple Framing":* This assumes that the beam and girder connections transfer only vertical shear reactions without bending moment when the structure is loaded with gravity load. The connections are allowed to rotate without restraint.
3. *Type 3, or "Semi-Rigid Framing":* This assumes that the connections can transfer vertical shear and also have the capacity to transfer some moment.

The Load and Resistance Factor Design (LRFD) Specifications (AISC, 1993) designate two types of construction in their provisions: Type FR (Fully Restrained) and Type PR (Partially Restrained). Type FR corresponds to ASD Type 1. Type PR includes ASD Types 2 and 3. If Type PR construction is used, the effect of connection flexibility must be taken into account in the analysis and design procedures.

The influence of partially restrained connections on structural analysis is not only in changing the moment distribution among beams and columns, but also in the increase of frame drift and accordingly the increase of the P–Δ effect in the frame analysis. Behavior and modeling of semi-rigid connections are presented in Chapter 2. Some practical methods of designing semi-rigid

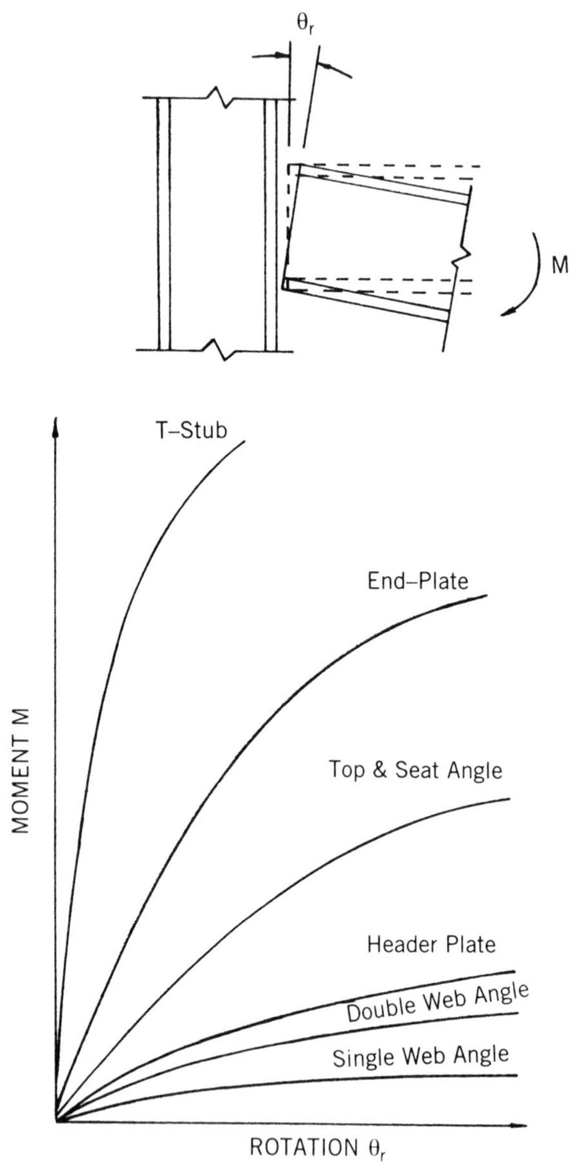

Fig. 1.3 Connection moment–rotation curves.

frames using the LRFD specifications with a direct second-order elastic analysis are described in detail in Chapter 4.

1.4 PLASTICITY IN FRAME DESIGN

In the engineering of steel frame structures, there is an increasing awareness of the value of analysis/design techniques that explicitly consider the structural response through the elastic and inelastic ranges and up to the maximum load limit state. This is evidenced by, and in part, as a result of the development of limit states design methodologies such as that of the AISC Load and Resistance Factor Design (LRFD) specification (AISC, 1993). Limit states design procedures are based on analysis at factored loads and limits of structural resistance, so there is no way to disguise the fact that a system will behave nonlinearly before its strength design conditions are reached (ECCS, 1984).

Typically, current limit states specifications refer to more sophisticated methods of analysis than older allowable stress design specifications. For example, the LRFD manual requires that for frames designed based on plastic analysis must account for the second-order P–Δ effects. For structures designed based on elastic load effects, the engineer is given the option to compute forces by a direct second-order analysis. The new Australian standard for structural steel design (AISC, 1990) refers explicitly to three levels of analysis: elastic analysis, plastic analysis, and advanced analysis. Advanced analysis is defined as one in which the relevant material properties, residual stresses, geometrical imperfections, second order effects, three-dimensional effects, erection procedures, and interaction with the foundations are accounted for.

In parallel with the development of limit states specifications, powerful personal computers and workstations are becoming readily available to design engineers. The advanced graphics and computing power of these new machines are making it possible to employ sophisticated analysis and design techniques that in the recent past were considered generally impractical for design office use. Methods of computer-aided analysis and design are becoming increasingly important, and less and less emphasis is being placed on simple hand procedures (White, Liew, and Chen, 1991, 1992). This does not discount the value of approximate analysis techniques, the importance of simple linear elastic analysis, and the need for simple code and specification procedures. There will always be a need for simple analysis/design procedures, and for many structures such techniques are sufficient. Furthermore, in the foreseeable future, the computing power available to design engineers is not expected to be great enough for all the possible effects on frame strength and stability to be considered explicitly in the analysis. However, we are rapidly approaching the time when the actual limit state behavior of a structural system can be considered, with *little* or *no extra effort* on the part of the engineer, through a *second-order inelastic analysis* method that is capable of representing what are typically the dominant behavioral effects. Of course, this means that the effects not considered appropriately in the analysis must still be accounted for in the design specifications, just as in the case of current specifications for elastic analysis/design.

The major benefit resulting from the use of second-order inelastic analysis/design is a more reliable consideration of frame stiffness, strength, and performance whenever more careful checks than attainable by elastic procedures are desired. Secondary benefits are a possible increase in the economy of steel construction and more uniform safety for some types of frames. These benefits are primarily due to the fact that elastic analysis does not explicitly account for any degradation in member stiffness and the corresponding *redistribution of forces* within the structural system prior to achievement of the design load. This degradation in stiffness is in general due to second-order effects associated with changes in the geometry of the structure (i.e., P–Δ and P–δ effects), as well as the plastification of the material within the structure (yielding of material). Although elastic analysis/design procedures consider the full inelastic strength and stability of individual frame members through rules and equations for member proportioning, the behavior of the members as part of the overall structural system is treated in an approximate way (e.g., through effective length factors). The load-carrying capacity of the structure is assumed to be reached when the ultimate (elastic or inelastic) strength/stability of one of its members is achieved by forces computed based on the assumption that the structure is elastic. In most cases, this results in a conservative design. But in some extreme cases, where member axial forces are large and the system is inherently flexible (e.g., possibly a structure that is loaded primarily by gravity loads and that has a large number of leaner columns and small lateral stiffness), it could be unconservative due to stability effects.

Much progress has been made in the development of nonlinear analysis techniques that can account for inelastic frame behavior up to the full application of the factored design load. However, many of these analysis approaches are still unsystematized, and, in general, improvements are still needed to make inelastic analysis practical for design office use. Most importantly, the implications of various analysis approaches with respect to the design of steel frames need to be studied such that suitable guidelines can be established for design practice. Appropriate computer-aided methodologies involving use of inelastic analysis need to be determined and outlined. These aspects are discussed in Chapter 5.

To clarify the terms used, several general types of inelastic methods of analysis are briefly described here. A full description is given in each relevant chapter. The term inelastic analysis is used here to refer to any method of analysis in which the effects of material yielding are accounted for. The inelastic analysis methods are categorized into the following two groups: *con-*

centrated plasticity (also commonly referred to as *plastic hinge*) and *spread of plasticity* (also called the *plastic zone*).

Concentrated Plasticity Analysis. The inelastic behavior is approximated by zero-length "plastic hinges" in the frame members. The member cross-section is assumed to be completely elastic until the capacity of the fully plastic section is reached. At this stage in the loading, the cross-section response is assumed to be perfectly plastic (no strain hardening). Therefore, another commonly used name for this type of analysis is *elastic-plastic analysis.*

Plastic-hinge analysis can be further subdivided into two categories, first-order and second-order elastic-plastic, based on whether second-order geometric effects are accounted for. For *first-order elastic-plastic* analysis, equilibrium is formulated on the undeformed geometry. As a result, the dominant factors that affect the strength of the structure are included in the analysis, but the stability aspects on structural performance are not considered. First-order elastic-plastic analysis predicts that the maximum capacity of the frame is reached when a plastic collapse mechanism is formed. Thus, this type of analysis predicts the same ultimate load capacity as conventional mechanism and statical methods of plastic analysis (i.e., first-order rigid-plastic methods) covered in standard texts on plastic analysis/design such as Beedle (1958), Heyman (1971), and Chen and Sohal (1995).

If equilibrium is formulated on the deformed structural configuration, the plastic-hinge analysis may be termed as *second-order elastic-plastic*. Elastic-plastic analysis only models yielding at zero-length plastic hinges, so this type of analysis does not accurately represent the second-order *inelastic* P–δ effects within individual members. However, in well-designed frames, these effects are usually expected to be a minor aspect of the overall structural response. The load-deflection behaviors for a structure analyzed by the first- and second-order plastic-hinge methods are illustrated in Fig. 1.1.

Spread-of-Plasticity Analysis. Because of the idealization that yielding occurs in concentrated plastic zones, the second-order elastic-plastic analysis may in some cases over predict the actual inelastic stiffness and strength of the structure. Although it is generally undesirable to allow formation of plastic hinges in the columns (i.e., a strong-column weak-beam design is typically sought), some distributed yielding ordinarily occurs in beam-columns prior to reaching the strength limit state. Inelastic column behavior at the strength limit state is inherent in current specifications for design of short to intermediate-length columns (Chen and Atsuta, 1976, 1977). The second type of inelastic analysis, referred to as *spread-of-plasticity analysis,* explicitly considers both geometric nonlinear effects as well as the explicit spread of plastic zones along the length and through the cross-section of the members. Therefore, it is capable of representing the effects of limited yielding in the beam-columns of a frame structure. The spread-of-plasticity analysis combines the theory of plasticity with the theory of stability and leads to a direct analysis of inelastic strength and stability for frame design. Three practical plastic-hinge-based analysis/design methods developed in recent years based on this advanced analysis technology are described in detail in Chapters 6–8 for rigid frames and in Chapter 9 for semi-rigid frames (Chen and White, 1992). Some education software and their manuals are given in Chapter 11. Second-order inelastic analysis and design of steel frames considering the beam-column panel-joint deformations are presented in Chapter 10.

1.5 SUMMARY

In the design of frame structures for strength and stability, engineers have long recognized the need to account in some way for inelastic-material behavior, the flexibility of connections, and second-order effects due to changes in geometry. However, until recently, accounting for these effects directly in analysis was infeasible or limited to simple structural systems. This is reflected in the common use of first-order elastic analysis with rigid or pin joints to estimate the distribution of forces within a structure, and the consideration of strength and stability effects solely through specification provisions for member design. Advancements in computer hardware, particularly engineering workstations, are making more sophisticated analysis methods feasible in design.

The main benefit of second-order inelastic analysis is, as in conventional plastic design, the accounting for inelastic redistribution of internal forces after the strength of certain members is reached. In addition, modern inelastic-analysis methods allow more rational consideration of the effects of connection flexibility on system strength and stability. It enables engineers to predict accurately all possible failure modes of a structural system they have designed when it is subjected to a given load combination.

A wide range of analysis methods may be formulated that account for inelastic performance of a structure. These methods range from two-dimensional simple plastic-hinge methods to more refined analysis in which the distribution of plastic zones is explicitly modeled. An argument is made for the use of simple plastic-hinge methods of analysis for proportioning of frame structures, and use of modified plastic-hinge concepts for final design checking. Two plastic-hinge-based methods for the design of steel frames are described in this book. The first one is based on the *notional-load plastic-hinge approach,* similar to the proposals put forward by Eurocode 3 (CEN, 1992). The other one is based on the *refined plastic-hinge approach,* similar to the concepts used previously in the development of the SSRC Column Curves adopted in the AISC ASD and AISC LRFD Specifications (AISC, 1989, 1993). The general validity of these two approaches has been confirmed by extensive comparisons with the results of tests on actual frames, as well as with the almost exact theoretical solutions based on the plastic-zone analysis. It is concluded that with modern developments in computers, both methods as given in details in Chapters 6–9 are feasible and practical, and are therefore suitable for adoption in practice.

The advanced analysis technology will serve the structural engineering profession into the next century with the state-of-the-art design methods. Will the advanced methods result in buildings built with lighter members? Not necessarily. Will buildings cost less to design or build? Not likely. Will it be more complicated to design than LRFD? No! What will it do then? The advanced methods described here will encourage designers to use more accurate analysis methods and computer programs. The advanced analysis can predict more accurately possible failure modes of a structure, exhibit a more uniform level of safety, and provide a better long-term serviceability and maintainability. It is the state-of-the-art design methods for the structural engineers for the twenty-first century.

REFERENCES

AISC (1989). *Allowable Stress Design and Plastic Design Specifications for Structural Steel Buildings,* Chicago: American Institute of Steel Construction.

AISC (1990). *Australian Limit States Specification,* AS4100, Sydney: Australian Institute of Steel Construction.

AISC (1993). *Load and Resistance Factor Design Specification for Structural Steel,* 2nd Ed., Chicago: American Institute of Steel Construction.

Beedle, L. S. (1958). *Plastic Design of Steel Frames,* New York: Wiley.

Beedle, L. S. (1991). *Stability Design of Metal Structures: A World View,* Bethlehem, PA: Structural Stability Research Council, Lehigh University.

Beedle, L. S. (1993). *Semi-Rigid Connections in Steel Frames,* New York: Council on Tall Buildings and Urban Habitat and McGraw-Hill, pp. 000–000.

CEN—European Committee for Standardization (1992). "Design of Steel Structures: Part 1—General Rules and Rules for Buildings," Vol. 1, February.

Chen, W. F., ed. (1987). "Joint Flexibility in Steel Frames," *Journal of Constructional Steel Research,* Vol. 8, Special Issue.

Chen, W. F., ed. (1988). "Steel Beam-to-Column Building Connections," *Journal of Constructional Steel Research,* Vol. 10, Special Issue.

Chen, W. F., and T. Atsuta (1976). *Theory of Beam-Columns, Vol. 1: In-Plane Behavior and Design,* New York: McGraw-Hill.

Chen, W. F., and T. Atsuta (1977). *Theory of Beam-Columns, Vol. 2: Space and Behavior and Design,* New York: McGraw-Hill.

Chen, W. F., and N. Kishi (1989). "Semi-Rigid Steel Beam-To-Column Connections: Data Base and Modeling," *Journal of Structural Engineering,* ASCE, Vol. 115, No. 1, pp. 105–119.

Chen, W. F., and E. M. Lui (1986). "Recent Developments in Structural Connections," in *Advances in Tall Buildings,* L. S. Beedle, ed., New York: Van Nostrand Reinhold, pp. 353–365.

Chen, W. F., and E. M. Lui (1987). *Structural Stability: Theory and Implementation,* New York: Elsevier.

Chen, W. F., and E. M. Lui (1991). *Stability Design of Steel Frames,* Boca Raton, FL: CRC Press.

Chen, W. F., and I. S. Sohal (1995). *Plastic Design and Second-Order Analysis of Steel Frames,* New York: Springer-Verlag.

Chen, W. F., and S. Toma, eds. (1994). *Advanced Analysis of Steel Frames: Theory, Software, and Applications,* Boca Raton, FL: CRC Press.

Chen, W. F., and D. W. White, eds. (1992). *Workshop and Monograph on Plastic Hinge Based Methods for Advanced Analysis and Design of Steel Frames,* Bethlehem, PA: Structural Stability Research Council, Lehigh University.

CSA (1989). *Limit States Design of Steel Structures,* CAN/CSA-S16, 1-M89, Toronto: Canadian Standards Association.

European Convention for Construction and Steelwork (1984). *Ultimate Limit State Calculation of Sway Frames with Rigid Joints,* Technical Working Group 8.2, Systems, Publication No. 33, Brussels.

Heyman, J. (1971). *Plastic Design of Frames,* Cambridge: Cambridge University Press.

SSRC (1988). *Guide to Stability Design Criteria for Metal Structures,* 4th ed., Galambos, T. V., ed. New York: Wiley.

White, D. W., R. J. Y. Liew, and W. F. Chen (1991). *Second-Order Inelastic Analysis for Frame Design: A Report to SSRC Task Group 29 on Recent Research and the Perceived State-of-the-Art,* Structural Engineering Report No. CE-STR-91-12, West Lafayette, IN: School of Civil Engineering, Purdue University.

White, D. W., R. J. Y. Liew, and W. F. Chen (1992, April 5). "Toward Advanced Inelastic Analysis in LRFD," *Proceedings of SSRC Task Group 29 Workshop on "Plastic Hinge Based Methods for Advanced Analysis and Design of Steel Frames,* Bethlehem, PA: Structural Stability Research Council, Lehigh University.

2

SEMI-RIGID CONNECTIONS

2.1 INTRODUCTION

Conventional analysis and the design of steel frameworks are performed using the assumption that the connections are either fully rigid or ideally pinned. The assumption of a fully rigid connection implies that no relative rotation of the connection occurs and the end moment of the beam is completely transferred to the column. On the other hand, the pinned connection implies that no restraint for rotation of the connection exists and the connection moment is always zero. However, as is evident from experimental observations, all beam-to-column connections used in current practice possess some stiffness that falls between the two extreme cases of fully rigid and ideally pinned.

The AISC LRFD Specifications (AISC, 1993) identify Type PR (Partially Restrained) construction as one of the two basic types of construction. In PR-type construction, the structural connections are assumed to offer some restraint to the members they connect. When the flexibility of connections is considered in the design of frames, the LRFD specification requires an explicit evaluation of the influence of the connection flexibility on frame stability by rational analytical or empirical means. However, the design specification provides for Type PR construction only broad principles for analysis and design. It is left to the engineer to implement these principles in a quantitative manner. Therefore, to establish the guideline for design of semi-rigid frames, it is necessary to know the behavior of actual beam-to-column connections and to formulate an appropriate connection model for use in the design and analysis of semi-rigid frames.

In this chapter, the behavior of semi-rigid connections is briefly discussed. The commonly used connection types are presented with their general behavior and characteristics. Several empirical and semi-empirical connection models based on curve-fitting techniques are reviewed. The main aim of this chapter is to provide moment–rotation characteristics and corresponding parameters of semi-rigid beam-to-column connections that can be used in the frame analysis as will be explained later in Chapter 3. To this end, the steel connection database developed by Chen and Kishi (1989), and by Kishi and Chen (1986a, 1986b) is introduced. This connection database is systematically controlled by the Steel Connection Data Bank (SCDB program). The SCDB program provides seven main functions, including routines for tabulating the experimental test data and the results calculated by three prediction equations.

2.2 BEHAVIOR OF SEMI-RIGID CONNECTIONS

A connection is a medium through which forces and moments are transmitted from one member to another. A beam-to-column connection is generally subject to axial force, shear force, bending moment, and torsion. For an in-plane study, the effect of torsion can be excluded. Furthermore, the effects of axial and shear forces are usually small compared to that of the bending moment. Consequently, for practical purposes, only the effect of moment on the rotational deformation of connections needs to be considered. As depicted in Fig. 2.1, the connection rotates by an amount θ_r when a moment M is applied. The angle θ_r corresponds to the relative rotation between the beam and the

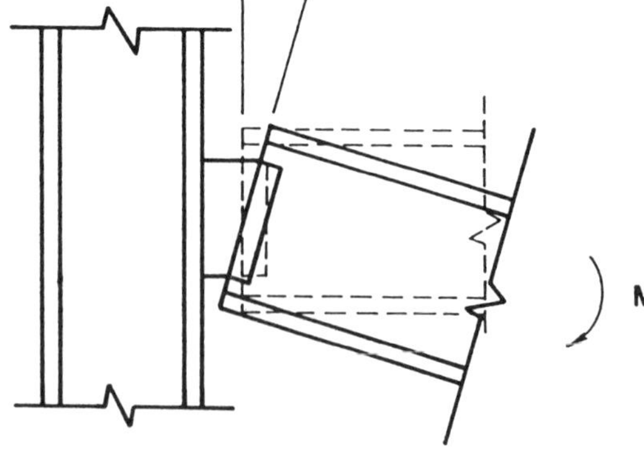

Fig. 2.1 Rotational deformation of a connection.

column at the connection. Therefore, the in-plane behavior of semi-rigid connection is represented by the M–θ_r relation, referred to as moment–rotation behavior. The moment–rotation behavior of a variety of commonly used semi-rigid connections is shown in Fig. 2.2. All types of connections exhibit nonlinear moment–rotation behavior that falls between the two extreme cases of ideally pinned and fully rigid connections. The single web-angle connection represents a very flexible joint, and the T-stub connection represents a rather rigid connection. The moment–rotation curves of all types of connections are nonlinear over the entire range of loading. The nonlinearity of connection behavior is due to a number of factors such as material discontinuity of the connection subassemblage, local yielding of some component part, local buckling of a plate element, and so on. Semi-rigid connections are generally very ductile, provided that no serious defects exist in the connection fabrication. Connection ductility is a significant quality as far as the ultimate ca-

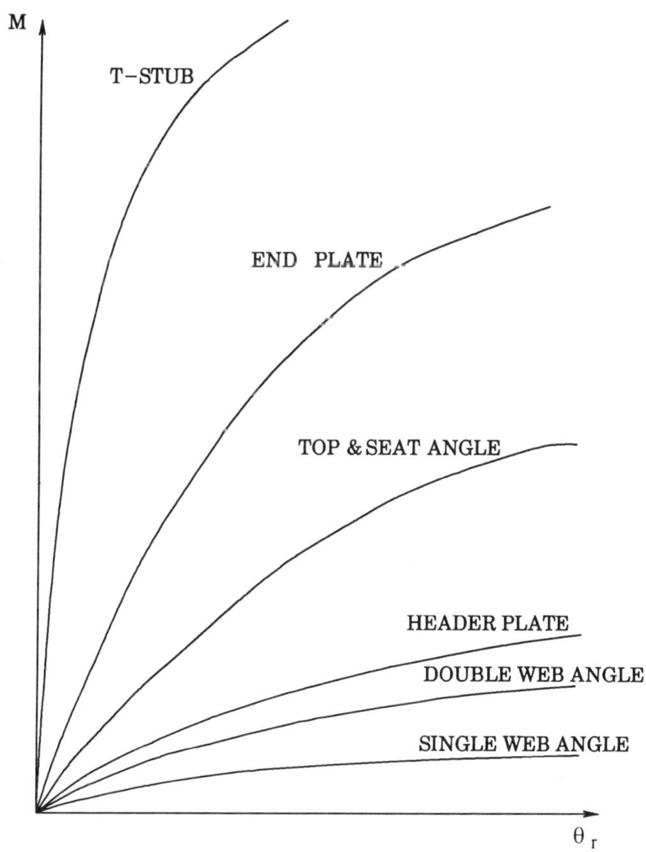

Fig. 2.2 Connection moment–rotation curves.

pacity of structures is concerned, particularly for those designed to resist against seismic forces.

In engineering practice, it is important to know in what cases the connections are assumed to be rigid, semi-rigid, or flexible. The proposals for the classification of connections have been presented independently by Eurocode 3 (CEN, 1992) and Bjorhvde et al. (1990), as illustrated in Figs. 2.3 and 2.4. The classification system by Bjorhovde et al. is intended for the case in which prior knowledge of the member details is not available. This implies that the

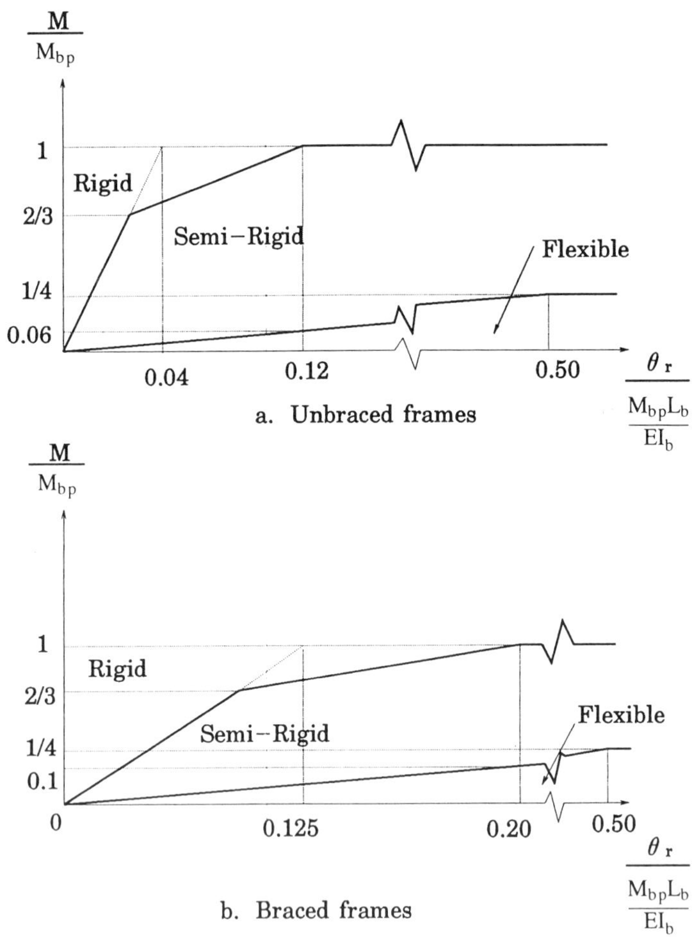

EI_b = bending rigidity of connected beam
L_b = beam length
M_{bp} = plastic moment capacity of connected beam

Fig. 2.3 EC3 classification system.

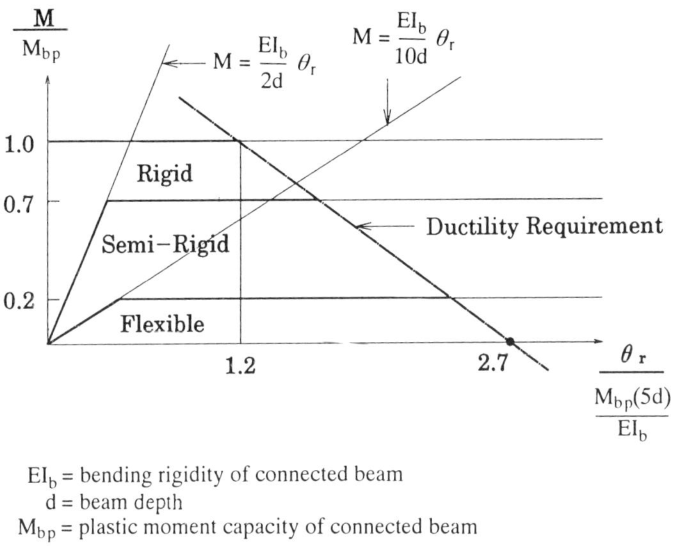

Fig. 2.4 Bjorhovede, Brazzetti, Colson classification system.

Bjorhovde's classification approach does not consider the overall behavior of frames. On the other hand, EC3 classification is based on the overall strength of frames. Thus, the EC3 classification is more rational if the layout and member details of the structural systems are known in advance. However, EC3 does not specify the ductility demand of connections, which is different from the proposal by Bjorhovde et al. Further, EC3 adopts an approximate method to evaluate the ultimate strength of frames. In addition to the strength limit states of the frames, the serviceability limit states should also be considered for classification.

As explained before, the existing classification systems are still considerably approximate in nature and their validity should be further examined in more detail. In fact, a precise elastic-plastic finite-displacement analysis (Goto, Miyashita, and Chen, 1994; Goto and Miyashita, 1995) showed that the boundary between rigid and semi-rigid presented either by EC3 or Bjohovde is rather strict in terms of the load-carrying capacity of frames.

2.3 TYPES OF SEMI-RIGID CONNECTIONS

2.3.1 Single Web-Angle Connections and Single Plate Connections

A single web-angle connection consists of an angle either bolted or welded to both the column and the beam web, as shown in Fig. 2.5a. On the other hand, a single plate connection uses the plate instead of the angle. This con-

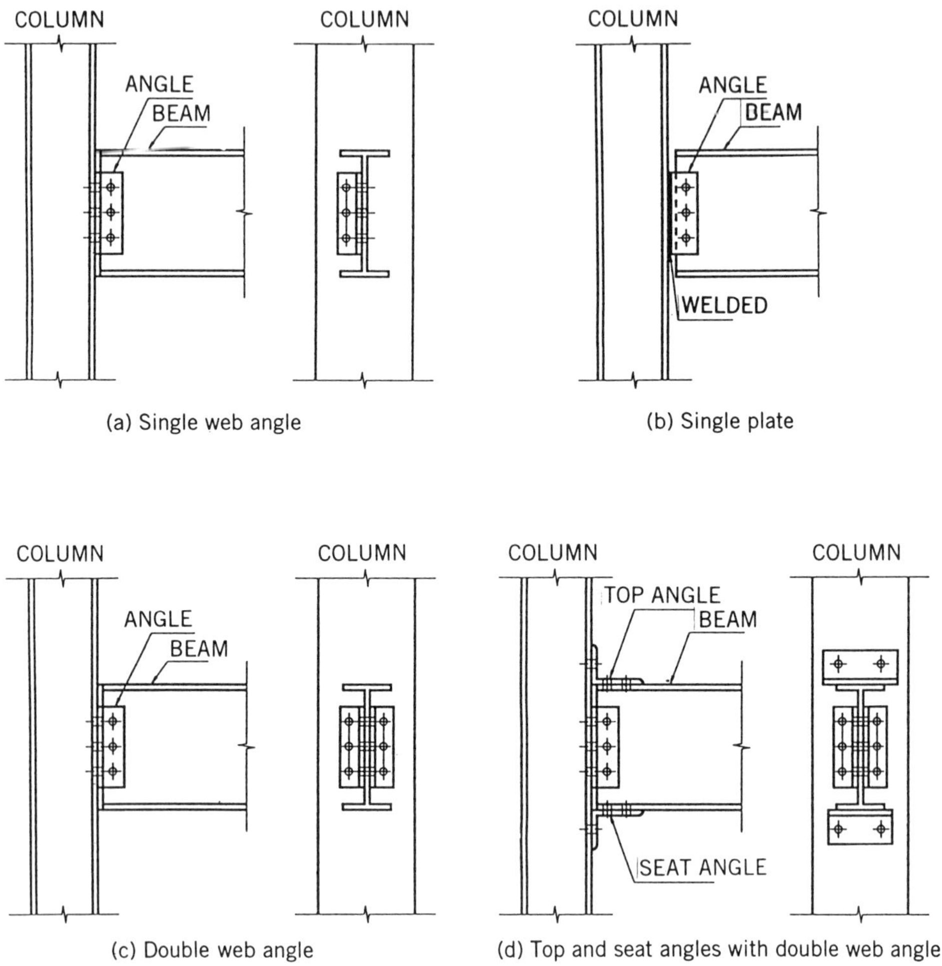

Fig. 2.5 Typical types of beam-column connections: (a) single web angle; (b) single plate; (c) double web angle; (d) top and seat angles with double-web angle; (e) top and seat angles; (f) extended end plate on the tension side only; (g) extended end plate on both sides; (h) flush end plate; and (i) header plate.

nection type requires less material than a single web-angle connection (Fig. 2.5b). Generally, the single web-angle connection has moment rigidity equal to about one-half of the double web-angle connection. The single plate connection has rigidity equal to or greater than the single web-angle connection because one side of the plate in the single plate connection is fully welded to the column flange.

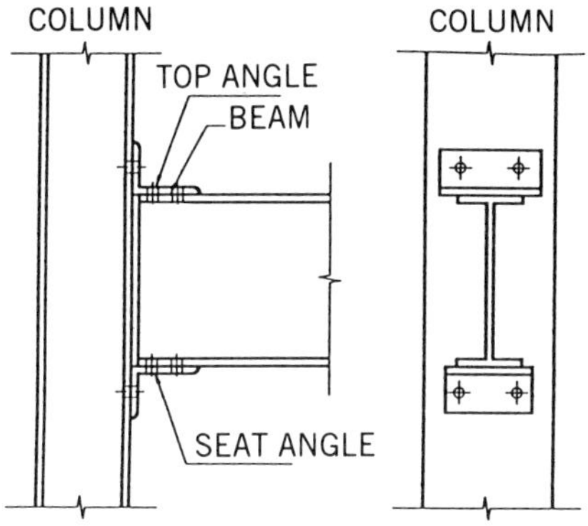

(e) Top and seat angles

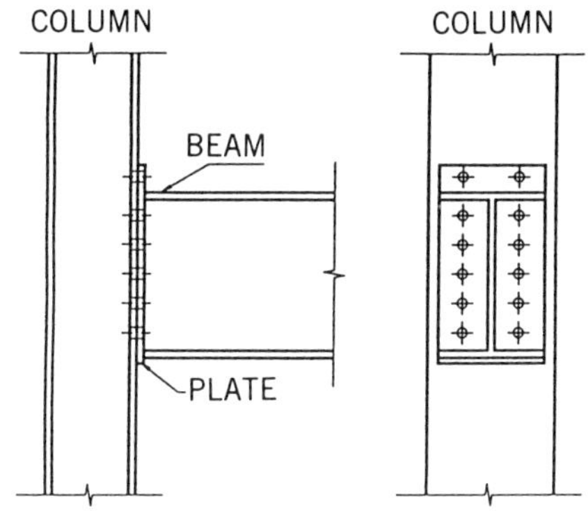

(f) Extended end plate on the tension side only

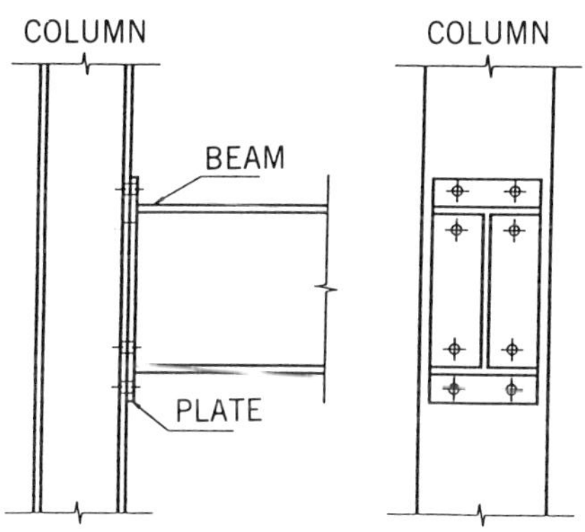

(g) Extended end plate on both sides

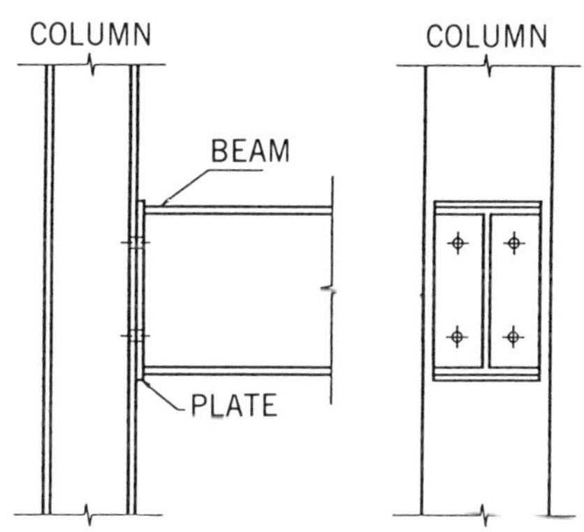

(h) Flush end plate

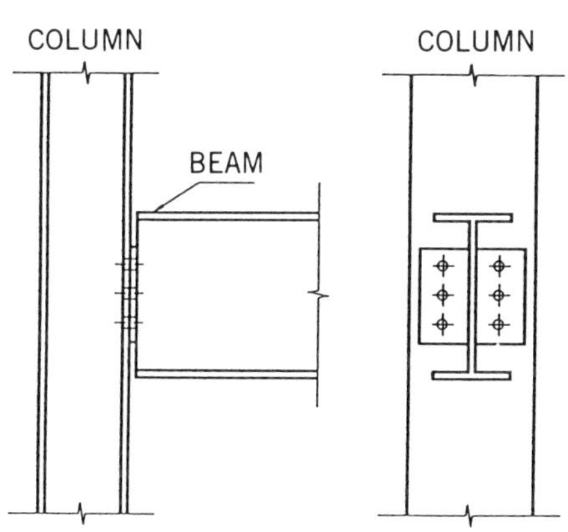

(i) Header plate

Fig. 2.5 *(Continued)*

2.3.2 Double Web-Angle Connections

A double web-angle connection consists of two angles, either bolted or riveted to both the column and the beam web, as shown in Fig. 2.5c. Rivets were used as fasteners in the earliest tests on double web-angle connections conducted by Rathbun (1936). In the 1950s, most specifications for the design of steel structures allowed the use of high-strength bolts in lieu of rivets. To clarify the effect of high-strength bolts on connection behavior when used in conjunction with rivets, Bell, Chesson, and Munse (1958) and Lewitt, Chesson, and Munse (1966) conducted experiments on riveted and bolted beam-to-column connections. Today, high-strength bolts are used popularly as fasteners for this type of connection. Although the connection rigidity of this type of connection is stiffer than those of single web-angle and single plate connections, the AISC ASD Specifications (AISC, 1989) consider this connection type as a Type 2 construction (simple connection or shear connection).

2.3.3 Top- and Seat-Angle Connection with Double Web Angle

This type of connection is a combination of a top- and seat-angle connection and a double web-angle connection. A typical top- and seat-angle connection with a double web angle is shown in Fig. 2.5d. Double web angles are used to improve the connection restraint characteristics of top- and seat-angle connections, and for shear transfer. This type of connection is considered as Type 3 framing of the AISC ASD Specifications (AISC, 1989), that is to say, a semi-rigid connection.

2.3.4 Top- and Seat-Angle Connection without Double Web Angle

A typical top- and seat-angle connection is shown in Fig. 2.5e. The AISC ASD Specifications (AISC, 1989) described this type of connection as follows: (1) the top angle is used to provide lateral support of the compression flange of the beam; and (2) the seat angle is to transfer only the vertical reaction of the beam to the column and should not give a significant restraining moment on the end of the beam. However, according to the experimental results, this connection can transfer not only the vertical reaction, but also some end moment of the beam to the column.

2.3.5 Extended End-Plate Connection and Flush End-Plate Connection

In general, an end plate is welded to the beam end along both the flanges and web in the fabricator's shop and bolted to the column in the field. The end-plate connection has been used extensively since the 1960s. The extended end-plate connections are classified into two types: as an end plate either extended on the tension side only or on both the tension and compression sides, as shown in Figs. 2.5f and 2.5g, respectively. A typical flush end-plate connection is shown in Fig. 2.5h. Because some end-plate connections are considered as Type FR construction rather than Type PR construction in AISC LRFD Specifications (AISC, 1993), they have often been used as means of transferring beam end moment to the column. The extended end-plate connection on both sides is preferred when the connection is subjected to moment reversal such as during severe earthquake loading. Although the flush end-plate connection is weaker than the extended end-plate connection, this connection type is often used in roof details. The behavior of an end-plate connection depends on whether the column flange near the connection is stiffened or not. The stiffeners of the column flanges act to prevent flexural deformation of the column flange, thereby influencing the behavior of the plate and fasteners.

2.3.6 Header-Plate Connection

A header-plate connection consists of an end plate, whose length is less than the depth of the beam, welded to the beam web and bolted to the column, as shown in Fig. 2.5i. The moment–rotation characteristics of this connection are similar to those of the double web-angle connection. Accordingly, a header-plate connection is used mainly to transfer the reaction of the beam to the column and is classified as a connection in Type 3 framing of the AISC ASD Specifications (AISC, 1989).

2.4 MODELING CONNECTIONS

2.4.1 General Remarks

A number of analytical studies of connection behavior using the finite-element method have been reported so far (Krishnamurthy et al., 1979; Patel and Chen, 1984; Driscoll, 1987; Kukreti, Murray, and Abolmaali, 1987). However, these approaches are unacceptable for practical use because cumbersome calculations are required to consider the material and geometrical nonlinearities. At the present, the most commonly used approach to describe the M–θ_r curve is to curve fit the experimental data with simple expressions. Several analytical models have been developed to represent connection flexibility using available experimental test data.

Early models used the initial connection stiffness as the key parameter in a linear M–θ_r model (Rathbun, 1936; Monforton and Wu, 1963; Lightfoot and LeMessurier, 1974). Although the linear model is very easy to use, it has a serious drawback. That is, the linear model is applicable only in a small range of the initial rotation. A closer approximation of true connection behavior can be obtained by using either the bilinear model (Tarpy and Cardinal, 1981; Lui and Chen, 1986) or the piecewise linear model. The piecewise linear model is composed of a series of straight-line segments. In these models, the abrupt changes in connection stiffness at the transition points make their practical use difficult. Jones, Kirby, and Nethercot (1980, 1981) proposed a cubic B-spline model to obtain a more suitable function. However, this model requires a large number of sampling data during the formulation process. Frye and Morris (1975) have reported a polynomial model to evaluate the behavior of several types of connections. In this model, the $M-\theta_r$ behavior is represented by an odd-power polynomial.

Lui and Chen (1986) used an exponential function to curve fit the experimental $M-\theta_r$ data. This model represents the monotonic nonlinear connection behavior well. However, if there are some sharp changes in the slope of the $M-\theta_r$ curve, this model cannot represent it adequately (Wu, 1989). Kishi and Chen (1986a, 1986b) refined the Lui–Chen exponential model to accommodate any sharp changes in the slope of the $M-\theta_r$ curve. The Kishi–Chen model, which is also referred to as the modified exponential, can be really used in lieu of the experimental $M-\theta_r$ data. Other exponential models have been reported by Yee and Melchers (1986) and Wu and Chen (1990). The Yee–Melchers model includes four parameters such as initial connection stiffness, strain-hardening connection stiffness, plastic moment capacity, and a constant. The Wu–Chen model is a three-parameter model whose independent parameters are initial connection stiffness, ultimate moment capacity, and a shape parameter.

The connection models using the power function, namely, power models, have been also reported. Colson and Louveau (1983) and Kishi and Chen (1990) proposed a similar model independently using three parameters: initial connection stiffness, ultimate moment capacity, and a shape parameter. These models have only three parameters, so they are not as accurate as the cubic-B-spline and/or the modified exponential models. Furthermore, the previous power models cannot estimate $M-\theta_r$ curve, which does not flatten out near final loading. However, the number of data required for this model is drastically reduced. Ang and Morris (1984) used a standardized Ramberg—Osgood function (Ramberg and Osgood, 1943) in the power form. The Ang–Morris model is composed of four parameters. Richard et al. (1980) analytically described the $M-\theta_r$ curve of single plate connections with a nondimensional power model. The procedure used to establish this model was based upon a nonlinear finite-element method in which nonlinear behavior of the bolts and connected plates was modeled by using a force–deformation relationship obtained from single shear bolt tests.

The connection models explained so far represent the behavior under monotonic loading. In addition to these models, the models under cyclic loading also have been presented in view of the necessity to analyze the behavior of semi-rigid frames subject to seismically induced load or alternating wind load (Cook, 1983; Goto, Suzuki, and Chen, 1991, 1993; Huang and Morris, 1991).

In the following section, some of the important connection models are discussed in more detail.

2.4.2 Behavior under Monotonic Loading

2.4.2.1 Frye–Morris Polynomial Model The most popular model for structural analysis is the polynomial function proposed by Frye and Morris (1975). The Frye–Morris model was developed based on a procedure formulated by Sommer (1969). They used the method of least squares to determine the constants of the polynomial. This model has the general form of

$$\theta_r = C_1(KM)^1 + C_2(KM)^3 + C_3(KM)^5 \qquad (2.4.1)$$

where K is a standardization parameter dependent upon the connection type and geometry, and C_1, C_2, and C_3 are curve-fitting constants.

The polynomial model represents the M–θ_r behavior reasonably well. The main drawback is that the nature of a polynomial is to peak and trough within a certain range. Then, the first derivative of this function, which indicates the tangent connection stiffness, may become negative at some value of the connection moment M. This is physically unacceptable. Furthermore, the negative stiffness makes structural analysis difficult if the load-control incremental method is used.

Following the procedure of Frye–Morris, Picard, Giroux, and Brun (1976) and Altman et al. (1982) developed prediction equations to describe the M–θ_r behavior for strap-angle connections and top- and seat-angle connections with double web angles, respectively. Goverdhan (1983) reestimated a size parameter in the standardization constant K for flush end-plate connections to get a good agreement with the M–θ_r curve obtained from experimental results. The curve-fitting constants C_1, C_2, and C_3 and the standardization constant K for each connection type are summarized in Table 2.1, and the size parameters for each type of connection are shown schematically in Fig. 2.6.

2.4.2.2 Modified Exponential Model The Chen–Lui exponential model has been refined to accommodate linear components by Kishi and Chen (1986a, 1986b) and is referred to herein as the modified exponential model.

This model is represented by a function of the following form:

$$M = M_0 + \sum_{j=1}^{m} A_j \left[1 - \exp\left(-\frac{|\theta_r|}{2j\alpha}\right)\right] + \sum_{k=1}^{n} R_k(|\theta_r| - |T_k|)H[|\theta_r| - |T_k|] \qquad (2.4.2)$$

where

A_j, R_k = curve-fitting parameters

$H[\theta]$ = Heaviside's step function (unity for $\theta \geq 0$, zero for $\theta < 0$)

TABLE 2.1 Curve-Fitting Constants and Standardization Constants for the Frye–Morris Polynomial Model (All Size Parameters are in Inches)

Connection Types	Curve-Fitting Constants	Standardization Constants
Single web-angle connection	$C_1 = 4.28 \times 10^{-3}$ $C_2 = 1.45 \times 10^{-9}$ $C_3 = 1.51 \times 10^{-16}$	$K = d_a^{-2.4} t_a^{-1.81} g^{0.15}$
Double web-angle connection	$C_1 = 3.66 \times 10^{-4}$ $C_2 = 1.15 \times 10^{-6}$ $C_3 = 4.57 \times 10^{-8}$	$K = d_a^{-2.4} t_a^{-1.81} g^{0.15}$
Top- and seat-angle with double web-angle connection	$C_1 = 2.23 \times 10^{-5}$ $C_2 = 1.85 \times 10^{-8}$ $C_3 = 3.19 \times 10^{-12}$	$K = d^{-1.287} t_t^{-1.128} t_w^{-0.415} l_t^{-0.694}$
Top- and seat-angle without double web-angle connection	$C_1 = 8.46 \times 10^{-4}$ $C_2 = 1.01 \times 10^{-4}$ $C_3 = 1.24 \times 10^{-8}$	$K = d^{-1.5} t^{-0.5} l_a^{-0.7} d_b^{-1.5}$
End-plate connection without column stiffeners	$C_1 = 1.83 \times 10^{-3}$ $C_2 = -1.04 \times 10^{-4}$ $C_3 = 6.38 \times 10^{-6}$	$K = d_g^{-2.4} t_p^{-0.4} d_b^{-1.5}$
End-plate connection with column stiffeners	$C_1 = 1.79 \times 10^{-3}$ $C_2 = -1.76 \times 10^{-4}$ $C_3 = 2.04 \times 10^{-4}$	$K = d_g^{-2.4} t_p^{-0.6}$
T-stub connection	$C_1 = 2.1 \times 10^{-4}$ $C_2 = 6.2 \times 10^{-6}$ $C_3 = -7.6 \times 10^{-9}$	$K = d^{-1.5} t^{-0.5} l_t^{-0.7} d_b^{-1.1}$
Header-plate connection	$C_1 = 5.10 \times 10^{-5}$ $C_2 = 6.20 \times 10^{-10}$ $C_3 = 2.40 \times 10^{-13}$	$K = t_p^{-1.6} g^{1.6} d_p^{-2.3} t_w^{-0.5}$

M_0 = initial connection moment
α = scaling factor for numerical stability
T_k = starting rotation of kth linear component given from the experimental M–θ_r curve

In general, $m = 6$ is adopted.

By using the linear interpolation technique for the original M–θ_r data, the weight function for each M–θ_r datum is nearly equal. Constants A_j and R_k for the exponential and linear terms of the function are determined by a linear regression analysis.

The instantaneous connection tangent stiffness R_{kt} at an arbitrary relative rotation $|\theta_r|$ can be evaluated by differentiating Eq. (2.4.2) with respect to $|\theta_r|$ as

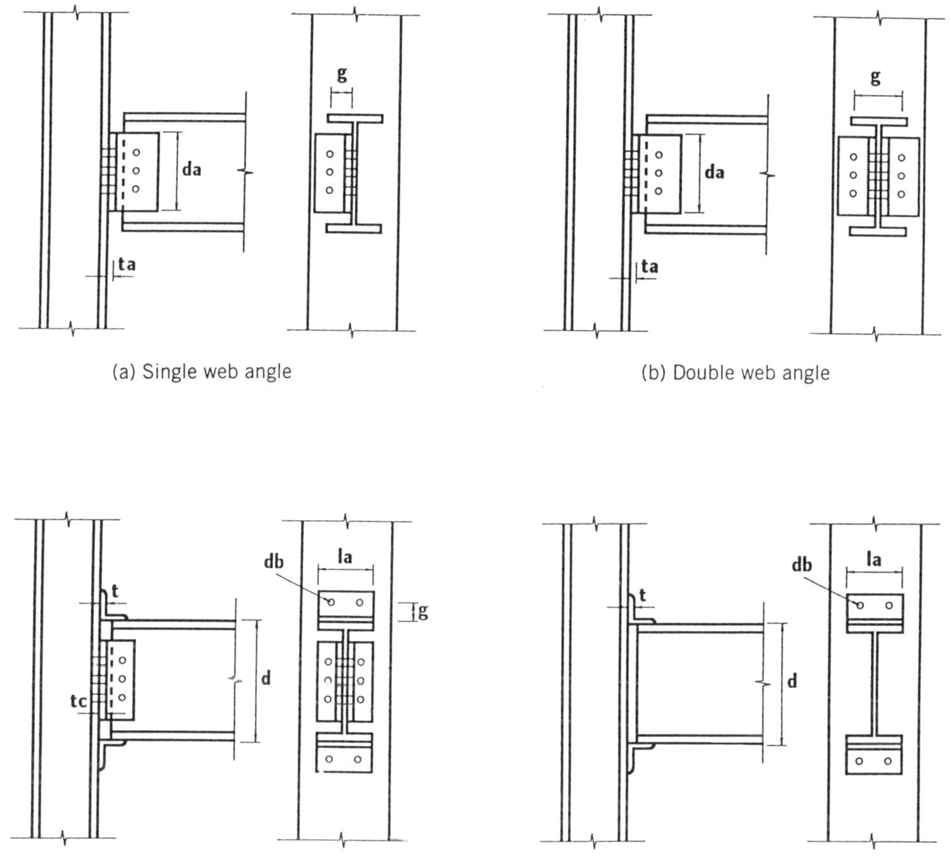

(a) Single web angle (b) Double web angle

(c) Top and seat angles with double web angle (d) Top and seat angles without double web angle

Fig. 2.6 Size parameters for various connection types of the Frye–Morris polynomial model: (a) single-web angle; (b) double-web angle; (c) top and seat angles without double-web angle; (d) top and seat angles with double-web angle; (e) end plate without column stiffeners; (f) end plate with column stiffeners; (g) header plate; and (h) T-stub.

$$R_{kt} = \left.\frac{dM}{d|\theta_r|}\right|_{|\theta_r|=|\theta_r|} = \sum_{j=1}^{m} \frac{A_j}{2j\alpha} \exp\left(-\frac{|\theta_r|}{2j\alpha}\right) + \sum_{k=1}^{n} R_k H[|\theta_r| - |T_k|] \quad (2.4.3)$$

Specifically, the initial stiffness R_{ki} is given by

$$R_{ki} = \left.\frac{dM}{d|\theta_r|}\right|_{|\theta_r|=0} = \sum_{j=1}^{m} \frac{A_j}{2j\alpha} + R_k H[|T_k|]\Big|_{k=1} \quad (2.4.4)$$

This model has the following merits:

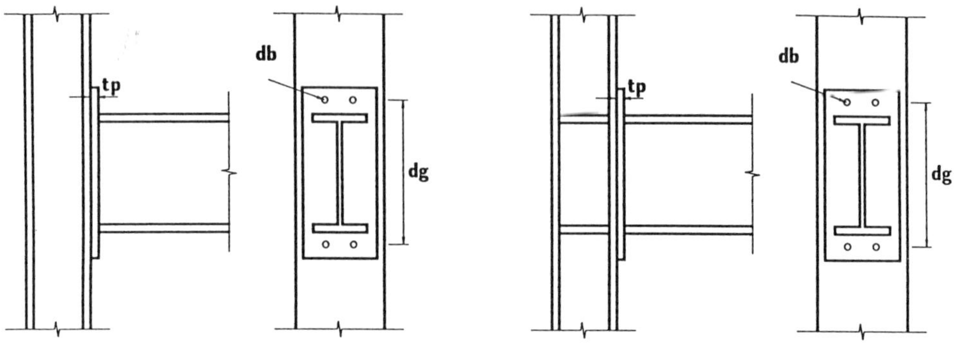

(e) End plate without column stiffeners

(f) End plate with column stiffeners

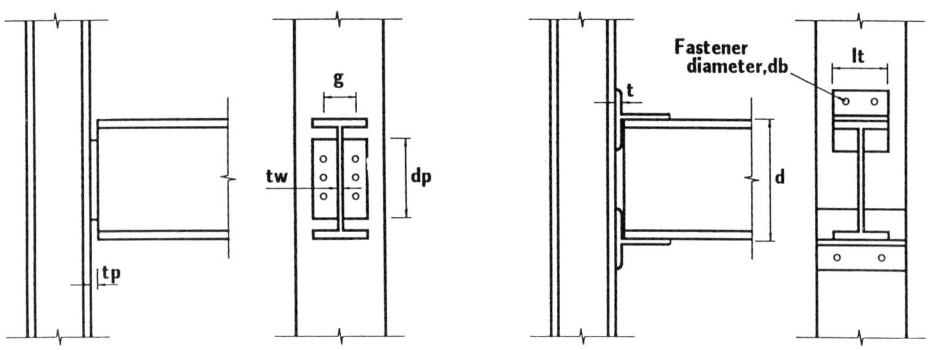

(g) Header plate

(h) T-stub

Fig. 2.6 (*Continued*)

1. The formulation is relatively simple and straightforward.
2. The abrupt change of the connection stiffness among the sampling data is only generated from inherent experimental characteristics.

The comparison between the Chen–Lui exponential model (1985) and the modified exponential model for numerical example of the test data, including a linear component, is shown in Fig. 2.7.

2.4.2.3 Three-Parameter Power Model The modified exponential model mentioned before is a curve-fitting equation obtained by using the least-mean-square technique for the experimental test data. From a different viewpoint, Chen and Kishi (1987) and Kishi et al. (1988a, 1988b) developed another

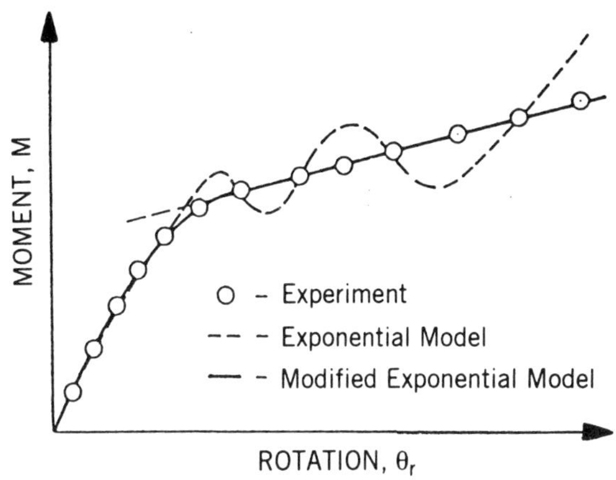

Fig. 2.7 Comparison between results by exponential and modified exponental models for M–θ data including a linear term.

procedure to predict the moment–rotation characteristics of steel beam-to-column connections. In this procedure, the initial connection stiffness and ultimate moment capacity of the connection are determined by a simple analytical model. By using these quantities so obtained, a three-parameter power model given by Richard and Abbott (1975) was adopted to represent the moment–rotation behavior of the connection.

The generalized form of this model is

$$M = \frac{R_{ki}\theta_r}{[1 + (\theta_r/\theta_0)^n]^{1/n}} \tag{2.4.5}$$

where

M_u = ultimate moment capacity
n = shape parameter
R_{ki} = initial connection stiffness
θ_0 = reference plastic rotation

Equation (2.4.5) has the shape shown in Fig. 2.8. From this figure, it is recognized that the larger the power index n, the steeper the curve. Shape parameter n can be determined by using the method of least squares for the differences between the predicted moments and the experimental test data.

This power model is an effective tool for designers to execute the second-order nonlinear structural analysis quickly and accurately. This is because the tangent connection stiffness R_{kt} and relative rotation θ_r can be determined directly from Eq. (2.4.5) without iteration, in which the tangent connection stiffness R_{kt} is

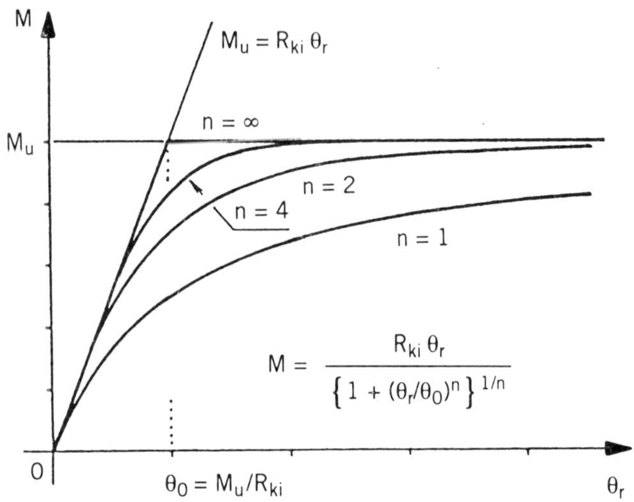

Fig. 2.8 Three-parameter power model.

$$R_{kt} = \frac{dM}{d\theta_r} = \frac{R_{ki}}{[1 + (\theta_r/\theta_0)^n]^{(n+1)/n}} \tag{2.4.6}$$

and the rotation θ_r is

$$\theta_r = \frac{M}{R_{ki}[1 - (M/M_u)]^n]^{1/n}} \tag{2.4.7}$$

The values of the initial connection stiffness R_{ki}, the ultimate moment capacity M_u, and shape parameter n of the three-parameter model are obtained for four types of connections: single/double web-angle connections and top and seat angle with/without double web-angle connections (Kishi et al., 1993). Herein, the procedures to determine the values of the three parameters are explained in some detail. The dimensions used for angles shown in Fig. 2.9 are defined as

g = distance between heel to center of fastener closest to web or flange of beam
$I_0 = t^3/12$ = geometrical moment of inertia
k = gauge distance from heel to top of fillet
l = angle length
$M_0 = \sigma_y t^2/4$ = pure elastic bending moment per unit length of angle
t = angle thickness

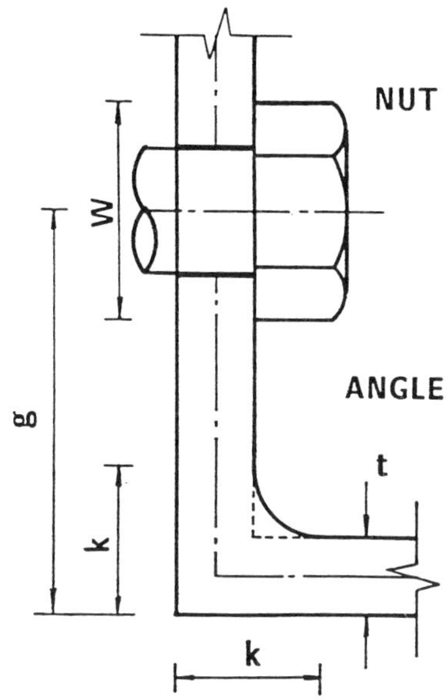

Fig. 2.9 Main parameters for an angle.

W = nut width
σ_y = yield stress of angle

Furthermore, we introduce the following nondimensional parameters:

$$\beta = \frac{g}{l}, \qquad \gamma = \frac{l}{t}, \qquad \delta = \frac{d}{t}, \qquad \kappa = \frac{k}{t}, \qquad \omega = \frac{W}{t}, \qquad \rho = \frac{t_w}{t_t}$$

in which d is the height of the beam and subscripts t and w denote the top angle and web angle, respectively.

The procedures to calculate the initial connection stiffness R_{ki} and the ultimate moment capacity M_u are summarized for the respective connections in the following.

Single/Double Web-Angle Connections. The values of R_{kiw} and M_{uw} for single web-angle connections are formulated as

$$\frac{R_{kiw}}{EI_{0w}} = \frac{12\alpha \cosh(\alpha\beta'_w)}{7.8[(\alpha\beta'_w \cosh(\alpha\beta'_w) - \sinh(\alpha\beta'_w)]} \tag{2.4.8}$$

$$\frac{M_{uw}}{M_{0w}t_w} = \frac{2\xi_w + 1}{3}\gamma_w^2 \tag{2.4.9}$$

where $\alpha = 4$ and β'_w is defined as

$$\beta'_w = \beta_w - \frac{1}{\gamma_w}\left(\kappa_w + \frac{\omega_w}{2}\right) \tag{2.4.10}$$

ξ_w in Eq. (2.4.9) is obtained by solving Eq (2.4.10), which is derived by combining the Drucker interaction equation between bending moment and shearing force with the Tresca yielding criterion as

$$\xi_w^4 + (\beta_w \gamma_w - \kappa_w)\xi_w - 1 = 0 \tag{2.4.11}$$

The nondimensional initial connection stiffness as a function of β'_w for single web-angle connections is shown in Fig. 2.10. The nondimensional ultimate connection moment is shown in Fig. 2.11, where β_w is taken as the abscissa and γ_w is varied from 5 to 20 with an increment of 5 for $\kappa_w = 1.5$ and 2.0.

The case of double web-angle connections can be obtained simply by doubling the values found from these figures.

Top- and Seat-Angle Connection without a Double Web Angle. Assuming that the center of rotation is located at the angle leg adjacent to the compression beam flange and the top angle acts as a cantilever beam to resist surcharged moment, the initial connection stiffness R_{kits} is obtained as

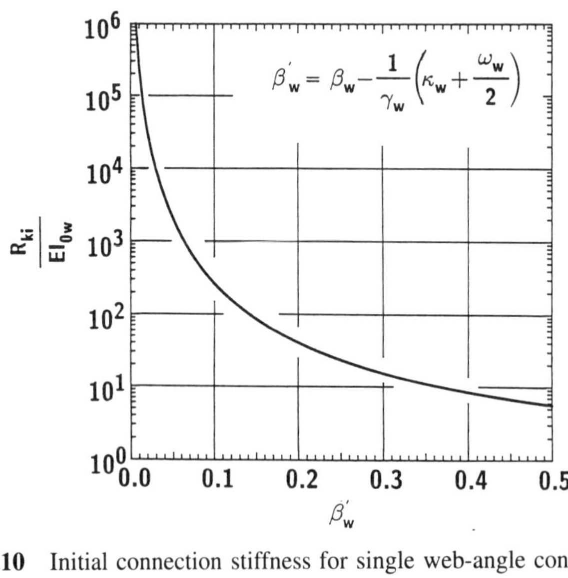

Fig. 2.10 Initial connection stiffness for single web-angle connections.

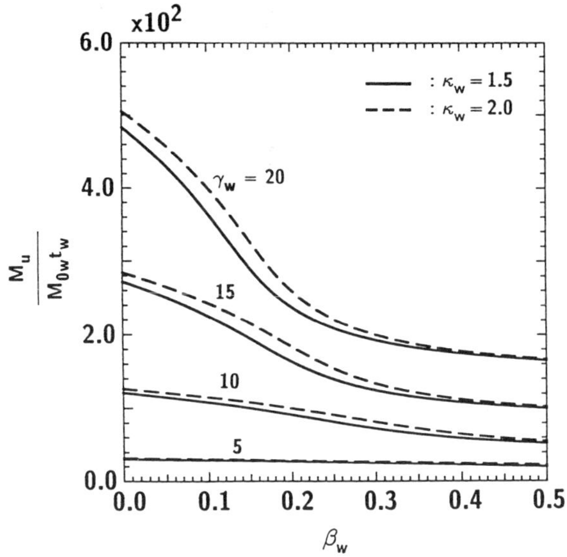

Fig. 2.11 Ultimate moment capacity for single web-angle connections.

$$\frac{R_{kits}}{EI_{0t}} = (1 + \delta_t)^2 D_{ts} \tag{2.4.12}$$

where D_{ts} is a function of β_t' and γ_t, and D_{ts} and β_t' are given by

$$D_{ts} = \frac{3}{\beta_t'[\gamma_t^2(\beta_t')^2 + 0.78]} \tag{2.4.13}$$

$$\beta_t' = \beta_t - \frac{1 + \omega_t}{2\gamma_t} \tag{2.4.14}$$

The ultimate moment capacity M_{uts} is obtained by assuming a simple failure mechanism. The equation for M_{uts} is given by

$$\frac{M_{uts}}{M_{0t}t_t} = \gamma_t[1 + \xi_t(1 + \beta_t^*) + 2(\kappa_t + \delta_t)] \tag{2.4.15}$$

where the variable ξ_t is a nondimensional ultimate shearing force acting at the plastic hinge. Here, as in the case of single web-angle connections, it is obtained by the following equation:

$$\xi_t^4 + \beta_t^* \xi_t - 1 = 0 \tag{2.4.16}$$

where β_t^* is defined as

$$\beta_t^* = \beta_t' \gamma_t - \kappa_t \tag{2.4.17}$$

The distribution of the coefficient D_{ts} for R_{kits} with respect to β_t' are shown in Fig. 2.12. where γ_t is varied from 5 to 40 with an increment of 5. Figure 2.13 is the chart for the nondimensional ultimate moment capacity M_{uts}. Figure 2.13a shows the results with variation of γ_t for $\delta_t = 40$ and 80, and Fig. 2.13b shows the variation of δ_t (10 to 80) for $\gamma_t = 5$ and 40.

Top- and Seat-Angle Connection with Double Web Angle. In this type of connection, the initial connection stiffness R_{ki} and the ultimate moment capacity M_u can be evaluated by separating the top- and seat-angle part and the double web-angle part as

$$\frac{R_{ki}}{EI_{0t}} = \frac{R_{kits}}{EI_{0t}} + \frac{R_{kiw}}{EI_{0t}} \tag{2.4.18a}$$

$$\frac{M_u}{M_{0t}t_t} = \frac{M_{uts}}{M_{0t}t_t} + \frac{M_{uw}}{M_{0t}t_t} \tag{2.4.18b}$$

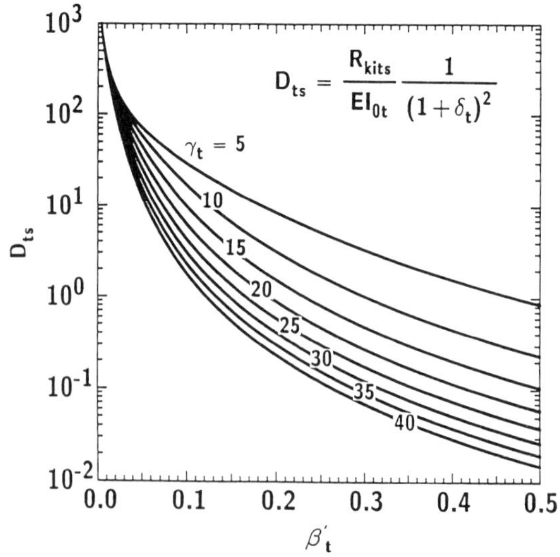

Fig. 2.12 Coefficient D_{ts} for R_{kits} of top and seat angle without double web-angle connections.

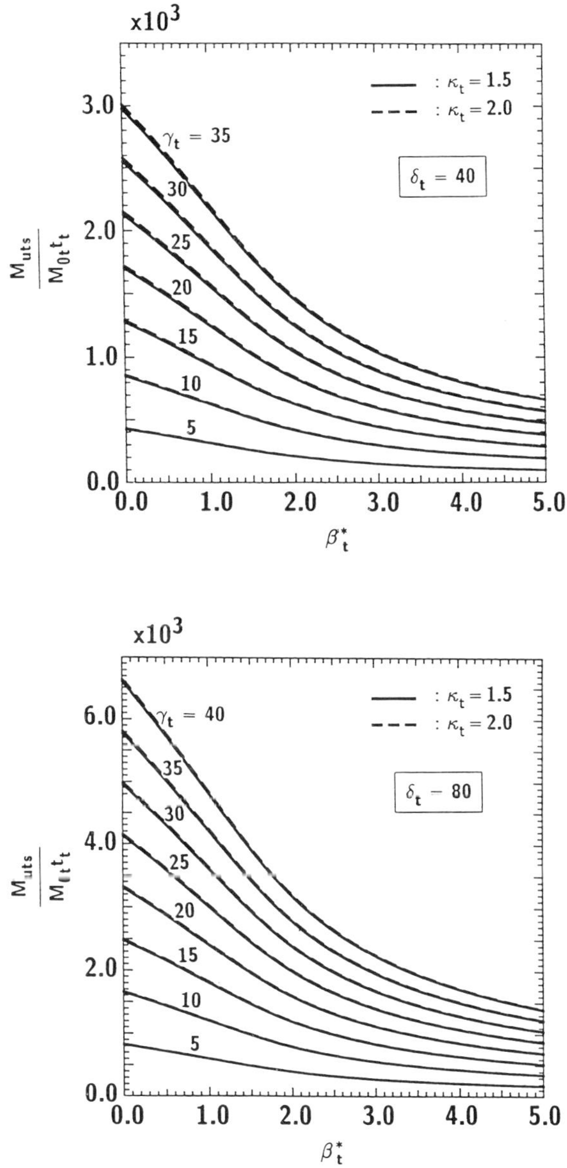

(a)

Fig. 2.13 Ultimate moment capacity for top and seat angle without double web-angle connections. (a) Varying γ_t for $\delta_t = 40$ and 80. (b) Varying γ_t for $\delta_t = 5$ and 40.

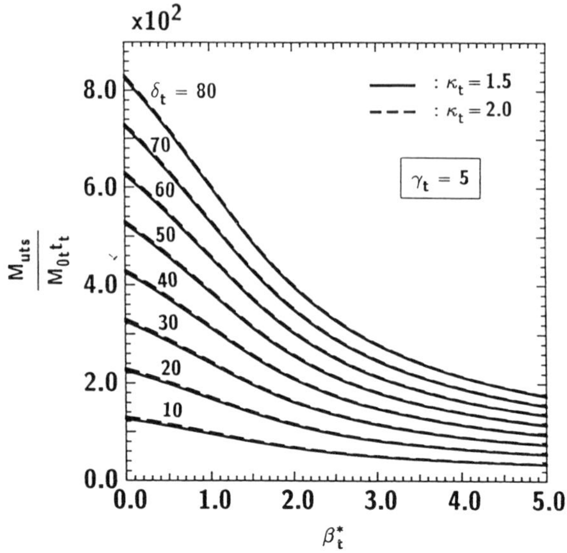

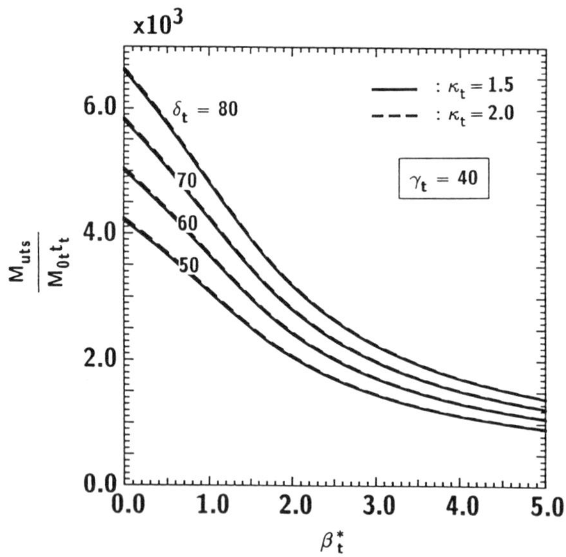

(b)

Fig. 2.13 (*Continued*)

For the top- and seat-angle parts of Eqs. (2.4.18a), (2.4.18b), (2.4.12), and (2.4.15) can be used to evaluate R_{kits}/EI_{0t} and $M_{uts}/M_0 t_t$.

As for the web-angle parts, the deformation pattern and failure mechanism of the present connection type are experimentally observed to be slightly different from those of the single/double web-angle connection. Thus, in lieu of Eqs. (2.4.8) and (2.4.9), Kishi et al. (1993) recommended the following formulas to calculate R_{kiw}/EI_{0t} and $M_{uw}/M_{0t}t_t$:

$$\frac{R_{kiw}}{EI_{0t}} = (1 + \delta_t)^2 \rho D_w \tag{2.4.19}$$

where D_w is

$$D_w = \frac{3}{2\beta'_w[\gamma_w^2(\beta'_w)^2 + 0.78]} \tag{2.4.20}$$

where β'_w is given by Eq. (2.4.14).

$$\frac{M_{uw}}{M_{0t}t_t} = \gamma_w(1 + \xi_w)\left[\frac{\xi_w - 1}{3(\xi_w + 1)}\gamma_w + \delta_w + \frac{1}{\rho}\right]\rho^3 \tag{2.4.21}$$

where ξ_w is obtained by solving Eq. (2.4.11).

The distributions of the coefficient D_w for R_{kiw}/EI_{0t} are given in Fig. 2.14.

M_{uw} has a total of five variables: β_w, δ_w, γ_w, ρ, and κ_w. By taking β_w as the abscissa, two types of charts to evaluate the value of $M_{uw}/M_{0t}t_t$ are shown in Figs. 2.15a and 2.15b. In both figures, γ_w is varied from 5 to 35 and/or 40 with an increment of 5.

Shape Parameter n. To determine the M–θ_r curves based on the three-parameter model, shape parameter n in Eq. (2.4.5) has to be calculated in addition to the initial stiffness R_{ki} and the ultimate moment capacity M_u. Herein, shape parameter n is calculated by the following procedures.

1. The shape parameter n for each experimental test data is numerically determined by the least-mean-square technique.
2. The shape parameter was assumed to be a linear function of $\log_{10} \theta_0$.
3. The constants of the linear function are calculated by the least-square-method for each connection type.

The empirical equations to calculate the shape parameter are summarized in Table 2.2.

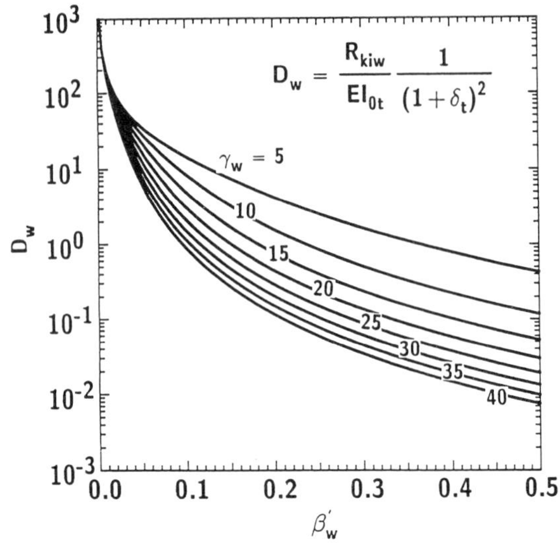

Fig. 2.14 Coefficient D_w for R_{kiw}/EI_{0t} of top and seat angle with double web-angle connections.

2.4.3 Behavior under Cyclic Loading

2.4.3.1 Independent Hardening Model The independent hardening model (Chen and Saleeb, 1982) may be one of the simplest connection models to express the behavior under cyclic loading. Because this model is simple to use, it is often implemented in the frame-analysis program (Goto, Suzuki, and Chen, 1989; Kishi et al., 1991). Cyclic moment–rotation relation based on this model is schematically shown in Fig. 2.16. In the case of unloading from point a on the initial loading curve, the connection unloads linearly down to $M = 0$, with the initial stiffness of R_{ki}. When the sign of the connection moment M changes at point b in the unloading process, the connection enters into the reverse loading process. In this case, the connection behavior, excluding the residual plastic rotation, resulting from the previous loading process is assumed to coincide with that of virgin connections under monotonic loading. The implementation of this model into frame analysis is explained in some detail in Section 3.4.

2.4.3.2 Kinematic Hardening Model A connection model considering kinematic hardening is presented by Huang and Morris (1991). The behavior of this connection model is illustrated in Fig. 2.17. To express the connection behavior in the initial monotonic loading, this model adopts the Richard–Abbott function in the form

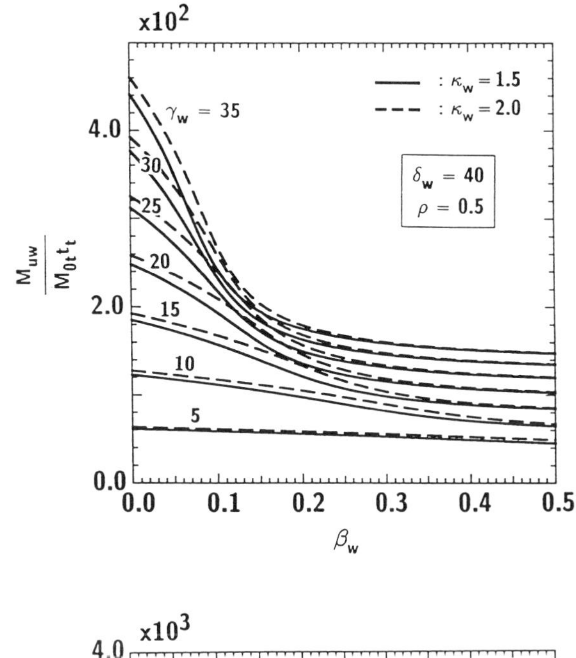

(a)

Fig. 2.15 Ultimate moment for the variation of top and seat angle with double web-angle connections. (a) $\rho = 0.5$ and 1.0 for $\delta_w = 40$. (b) $\rho = 0.5$ and 1.0 for $\delta_w = 80$.

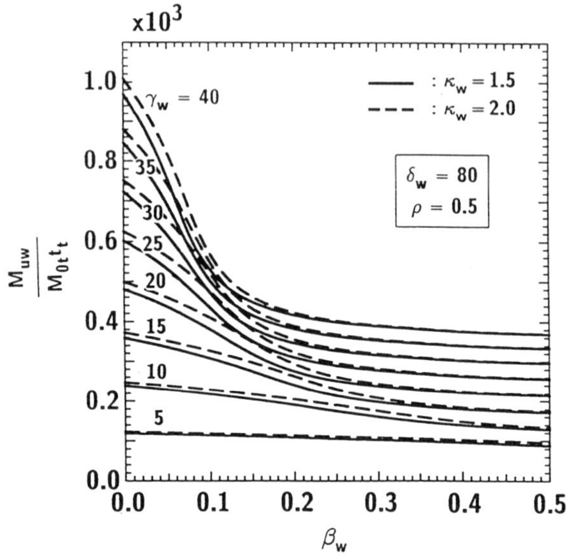

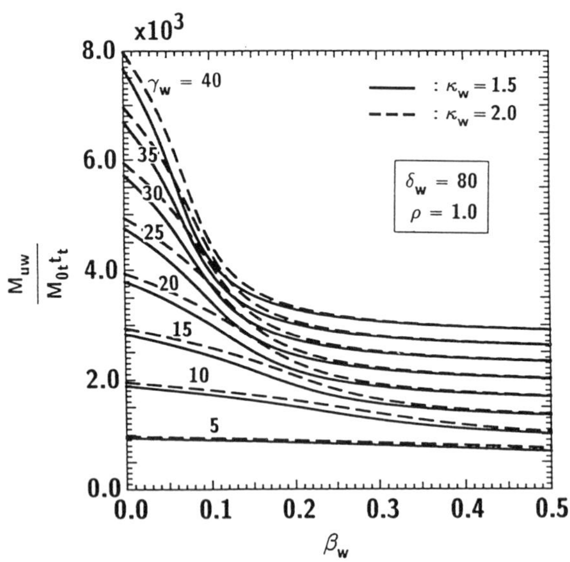

(b)

Fig. 2.15 *(Continued)*

TABLE 2.2 Empirical Equations for Shape Parameter n

Type No.	Connection Type	n
1	Single web-angle connection	$0.520 \log_{10} \theta_0 + 2.291$ $(\log_{10} \theta_0 > -3.073)$ 0.695 $(\log_{10} \theta_0 \leq -3.073)$
2	Double web-angle connection	$1.322 \log_{10} \theta_0 + 3.952$ $(\log_{10} \theta_0 > -2.582)$ 0.573 $(\log_{10} \theta_0 \leq -2.582)$
3	Top- and seat-angle connection (without double web-angle)	$1.398 \log_{10} \theta_0 + 4.631$ $(\log_{10} \theta_0 > -2.721)$ 0.827 $(\log_{10} \theta_0 \leq -2.721)$
4	Top- and seat-angle connection (with double web-angle)	$2.003 \log_{10} \theta_0 + 6.070$ $(\log_{10} \theta_0 > -2.880)$ 0.302 $(\log_{10} \theta_0 \leq -2.880)$

$$M = R_{ki}\theta_r \left(\frac{1 - R_b/R_{ki}}{\{1 + [(1 - R_b/R_{ki})\theta_r/\theta_0]^n\}^{1/n}} + \frac{R_b}{R_{ki}} \right) \tag{2.4.22}$$

where

n = shape parameter
R_b = slope of the asymptotic line

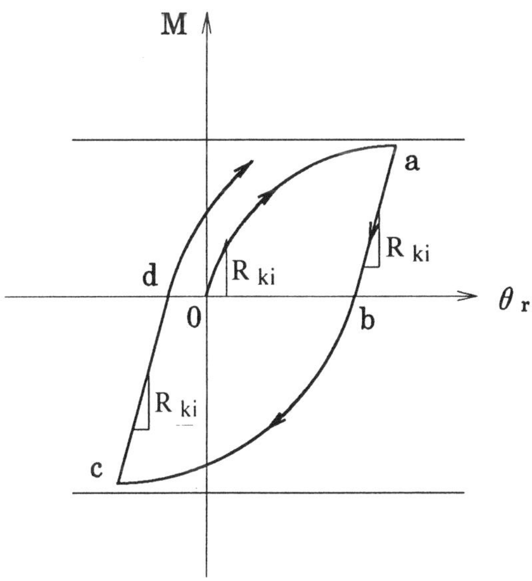

Fig. 2.16 Independent hardening model.

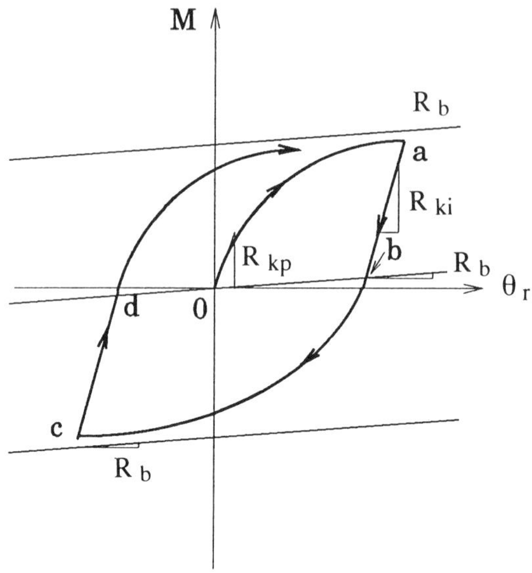

Fig. 2.17 Kinematic hardening model.

R_{ki} = initial connection stiffness
θ_0 = reference plastic rotation

The preceding function is different from Eq. (2.4.5) in the point that the slope of the asymptotic line is not zero. In other words, Eq. (2.4.22) coincides with Eq. (2.4.5) when R_b is zero.

In the event of a moment reversal at point *a* on the curve in Fig. 2.17, the linear relationship with the slope of the initial connection stiffness R_{ki} is applied until it intersects a line through point 0 with a slope of R_b. If the moment reversal continues, a double mirror image of the initial Richard–Abbot function is applicable.

2.4.3.3 Bounding Surface Model with Internal Variables The previously mentioned independent and kinematic hardening models are simple to use in frame analysis. Although these models can take care of the moment–rotation relation under one cycle of loading, unloading, and reverse loading, the connection behavior under the repetition of this loading cycle may not be expressed with acceptable accuracy. As shown in Fig. 2.18, one problem associated with the independent and kinematic hardening models is that the moment–rotation relation is completely different according to whether the connection is reloaded from the region between *a* and *b* or from the region between *b* and *c*. To overcome this deficiency, a bounding surface model with

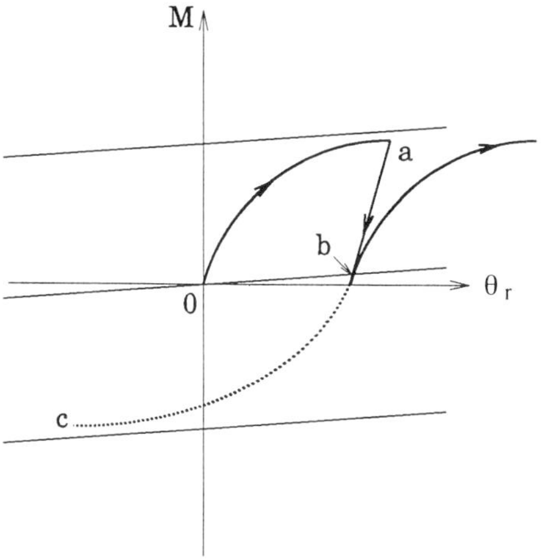

(a) Reloading from the region between b and c

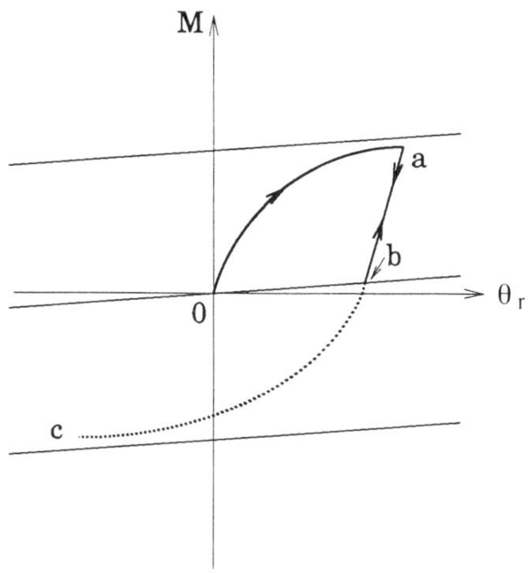

(b) Reloading from the region between a and b

Fig. 2.18 Problems associated with the kinematic hardening model. (a) Reloading from the region between *b* and *c*. (b) Reloading from the region between *a* and *b*.

internal variables (Dafalias and Popov, 1976) may be used (Cook, 1983; Goto et al., 1991, 1993).

In this model, the moment–rotation relation is defined in incremental form:

$$\Delta M = R_{kt} \Delta \theta_r \tag{2.4.23}$$

where R_{kt} is the tangent stiffness of connections. R_{kt} is expressed in terms of the initial stiffness R_{ki} and the plastic tangent stiffness R_{kp} as

$$R_{kt} = \frac{R_{ki} R_{kp}}{R_{ki} + R_{kp}} \tag{2.4.24}$$

The plastic tangent stiffness R_{kp} is expressed by using the plastic internal variables δ and δ_{in} as

$$R_{kp} = R_b + h \frac{\delta}{\delta_{in} - \delta} \tag{2.4.25}$$

where h is the hardening shape parameter, R_b is the slope of the bounding line, δ is the distance of the current moment state from the corresponding bound, and δ_{in} is the value of δ at the initiation for each loading process.

These quantities are schematically shown in Fig. 2.19 using the moment–plastic rotation ($M–\theta_r^p$) curve.

2.5 CONNECTION DATABASE

2.5.1 General Remarks

Full-scale experiments are the only way to obtain the real moment–rotation characteristics of semi-rigid connections. Wilson and Moore (1917) performed the first experiment. Since then, extensive tests have been conducted. Jones et al. (1980, 1983) collected a total of 323 tests from 29 studies. Including these test data, Nethercot (1985a, 1985b) examined and evaluated more than 800 individual tests from over 70 experimental papers. Goverdhan (1983) collected a total of 230 experimental moment–rotation curves and digitized them to form the database of connection behavior.

The database (Kishi and Chen, 1986a, 1986b) presented herein was compiled by expanding and extending Goverdhan's collection to a total of 303 tests. The literature survey encompassed experimental data published from 1936 (Rathbun) to 1985, on riveted, bolted, and welded connections. All of the 303 tests collected so far are classified into seven types, as shown in Table 2.3. The references and number of experimental curves of each connection

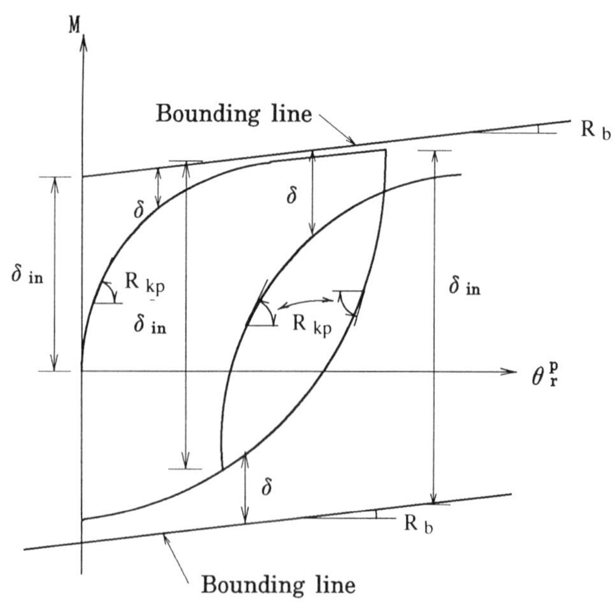

Fig. 2.19 Bounding surface model with internal variables.

TABLE 2.3 Connection Types

Type Number	Connection Type
1	Single web-angle connections
2	Double web-angle connections
3	Top- and seat-angle connections with double web angle
4	Top- and seat-angle connections
5	Extended end-plate connections
6	Flush end-plate connections
7	Header-plate connections

type are listed in Table 2.4. These test data are stored in the computer as the steel beam-to-column connection database at the Purdue University Computer Center. The items from each test as registered in the database are as follows:

1. Connection type and fastening mode
2. Author, test I.D., and tested country
3. Material properties and size of fasteners
4. Material strength of the angles designed for connection elements
5. All parameters used in beam-to-column connections
6. Moment–rotation test data

Numerical values registered in the database are standardized in U.S. customary units by executing a subprogram as the preprocessor. The connection type is classified as shown in Table 2.3 and the fastening mode as shown in Table 2.5. The notations of major parameters for each connection type are identified in the next section.

2.5.2 Parameter Definition for Connection Type

2.5.2.1 Single Web-Angle Connections and Single Plate Connections Collected experimental moment–rotation data for a single web-angle with bolted beam-to-column connections are tabulated in Table 2.6. A single plate and a single web angle with bolted to the beam and welded to the column connections are tabulated in Table 2.7. The notations for major parameters of these types of connections as shown in Figs. 2.20 and 2.21 are identified as follows:

cl = distance from center of lower fastener holes to bottom edge of angle in beam web

cu = distance from center of upper fastener holes to top edge of angle in beam web

TABLE 2.4 Authors and Numbers of Experimental M–θ_r Curves for Each Connection Type

References for Experimental Curves	Number of Tests
Single Web-Angle Connections, Single Plate Connections	
Lipson, 1968	30
Thompson et al., 1970	12
Lipson, 1977	8
Richard et al., 1982	4
Double Web-Angle Connections	
Rathbun, 1936	7
Bell et al., 1958	4
Lewitt et al., 1966	6
Sommer, 1969	4
Thompson et al., 1970	48
Bose, 1981	1
Top- and Seat-Angle Connections with Double Web Angle	
Rathbun, 1936	3
Azizinamini et al., 1985	20
Top- and Seat-Angle Connections without Double Web Angle	
Rathbun, 1936	3
Hechtman et al., 1947	12
Maxwell et al., 1981	12
Marley et al., 1982	26
Extended End-Plate Connections	
Johnson et al., 1960	1
Sherbourne, 1961	5
Bailey, 1970	26
Surtees et al., 1970	6
Packer et al., 1977	3
Ioannides, 1978	6
Dews, 1979	3
Grundy et al., 1980	2
Johnstone et al., 1981	8
Flush End-Plate Connections	
Ostrander, 1970	24
Header-Plate Connections	
Sommer, 1969	20

TABLE 2.5 Fastening-Mode Patterns

Pattern Number, N	Fastening-Mode Pattern
1	All riveted
2	All bolted
3	Riveted to beam and bolted to column
4	Bolted to beam and riveted to column
5	Riveted to beam and welded to column
6	Bolted to beam and welded to column
7	Welded to beam and riveted to column
8	Welded to beam and bolted to column
9	All riveted without column stiffeners
10	All riveted with column stiffeners
11	All bolted without column stiffeners
12	All bolted with column stiffeners

gb = distance from angle's heel to center of bolt holes in leg adjacent to beam web face
gc = distance from angle's heel to center of bolt holes in leg adjacent to column face
ll = distance from bottom edge of angle to compression flange of beam
lp = angle height
lu = distance from top edge of angle to tension flange of beam
nb = total number of fasteners in beam
nc = total number of fasteners in column
pb = distance between two rows of fasteners in beam web
pc = distance between two rows of fasteners in column
qb = distance between two lines of fasteners in beam web
qc = distance between two lines of fasteners in column
ta = angle thickness

2.5.2.2 Double Web-Angle Connections Collected experimental moment–rotation data for double web-angle connections are tabulated in Table 2.8. The notations for major parameters of this connection type as shown in Fig. 2.22 are identified as follows:

cl = distance from center of lower fastener holes to bottom edge of angle in beam web
cu = distance from center of upper fastener holes to top edge of angle in beam web
gb = distance from angle's heel to center of bolt holes in leg adjacent to beam web face
gc = distance from angle's heel to center of bolt holes in leg adjacent to column face

TABLE 2.6 Collected Test Data for Single Web-Angle with Bolted Beam-to-Column Connections

No.	Author	Test I.D.	Angle	*l* (in.)	*gc* (in.)	*pc* (in.)	*gb* (in.)	*db*	*nc*	*nb*
1	Lipson, 1968	AA-2/1	4 × 3.5 × 1/4	5.50	2.56	2.25	3.00	3/4	1 × 2	1 × 2
2		AA-2/2	4 × 3.5 × 1/4	5.50	2.56	2.25	3.00	3/4	1 × 2	1 × 2
3		AA-3/1	4 × 3.5 × 1/4	8.50	2.56	2.25	3.00	3/4	1 × 3	1 × 3
4		AA-3/2	4 × 3.5 × 1/4	8.50	2.56	2.25	3.00	3/4	1 × 3	1 × 3
5		AA-4/1	4 × 3.5 × 1/4	11.50	2.56	2.25	3.00	3/4	1 × 4	1 × 4
6		AA-4/2	4 × 3.5 × 1/4	11.50	2.56	2.25	3.00	3/4	1 × 4	1 × 4
7		AA-5/1	4 × 3.5 × 1/4	14.50	2.56	2.25	3.00	3/4	1 × 6	1 × 6
8		AA-5/2	4 × 3.5 × 1/4	14.50	2.56	2.25	3.00	3/4	1 × 6	1 × 6
9		AA-6/1	4 × 3.5 × 1/4	17.50	2.56	2.25	3.00	3/4	1 × 6	1 × 6
10		AA-6/2	4 × 3.5 × 1/4	17.50	2.56	2.25	3.00	3/4	1 × 6	1 × 6
11		BB-4/1	4 × 3.5 × 5/16	11.50	2.56	2.25	3.00	3/4	1 × 4	1 × 4
12		BB-4/2	4 × 3.5 × 5/16	11.50	2.56	2.25	3.00	3/4	1 × 4	1 × 4
13		B2-4	4 × 3.5 × 5/16	11.50	2.56	2.25	3.00	3/4	1 × 4	1 × 4
14		B3-4	4 × 3.5 × 5/16	11.50	2.56	2.25	3.00	3/4	1 × 4	1 × 4
15		C-4	5 × 3.5 × 5/16	11.50	1.94	2.25	3.00	3/4	1 × 4	1 × 4
16		D-4	4 × 3.5 × 5/16	11.50	1.94	3.25	3.00	3/4	1 × 4	1 × 4

TABLE 2.7 Collected Test Data for Single Web-Angle with Bolted to Beam and Welded to Column Connections

No.	Author	Test I.D.	Angle	l (in.)	pc (in.)	db	nb
1	Lipson, 1968	P1-2/1	—	5.50	2.50	3/4	1 × 2
2		P1-2/2	—	5.50	2.50	3/4	1 × 2
3		P1-3/1	—	8.50	2.50	3/4	1 × 3
4		P1-3/2	—	8.50	2.50	3/4	1 × 3
5		P1-4/1	—	11.50	2.50	3/4	1 × 4
6		P1-4/2	—	11.50	2.50	3/4	1 × 4
7		P1-5/1	—	14.50	2.50	3/4	1 × 5
8		P1-5/2	—	14.50	2.50	3/4	1 × 5
9		P1-6/1	—	17.50	2.50	3/4	1 × 6
10		P1-6/2	—	17.50	2.50	3/4	1 × 6
11		Q-2/1	4 × 3.5 × 5/16	4.75	2.50	3/4	1 × 2
12		Q-3/1	4 × 3.5 × 5/16	8.00	2.50	3/4	1 × 3
13		Q-4/1	4 × 3.5 × 5/16	12.00	2.50	3/4	1 × 4
14		Q-5/1	4 × 3.5 × 5/16	14.50	2.50	3/4	1 × 5
15	Lipson, 1977	A1-5	4 × 3 × 3/8	14.50	2.50	3/4	1 × 5
16		A1-6	4 × 3 × 3/8	17.50	2.50	3/4	1 × 6
17		A1-7	4 × 3 × 3/8	20.50	2.50	3/4	1 × 7
18		A1-8	4 × 3 × 3/8	23.50	2.50	3/4	1 × 8
19		A1-9	4 × 3 × 3/8	26.50	2.50	3/4	1 × 9
20		A1-10	4 × 3 × 3/8	29.50	2.50	3/4	1 × 10
21		A1-11	4 × 3 × 3/8	32.50	2.50	3/4	1 × 11
22		A1-12	4 × 3 × 3/8	35.50	2.50	3/4	1 × 12
23	Thompson et al., 1970	F-1 alt	—	11.50	2.25	3/4	2 × —
24		F-1 alt	—	11.50	2.25	3/4	2 × —
25		G-1 alt	—	11.50	2.25	3/4	2 × —
26		G-1 alt	—	11.50	2.25	3/4	2 × —
27		H-1 alt	—	11.50	2.25	3/4	2 × —
28		H-1 alt	—	11.50	2.25	3/4	2 × —

TABLE 2.7 Collected Test Data for Single Web-Angle with Bolted to Beam and Welded to Column Connections (*Continued*)

No.	Author	Test I.D.	Angle	l (in.)	pc (in.)	db	nb
29		F-2-1 alt	—	11.50	2.25	3/4	2 × —
30		F-2-1 alt	—	11.50	2.25	3/4	2 × —
31		F-2-2 alt	—	11.50	2.25	3/4	2 × —
32		F-2-2 alt	—	11.50	2.25	3/4	2 × —
33		F-2-3 alt	—	11.50	2.25	3/4	2 × —
34		F-2-3 alt	—	11.50	2.25	3/4	2 × —
35	Richard et al., 1982	RGKL2	—	6.00	—	3/4	1 × 2
36		RGKL3	—	9.00	—	3/4	1 × 3
37		RGKL5	—	15.00	—	3/4	1 × 5
38		RGKL7	—	21.50	—	7/8	1 × 7

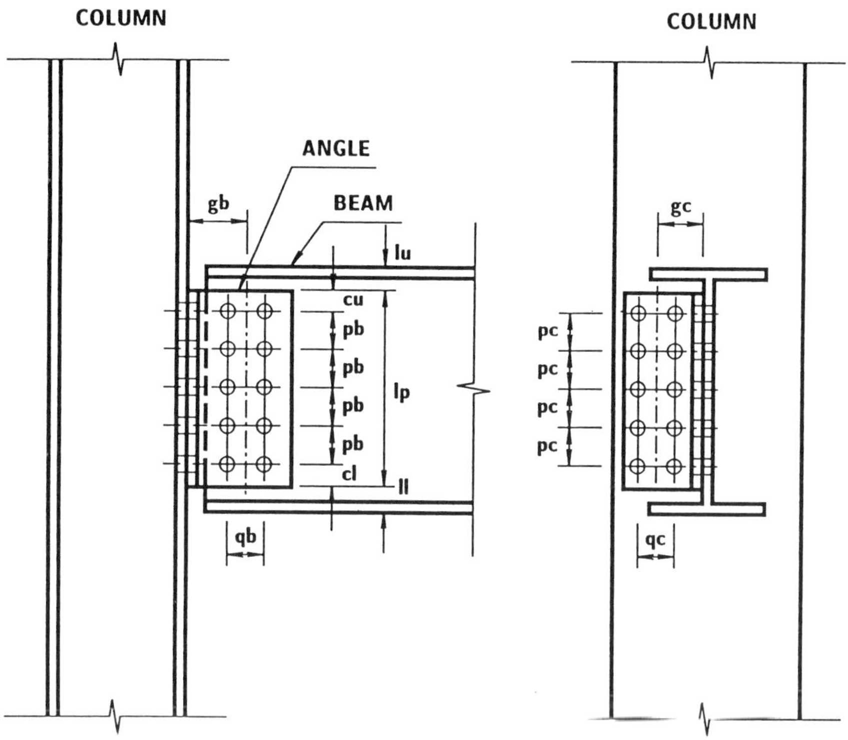

Fig. 2.20 Size parameters for single web-angle connections.

ll = distance from bottom edge of angle to compression flange of beam
lp = angle height
lu = distance from top edge of angle to tension flange of beam
nb = total number of fasteners in beam
nc = total number of fasteners in column
pb = distance between two rows of fasteners in beam web
pc = distance between two rows of fasteners in column
qb = distance between two lines of fasteners in beam web
qc = distance between two lines of fasteners in column
ta = angle thickness

2.5.2.3 Top- and Seat-Angle Connections with Double Web Angle The experimental studies on this connection type were mainly performed by Altman et al. (1982) and Azizinamini et al. (1985) at the University of South Carolina. The test program included specimen subject to both cyclic loading as well as static loading. However, cyclic loading test data are excluded from the database, as listed in Table 2.9. The notations for major parameters of this connection type as shown in Fig. 2.23 are as follows:

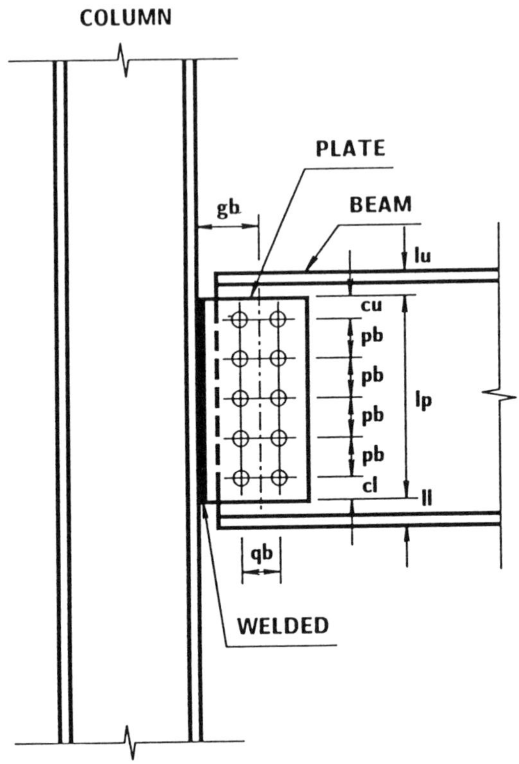

Fig. 2.21 Size parameters for single plate connections.

On the Web Angle

cl = distance from center of lower fastener holes to bottom edge of angle in beam web
cu = distance from center of upper fastener holes to top edge of angle in beam web
gb = distance from angle's heel to center of bolt holes in leg adjacent to beam web face
gc = distance from angle's heel to center of bolt holes in leg adjacent to column face
ll = distance from bottom edge of angle to compression flange of beam
lp = angle height
lu = distance from top edge of angle to tension flange of beam
nb = total number of fasteners in beam
nc = total number of fasteners in column
pb = distance between two rows of fasteners in beam web
pc = distance between two rows of fasteners in column

TABLE 2.8 Collected Test Data for Double Web-Angle Connections

No.	Author	Test I.D.	Angle	l (in.)	gc (in.)	pc (in.)	gb (in.)	db	nc	nb
1	Rathbun, 1936	A-1	6 × 4 × 3/8	2.50	2.63	3.50		7/8	1 × 1	2 × 1
2		A-2	6 × 4 × 3/8	6.00	2.63	3.50	3.00	7/8	1 × 2	2 × 2
3		A-3	6 × 6 × 3/8	6.00	2.63	3.50	3.00	7/8	2 × 2	2 × 2
4		A-4	4 × 3.5 × 3/8	9.00	2.56	2.25	3.00	7/8	1 × 3	1 × 3
5		A-5	6 × 6 × 3/8	9.00	3.56	3.50	3.00	7/8	2 × 3	2 × 3
6		A-6	4 × 3.5 × 3/8	15.00	2.50	2.25	3.00	7/8	1 × 5	1 × 5
7		A-7	6 × 6 × 3/8	15.00	3.50	3.50	3.00	7/8	2 × 5	2 × 5
8	Bell et al., 1958	FK-4A	6 × 4 × 3/8	11.50	2.57	4.50	3.00	M20	1 × 4	2 × 4
9		FK-4B	6 × 4 × 3/8	11.50	2.57	4.50	3.00	M20	1 × 4	2 × 4
10		FK-4C	6 × 4 × 3/8	11.50	2.57	4.50	3.00	M20	1 × 4	2 × 4
11		FK-4R	6 × 4 × 3/8	11.50	2.57	4.50	3.00	M20	1 × 4	2 × 4
12	Lewitt et al., 1966	FK-3	6 × 4 × 3/8	8.50	2.63	3.50	3.00	3/4	1 × 3	2 × 3
13		FK-4AB-M	6 × 4 × 3/8	11.50	2.07	3.50	3.00	3/4	2 × 4	2 × 4
14		FK-4P	6 × 4 × 3/8	11.50	2.07	3.50	3.00	3/4	2 × 4	2 × 4
15		WK-4	6 × 4 × 3/8	11.50	2.57	3.50	3.00	3/4	1 × 4	2 × 4
16		FB-4	6 × 3.5 × 3/8	11.50	2.57	2.25	3.00	3/4	1 × 4	1 × 4
17		FK-5	6 × 4 × 7/16	14.50	2.55	3.50	3.00	3/4	1 × 5	2 × 5
18	Bose, 1981	B-1	150 × 90 × 15	15.75	2.38	3.54	2.95	M20	1 × 5	2 × —
19	Thompson et al., 1970	A1-1 alt	4 × 3.5 × 5/16	11.50	2.55	2.25	—	3/4	— × —	— × 2
20		A1-1 alt	4 × 3.5 × 5/16	11.50	2.55	2.25	—	3/4	— × —	— × 2
21		A1-2 alt	4 × 3.5 × 5/16	11.50	2.55	2.25	—	3/4	— × —	— × 2
22		A1-2 alt	4 × 3.5 × 5/16	11.50	2.55	2.25	—	3/4	— × —	— × 2
23		A1-3 alt	4 × 3.5 × 5/16	11.50	2.55	2.25	—	3/4	— × —	— × 2
24		A1-3 alt	4 × 3.5 × 5/16	11.50	2.55	2.25	—	3/4	— × —	— × 2
25		B1-1 alt	4 × 3.5 × 5/16	11.50	2.55	2.25	—	3/4	— × —	— × 2
26		B1-1 alt	4 × 3.5 × 5/16	11.50	2.55	2.25	—	3/4	— × —	— × 2
27		B1-2 alt	4 × 3.5 × 1/2	11.50	2.55	2.25	—	3/4	— × —	— × 2

TABLE 2.8 Collected Test Data for Double Web-Angle Connections (*Continued*)

No.	Author	Test I.D.	Angle	l (in.)	gc (in.)	pc (in.)	gb (in.)	db	nc	nb
28		B1-2 alt	4 × 3.5 × 1/2	11.50	2.55	2.25	—	3/4	— × —	— × 2
29		B1-3 alt	4 × 3.5 × 1/2	11.50	2.55	2.25	—	3/4	— × —	— × 2
30		B1-3 alt	4 × 3.5 × 1/2	11.50	2.55	2.25	—	3/4	— × —	— × 2
31		D1-1 alt	4 × 3.5 × 5/16	11.50	1.80	2.25	—	3/4	— × —	— × 2
32		D1-1 alt	4 × 3.5 × 5/16	11.50	1.80	2.25	—	3/4	— × —	— × 2
33		D1-2 alt	4 × 3.5 × 5/16	11.50	1.80	2.25	—	3/4	— × —	— × 2
34		D1-2 alt	4 × 3.5 × 5/16	11.50	1.80	2.25	—	3/4	— × —	— × 2
35		D1-3 alt	4 × 3.5 × 5/16	11.50	1.80	2.25	—	3/4	— × —	— × 2
36		D1-3 alt	4 × 3.5 × 5/16	11.50	1.80	2.25	—	3/4	— × —	— × 2
37		E1-1 alt	4 × 3.5 × 1/2	11.50	1.80	2.25	—	3/4	— × —	— × 2
38		E1-1 alt	4 × 3.5 × 1/2	11.50	1.80	2.25	—	3/4	— × —	— × 2
39		E1-2 alt	4 × 3.5 × 1/2	11.50	1.80	2.25	—	3/4	— × —	— × 2
40		E1-2 alt	4 × 3.5 × 1/2	11.50	1.80	2.25	—	3/4	— × —	— × 2
41		E1-3 alt	4 × 3.5 × 1/2	11.50	1.80	2.25	—	3/4	— × —	— × 2
42		E1-3 alt	4 × 3.5 × 1/2	11.50	1.80	2.25	—	3/4	— × —	— × 2
43		A2-1 alt	4 × 3.5 × 5/16	11.50	2.55	2.25	—	3/4	— × —	— × 2
44		A2-1 alt	4 × 3.5 × 5/16	11.50	2.55	2.25	—	3/4	— × —	— × 2
45		A2-2 alt	4 × 3.5 × 5/16	11.50	2.55	2.25	—	3/4	— × —	— × 2
46		A2-2 alt	4 × 3.5 × 5/16	11.50	2.55	2.25	—	3/4	— × —	— × 2
47		A2-3 alt	4 × 3.5 × 5/16	11.50	2.55	2.25	—	3/4	— × —	— × 2
48		A2-3 alt	4 × 3.5 × 5/16	11.50	2.55	2.25	—	3/4	— × —	— × 2
49		B2-1 alt	4 × 3.5 × 5/16	11.50	1.80	2.25	—	3/4	— × —	— × 2

50		B2-1 alt	4 × 3.5 × 5/16	11.50	1.80	2.25	—	3/4	— × —	— × 2
51		B2-2 alt	4 × 3.5 × 5/16	11.50	1.80	2.25	—	3/4	— × —	— × 2
52		B2-2 alt	4 × 3.5 × 5/16	11.50	1.80	2.25	—	3/4	— × —	— × 2
53		B2-3 alt	4 × 3.5 × 5/16	11.50	1.80	2.25	—	3/4	— × —	— × 2
54		B2-3 alt	4 × 3.5 × 5/16	11.50	1.80	2.25	—	3/4	— × —	— × 2
55		C2-1 alt	4 × 3.5 × 1/2	11.50	2.55	2.25	—	3/4	— × —	— × 2
56		C2-1 alt	4 × 3.5 × 1/2	11.50	2.55	2.25	—	3/4	— × —	— × 2
57		C2-2 alt	4 × 3.5 × 1/2	11.50	2.55	2.25	—	3/4	— × —	— × 2
58		C2-2 alt	4 × 3.5 × 1/2	11.50	2.55	2.25	—	3/4	— × —	— × 2
59		C2-3 alt	4 × 3.5 × 1/2	11.50	2.55	2.25	—	3/4	— × —	— × 2
60		C2-3 alt	4 × 3.5 × 1/2	11.50	2.55	2.25	—	3/4	— × —	— × 2
61		D2-1 alt	4 × 3.5 × 1/2	11.50	1.80	2.25	—	3/4	— × —	— × 2
62		D2-1 alt	4 × 3.5 × 1/2	11.50	1.80	2.25	—	3/4	— × —	— × 2
63		D2-2 alt	4 × 3.5 × 1/2	11.50	1.80	2.25	—	3/4	— × —	— × 2
64		D2-2 alt	4 × 3.5 × 1/2	11.50	1.80	2.25	—	3/4	— × —	— × 2
65		D2-3 alt	4 × 3.5 × 1/2	11.50	1.80	2.25	—	3/4	— × —	— × 2
66		D2-3 alt	4 × 3.5 × 1/2	11.50	1.80	2.25	—	3/4	— × —	— × 2
67	Sommer, 1969	TEST21	3.5 × 3 × 3/8	9.00	1.83		3.00	3/4	1 × 3	
68		TEST22	3.5 × 3 × 3/8	12.00	1.83		3.00	3/4	1 × 4	
69		TEST23	4 × 3 × 3/8	15.00	2.53		3.00	3/4	1 × 5	
70		TEST24	4 × 3 × 3/8	18.00	2.53		3.00	3/4	1 × 6	

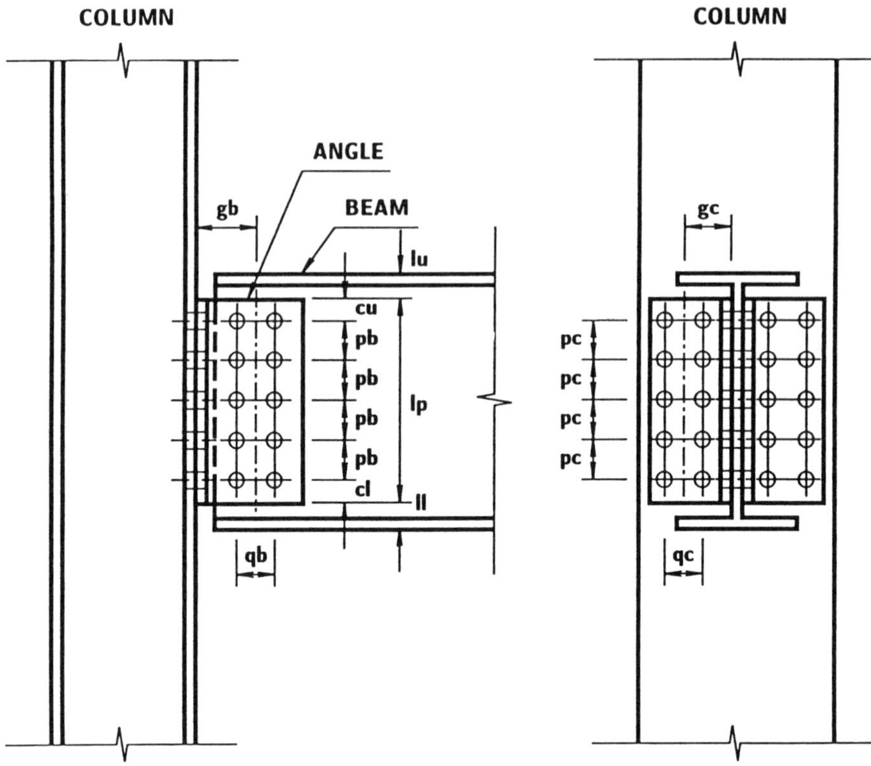

Fig. 2.22 Size parameters for double web-angle connections.

qb = distance between two lines of fasteners in beam web
qc = distance between two lines of fasteners in column
ta = web-angle thickness

On the Top and Seat Angles

gs = gauge from seat-angle heel to center of fastener holes in leg under compression beam flange
gs' = gauge from seat-angle heel to center of fastener holes in leg adjacent to column face
gt = gauge from top-angle heel to center of fastener holes in leg seating on tension beam flange
gt' = gauge from top-angle heel to center of fastener holes in leg adjacent to column face
ls = width of seat angle along column
lt = width of top angle along column
ns = total number of fasteners in compression beam flange

TABLE 2.9 Collected Test Data for Top- and Seat-Angle with Double Web-Angle Connections

No.	Author	Test I.D.	Beam	Column	Flange Angle	Web Angle
1	Azizinamini et al., 1985	8S1	W8 × 21	W12 × 58	6 × 3.5 × 5/16 × 6.0	4 × 3.5 × 1/4 × 5.5
2		8S2	W8 × 21	W12 × 58	6 × 3.5 × 3/8 × 6.0	4 × 3.5 × 1/4 × 5.5
3		8S3	W8 × 21	W12 × 58	6 × 3.5 × 5/16 × 8.0	4 × 3.5 × 1/4 × 5.5
4		8S4	W8 × 21	W12 × 58	6 × 6.0 × 3/8 × 6.0	4 × 3.5 × 1/4 × 5.5
5		8S5	W8 × 21	W12 × 58	6 × 4.0 × 3/8 × 8.0	4 × 3.5 × 1/4 × 5.5
6		8S6	W8 × 21	W12 × 58	6 × 4.0 × 5/16 × 6.0	4 × 3.5 × 1/4 × 5.5
7		8S7	W8 × 21	W12 × 58	6 × 4.0 × 3/8 × 6.0	4 × 3.5 × 1/4 × 5.5
8		8S8	W8 × 21	W12 × 58	6 × 3.5 × 5/16 × 6.0	4 × 3.5 × 1/4 × 5.5
9		8S9	W8 × 21	W12 × 58	6 × 3.5 × 3/8 × 6.0	4 × 3.5 × 1/4 × 5.5
10		8S10	W8 × 21	W12 × 58	6 × 3.5 × 1/2 × 6.0	4 × 3.5 × 1/4 × 5.5
11		14S1	W14 × 38	W12 × 96	6 × 4.0 × 3/8 × 8.0	4 × 3.5 × 1/4 × 8.5
12		14S2	W14 × 38	W12 × 96	6 × 4.0 × 1/2 × 8.0	4 × 3.5 × 1/4 × 8.5
13		14S3	W14 × 38	W12 × 96	6 × 4.0 × 3/8 × 8.0	4 × 3.5 × 1/4 × 5.5
14		14S4	W14 × 38	W12 × 96	6 × 4.0 × 3/8 × 8.0	4 × 3.5 × 3/8 × 8.5
15		14S5	W14 × 38	W12 × 96	6 × 4.0 × 3/8 × 8.0	4 × 3.5 × 1/4 × 8.5
16		14S6	W14 × 38	W12 × 96	6 × 4.0 × 1/2 × 8.0	4 × 3.5 × 1/4 × 8.5
17		14S8	W14 × 38	W12 × 96	6 × 4.0 × 5/8 × 8.0	4 × 3.5 × 1/4 × 8.5
18		14S9	W14 × 38	W12 × 96	6 × 4.0 × 1/2 × 8.0	4 × 3.5 × 1/4 × 8.5
19	Azizinamini et al., 1985	14WS1	W14 × 38	W12 × 96	6 × 4.0 × 3/8 × 8.0	4 × 3.5 × 1/4 × 8.5
20		14WS2	W14 × 38	W12 × 96	6 × 4.0 × 1/2 × 8.0	4 × 3.5 × 1/4 × 8.5
21	Rathbun, 1936	B-11	W12 × 31.8	—	6 × 6.0 × 3/8 × 9 (TOP)	4 × 3.5 × 3/8 × 9
22		B-12	W12 × 31.8	—	6 × 6.0 × 3/8 × 14 (TOP)	4 × 3.5 × 3/8 × 9

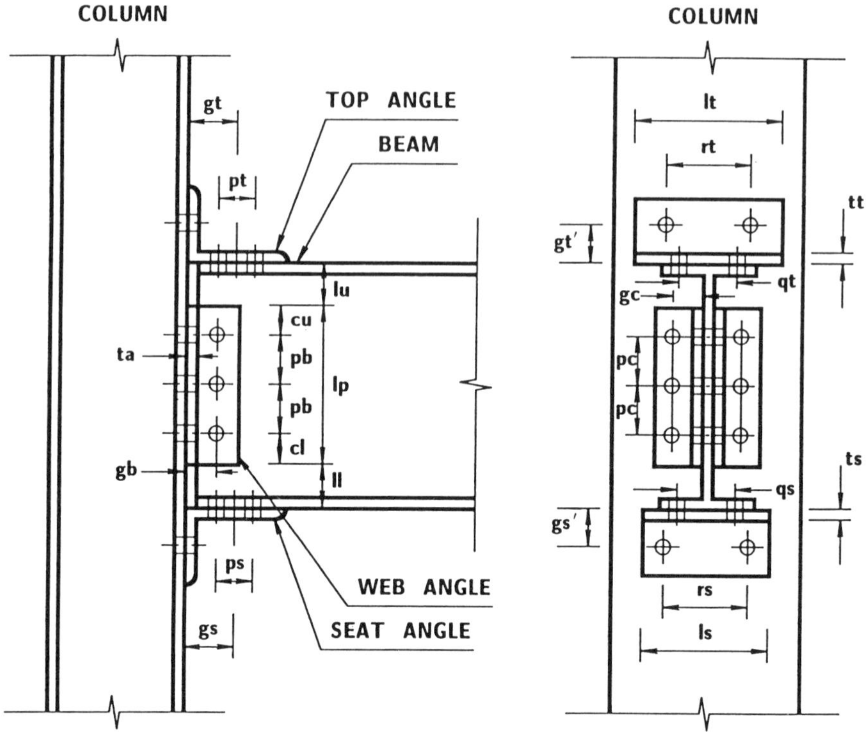

Fig. 2.23 Size parameters for top- and seat-angle connections with double web angle.

ns' = total number of fasteners in leg adjacent to column face in compression side

nt = total number of fasteners in tension beam flange

nt' = total number of fasteners in leg adjacent to column face in tension side

ps = distance between two rows of fasteners in leg under compression beam flange

ps' = distance between two rows of fasteners in leg adjacent to column face in compression side

pt = distance between two rows of fasteners in leg seating on tension beam flange

pt' = distance between two rows of fasteners in leg adjacent to column face in tension side

qs = distance between two inner lines of fasteners in leg under compression beam flange

$qs2$ = distance between outer and inner lines of fasteners in leg under compression beam flange where fastener line number exceeds 3

qt = distance between two inner lines of fasteners in leg seating on tension flange
$qt2$ = distance between outer and inner lines of fasteners in leg seating on tension beam flange where fastener line number exceeds 3
rs = distance between two inner lines of fasteners in leg adjacent to column face in compression side
$rs2$ = distance between outer and inner lines of fasteners in leg adjacent to column face in compression side where fastener line number exceeds 3
rt = distance between two inner lines of fasteners in leg adjacent to column face in tension side
$rt2$ = distance between outer and inner lines of fasteners in leg adjacent to column face in tension side where fastener line exceeds 3
ts = seat-angle thickness
tt = top-angle thickness

2.5.2.4 Top- and Seat-Angle Connections The collected test data for this connection type are listed in Table 2.10. The notations for major parameters of this connection type as shown in Fig. 2.24 are as follows:

gs = gauge from seat-angle heel to center of fastener holes in leg under compression beam flange
gs' = gauge from seat-angle heel to center of fastener holes in leg adjacent to column face
gt = gauge from top-angle heel to center of fastener holes in leg seating on tension beam flange
gt' = gauge from top-angle heel to center of fastener holes in leg adjacent to column face
ls = width of seat angle along column
lt = width of top angle along column
ns = total number of fasteners in compression beam flange
ns' = total number of fasteners in leg adjacent to column face in compression side
nt = total number of fasteners in tension beam flange
nt' = total number of fasteners in leg adjacent to column face in tension side
ps = distance between two rows of fasteners in leg under compression beam flange
ps' = distance between two rows of fasteners in leg adjacent to column face in compression side
pt = distance between two rows of fasteners in leg seating on tension beam flange
pt' = distance between two rows of fasteners in leg adjacent to column face in tension side

TABLE 2.10 Collected Test Data for Top- and Seat-Angle Connections

No.	Author	Test I.D.	Beam	Column	Angle	*lt* (in.)	*gt* (in.)	*tt* (in.)
1	Rathbun, 1936	B-8	W12 × 31.8	—	6 × 6 × 3/8 × 6 (TOP)	6.00	3.50	0.38
2		B-9	W12 × 31.8	—	6× 6 × 3/8 × 8 (TOP)	6.00	3.50	0.38
3		B-10	W12 × 31.8	—	6 × 6 × 3/8 × 14 (TOP)	6.00	3.50	0.38
4	Hechtmann et al., 1947	NO 2	W12 × 25	W10 × 49	6 × 6 × 1/2 × 6.75 (TOP)	4.00	3.50	0.63
5		NO 5	W18 × 47	W12 × 65	6 × 6 × 7/8 × 7.5 (TOP)	4.00	3.50	0.50
6		NO 9	W18 × 47	W12 × 65	6 × 6 × 7/8 × 7.5 (TOP)	4.00	3.50	0.63
7		NO 10	W18 × 47	W12 × 65	6 × 6 × 7/8 × 7.5 (TOP)	4.00	3.50	0.75
8		NO 11	W18 × 47	W14 × 58	6 × 6 × 7/8 × 7.5 (TOP)	4.00	3.50	0.50
9		NO 16	W12 × 25	W10 × 49	6 × 6 × 1/2 × 6.75 (TOP)	4.00	4.75	0.50
10		NO 17	W12 × 25	W10 × 49	6 × 6 × 1/2 × 6.75 (TOP)	4.00	3.50	0.50
11		NO 18	W12 × 50	W10 × 49	6 × 6 × 1/2 × 8 (TOP)	4.00	3.50	0.50
12		NO 20	W14 × 34	W12 × 65	6 × 6 × 5/8 × 7.25 (TOP)	4.00	3.50	0.63
13		NO 22	W16 × 40	W12 × 65	6 × 6 × 3/4 × 7.25 (TOP)	4.00	3.50	0.63
14		NO 23	W18 × 40	W14 × 58	6× 6 × 3/4 × 7.25 (TOP)	4.00	3.50	0.63
15		NO 24	W18 × 47	W12 × 65	6 × 6 × 7/8 × 7.5 (TOP)	4.00	3.50	0.63
16	Marley et al., 1982	A-1/4-1	W5 × 16	W5 × 16	4 × 4 × 1/4 × 5.0	4.00	—	0.25
17		A-1/4-2	W5 × 16	W5 × 16	4 × 4 × 1/4 × 5.0	4.00	—	0.25
18		B-1/4-1	W5 × 16	W5 × 16	4 × 4 × 1/4 × 5.0	4.00	—	0.25
19		B-1/4-2	W5 × 16	W5 × 16	4 × 4 × 1/4 × 5.0	4.00	—	0.25
20		C1-1/4-2	W5 × 16	W5 × 16	4 × 4 × 1/4 × 5.0	4.00	—	0.25
21		C2-1/4-1	W5 × 16	W5 × 16	4 × 4 × 1/4 × 5.0	4.00	—	0.25
22		C2-1/4-2	W5 × 16	W5 × 16	4 × 4 × 1/4 × 5.0	4.00	—	0.25
23		D1-1/4-1	W5 × 16	W5 × 16	4 × 4 × 1/4 × 5.0	4.00	—	0.25
24		D1-1/4-2	W5 × 16	W5 × 16	4 × 4 × 1/4 × 5.0	4.00	—	0.25
25		D2-1/4-1	W5 × 16	W5 × 16	4 × 4 × 1/4 × 5.0	4.00	—	0.25
26		D2-1/4-2	W5 × 16	W5 × 16	4 × 4 × 1/4 × 5.0	4.00	—	0.25
27		D3-1/4-1	W5 × 16	W5 × 16	4 × 4 × 1/4 × 5.0	4.00	—	0.25

28		D3-1/4-2	W5 × 16	W5 × 16	4 × 4 × 1/4 × 5.0	4.00	—	0.25
29		A-1/2-1	W5 × 16	W5 × 16	4 × 4 × 1/4 × 5.0	4.00	—	0.25
30		B-1/2-1	W5 × 16	W5 × 16	4 × 4 × 1/4 × 5.0	4.00	—	0.25
31		B-1/2-2	W5 × 16	W5 × 16	4 × 4 × 1/4 × 5.0	4.00	—	0.25
32		C1-1/2-1	W5 × 16	W5 × 16	4 × 4 × 1/4 × 5.0	4.00	—	0.25
33		C1-1/2-2	W5 × 16	W5 × 16	4 × 4 × 1/4 × 5.0	4.00	—	0.25
34		C2-1/2-1	W5 × 16	W5 × 16	4 × 4 × 1/4 × 5.0	4.00	—	0.25
35		C2-1/2-2	W5 × 16	W5 × 16	4 × 4 × 1/4 × 5.0	4.00	—	0.25
36		D1-1/2-1	W5 × 16	W5 × 16	4 × 4 × 1/4 × 5.0	4.00	—	0.25
37		D1-1/2-2	W5 × 16	W5 × 16	4 × 4 × 1/4 × 5.0	4.00	—	0.25
38		D2-1/2-1	W5 × 16	W5 × 16	4 × 4 × 1/4 × 5.0	4.00	—	0.25
39		D2-1/2-2	W5 × 16	W5 × 16	4 × 4 × 1/4 × 5.0	4.00	—	0.25
40		D3-1/2-1	W5 × 16	W5 × 16	4 × 4 × 1/4 × 5.0	4.00	—	0.25
41		D3-1/2-2	W5 × 16	W5 × 16	4 × 4 × 1/4 × 5.0	4.00	—	0.25
42	Maxwell et al., 1981	A1	457 × 191 × UB	305 × 305 × UC	150 × 90 × 10 × 150 mm	3.54	—	0.39
43		A2	457 × 191 × UB	305 × 305 × UC	150 × 90 × 10 × 200 mm	3.54	—	0.39
44		A3	457 × 191 × UB	305 × 305 × UC	150 × 90 × 10 × 150 mm	3.54	—	0.39
45		A4	457 × 191 × UB	305 × 305 × UC	150 × 90 × 10 × 200 mm	3.54	—	0.39
46		B1	457 × 191 × UB	305 × 305 × UC	150 × 90 × 12 × 150 mm	3.54	—	0.47
47		B2	457 × 191 × UB	305 × 305 × UC	150 × 90 × 12 × 200 mm	3.54	—	0.47
48		B3	457 × 191 × UB	305 × 305 × UC	150 × 90 × 12 × 150 mm	3.54	—	0.47
49		B4	457 × 191 × UB	305 × 305 × UC	150 × 90 × 12 × 200 mm	3.54	—	0.47
50		C1	457 × 191 × UB	305 × 305 × UC	150 × 90 × 15 × 150 mm	3.54	—	0.59
51		C2	457 × 191 × UB	305 × 305 × UC	150 × 90 × 15 × 200 mm	3.54	—	0.59
52		C3	457 × 191 × UB	305 × 305 × UC	150 × 90 × 15 × 150 mm	3.54	—	0.59
53		C4	457 × 191 × UB	305 × 305 × UC	150 × 90 × 15 × 200 mm	3.54	—	0.59

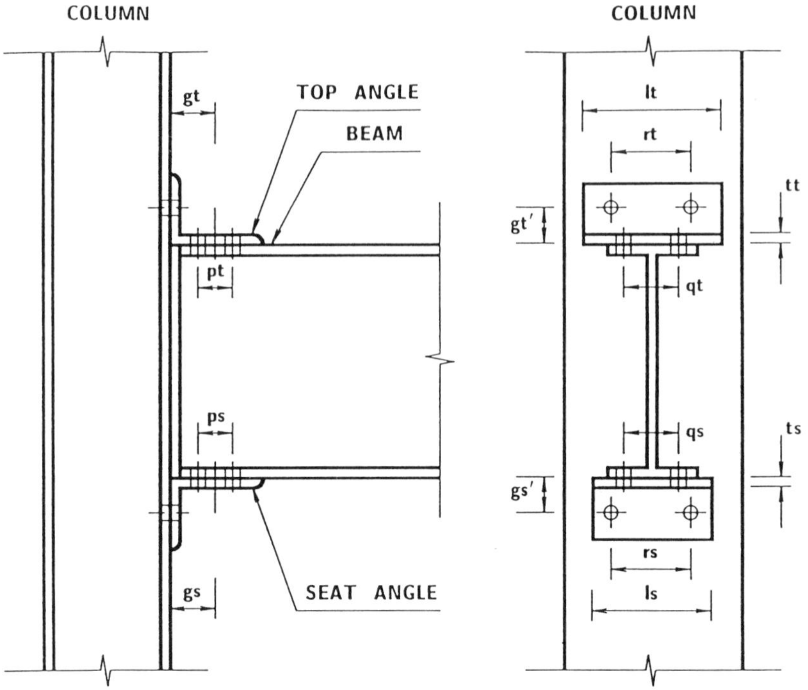

Fig. 2.24 Size parameters for top- and seat-angle connections without double web angle.

qs = distance between two inner lines of fasteners in leg under compression beam flange

$qs2$ = distance between outer and inner lines of fasteners in leg under compression beam flange where fastener line number exceeds 3

qt = distance between two inner lines of fasteners in leg seating on tension flange

$qt2$ = distance between outer and inner lines of fasteners in leg seating on tension beam flange where fastener line number exceeds 3

rs = distance between two inner lines of fasteners in leg adjacent to column face in compression side

$rs2$ = distance between outer and inner lines of fasteners in leg adjacent to column face in compression side where fastener line number exceeds 3

rt = distance between two inner lines of fasteners in leg adjacent to column face in tension side

$rt2$ = distance between outer and inner lines of fasteners in leg adjacent to column face in tension side where fastener line exceeds 3

ts = seat-angle thickness
tt = top-angle thickness

2.5.2.5 *Extended End-Plate Connections* The type of extended end-plate connections includes both cases of extended on the tension and compression sides and extended on the tension side only. Collected experimental moment–rotation data for extended end-plate connections are tabulated in Table 2.11. The notations for major parameters of this connection type shown in Fig. 2.25 are identified as follows:

ct = distance from tension-side edge of plate to center of outer tension-side fastener holes
li = distance from center of lower tension-side fastener holes to center of upper compression-side fastener holes
pt = distance between two rows of tension-side fasteners

The case of end-plate connections extended only on the tension side.

cc = distance from center of compression-side flange of beam to compression-side edge of plate
pc = distance from inner row of fasteners on compression side to center of compression-side flange of beam

The case of end-plate connections extended on both tension and compression sides.

cc = distance from outer row of fasteners on compression side to the compression-side edge of plate
gc = distance between two inner lines of compression-side fasteners
*gc*2 = distance from outer line to inner line of compression-side fasteners where fastener line number exceeds 3
gi = distance between two inner lines of central fasteners
*gi*2 = distance from outer line to inner line of central fasteners where fastener line number exceeds 3
gt = distance between two inner lines of tension-side fasteners
*gt*2 = distance from outer line to inner line of tension-side fasteners where fastener line number exceeds 3
lp = plate length
pc = distance between two rows of compression-side fasteners
pi = distance between two inner rows of central fasteners
pic = distance from inner row of fasteners on compression side to the lower row of central fasteners
pit = distance from inner row of fasteners on tension side to upper row of central fasteners
tp = plate thickness

TABLE 2.11 Collected Test Data for Extended End-Plate Connections

No.	Author	Test I.D.	Beam	Column	*lp* (in.)	*tp* (in.)	*g* (in.)	*p* (in.)	*db*
1	Ioannides, 1978	TEST 1	W14 × 22	W8 × 35	17.25	0.63	5.50	3.50	3/4″
2		TEST 2	W18 × 35	W10 × 49	21.50	0.75	5.50	3.75	7/8″
3		TEST 3	W24 × 55	W14 × 48	27.25	0.88	5.50	4.00	1″
4		TEST 4	W14 × 22	W8 × 35	17.25	0.88	5.50	3.50	3/4″
5		TEST 5	W18 × 35	W10 × 49	21.50	1.25	5.50	3.75	7/8″
6		TEST 6	W24 × 55	W14 × 48	27.25	1.25	5.50	4.00	1″
7	Dews, 1979	TEST 1	W14 × 22	W8 × 31	16.81	0.97	5.50	3.50	3/4″
8		TEST 2	W14 × 22	W8 × 31	16.92	0.94	5.53	3.50	3/4″
9		TEST 3	W16 × 26	W10 × 33	19.14	1.56	5.38	3.75	1″
10	Sherbourne, 1961	TEST A1	15 × 5 RSJ 42	8 × 8 UC 35	14.50	1.25	4.00	3.50	3/4″
11	Bailey, 1970	TEST A1-L	12 × 5 RSJ 32	8 × 8 UC 48	—	1.00	4.00	5.50	7/8″
12		TEST A1-R	12 × 5 RSJ 32	8 × 8 UC 48	—	1.00	4.00	4.50	7/8″
13		TEST A2-L	10 × 4 UB 17	8 × 8 UC 40	—	0.75	4.00	4.38	5/8″
14		TEST A2-R	10 × 4 UB 17	8 × 8 UC 40	—	0.75	4.00	4.38	5/8″
15		TEST A3-L	10 × 4 UB 19	8 × 8 UC 40	—	0.75	4.00	4.38	3/4″
16		TEST A3-R	10 × 4 UB 19	8 × 8 UC 40	—	0.75	4.00	4.38	3/4″
17	Surtees et al., 1970	TEST C1	12 × 5 UB 25	8 × 8 UC 48	—	0.75	3.50	4.50	3/4″
18		TEST C2	15 × 6 UB 40	10 × 10 UC 60	—	0.75	3.50	4.50	3/4″
19		TEST C3	15 × 6 UB 40	10 × 10 UC 60	—	0.75	3.50	4.50	1″
20		TEST C4	15 × 6 UB 40	10 × 10 UC 60	—	1.00	3.50	4.50	1″
21		TEST C5	15 × 6 UB 40	10 × 10 UC 60	—	1.00	3.50	4.50	1″
22	Packer et al., 1977	TEST J1	254× 102 UB 2	152× 152 UC 3	9.76	0.59	3.39	3.82	M16
23		TEST J2	254 × 102 UB 2	152 × 152 UC 3	9.75	0.59	3.39	3.81	M16
24		TEST J3	254 × 102 UB 2	152 × 152 UC 2	9.76	0.59	3.39	3.82	M16
25	Grundy et al., 1980	T1	610 UB 113	310 UC 240	—	1.00	3.50	3.75	7/8″
26		T2	610 UB 113	310 UC 240	—	1.25	3.50	3.75	7/8″
27	Sherbourne, 1961	TEST A2	15 × 5 RSJ 42	8 × 8 UC 35	14.50	1.25	4.00	3.50	3/4″
28		TEST A3	15 × 5 RSJ 42	8 × 8 UC 35	14.50	0.75	4.00	3.50	3/4″

29		TEST B1	15 × 5 RSJ 42	8 × 8 UC 35	17.00	1.00	4.00	5.00	7/8″
30		TEST B2	15 × 5 RSJ 42	8 × 8 UC 35	17.00	0.75	4.00	5.00	7/8″
31	Johnson et al., 1960	TEST 5	10 × 4.5 RSJ	8 × 8 UC 45	—	0.50	3.50	3.75	3/4″
32	Bailey, 1970	B4-L	12 × 5 UB 32	8 × 8 UC 40	—	1.25	4.00	4.50	7/8″
33		B4-R	12 × 5 UB 32	8 × 8 UC 40	—	1.25	4.00	4.50	7/8″
34		B5-L	14 × 5 UB 26	8 × 8 UC 40	—	1.00	4.00	4.25	3/4″
35		B5-R	14 × 5 UB 26	8 × 8 UC 40	—	1.00	4.00	4.25	3/4″
36		B6-L	14 × 6.75 UB	8 × 8 UC 40	—	1.38	4.00	4.31	1″
37		B6-R	14 × 6.75 UB	8 × 8 UC 40	—	1.38	4.00	4.31	1″
38		B7-L	12 × 6.5 UB	8 × 8 UC 40	—	1.25	4.00	4.50	1″
39		B7-R	12 × 6.5 UB	8 × 8 UC 40	—	1.25	4.00	4.50	1″
40		B8-L	8 × 5.25 UB 2	8 × 8 UC 40	—	0.75	4.00	4.50	3/4″
41		B8-R	8 × 5.25 UB 2	8 × 8 UC 40	—	0.75	4.00	4.50	3/4″
42		C9-L	8 × 5.25 UB 2	8 × 8 UC 40	—	0.75	4.00	4.50	5/8″
43		C9-R	8 × 5.25 UB 2	8 × 8 UC 40	—	0.75	4.00	4.50	5/8″
44		C10-L	12 × 5 UB 32	8 × 8 UC 40	—	0.81	4.00	4.50	3/4″
45		C10-R	12 × 5 UB 32	8 × 8 UC 40	—	0.81	4.00	4.50	3/4″
46		C11-L	14 × ;5 UB 26	8 × 8 UC 40	—	0.88	4.00	4.25	5/8″
47		C11-R	14 × 5 UB 26	8 × 8 UC 40	—	0.88	4.00	4.25	5/8″
48		C12-L	12 × 6.5 UB	8 × 8 UC 40	—	1.06	4.00	4.38	7/8″
49		C12-R	12 × 6.5 UB	8 × 8 UC 40	—	1.06	4.00	4.38	7/8″
50		C13-L	14 × 6.75 UB	8 × 8 UC 40	—	1.25	4.00	4.63	1″
51		C13-R	14 × 6.75 UB	8 × 8 UC 40	—	1.25	4.00	4.63	1″
52	Surtees et al., 1970	TEST C6	18 × 6 UB 55	10 × 10 UC 89	—	1.13	6.00	6.00	11/8″
53	Johnstone et al., 1981	TEST 1-L	310 UB 46	250 UC 89	—	1.26	5.51	7.01	M30
54		TEST 1-R	310 UB 46	250 UC 89	—	1.26	5.51	7.01	M30
55		TEST 2-L	310 UB 46	250 UC 89	—	1.26	5.51	7.01	M30
56		TEST 2-R	310 UB 46	250 UC 89	—	1.26	5.51	7.01	M30
57		TEST 3-L	310 UB 46	250 UC 89	—	0.94	5.12	4.02	M24
58		TEST 3-R	310 UB 46	250 UC 89	—	0.94	5.12	4.02	M24
59		TEST 4-L	310 UB 46	250 UC 89	—	0.63	5.12	4.53	M24
60		TEST 4-R	310 UB 46	250 UC 89	—	0.63	5.12	4.53	M24

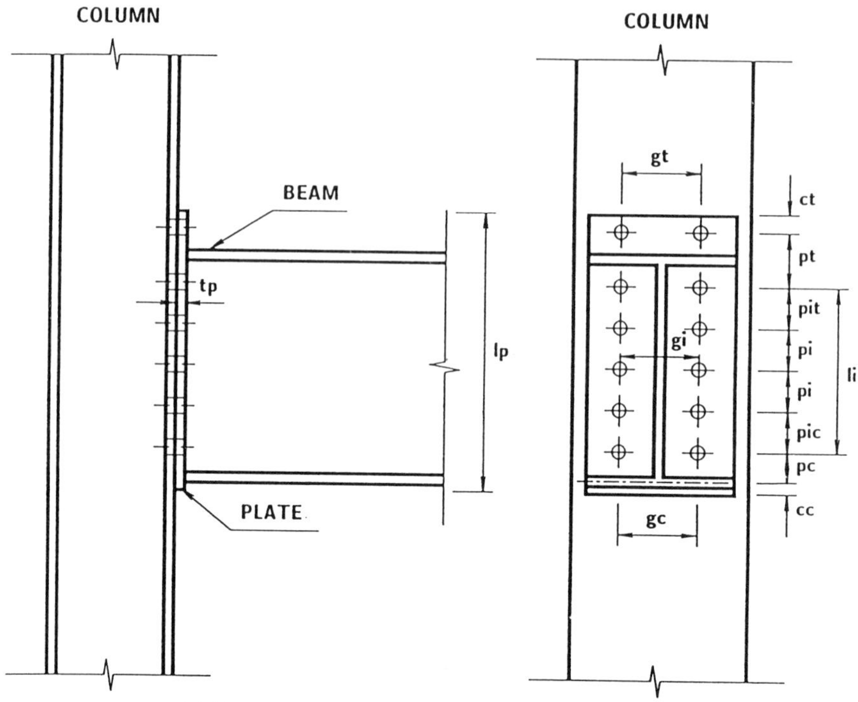

(a) Extended end plate on the tension side only

Fig. 2.25 Size parameters for extended end-plate connections. (a) Extended end plate on the tension side only. (b) Extended end plate on the tension and compression sides.

2.5.2.6 Flush End-Plate Connections Collected experimental moment–rotation data for flush-end connections are tabulated in Table 2.12. The notations for major parameters of this connection type as shown in Fig. 2.26 are identified as follows:

- cc = distance from outer surface of compression-side flange of beam to compression-side edge of plate
- ct = distance from tension-side edge of plate to outer surface of tension-side flange of the beam
- gc = distance between two inner lines of compression-side fasteners
- $gc2$ = distance from outer line to inner line of compression-side fasteners where fastener line number exceeds 3
- gi = distance between two inner lines of central fasteners
- $gi2$ = distance from outer line to inner line of central fasteners where fastener line number exceeds 3

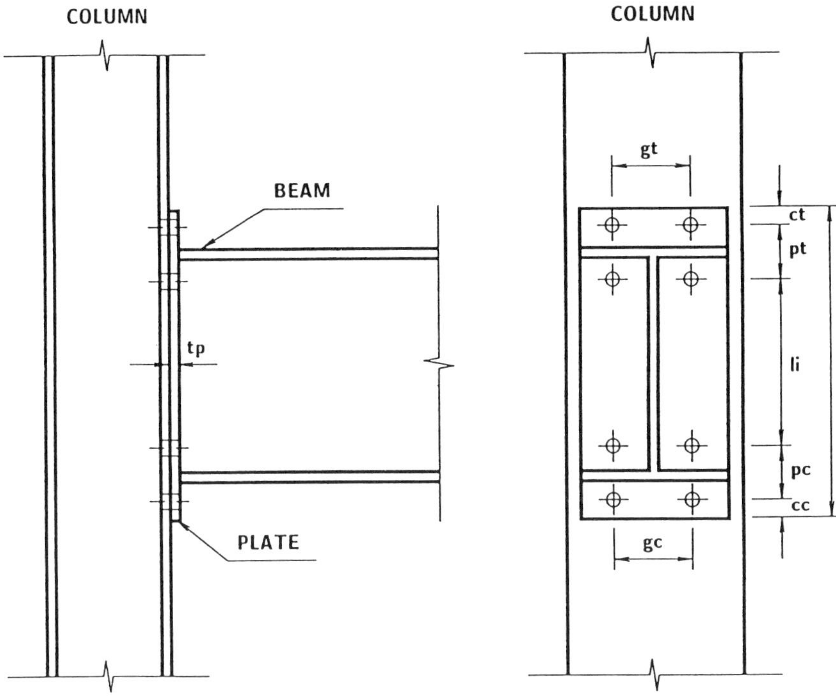

(b) Extended end plate on the tension and compression sides

Fig. 2.25 *(Continued)*

gt = distance between two inner lines of tension-side fasteners

*gt*2 = distance from outer line to inner line of tension-side fasteners where fastener line number exceeds 3

li = distance from center of tension-side fastener holes to center of compression-side fastener holes

lp = plate length

pc = distance from row of compression-side fasteners to outer surface of compression-side flange of beam

pi = distance between two inner rows of central fasteners

pic = distance from row of compression-side fasteners to row of central fasteners

pit = distance from the row of tension-side fasteners to upper row of central fasteners

pt = distance from outer surface of the tension-side flange to row of tension-side fasteners

tp = plate thickness

TABLE 2.12 Collected Test Data for Flush End-Plate Connections

No.	Author	Test I.D.	Beam	Column	*li* (in.)	*g* (in.)	*tp* (in.)	*db* (in.)
1	Ostrander, 1970	TEST 1	W10 × 21	W8 × 28	5.00	3.50	0.50	3/4
2		TEST 3	W10 × 21	W8 × 28	5.00	3.50	0.38	3/4
3		TEST 4	W10 × 21	W8 × 28	5.00	3.50	0.25	3/4
4		TEST 9	W10 × 21	W8 × 28	5.00	3.50	0.75	3/4
5		TEST 11	W12 × 27	W8 × 40	7.00	4.00	0.38	3/4
6		TEST 12	W12 × 27	W8 × 40	7.00	4.00	0.50	3/4
7		TEST 13	W12 × 27	W8 × 40	7.00	4.00	0.63	3/4
8		TEST 17	W12 × 27	W8 × 24	7.00	4.00	0.38	3/4
9		TEST 18	W12 × 27	W8 × 24	7.00	4.00	0.50	3/4
10		TEST 19	W12 × 27	W8 × 24	7.00	4.00	0.63	3/4
11		TEST 23	W12 × 27	W8 × 48	7.00	4.00	0.63	3/4
12		TEST 2	W10 × 21	W8 × 28	5.00	3.50	0.50	3/4
13		TEST 5	W10 × 21	W8 × 28	5.00	3.50	0.50	3/4
14		TEST 6	W10 × 21	W8 × 28	5.00	3.50	0.38	3/4
15		TEST 7	W10 × 21	W8 × 28	5.00	3.50	0.25	3/4
16		TEST 8	W10 × 21	W8 × 28	5.00	3.50	0.25	3/4
17		TEST 10	W10 × 21	W8 × 28	5.00	3.50	0.75	3/4
18		TEST 14	W12 × 27	W8 × 40	7.00	4.00	0.38	3/4
19		TEST 15	W12 × 27	W8 × 40	7.00	4.00	0.50	3/4
20		TEST 16	W12 × 27	W8 × 40	7.00	4.00	0.63	3/4
21		TEST 20	W12 × 27	W8 × 24	7.00	4.00	0.38	3/4
22		TEST 21	W12 × 27	W8 × 24	7.00	4.00	0.50	3/4
23		TEST 22	W12 × 27	W8 × 24	7.00	4.00	0.63	3/4
24		TEST 24	W12 × 27	W8 × 48	7.00	4.00	0.63	3/4

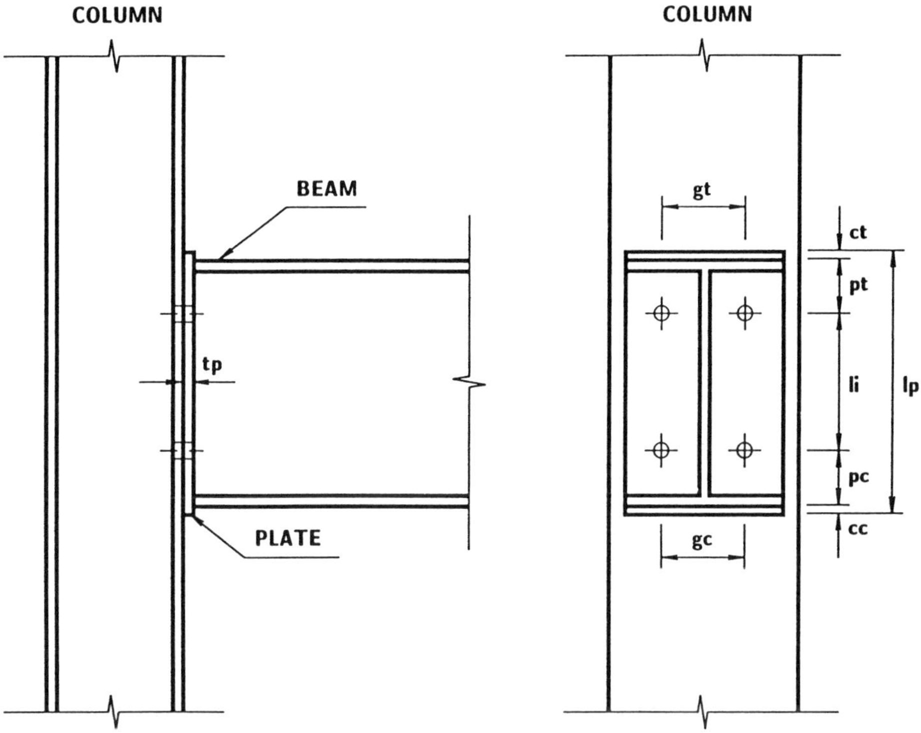

Fig. 2.26 Size parameters for flush end-plate connections.

2.5.2.7 Header-Plate Connections All of the collected moment–rotation tests on header-plate connections were performed by Sommer (1969) at the University of Toronto. He tested 20 specimens listed in Table 2.13. The notations for major parameters of this connection type as shown in Fig. 2.27 are identified as follows:

cc = distance from center of lower fastener holes to bottom edge of plate
ct = distance from center of upper fastener holes to top edge of plate
g = distance between two lines of fasteners
lc = distance from compression beam flange to bottom edge of plate
lp = plate depth
lt = distance from tension beam flange to top edge of plate
n = total number of fasteners
p = distance between two rows of fasteners
tbw = beam web thickness
tp = plate thickness

TABLE 2.13 Collected Test Data for Header-Plate Connections

No.	Author	Test I.D.	Beam	Column	*l* (in.)	*g* (in.)	*tp* (in.)	*db* (in.)
1	Sommer, 1969	TEST 5	W18 × 45	—	15.00	4.00	0.25	3/4
2		TEST 6	W24 × 76	—	9.00	4.00	0.25	3/4
3		TEST 7	W24 × 76	—	12.00	4.00	0.25	3/4
4		TEST 8	W24 × 76	—	15.00	4.00	0.25	3/4
5		TEST 9	W24 × 76	—	18.00	4.00	0.25	3/4
6		TEST 10	W18 × 45	—	9.00	4.00	0.38	3/4
7		TEST 11	W18 × 45	—	12.00	4.00	0.38	3/4
8		TEST 12	W18 × 45	—	15.00	4.00	0.38	3/4
9		TEST 13	W24 × 76	—	9.00	4.00	0.38	3/4
10		TEST 14	W24 × 76	—	12.00	4.00	0.38	3/4
11		TEST 15	W24 × 76	—	15.00	5.50	0.38	3/4
12		TEST 16	W24 × 76	—	18.00	5.50	0.38	3/4
13		TEST 17	W24 × 76	—	18.00	5.50	0.38	3/4
14		TEST 18	W24 × 76	—	15.00	5.50	0.25	3/4
15		TEST 19	W24 × 76	—	12.00	5.50	0.50	3/4
16		TEST 20	W24 × 76	—	15.00	5.50	0.50	3/4
17		TEST 25	W18 × 45	—	12.00	4.00	0.25	3/4
18		TEST 26	W18 × 45	—	9.00	4.00	0.25	3/4
19		TEST 27	W12 × 27	—	9.00	4.00	0.25	3/4
20		TEST 28	W12 × 27	—	6.00	4.00	0.25	3/4

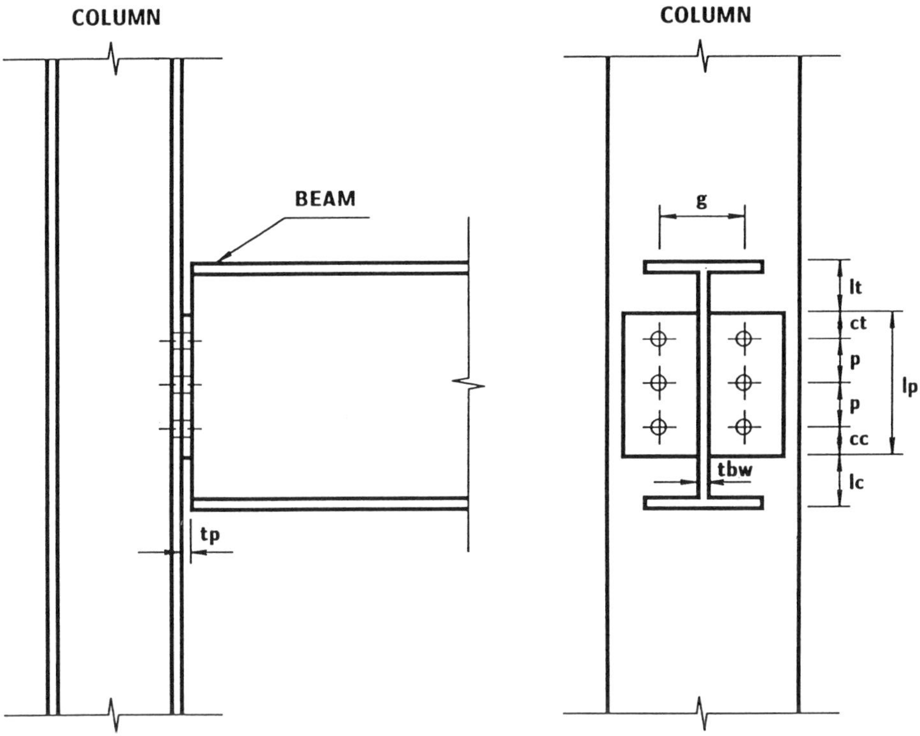

Fig. 2.27 Size parameters for header-plate connections.

2.6 STEEL CONNECTION DATA BANK PROGRAM

2.6.1 Outline of SCDB

The experimental connection test data collected in the previous section can be controlled by the Steel Connection Data Bank (SCDB) program. The first version of the SCDB program has seven main routines. To be controlled by any workstation with a FORTRAN compiler, the plotter routines have been removed. The new version of the SCDB program for this book consists of the following main routines:

1. Transforming the outputted unit system from the original U.S. customary units to the MKS system.
2. Setting and printing the selected test data.
3. Setting and printing general tables of test data concerning connection type and mode.
4. Determining and printing the values of three prediction curves for selected test data.

In this program, three prediction equations are prepared: the Frye–Morris polynomial model, the modified exponential model, and the three-parameter power model. The prediction curves obtained from the three-parameter power model are limited to the angle types of connections. In the three-parameter power model, the initial connection stiffness R_{ki} and the ultimate connection moment M_u are based on the values determined from the simple mechanics and mechanism formulated by Kishi and Chen (1990). The shape parameter n is determined numerically by applying the least-mean-square technique for the connection moments between experimental test data and Eq. (2.4.5). The validity of using either the three-parameter power model or the modified exponential model to fit the experimental curves for practical use is discussed by Kishi et al. (1991). It was concluded that the three-parameter power model can replace the experimental test data in describing adequately the connection M–θ_r behavior for practical use.

Depending on the users demand, it is possible to execute the SCDB program together with the previously mentioned routines. This program enables the user to make comparative studies of the role of different joint parameters on M–θ_r behavior.

2.6.2 User's Manual for the SCDB Program

The SCDB program is written in FORTRAN, controlled by the UNIX system, and is composed of four files (scdb1.f–scdb4.f). These files are in a diskette in the source list. A Makefile is also included to provide a specification on how an executable program can be built using the make utility of the UNIX system. The user can compile them and make the file executable. This program has been tested in computing environments: a Sun SPARCstation IPX using a Sun FORTRAN 77 compiler and a Sony RISC NEWS using RISC NEWS FORTRAN. A hard disk is generally required. To execute the SCDB program, the files shown in Table 2.14 have to be prepared. Note that scdb.d and BANK/no1–BANK/no7 in the table are files for input, and bank.d is for output. The files of BANK/no1–BANK/no7 are the files for the original database of each type of connection, which are also provided in the diskette. File scdb.d should be defined in the same directory with an executable program, and input data should be created according to user's purpose as follows:

TABLE 2.14 List of Files Used in the SCDB Program

File Name	Contents
scdb.d	Input data file
BANK/no1–BANK/no7	Original database of each connection type
bank.d	Output file
temp	Working file

Lines 1 to 7 (7a50)
File names with full pass of files BANK/no1 to BANK/no7.

Line 8 (4i5)
idsi: Index for unit name of outputted data
0: U.S. customary units
1: MKS units, except dimensions (cm) of connection
icnct: Identification number of the connection type referred to in Table 2.3
ilist: Flag for output
0: Print data bank list
n: Print the table concerning the *n*th fastening mode referred to in Table 2.15. In this table, the cases of types of end-plate connections are taken from number 9 through 12 or numbers 16 and 17
nterm: Number of terms considered in the modified exponential curve-fitting equation. The case of nterm = 6 gives a good result.
0: Default set to 6

Line 9 (i5)
ngroup: Number of calculating group of experimental test data for the type of connections given by icnct in Line 8.

Line 10 (10i5)
istart(i), i=1, ngroup:
Start the sequential number of test data in the *i*th group specified by the preceding variable ngroup

TABLE 2.15 Fastening-Mode Patterns for Making Table

Pattern Number, *N*	Fastening Mode Pattern
1	All riveted
2	All bolted
3	Riveted to beam and bolted to column
4	Bolted to beam and riveted to column
5	Riveted to beam and welded to column
6	Bolted to beam and welded to column
7	Welded to beam and riveted to column
8	Welded to beam and bolted to column
9	All riveted without column stiffeners
10	All riveted with column stiffeners
11	All bolted without column stiffeners
12	All bolted with column stiffeners
13	Riveted/bolted to beam and column (1, 2, 3, 4)
14	Riveted/bolted to beam and welded to column (5, 6)
15	Riveted/bolted to column and welded to beam (7, 8)
16	Riveted/bolted without column stiffener (9, 11)
17	Riveted/bolted with column stiffener (10, 12)
20	All modes included

TABLE 2.16 Content of scdb.d for Example 1

```
BANK/no1
BANK/no2
BANK/no3
BANK/no4
BANK/no5
BANK/no6
BANK/no7
    0    2    0    0
    1
    1
    1
```

Line 11 (10i5)
iend(i), i=1, ngroup:
Finish the sequential number of test data in the *i*th group specified by the preceding variable ngroup.

2.6.3 Examples

Example 1: Printing the Bank List The first example is to print a bank list: connection parameters, experimental $M\text{–}\theta_r$ data, parameters for each prediction equation, and the comparison of the values obtained from experimental test data with the values of prediction equations. The input file, named scdb.d for this example, is given in Table 2.16, which is for making a bank list on the experimental test data of sequential number 1 for the type of double web-angle connections. The content of output file bank.d is shown in Table 2.18 and it can be used to verify the SCDB program upon installation

TABLE 2.17 Content of scdb.d for Example 2

```
BANK/no1
BANK/no2
BANK/no3
BANK/no4
BANK/no5
BANK/no6
BANK/no7
    0    2    3    0
    1
    1
   70
```

TABLE 2.18 Content of bank.d for Example 1

II - 1

Connection type : Double web-angle connections
Mode : All riveted

Tested by : J. C. Rathbun (1936) U.S.A.
Test Id. : A-1

Column : --
Beam : W6 × 12.5
Angle : 6 × 4 × 3/8

Fasteners: -- -X-7/8″D
15/16″ Oversize holes
Material : G40.21
Fy = 44.00 ksi
Fu = 64.80 ksi

Major parameters

lp = 2.5000″	lu = 1.7500″	ll = 1.7500″	ta = 0.3750″
gb = 3.5000″	gc = 2.6250″	cu = 1.2500″	cl = 1.2500″
qb = 2.5000″	nb = 2 × 1	nc = 1 × 1	

Remark 1) CONNECTION ANGLE RIVETED TO 1/2″ MOUNING PLATE.
2)

No	Moment (k-in)	Rotation (radians) × 1/1000	No	Moment (k-in)	Rotation (radians) × 1/1000
1	0.0	0.00	26	17.1	19.17
2	0.6	0.60	27	17.6	19.94
3	1.2	1.20	28	18.0	20.85
4	1.9	1.82	29	18.5	21.76
5	2.5	2.44	30	19.3	23.10
6	3.2	3.12	31	19.5	23.98
7	4.0	3.81	32	19.8	24.85
8	4.6	4.41	33	20.0	25.78
9	5.3	5.01	34	20.2	26.71
10	6.2	6.08	35	20.6	28.15
11	7.1	7.02	36	20.7	29.00
12	7.9	7.79	37	20.9	29.84
13	8.7	8.55			
14	9.3	9.31			
15	10.0	10.06			
16	10.8	10.86			
17	11.7	11.80			
18	12.4	12.67			
19	13.2	13.54			
20	13.7	14.21			
21	14.2	14.88			
22	14.9	15.85			
23	15.6	16.82			
24	16.1	17.61			
25	16.6	18.40			

TABLE 2.18 (*Continued*)

Moment-rotation prediction equations (R : × 1/1000 radians)

Frye and Morris polynomial model : R = Sum (Ai × (K*Bm)**Pi × 10**Qi)

xd = 2.500000″	g = 5.500000″	t = 0.375000″			
A1 = 3.660000	A2 = 1.150000	A3 = 4.570000	K = 0.845282		
P1 = 1	P2 = 3	P3 = 5	Q1 = −1	Q2 = −3	Q3= −5

Modified exponential model :

Bm = Sum (Ai × (1 − exp(− R/(2*i*C)))) + Sum (RKj × (R − Rj0)) + Bm0

C = 0.39904000E+00 Bmo= 0.00000000E+00 Nexp= 6 Nliner= 0

Ai = −0.56289585E+01 0.10550337E+03 −0.69927898E+03 0.19892424E+04 −0.24849136E+04 0.11169381E+04

Power model : Bm = (Rki × R) / (1 + (R/RO)***rn)**(1/rn)

rn = 1.807 rki = 0.13188761E+01 rmu = 0.24538170E+02

No	Rotation (radians) × 1/1000	Moment (kip-inch) Expri.	Poly.	M. Expo.	P. Model.	Connection stiffness (kip-inch) × 1000 Expri.	Poly.	M. Expo.	P. Model.

1	0.00	0.0	0.0	0.0	0.0	0.1019E+01	0.3232E+01	0.6440E+00	0.1319E+01
3	1.20	1.2	3.7	1.2	1.6	0.1024E+01	0.2799E+01	0.1109E+01	0.1304E+01
5	2.44	2.5	6.5	2.5	3.2	0.1047E+01	0.1736E+01	0.1017E+01	0.1268E+01
7	3.81	4.0	8.4	3.9	4.9	0.1073E+01	0.1064E+01	0.1027E+01	0.1210E+01
9	5.01	5.3	9.5	5.2	6.3	0.1023E+01	0.7775E+00	0.1034E+01	0.1148E+01
11	7.02	7.1	10.7	7.2	8.5	0.9703E+00	0.5358E+00	0.9863E+00	0.1031E+01
13	8.55	8.7	11.5	8.7	10.0	0.9527E+00	0.4344E+00	0.9390E+00	0.9378E+00
15	10.06	10.0	12.1	10.1	11.3	0.9321E+00	0.3676E+00	0.9037E+00	0.8478E+00
17	11.80	11.7	12.7	11.6	12.7	0.9088E+00	0.3133E+00	0.8700E+00	0.7494E+00
19	13.54	13.2	13.2	13.1	13.9	0.8136E+00	0.2737E+00	0.8293E+00	0.6591E+00
21	14.88	14.2	13.5	14.2	14.8	0.7470E+00	0.2498E+00	0.7872E+00	0.5959E+00
23	16.82	15.6	14.0	15.6	15.9	0.6690E+00	0.2222E+00	0.7081E+00	0.5143E+00
25	18.40	16.6	14.3	16.7	16.6	0.6361E+00	0.2040E+00	0.6325E+00	0.4564E+00
27	19.94	17.6	14.6	17.6	17.3	0.5832E+00	0.1893E+00	0.5538E+00	0.4064E+00
29	21.76	18.5	15.0	18.5	18.0	0.5405E+00	0.1747E+00	0.4619E+00	0.3552E+00
31	23.98	19.5	15.3	19.4	18.7	0.2999E+00	0.1598E+00	0.3580E+00	0.3023E+00
33	25.78	20.0	15.6	20.0	19.2	0.2456E+00	0.1496E+00	0.2845E+00	0.2660E+00
35	28.15	20.6	16.0	20.6	19.8	0.2234E+00	0.1383E+00	0.2048E+00	0.2259E+00
37	29.84	20.9	16.2	20.9	20.2	0.2279E+00	0.1312E+00	0.1597E+00	0.2018E+00

TABLE 2.19 Content of bank.d for Example 2

No	Author	Test Id.	Angle	lp(in)	gc(in)	gb(in)	pc(in)	db	nc	nb
			II. Double web-angle connections Riveted-to-beam and bolted-to-column							
1	W. C. Bell et al. (1958)	FK-4A	6 × 4 × 3/8	11.5000	2.5730	4.5000	3.0000	3/4″	1 × 4	2 × 4
2		FK-4B	6 × 4 × 3/8	11.5000	2.5730	4.5000	3.0000	3/4″	1 × 4	2 × 4
3		FK-4C	6 × 4 × 3/8	11.5000	2.5730	4.5000	3.0000	3/4″	1 × 4	2 × 4
4	C. W. Lewitt et al. (1966)	FK-3	6 × 4 × 3/8	8.5000	2.6320	3.5000	3.0000	3/4″	1 × 3	2 × 3
5		FK-4P	6 × 4 × 3/8	11.5000	2.0710	3.5000	3.0000	3/4″	2 × 4	2 × 4
6		WK-4	6 × 4 × 3/8	11.5000	2.5710	3.5000	3.0000	3/4″	1 × 4	2 × 4
7		FB-4	6 × 3.5 × 3/8	11.5000	2.5710	2.2500	3.0000	3/4″	1 × 4	1 × 4
8		FK-5	6 × 4 × 7/16	14.5000	2.5500	3.5000	3.0000	3/4″	1 × 5	2 × 5

in a computer system. In the diskette, the input and output files for this example are found in files of scdb_ex1.d and bank_ex1.d, respectively.

Example 2: Printing General Tables of Test Data The second example is to print the general table of experimental test data concerning connection type and mode. In this example, a general table is made of test data concerning double web-angle connections with riveted-to-beam and bolted-to-column detailings. Input file scdb.d for this example is given Table 2.17. Output file scbd.d, which is obtained from the SCDB program, is shown in Table 2.19. Those files are provided in the diskette in files of scdb_ex2.3 and bank_ex2.d, respectively.

REFERENCES

AISC (1989). *Allowable Stress Design and Plastic Design Specifications for Structural Steel Buildings,* 9th Ed., Chicago: American Institute of Steel Construction.

AISC (1993). *Load and Resistance Factor Design Specifications for Structural Steel Buildings,* 2nd Ed., Chicago: American Institute of Steel Construction.

Altman, W. G., Jr., A. Azizinamini, J. H. Bradburn, and J. B. Radziminski (1982). *Moment-Rotation Characteristics of Semi-Rigid Steel Beam-to-Column Connection,* Columbia: Department of Civil Engineering, University of South Carolina.

Ang, K. M., and G. A. Morris (1984). "Analysis of Three-Dimensional Frame with Flexible Beam-Column Connections," *Canadian Journal of Civil Engineering,* Vol. 11, pp. 245–254.

Azizinamini, A., J. H. Bradburn, and J. B. Radziminski (1985). *Static and Cyclic Behavior of Semi-Rigid Steel Beam-Column Connections,* Columbia: Department of Civil Engineering, University of South Carolina.

Bailey, J. R. (1970). "Strength and Rigidity of Bolted Beam-to-Column Connections," *Proceedings of the Conference on Joints in Structures,* Vol. 1, Paper 4, Sheffield, U.K.: University of Sheffield.

Bell, W. G., E. Chesson, Jr., and W. H. Munse (1958). *Static Tests of Standard Riveted and Bolted Beam-to-Column Connections,* Urbana: University of Illinois Engineering Experiment Station.

Bjorhovde, R., J. Brozzetti, and A. Colson (1990). "A Classification System for Beam-to-Column Connections," *Journal of Structural Engineering,* ASCE, Vol. 116, No. ST11, pp. 3059–3076.

Bose, B. (1981). "Moment-Rotation Characteristics of Semi-Rigid Joints in Steel Structures," *Journal of Institute Engineers* (India), Part CI, Civil Engineering Division, Vol. 62, No. 2, pp. 128–132.

CEN—European Committee for Standardization (1992). *Design of Steel Structures: Part 1—General Rules for Buildings,* Vol. 1, February.

Chen, W. F., and N. Kishi (1987). *Moment-Rotation Relation of Top- and Seat-Angle Connections,* CE-STR-87-4, West Lafayette, IN: School of Civil Engineering, Purdue University. See also *Proceedings of the International Colloquium on Bolted and Special Connections,* Moscow, May 15–20, 1989.

Chen, W. F., and N. Kishi (1989). "Semirigid Steel Beam-to-Column Connections: Data Base and Modeling," *Journal of Structural Engineering,* ASCE, Vol. 115, No. ST1, pp. 105–119.

Chen, W. F., and E. M. Lui (1985). "Column with End Restraint and Bending in Load and Resistance Factor Design," *AISC Engineering Journal,* Vol. 22, No. 4, pp. 105–132.

Chen, W. F., and A. F. Saleeb (1982). *Uniaxial Behavior and Modeling in Plasticity,* Structural Engineering Report No. CE-STR-82-35, West Lafayette, IN: School of Civil Engineering, Purdue University.

Colson, A., and J. M. Louveau (1983). "Connections Incidence on the Inelastic Behavior of Steel Structures," *Proceedings of the Euromech Colloquium,* October, p. 174.

Cook, N. E. (1983). "Strength of Flexibly-Connected Steel Frames under Load Histories," Ph.D. thesis, University of Colorado, Boulder.

Dafalias, Y. F., and E. P. Popov (1976). "Plastic Internal Variables Formalism of Cyclic Plasticity," *Journal of Applied Mechanics,* Vol. 43,pp. 645–651.

Dews, R. J. (1979). "Experimental Test Results on Experimental End-Plate Moment Connections," M.Sc. thesis, Vanderbilt University, Nashville.

Driscoll, G. C. (1987). "Elastic-Plastic Analysis of Top- and Seat-Angle Connections," *Journal of Constructional Steel Research,* Vol. 18, pp. 119–136.

Frye, M. J., and G. A. Morris (1975). "Analysis of Flexibly Connected Steel Frames," *Canadian Journal of Civil Engineers,* Vol. 2, pp. 280–291.

Goto, Y., and S. Miyashita (1995). "Validity of Classification System of Semi-Rigid Connections," in *Behavior of Flexible Joints and Dynamic Analysis of Steel Frames with Semi-Rigid Connections,* edited by S. L. Chan, Special Issue of *Journal of Engineering Structures,* London: Butterworth.

Goto, Y., S. Miyashita, and W. F. Chen (1994). "Classification of Semi-Rigid Frames Based on the Elastic-Plastic Mechanical Behavior of Frames," *Proceedings of the ASCE Structures Congress XII,* Atlanta, edited by N. C. Baker and B. J. Goodno, Vol. 2, pp. 1191–1196, ASCE, New York.

Goto, Y., S. Suzuki, and W. F. Chen (1989). "Bifurcation and Limit-Load Instability of Flexibly Jointed Frames," *Proceedings of the International Colloquium on Bolted and Special Structural Joints,* Moscow, May, pp. 80–90.

Goto, Y., S. Suzuki, and W. F. Chen (1991). "Analysis of Critical Behavior of Semi-Rigid Frames with or without Load History in Connections," *International Journal of Solids and Structures,* Vol. 27, No. 4, pp. 467–483.

Goto, Y., S. Suzuki, and W. F. Chen (1993). "Stability Behavior of Semi-Rigid Sway Frames," *Engineering Structures,* Vol. 15, No. 3, pp. 209–219.

Goverdhan, A. V. (1983). "A Collection of Experimental Moment-Rotation Curves and Evaluation of Prediction Equations for Semi-Rigid Connections," M.Sc. thesis, Vanderbilt University, Nashville.

Grundy, P., I. R. Thomas, and I. D. Bennetts (1980). "Beam-to-Column Moment Connections," *Journal of Structural Engineering,* ASCE, Vol. 106, No. ST1, pp. 313–330.

Hechtman, R. A., and B. G. Johnston (1947). *Riveted Semi-Rigid Beam-to-Column Building Connections,* Progress Report No. 1, Bethlehem, PA: AISC Research, Lehigh University.

Ioannides, S. A. (1978). "Flange Behavior in Bolted End-Plate Moment Connections," Ph.D. thesis, Vanderbilt University, Nashville.

Huang, J., and G. Morris (1991). "Analysis of Flexibly Connected Frames under Non-proportional Loading," *Proceedings of the Second International Workshop on Connections in Steel Structures,* R. Bjorhovde et al., eds. April 10–12, Pittsburgh. Chicago: AISC.

Johnson, L. G., J. C. Cannon, and L. A. Spooner (1960). "High Tensile Preloaded Bolted Joints for Development of Full Plastic Moments," *British Welding Journal,* No. 7, pp. 560–569.

Johnstone, N. D., and W. R. Walpole (1981). *Bolted End-Plate Beam to Column Connections under Earthquake Type Loading,* Research Report 81-7, Christchurch, New Zealand: Department of Civil Engineering, University of Canterbury.

Jones, S. W., P. A. Kirby, and D. A. Nethercot (1980). "Effect of Semi-Rigid Connections on Steel Column Strength," *Journal of Constructional Steel Research,* Vol. 1, No. 1, pp. 38–46.

Jones, S. W., P. A. Kirby, and D. A. Nethercot (1981). "Modeling of Semi-Rigid Connection Behavior and Its Influence on Steel Column Behavior," in *Joints in Structural Steelwork,* J. H. Howlett, W. M. Jenkins, and R. Stainsby, eds., London: Pentech Press, pp. 5.73–5.78.

Jones, S. W., P. A. Kirby, and D. A. Nethercot (1983). "The Analysis of Frames with Semi-Rigid Connections—A State-of-the-Art Report," *Journal of Constructional Steel Research,* Vol. 3, No. 2, pp. 2–13.

Kishi, N., and W. F. Chen (1986a). *Steel Connection Data Bank Program,* CE-STR-86-18, West Lafayette, IN: School of Civil Engineering, Purdue University. See also *Proceedings of the Sessions at Structures Congress Related to Materials and Member Behavior,* Orlando, FL, August 17–20, New York: ASCE, 1987, pp. 89–106.

Kishi, N., and W. F. Chen (1986b). *Data Base of Steel Beam-to-Column Connections,* Structural Engineering Report No. CE-STR-86-26, 2 vols., West Lafayette, IN: School of Civil Engineering, Purdue University.

Kishi, N., and W. F. Chen (1990). "Moment-Rotation Relations of Semi-Rigid Connections with Angles," *Journal of Structural Engineering,* ASCE, Vol. 116, No. ST7, pp. 1813–1834.

Kishi, N., W. F. Chen, Y. Goto, and K. G. Matsuoka (1991). "Applicability of Three-Parameter Power Model to Structural Analysis of Flexibly Jointed Frames," *Proceedings of Mechanics Computing in 1990's and Beyond,* H. Adeli and R. L. Sierakowski, eds., Columbus, Ohio, May 20–22, 233–237.

Kishi, N., W. F. Chen, Y. Goto, and K. G. Matsuoka (1993). "Design Aid of Semi-Rigid Connections for Frame Analysis," *AISC Engineering Journal,* 3rd Quarter, pp. 90–107.

Kishi, N., W. F. Chen, K. G. Matsuoka, and S. G. Nomachi (1988a). "Moment-Rotation Relation of Top- and Seat-Angle with Double Web-Angle Connections," *Proceedings of the State-of-the-Art Workshop on Connections and the Behavior, Strength and Design of Steel Structures,* R. Bjorhovde, J. Brozzetti, and A. Colson, eds., Ecole Normale Superieure de Cachan, France, May 25–27, 1987, London: Elsevier, pp. 121–134.

Kishi, N., W. F. Chen, K. G. Matsuoka, and S. G. Nomachi (1988b). "Moment-Rotation Relation of Single/Double Web-Angle Connections," *Proceedings of the State-of-the-Art Workshop on Connections and the Behavior, Strength and Design of Steel Structures,* R. Bjorhovde, J. Brozzetti, and A. Colson, eds., Ecole Normale Superieure de Cachan, France, May 25–27, 1987, London: Elsevier, pp. 135–149.

Krishnamurthy, N., H. T. Huang, P. K. Jefferey, and L. K. Avery (1979). "Analytical $M-\theta$ Curves for End-Plate Connections," *Journal of Structural Engineering,* ASCE, Vol. 105, No. ST1, pp. 133–145.

Kukreti, A. R., T. M. Murray, and A. Abolmaali (1987). "End-Plate Connection Moment-Rotation Relationship," *Journal of Constructional Steel Research,* Vol. 8, pp. 137–157.

Lewitt, C. W., E. Chesson, Jr., and W. H. Munse (1966). *Restraint Characteristics of Flexible Riveted and Bolted Beam-to-Column Connections,* Urbana: Department of Civil Engineering, University of Illinois.

Lightfoot, E., and A. P. LeMessurier (1974). "Elastic Analysis of Frameworks with Elastic Connections," *Journal of Structural Engineering,* ASCE, Vol. 100, No. ST6, pp. 1297–1309.

Lipson, S. L. (1968). "Single-Angle and Single-Plate Beam Framing Connections," *Proceedings of the Canadian Structural Engineering Conference,* Toronto, pp. 141–162.

Lipson, S. L. (1977). "Single Angle Welded-Bolted Connections," *Journal of Structural Engineering,* ASCE, Vol. 103, No. ST3, pp. 559–571.

Lui, E. M., and W. F. Chen (1986). "Analysis and Behavior of Flexibly Jointed Frames," *Engineering Structures,* Vol. 8, No. 2, pp. 107–118.

Marley, M. J., and K. H. Gerstle (1982). *Analysis and Tests of Flexibly-Connected Steel Frames,* Report to AISC under Project 199, Chicago: American Institute of Steel Construction.

Maxwell, S. M., et al. (1981). "A Realistic Approach to the Performance and Application of Semi-Rigid Jointed Structures," in *Joints in Structural Steelwork,* edited by J. H. Howeltt, W. M. Jenkins, and R. Stainsby, London: Pentech Press, pp. 2.71–2.98.

Monforton, A.R., and T. S. Wu (1963). "Matrix Analysis of Semi-Rigidly Connected Frames," *Journal of Structural Engineering,* ASCE, Vol. 87, No. ST6, pp. 13–42.

Nethercot, D. A. (1985a). "Utilization of Experimentally Obtained Connection Data in Assessing the Performance of Steel Frames," in *Connection Flexibility and Steel Frames,* edited by W. F. Chen, New York: Structural Division, ASCE.

Nethercot, D. A. (1985b). *Steel Beam-to-Column Connections—A Review of Test Data,* London: CIRIA.

Ostrander, J. R. (1970). "An Experimental Investigation of End-Plate Connections," M.Sc. thesis, University of Saskatchewan.

Packer, J. A., and L. J. Morris (1977). "A Limit State Design Method for the Tension Region of Bolted Beam-Column Connections," *The Structural Engineer,* Vol. 55, No. 10, pp. 446–458.

Patel, K. V., and W. F. Chen (1984). "Nonlinear Analysis of Steel Moment Connections," *ASCE,* Vol. 110, No. ST8, pp. 1861–1874.

Picard, A., Y.-M. Giroux, and P. Brun (1976). Discussion of "Analysis of Flexibly Connected Steel Frames," by M. J. Frye and G. A. Morris, *Canadian Journal of Civil Engineering,* Vol. 3, No. 2, pp. 350–352.

Ramberg, W., and W. R. Osgood (1943). *Description of Stress-Strain Curves by Three Parameters,* Technical Note 902, Washington, DC: National Advisory Committee for Aeronautics.

Rathbun, J. C. (1936). "Elastic Properties of Riveted Connections," *Transactions of the ASCE,* Paper No. 1933, Vol. 101, pp. 524–563.

Richard, R. M., and B. J. Abbott (1975). "Versatile Elastic-Plastic Stress–Strain Formula," *Journal of Engineering Mechanics Division, ASCE,* Vol. 101, No. EM4, pp. 511–515.

Richard, R. M., P. E. Gillet, J. D. Kriegh, and B. A. Lewis (1980). "The Analysis and Design of Single Plate Framing Connections," *AISC Engineering Journal,* 2nd Quarter, pp. 38–52.

Richard, R. M., J. D. Kriegh, and D. E. Hormby (1982). "Design of Single Plate Framing Connections with A307 Bolts," *AISC Engineering Journal,* 4th Quarter, pp. 209–213.

Sherbourne, A. N. (1961). "Bolted Beam-to-Column Connections," *The Structural Engineers,* No. 39, pp. 203–210.

Simitses, G. J., and A. S. Vlahinos (1984). "Elastic Stability of Rigidly and Semi-Rigidly Connected Unbraced Frames," in *Steel Frame Structures,*R. Narayanan, ed., London: Elsevier, pp. 115–152.

Sommer, W. H. (1969). "Behavior of Welded-Header-Plate Connections," M.Sc. thesis, University of Toronto.

Surtees, J. O., and A. J. Mann (1970). "End Plate Connection in Plastically Designed Structures," *Proceedings of the Conference on Joints in Structures,* Vol. 1, Paper 5, Sheffield, U.K.: University of Sheffield.

Tarpy, T. S., and J. W. Cardinal (1981). "Behavior of Semi-Rigid Beam to-Column End Plate Connections," in *Joints in Structural Steelwork,* J. H. Howlett, W. M. Jenkins, and R. Stainsby, eds., London: Pentech Press.

Thompson, L. E., R. J. McKee, and D. A. Visintainer (1970). *An Investigation of Rotation Characteristics of Web Shear Framed Connections Using A-36 and A-441 Steel,* Rolla: Department of Civil Engineering, University of Missouri.

Wilson, W. M., and H. F. Moore (1917). *Tests to Determine the Rigidity of Riveted Joints in Steel Structures,* Bulletin No. 104, Urbana: Engineering Experiment Station, University of Illinois.

Wu, F. S. (1989). "Semi-Rigid Connections in Steel Frames," Ph.D. thesis, Purdue University, West Lafayette, IN.

Wu, F. S., and W. F. Chen (1990). "A Design Model for Semi-Rigid Connections," *Engineering Structures,* Vol. 12, No. 2, pp. 88–97.

Yee, Y. L., and R. E. Melchers (1986). "Moment–Rotation Curves for Bolted Connections," *Journal of Structural Engineering,* ASCE, Vol. 112, No. ST3, pp. 615–635.

3

SECOND-ORDER ELASTIC ANALYSIS FOR SEMI-RIGID FRAMES

3.1 INTRODUCTION

The Load and Resistance Factor Design (LRFD) Specification (AISC, 1993) designates two types of constructions in their provisions: Type FR (Fully Restrained) and Type PR (Partially Restrained). Type FR construction corresponds to the so-called "rigid frame." Type PR construction, which assumes that the connections of beams and girders possess an insufficient rigidity to hold the original angles between intersecting members virtually unchanged, is usually referred to as the "semi-rigid frame."

The 1993 AISC LRFD specification recommends the following interaction equation for the design of beam-columns.

For $(P_u/\phi_c P_n) \geq 0.2$:

$$\frac{P_u}{\phi_c P_n} + \frac{8}{9}\left(\frac{M_{ux}}{\phi_b M_{nx}} + \frac{M_{uy}}{\phi_b M_{ny}}\right) \leq 1.0 \tag{3.1.1a}$$

For $(P_u/\phi_c P_n) < 0.2$:

$$\frac{P_u}{2\phi_c P_n} + \frac{M_{ux}}{\phi_b M_{nx}} + \frac{M_{uy}}{\phi_b M_{ny}} \leq 1.0 \tag{3.1.1b}$$

where P_n is the axial compression capacity of the axially loaded column; M_{nx} and M_{ny} are the moment resisting capacities of the laterally unsupported beam bent about the x- and y-axes, respectively; ϕ_c is the column resisting factor (= 0.85); ϕ_b is the beam resisting factor (= 0.90); P_u is the required axial strength; and M_{ux} and M_{ux} are the required flexural strengths of the members about the x- and y-axes, respectively, calculated by considering only the geometrical nonlinearity of structures known as the second-order effect.

In the design of Type FR construction, the specification uses two modifiers, B_1- and B_2-factors, to consider the second-order effect. These amplification factors are directly applied to the result of a first-order analysis.

In contrast, for Type PR construction, the AISC LRFD specification recommends no specific design method of analysis. If a linear relationship could be assumed between the joint moment and the joint rotation, the B_1- and B_2-factor method can also be applied to the analysis of Type PR construction as long as the connection flexibility is considered in the first-order analysis. However, experiments have shown that connection moment–rotation curves are generally always nonlinear over the entire range. It is therefore necessary to calculate the required flexural strengths M_{ux}, M_{uy} in a PR construction by considering both the material nonlinearity of connections and the geometrical nonlinearity of structures.

In order to calculate the required flexural strengths for Type PR construction, this chapter presents a versatile computer-based method of second-order elastic analysis that is numerically stable and also can be easily implemented in a computer. In this development, special attention has been paid to the necessary simplification of the analytical procedures without losing numerical stability and accuracy of a rigorous solution.

For the ease of computer implementation, the direct stiffness matrix method is adopted here as a numerical procedure in the discrete analysis. The stiffness equations for beam and column members were first derived in closed form from the governing differential equations of a moderately large rotation (Goto and Chen, 1987a; Suzuki, and Chen, 1991a, 1991b). This approximation is valid for the present applications, because extremely large displacements seldom occur in the design of framed structures. This derivation procedure of the stiffness equation is presented in Section 3.3, followed by a brief description of the implementation of connection models and an iterative method to solve the nonlinear stiffness equations (Goto and Chen, 1987b).

Modeling of connections is an important subject that strongly influences the accuracy of analysis with respect to Type PR construction. Although several practical connection models have been developed in recent years, no single model has yet been widely accepted that can adequately describe the overall behavior of various connections used in engineering practice. Herein, we select the most appropriate connection model from the comprehensive connection data bank explained in Chapter 2.

The analysis program, which is basically the same as FLFRM (Goto and Chen, 1986), is introduced and developed in Sections 3.8 and 3.9 with some refinement made to the original program (Kishi et al. 1993).

Last but not least, when we use the interaction equation of Eq. (3.1.1a) or (3.1.2b) in the design of the Type PR construction, we have to further evaluate the column effective length factor K by taking into account the effect of semi-rigid connections. The determination of the K-factor for semi-rigid frames is discussed later in Chapter 4.

3.2 SECOND-ORDER THEORY FOR IN-PLANE FRAMES UNDER MODERATELY LARGE ROTATIONS

The governing differential equations for a second-order analysis are formulated through the principle of virtual work, introducing the usual beam assumptions. In this way, accurate equilibrium equations consistent with the compatibility equations can be easily derived by a purely mathematical manipulation. This method originally proposed by Washizu (1968) was refined extensively by Nishino Kasemset, and Lee (1973), Nishino, Kurakata, and Goto (1975), Nishino and Hasegawa (1979), and Goto et al. (1985) in a very rigorous manner to formulate the geometrically nonlinear theories of thin-walled members.

Consider a plane frame member 1–2, as shown in Fig. 3.1, subjected to a uniformly distributed lateral force p_y acting perpendicular to the member axis before deformation. A Cartesian coordinate system (x, y) as well as displacement components (u, v) are defined at the initial configuration of the member.

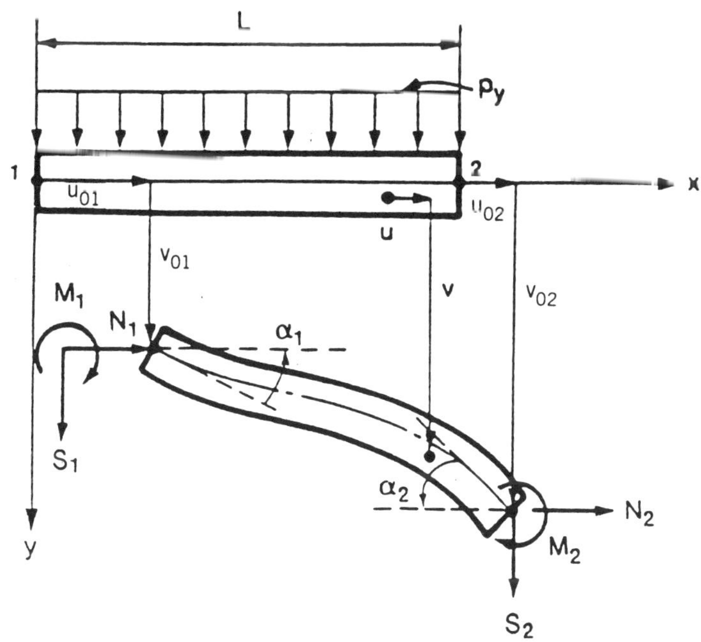

Fig. 3.1 Coordinate system for a member.

First, we derive the nonlinear axial strain that will be used in the equation of virtual work (Goto and Chen, 1987a, 1989). For the present design applications, the conditions of small strains as well as moderately large rotations are introduced herein to simplify the nonlinear strain.

Consider an infinitesimal fiber dx of the member parallel to the centroidal axis, as shown in Fig. 3.2. The axial component of the Green strain tensor is given by

$$e_{xx} = \frac{\partial u}{\partial x} + \frac{1}{2}\left(\frac{\partial u}{\partial x}\right)^2 + \frac{1}{2}\left(\frac{\partial v}{\partial x}\right)^2 \tag{3.2.1}$$

e_{xx} can be related to the corresponding component ε_x of the so-called engineering strain as

$$e_{xx} = \varepsilon_x(1 + \tfrac{1}{2}\varepsilon_x) \tag{3.2.2}$$

where

$$\varepsilon_x = \left[\left(1 + \frac{\partial u}{\partial x}\right)^2 + \left(\frac{\partial v}{\partial x}\right)^2\right]^{1/2} - 1 \tag{3.2.3}$$

Under the conditions of small strains mathematically expressed as

$$e_{xx} << 1 \qquad \text{or} \qquad \varepsilon_x << 1 \tag{3.2.4}$$

the Green strain e_{xx} coincides with the engineering strain ε_x. Thus, in the customary design analysis of frames, it is unnecessary to take into account the difference of strains according to their definitions.

The nonlinear strain e_{xx} is simplified by introducing the condition of moderately large rotations. This condition expressed by

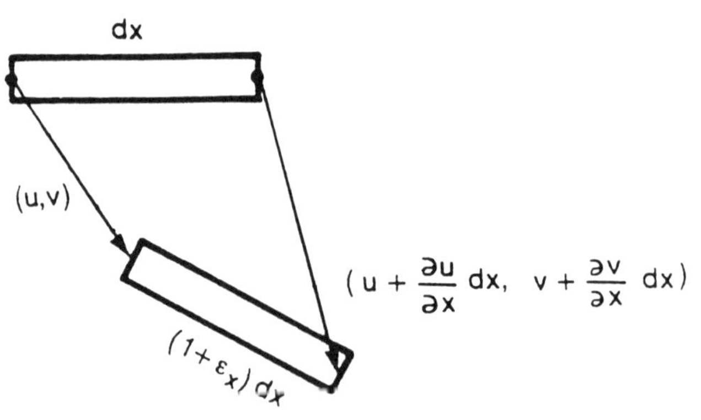

Fig. 3.2 Deformation of an infinitesimal longitudinal fiber.

$$\left(\frac{\partial v}{\partial x}\right)^2 << 1 \tag{3.2.5}$$

is less restrictive when compared with the condition of small rotation, that is, $(\partial v/\partial x) << 1$.

By using Eq. (3.2.5), the order of magnitude of $\partial u/\partial x$ can be known. For this purpose, Eq. (3.2.1) is solved for $\partial u/\partial x$ as

$$\frac{\partial u}{\partial x} = -1 + (1 + \lambda)^{1/2} \tag{3.2.6}$$

$$\lambda = 2e_{xx} - \left(\frac{\partial v}{\partial x}\right)^2 \tag{3.2.7}$$

From Eqs. (3.2.4), (3.2.5), and (3.2.6), $\partial u/\partial x$ is known to have the same order of magnitude as λ. This implies that

$$\frac{\partial u}{\partial x} = O(\lambda) < < 1 \tag{3.2.8}$$

The use of Eq. (3.2.8) reduces Eq. (3.2.1) to

$$e_{xx} = \frac{\partial u}{\partial x} + \frac{1}{2}\left(\frac{\partial v}{\partial x}\right)^2 \tag{3.2.9}$$

Herein, we introduce the usual beam assumptions, that is, the assumptions of no change of cross-sectional shapes and the Bernoulli–Euler hypothesis where the transverse plane is assumed to be plane and normal to the beam axis throughout deformation. Based on these two assumptions, the displacement field can be expressed as follows using the displacement components (u_0, v_0) on the centroidal axis.

$$u = u_0 - y \sin \alpha \tag{3.2.10a}$$

$$v = v_0 - y(1 - \cos \alpha) \tag{3.2.10b}$$

where

$$\sin \alpha = \frac{dv_0/dx}{[(1 + du_0/dx)^2 + (dv_0/dx)^2]^{1/2}} \tag{3.2.11a}$$

$$\cos \alpha = \frac{1 + du_0/dx}{[(1 + du_0/dx)^2 + (dv_0/dx)^2]^{1/2}} \tag{3.2.11b}$$

Equations (3.2.10a) and (3.2.10b) can be simplified by the conditions (3.2.5) and (3.2.8) as

$$u = u_0 - y\frac{dv_0}{dx} \tag{3.2.12a}$$

$$v = v_0 \tag{3.2.12b}$$

Note that u_0 and v_0 are both functions of x.

By substituting Eqs. (3.2.12a) and (3.2.12b) into Eq. (3.2.9), the axial component of the strain tensor is expressed by the displacement on the centroidal axis:

$$e_{xx} = \frac{du_0}{dx} - y\frac{d^2v_0}{dx^2} + \frac{1}{2}\left(\frac{dv_0}{dx}\right)^2 \tag{3.2.13}$$

This nonlinear strain–displacement relationship is what is customarily used in a second-order finite-element analysis for plane frames (Connor, Logcher, and Chan, 1968; SSRC, 1988, Mallet and Marcal, 1968). To obtain the closed-form solution corresponding to the customary finite-element solution, the governing differential equations are derived through the principle of virtual work.

By using the strain tensor component e_{xx} of Eq. (3.2.13) and the corresponding stress component σ_{xx} in addition to the components of external forces and displacements, as shown in Fig. 3.1, the equation of virtual work for the plane member is given by

$$\delta\Pi = \int_0^L\int_A \sigma_{xx}\,\delta e_{xx}\,dA\,dx - \int_0^L p_y\,\delta v_0 - [n_x(N_i\,\delta u_{0i} + S_i\,\delta v_{0i} + M_i\,\delta v'_{0i})]_1^2 = 0 \tag{3.2.14}$$

in which $\int_A dA$ is integration over the cross-sectional area; $\int_0^L dx$ is the integration over the length of the member; and n_x has the values of -1 and 1, respectively, at nodes 1 and 2.

The virtual strain δe_{xx} can be calculated by taking the variation of Eq. (3.2.13):

$$\delta e_{xx} = \delta u'_0 - y\,\delta v''_0 + v'_0\,\delta v'_0 \tag{3.2.15}$$

in which the notation $(.)'$ is introduced for simplicity to express the differentiation with respect to x. Substituting Eq (3.2.15) into Eq. (3.2.14) and integrating by parts lead to

$$\delta\Pi = [(N - n_x N_i)\ \delta u_{0i} + (Nv_0' + M' - n_x S_i)\ \delta v_{0i} - (M + n_x M_i)\ \delta v_0']_1^2 - \int_0^L \{N'\ \delta u_0 + [(Nv_0' + M')' + p_y]\ \delta v_0\}\ dx \tag{3.2.16}$$

in which

$$N = \int_A \sigma_{xx}\ dA \tag{3.2.17a}$$

$$M = \int_A \sigma_{xx} y\ dA \tag{3.2.17b}$$

These stress resultants correspond to axial force and bending moment, respectively.

Equilibrium equations and the associated boundary conditions are obtained from the necessary and sufficient conditions for Eq. (3.2.16) to hold for any virtual displacements. The terms in the largest brackets under the integral sign in Eq. (3.2.16) yield the equilibrium equations. Thus,

$$N' = 0 \tag{3.2.18a}$$

$$(Nv_0' + M')' + p_y = 0 \tag{3.2.18b}$$

and the integrated terms give the associated boundary conditions at nodes 1 and 2 as

$$u_0 = u_{0i} \quad \text{or} \quad N = n_x N_i \tag{3.2.19a}$$

$$v_0 = v_{0i} \quad \text{or} \quad Nv_0' + M' = n_x S_i \tag{3.2.19b}$$

$$v_0' = \alpha_i \quad \text{or} \quad M = -n_x M_i \quad (i = 1, 2) \tag{3.2.19c}$$

where S_i and α_i are vertical forces and end rotations of the member, respectively.

The stress resultant–displacement relations are obtained by substituting Eq. (3.2.13) into Eqs. (3.2.17a) and (3.2.17b). If the x-axis is selected such that it coincides with the centroidal axis of the member, these relations are simplified as

$$N = EA\ [u_0' + \tfrac{1}{2}(v_0')^2] \tag{3.2.20a}$$

$$M = -EIv_0'' \tag{3.2.20b}$$

Equations (3.2.18) to (3.2.20) are the consistent governing equations of beam-columns derived from the theorem of virtual work. It should be noted that in these equations, the effect known as bowing or curvature shortening is included in Eq. (3.2.20a). We refer here to the theory with these governing equations as the *nonlinear beam-column theory.* This theory yields a symmetric tangent stiffness matrix consistent with the energy theory of elastic solids (Goto et al., 1991b). If the bowing effect is ignored in Eq. (3.2.20a), this equation is reduced to

$$N = EAu_0' \tag{3.2.21}$$

The beam-column theory using Eq. (3.2.21) instead of Eq. (3.2.20a) is commonly used. Because Eq. (3.2.21) does not include nonlinear terms resulting from the bowing effect, this theory is referred to as the *linearized beam-column theory.* Due to the exclusion of the nonlinear term, the linearized beam-column theory is mathematically easy to handle, but the tangent stiffness matrix derived from this theory becomes asymmetric, unless the prebuckling deformation or the increment of the axial force is ignored. This is against the elastic theory in the a strict sense. As demonstrated by Goto et al. (1991b), the difference between the two theories in some cases can be significant for critical behaviors such as bifurcation and limit-load instability. However, this difference is negligible for most design analyses of rectangular frames either rigidly or semi-rigidly connected (Goto and Chen, 1987a, 1987b), and, therefore, satisfactory results for design purposes can be generally obtained by using the linearized beam-column theory. Although both theories are implemented in the analysis program introduced in Sections 3.8 and 3.9, the linearized beam-column theory requires less time to achieve convergent solutions and is therefore recommended here for general use in the customary design analysis.

As can be seen from the assumptions of Eqs. (3.2.4) and (3.2.5) adopted in the derivation, the preceding theories should be used primarily for the case when frames undergo moderately large rotation, as in the design analysis. To deal with the case of extremely large rotations, the beam-column theories can be approximately used within the corotational framework (Saafan, 1963; Oran, 1973).

The exact geometrically nonlinear theories for two-dimensional beam-columns are reported by Nishino et al. (1975) and Nishino and Hasegawa (1979), and their general closed-form solutions using elliptic integrals are also given by Goto, Yamashita, and Matsuura (1987) and Goto, Yoshimitsu, and Obata (1990). These solutions can be used as benchmark checks for approximate numerical analyses. The accuracy of the linearized and the nonlinear beam-column theories is numerically shown later in Section 3.7.

3.3 STIFFNESS EQUATIONS FOR A BEAM-COLUMN MEMBER

Closed-form stiffness equations for a beam-column member are derived from the governing equations in Section 3.2. Although it is possible and easier to obtain stiffness equations based on finite-element techniques, the stiffness equations so derived are approximate and it is necessary to divide a member into finite elements to achieve more accurate solutions. This approach normally increases the number of input data, computational time, and required computer capacity. Furthermore, the finite-element method using the nonlinear beam-column theory (Mallet and Marcal, 1968) sometimes cannot yield accurate solutions when the rotations become relatively large (Byskov, 1989). To avoid these inconveniences, we use the closed-form stiffness equations as derived by Goto and Chen (1987a, 1987b) and Goto et al. (1991a, 1991b).

From the analytical solutions obtained earlier, stiffness equations for a beam-column member can be derived. The coefficients of these equations are expressed either by hyperbolic functions or trigonometric functions according to whether the axial force is tensile or compressive. However, if we use the previous functions, the coefficients of the stiffness equations become indefinite when the axial force is zero. This singularity causes numerical instability in the case when the axial force approaches zero. To circumvent this situation as well as to avoid the use of different functions for compressive and tensile forces, a power-series expansion is used (Goto and Chen, 1987a). As a result, we can remove the singularity and also unify the expressions.

Consider the following power-series expansions for the trigonometric and the hyperbolic functions:

$$\left.\begin{array}{l}\sin(\gamma L)\\ \sinh(\gamma L)\end{array}\right\} = \gamma L + \gamma L \sum_{n=1}^{\infty} \frac{1}{(2n+1)!} (-\tilde{N}_1)^n \tag{3.3.1a}$$

$$\left.\begin{array}{l}\cos(\gamma L)\\ \cosh(\gamma L)\end{array}\right\} = 1 + \sum_{n=1}^{\infty} \frac{1}{(2n)!} (-\tilde{N}_1)^n \tag{3.3.1b}$$

$$\tilde{N}_1 = \frac{N_1 L^2}{EI} = \frac{\pi^2 N_1}{P_e} \tag{3.3.1c}$$

$$P_e = \frac{\pi^2 EI}{L^2} \tag{3.3.1d}$$

$$\gamma = \left(\frac{|N_1|}{EI}\right)^{1/2} \tag{3.3.1e}$$

where EI is the sectional rigidity and L is the length of the member. In the preceding equations, the axial force N_1 of member 1,2 is taken as positive for a tensile and negative for a compressive force.

The closed-form stiffness equations obtained from the nonlinear beam-column theory and the linearized beam-column theory are given, respectively, as follows (Goto et al., 1991b).

The nonlinear theory:

$$\tilde{F}_i = \sum_{j=1}^{6} \overline{K}_{ij} (\tilde{N}_1)\, \tilde{d}_j + \sum_{j=1}^{6} \sum_{k=1}^{6} \overline{K}_{ijk} (\tilde{N}_1)\, \tilde{d}_j \tilde{d}_k + \tilde{p}_y \sum_{j=1}^{6} \overline{L}_{ij} (\tilde{N}_1)\, \tilde{d}_j + \tilde{p}_y \overline{L}_i(\tilde{N}_1) + \tilde{p}_y^2 \overline{C}_i(\tilde{N}_1) \tag{3.3.2}$$

The linearized theory:

$$\tilde{F}_i = \sum_{j=1}^{6} \overline{K}_{ij} (\tilde{N}_1)\tilde{d}_j + \tilde{p}_y \overline{L}_i(\tilde{N}_1) \tag{3.3.3}$$

where

$$^t\{\tilde{F}_i\} = (\tilde{N}_1, \tilde{S}_1, \tilde{M}_1, \tilde{N}_2, \tilde{S}_2, \tilde{M}_2) \tag{3.3.4a}$$

$$^t\{\tilde{d}_i\} = (\tilde{u}_1, \tilde{v}_1, \alpha_1, \tilde{u}_2, \tilde{v}_2, \alpha_2) \tag{3.3.4b}$$

$$\tilde{N}_\beta = N_\beta L^2/EI \tag{3.3.5a}$$

$$\tilde{S}_\beta = S_\beta L^2/EI \tag{3.3.5b}$$

$$\tilde{M}_\beta = M_\beta L/EI \tag{3.3.5c}$$

$$\tilde{p}_y = p_y L^3/EI \tag{3.3.5d}$$

$$\tilde{u}_\beta = u_{0\beta}/L \tag{3.3.5e}$$

$$\tilde{v}_\beta = v_{0\beta}/L \qquad (\beta = 1, 2) \tag{3.3.5f}$$

The coefficients of the stiffness equation, $\overline{K}_{ij}(\tilde{N}_1)$, $\overline{K}_{ijk}(\tilde{N}_1)$, $\overline{L}_{ij}(\tilde{N}_1)$, $\overline{L}_i(\tilde{N}_1)$, and $\overline{C}_i(\tilde{N}_1)$, are functions of the axial force $\tilde{N}_1$. Specifically, $\overline{K}_{ij}$ and $\overline{K}_{ijk}$ are symmetric with respect to subscripts (i,j) and (j,k), respectively. All the foregoing coefficients are shown in the following:

$$[\overline{K}_{ij}] = \begin{bmatrix} AL^2/I & 0 & 0 & -AL^2/I & 0 & 0 \\ & 12\phi_1 & 6\phi_2 & 0 & -12\phi_1 & 6\phi_2 \\ & & 4\phi_3 & 0 & -6\phi_2 & 6\phi_4 \\ & & & AL^2/I & 0 & 0 \\ & \text{sym.} & & & 12\phi_1 & -6\phi_2 \\ & & & & & 4\phi_3 \end{bmatrix} \tag{3.3.6}$$

where ϕ_α (α = 1 to 4) are stability functions and their series expansions with respect to the axial force $\tilde{N}_1$ are expressed in the following:

$$\phi_1 = \frac{1 + \sum_{n=1}^{\infty} \frac{1}{(2n+1)!} (-\tilde{N}_1)^n}{12\phi} \tag{3.3.7a}$$

$$\phi_2 = \frac{\frac{1}{2} + \sum_{n=1}^{\infty} \frac{1}{(2n+2)!} (-\tilde{N}_1)^n}{6\phi} \tag{3.3.7b}$$

$$\phi_3 = \frac{\frac{1}{3} + \sum_{n=1}^{\infty} \frac{2(n+1)}{(2n+3)!} (-\tilde{N}_1)^n}{4\phi} \tag{3.3.7c}$$

$$\phi_4 = \frac{\frac{1}{6} + \sum_{n=1}^{\infty} \frac{1}{(2n+3)!} (-\tilde{N}_1)^n}{2\phi} \tag{3.3.7d}$$

where ϕ is defined as

$$\phi = \frac{1}{12} + \sum_{n=1}^{\infty} \frac{2(n+1)}{(2n+4)!} (-\tilde{N}_1)^n \tag{3.3.8}$$

$$\begin{aligned} \overline{K}_{1jk} = -\overline{K}_{4jk} &= f_1 A_j A_k + f_2 \overline{K}_{2j} \overline{K}_{2k} + f_3 \overline{K}_{3j} \overline{K}_{3k} \\ &\quad + f_4 (A_j \overline{K}_{2k} + A_k \overline{K}_{2j})/2 + f_5 (A_j \overline{K}_{3k} + A_k \overline{K}_{3j})/2 \\ &\quad + f_6 (\overline{K}_{2j} \overline{K}_{3k} + \overline{K}_{3j} \overline{K}_{2k})/2 \end{aligned} \tag{3.3.9a}$$

$$\overline{K}_{2jk} = \overline{K}_{3jk} = \overline{K}_{5jk} = \overline{K}_{6jk} = 0 \tag{3.3.9b}$$

where ${}^t\{A_i\} = (0\ 0\ 1\ 0\ 0\ 0)$. f_b (b = 1 to 10) are expressed by the power series with respect to the axial force:

$$f_1 = 1 + \frac{1}{8} \sum_{n=1}^{\infty} \frac{4^{n+1}(-\tilde{N}_1)^n}{(2n+1)!} \tag{3.3.10a}$$

$$f_2 = \frac{1}{20} + 2\sum_{n=1}^{\infty} \frac{4^{n+1}(-\tilde{N}_1)^n}{(2n+5)!} - 2\sum_{n=1}^{\infty} \frac{(-\tilde{N}_1)^n}{(2n+5)!} \tag{3.3.10b}$$

$$f_3 = \frac{1}{3} + \frac{1}{2} \sum_{n=1}^{\infty} \frac{4^{n+1}(-\tilde{N}_1)^n}{(2n+3)!} \tag{3.3.10c}$$

$$f_4 = \frac{1}{3} - 2\sum_{n=1}^{\infty} \frac{(-\tilde{N}_1)^n}{(2n+3)!} + \sum_{n=1}^{\infty} \frac{4^{n+1}(-\tilde{N}_1)^n}{(2n+3)!} \tag{3.3.10d}$$

$$f_5 = -1 - \frac{1}{2}\sum_{n=1}^{\infty} \frac{4^{n+1}(-\tilde{N}_1)^n}{(2n+2)!} \tag{3.3.10e}$$

$$f_6 = -\frac{1}{4} + \sum_{n=1}^{\infty} \frac{2}{(2n+4)!}(-\tilde{N}_1)^n - 2\sum_{n=1}^{\infty} \frac{4^{n+1}}{(2n+4)!}(-\tilde{N}_1)^n \tag{3.3.10f}$$

$$f_7 = \frac{1}{252} - 4\sum_{n=1}^{\infty} \frac{(n+3)(-\tilde{N}_1)^n}{(2n+7)!} + 2\sum_{n=1}^{\infty} \frac{4^{n+2}(-\tilde{N}_1)^n}{(2n+7)!} \tag{3.3.10g}$$

$$f_8 = \frac{1}{12} + 2\sum_{n=1}^{\infty} \frac{1}{(2n+4)!}(-\tilde{N}_1)^n - \tilde{N}_1\left[\frac{1}{6} + \sum_{n=1}^{\infty} \frac{1}{(2n+3)!}(-\tilde{N}_1)^n\right]^2 \tag{3.3.10h}$$

$$f_9 = \left[\frac{1}{6} + \sum_{n=1}^{\infty} \frac{1}{(2n+3)!}(-\tilde{N}_1)^n\right]^2 \tag{3.3.10i}$$

$$f_{10} = -\frac{1}{15} + 4\sum_{n=1}^{\infty} \frac{n+2}{(2n+5)!}(-\tilde{N}_1)^n - \sum_{n=1}^{\infty} \frac{4^{n+2}}{(2n+5)!}(-\tilde{N}_1)^n \tag{3.3.10j}$$

$$\begin{aligned}\overline{L}_{1j} = -\overline{L}_{4j} = &-f_2\overline{K}_{2j} - f_3\overline{K}_{3j}/\{2(2\phi_3 + \phi_4)\} - f_4A_j/2 \\ &- f_5A_j/\{4(2\phi_3 + \phi_4)\} - f_6\overline{K}_{3k}/2 - f_6\overline{K}_{2j}/\{4(2\phi_3 + \phi_4)\} \\ &+ f_8A_j + f_9\overline{K}_{2j} + f_{10}\overline{K}_{3j}\end{aligned} \tag{3.3.11a}$$

$$\overline{L}_{2j} = \overline{L}_{3j} = \overline{L}_{5j} = \overline{L}_{6j} = 0 \tag{3.3.11b}$$

$$\overline{L}_1 = \overline{L}_4 = 0, \qquad \overline{L}_2 = \overline{L}_5 = -1/2, \qquad \overline{L}_3 = -\overline{L}_6 = -1/\{4(2\phi_3 + \phi_4)\} \tag{3.3.12a–c}$$

$$\begin{aligned}\overline{C}_1 = -\overline{C}_4 = &f_2/4 + f_3/\{48(2\phi_3 + \phi_4)\} + f_6/\{8(2\phi_3 + \phi_4)\} \\ &+ f_7 - f_9/2 - f_{10}/\{4(2\phi_3 + \phi_4)\}\end{aligned} \tag{3.3.13a}$$

$$\overline{C}_2 = \overline{C}_3 = \overline{C}_5 = \overline{C}_6 = 0 \tag{3.3.13b}$$

Functions ϕ_i and f_i as given in Eqs. (3.3.8) and (3.3.10) are expressed mathematically by infinite-series expansions in order to unify the different expressions according to the sign of the axial force as well as to avoid the numerical instability when the axial force approaches zero. However, for practical numerical calculations, the necessary numbers of terms to be adopted in the series expansions are the first 10 terms for ϕ_i and the first 14 terms for f_i (Goto and Chen, 1987a). These numbers are sufficient to achieve numerical convergence.

Another form of series expansion is also possible to express the stability functions ϕ_i by directly applying a Taylor-series expansion as

$$\phi_1 = 1 - \frac{1}{10}\tilde{N}_1 - \frac{3}{25{,}000}\tilde{N}_1^2 + \cdots \tag{3.3.14a}$$

$$\phi_2 = 1 - \frac{1}{60}\tilde{N}_1 - \frac{3}{25{,}000}\tilde{N}_1^2 + \cdots \tag{3.3.14b}$$

$$\phi_3 = 1 - \frac{1}{30}\tilde{N}_1 - \frac{11}{25{,}000}\tilde{N}_1^2 + \cdots \tag{3.3.14c}$$

$$\phi_4 = 1 + \frac{1}{60}\tilde{N}_1 - \frac{13}{25{,}000}\tilde{N}_1^2 + \cdots \tag{3.3.14d}$$

If we substitute the series expansion given by Eqs. (3.3.14a) to (3.3.14d) into the stiffness equation of Eq. (3.3.3) and retain the first two terms of the stiffness matrix and the first term of the distributed-force vector, respectively, with respect to the expansion, the following approximate stiffness equation can be obtained.

$$\begin{Bmatrix} \tilde{N}_1 \\ \tilde{S}_1 \\ \tilde{M}_1 \\ \tilde{N}_2 \\ \tilde{S}_2 \\ \tilde{M}_2 \end{Bmatrix} = \left[\begin{bmatrix} AL^2/I & 0 & 0 & -AL^2/I & 0 & 0 \\ & 12 & 6 & 0 & -12 & 6 \\ & & 4 & 0 & -6 & 2 \\ & & & AL^2/I & 0 & 0 \\ & \text{sym.} & & & 12 & -6 \\ & & & & & 4 \end{bmatrix} - \tilde{N}_1 \begin{bmatrix} 0 & 0 & 0 & 0 & 0 & 0 \\ & 6/5 & 1/10 & 0 & -6/5 & 1/10 \\ & & 2/15 & 0 & -1/10 & -1/30 \\ & & & 0 & 0 & 0 \\ & \text{sym.} & & & 6/5 & -1/10 \\ & & & & & 2/15 \end{bmatrix} \right] \begin{Bmatrix} \tilde{u}_1 \\ \tilde{v}_1 \\ \alpha_1 \\ \tilde{u}_2 \\ \tilde{v}_2 \\ \alpha_2 \end{Bmatrix} + \begin{Bmatrix} 0 \\ -\tilde{p}_y/2 \\ -\tilde{p}_y/12 \\ 0 \\ -\tilde{p}_y/2 \\ \tilde{p}_y/12 \end{Bmatrix} \tag{3.3.15}$$

The preceding equation is the most commonly used stiffness equation (Allen and Bulson, 1980; White, 1985) in general practice. The same equation can be obtained by a finite-element approximation that assumes polynominal displacement functions (Allen and Bulson, 1980). Equation (3.3.15) has a simpler form and is suitable for eigenvalue analysis to obtain the buckling load and the buckling mode. However, due to the truncations of nonlinear terms with respect to $\tilde{N}_1$, Eq. (3.3.15) loses its accuracy when $|\tilde{N}_1|$ becomes large. In such a case, a member must be divided into smaller elements to keep $\tilde{N}_1$ a small quantity.

In the present analysis, the secant stiffness method, explained later in Section 3.6, is employed as a numerical iterative procedure and the stiffness equations given by Eqs. (3.3.2), (3.3.3), and (3.3.15) are adequate for the present purpose of design analysis. However, as displacements become large, the secant stiffness method loses its accuracy and it becomes difficult to obtain convergent solutions. In such a case, iterative procedures using the tangent stiffness such as the Newton–Raphson method combined with the arc-length controls (Riks, 1979) have to be used. Furthermore, in order to analyze the critical behaviors of semi-rigid frames known as *limit-load instability* and *bifurcation,* the tangent stiffness is essential, as explained by Goto et al. (1991a, 1991b). The closed-form tangent stiffness corresponding to both stiffness equations, Eqs. (3.3.2) and (3.3.3), are explicitly shown by Goto et al. (1991b). In the case of the *linearized beam-column theory,* the tangent stiffness equations coincide with the stiffness equations, Eq. (3.3.3) or (3.3.15), if the prebuckling deformations are ignored. This is the reason why Eq. (3.3.3) or (3.3.15) is commonly used to calculate the bifurcation load in the practical applications where the prebuckling deformations are usually negligibly small.

3.4 MODELING SEMI-RIGID CONNECTIONS

The flexible connection is represented herein by a discrete nonlinear rotational spring, whose moment–rotation (M–θ_r) behavior under a monotonically increasing load is symbolically expressed by

$$M = k_c(\theta_r, M)\theta_r \tag{3.4.1}$$

where $k_c(\theta_r, M)$, a function of θ_r and M, corresponds to the secant stiffness of the connection. It is assumed here that the moment–rotation relationship given by Eq. (3.4.1) is known for each connection in the frame either numerically or in some forms of function. The connection models considered in the present analysis are summarized in Table 3.1. The details of these models are explained in Chapter 2.

In the case of unloading and reverse loading, the independent hardening model (Chen and Saleeb, 1982) is adopted for simplicity to express the connection behavior.

In the unloading process, the connection is assumed to unload linearly up to $M = 0$, following the initial stiffness R_{ki} of the connection, as shown in Fig. 3.3a. Hence, the moment–rotation relationship on the unloading path initiating from the point (M_0, θ_{ro}) is given by

TABLE 3.1 Connection Models

Model	Form of Function	Parameters
Modified exponential (Kishi and Chen, 1986)	$M = M_0 t \sum_{i=1}^{m} A_i \left[1 - \exp\left(-\frac{\theta_r}{2i\alpha}\right)\right] + \sum_{j=1}^{n} R_j H(\theta_r - T_j)(\theta_r - T_j)$	A_i, R_j = curve-fitting constants $H(x)$ = Heaviside's step function T_j = starting rotation of linear components α = scaling factor
Polynomial (Frye and Morris, 1975)	$\theta_r = C_1(KM)^{p_1} \times 10^{n_1} + C_2(KM)^{p_2} \times 10^{n_2} + C_3(KM)^{p_3} \times 10^{n_3}$	C_i = curve-fitting constants K = standardization parameter
Three-parameter power (Kishi and Chen, 1990)	$M = \frac{R_{ki}\theta_r}{[1 + \theta_r/(M_u/R_{ki})^n]^{1/n}}$	M_u = ultimate moment capacity n = shape parameter R_{ki} = initial stiffness

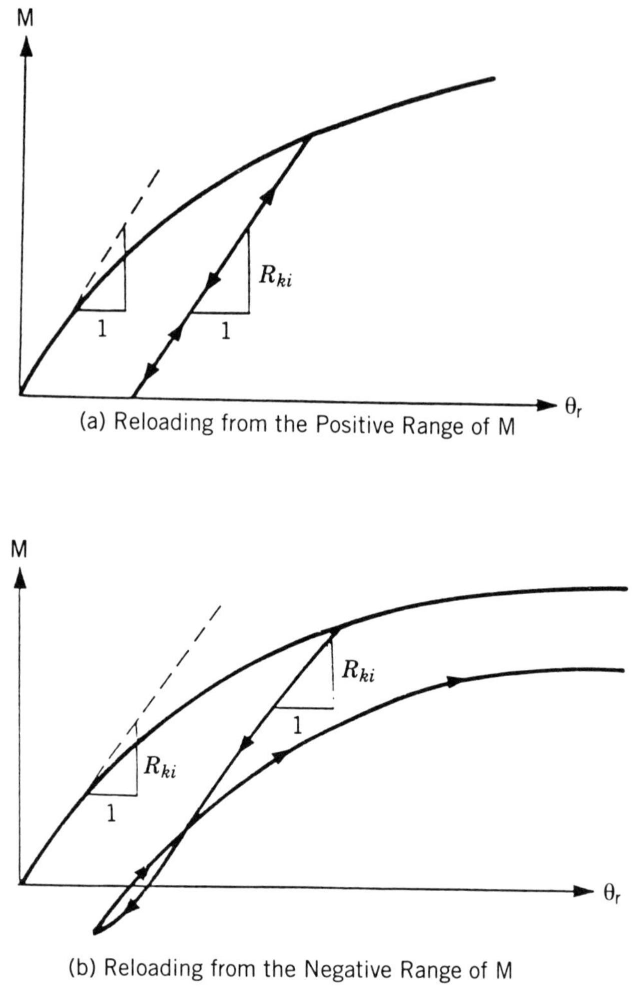

Fig. 3.3 Problems associated with the independent hardening model. Reloading from (a) the positive range of M and from (b) the negative range of M.

$$M = R_{ki}\,(\theta_r - \theta_{ro}) + M_0 \tag{3.4.2}$$

When the sign of connection moment M changes in the unloading process, the connection enters into the reverse-loading process. In this case, the connection behavior, excluding the residual plastic rotation θ_r^p rsulted from the previous loading process, is assumed as follows to coincide with that of virgin connections under monotonic loading.

$$M = k_c(|\theta_r - \theta_r^p|, M)(\theta_r - \theta_r^p) \tag{3.4.3}$$

Equations (3.4.1), (3.4.2), and (3.4.3) can be generally expressed as

$$M = k_c\,\theta_r + M_c \tag{3.4.4}$$

On the loading path:

$$k_c = k_c(\theta_r, M), \qquad M_c = 0 \tag{3.4.5a, b}$$

On the unloading path:

$$k_c = R_{ki}, \qquad M_c = M_0 - R_{ki}\,\theta_{ro} \tag{3.4.6a, b}$$

On the reverse-loading path:

$$k_c = k_c(|\theta_r - \theta_r^p|, M), \qquad M_c = -k_c\,\theta_r^p \tag{3.4.7a, b}$$

As for the connection behavior under cyclic loading, only few experimental data are known. Thus, it is difficult to use the curve-fitting technique to represent the moment–rotation relation. In the present formulation, the independent hardening model is used for simplicity to represent the inelastic behavior. Although this model can take care of the moment- rotation relation under one cycle of loading, unloading, and reverse loading, the connection behavior under the repetition of the previous loading cycle may not be expressed with acceptable accuracy. As explained in Section 2.4.3, one problem associated with the independent hardening model is that the moment–rotation relation is completely different according to whether the connection is reloaded from the positive range of M or from the negative range (Fig. 3.3). To overcome the deficiency of the independent hardening model, Goto et al. (1991a, 1993) introduced the bounding surface model (Dafalias and Popov, 1976) to describe the cyclic behavior after the first unloading occurs.

3.5 MODIFIED SECANT STIFFNESS EQUATION WITH CONNECTION FLEXIBILITY

In order to reduce the number of degrees of freedom in the numerical analysis based on the secant stiffness method, a modified secant stiffness equation for a beam-column member with flexible connections at member ends (Fig. 3.4) is derived. The derivation procedure is made herein specifically for a member with connections at both ends, as illustrated in Fig. 3.5. The modified stiffness equation with the flexible connection at either end of a member can be obtained in a similar manner.

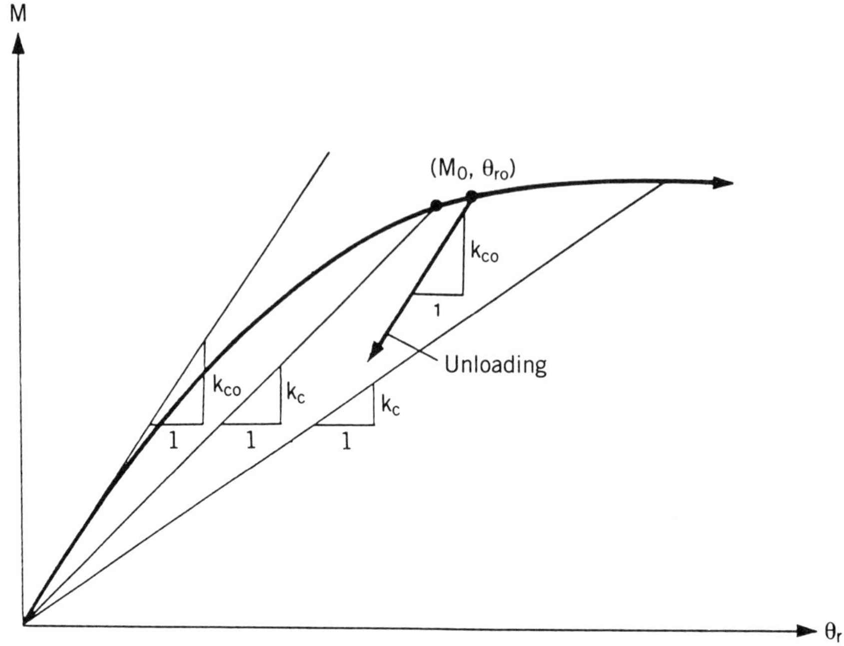

Fig. 3.4 Connection behavior.

To derive the modified stiffness equation, the rotational degrees of freedom of α_1 and α_2 for the beam-column member are eliminated. Relative rotation θ_{rm} at node m is expressed as follows by the rotational degree of freedom of the beam-column member (α_m) as well as that of the connection (α_{cm}).

$$\theta_{rm} = \alpha_{cm} - \alpha_m \qquad (m = 1, 2) \tag{3.5.1}$$

where subscript m indicates a quantity at node m.

Fig. 3.5 Beam-column member with flexible connections at both ends.

With Eqs. (3.4.4) and (3.5.1), the rotational degree α_m at the end of the beam-column member is given by

$$\alpha_m = \alpha_{cm} - \tilde{M}_m/\tilde{k}_{cm} + \tilde{M}_{cm}/\tilde{k}_{cm} \tag{3.5.2}$$

$$\tilde{M}_{cm} = M_{cm}L/EI, \qquad \tilde{k}_{cm} = k_{cm}L/EI \tag{3.5.3}$$

For simplicity, the stiffness equations of Eqs. (3.3.2) and (3.3.3) are assumed to be expressed in the secant stiffness form as

$$\{\tilde{F}_i\} = [\overline{K}^s_{ij}]\{\tilde{d}_j\} + \{\tilde{p}^e_i\} \tag{3.5.4}$$

where the secant stiffness matrix $[\overline{K}^s_{ij}]$ as well as the equivalent member fixed-end force vector $\{\tilde{p}^e_i\}$ can be expressed by the following:

The nonlinear theory:

$$\overline{K}^s_{ij} = \overline{K}_{ij}(\tilde{N}_1) + \sum_{k=1}^{6} \overline{K}_{ijk}(\tilde{N}_1)\tilde{d}_k + \tilde{p}_y\overline{L}_{ij} \tag{3.5.5}$$

$$\tilde{p}^e_i = \tilde{p}_y\overline{L}_i(\tilde{N}_1) + \tilde{p}^2_y\overline{C}_i(\tilde{N}_1) \tag{3.5.6}$$

The linearized theory:

$$\overline{K}^s_{ij} = \overline{K}_{ij}(\tilde{N}_1) \tag{3.5.7}$$

$$\tilde{p}^e_i = \tilde{p}_y\overline{L}_i(\tilde{N}_1) \tag{3.5.8}$$

Substituting Eq. (3.5.2) into Eq. (3.5.4) and solving the equation with respect to $\tilde{M}_m$ ($m = 1, 2$), the modified stiffness equation for a beam-column member with flexible connections at both ends is obtained as

$$\{\tilde{F}^c_i\} = [\overline{K}^c_{ij}]\{\tilde{d}^c_j\} + \{\tilde{p}^{ec}_i\} \tag{3.5.9}$$

3.6 SOLUTION PROCEDURE FOR NONLINEAR STIFFNESS EQUATIONS

The load-control secant stiffness method is used here to solve the structural stiffness equation that is nonlinear in terms of the displacements as well as the axial force. This is because the secant stiffness method is not only simple in computer implementation, but also stable regardless of the shape of the connection's moment–rotation curves, as long as the displacements are not very large. Indeed, the iterative procedure using the tangent stiffness, such as the Newton–Raphson method, guarantees the second-order convergence, but this method requires the calculations of tangent stiffness as well as as unbalanced force. Hence, the programming for the tangent stiffness method becomes more complicated than that of the secant stiffness method that is almost like a simple repetition of the customary first-order analysis. Furthermore, the tangent stiffness is much influenced by a local variation of the connection moment–rotation curves, and the iteration using tangent stiffness may encounter numerical difficulty if the slope of the connection curves changes abruptly or irregularly. Such a problem can occur when the connection moment–rotation curves are expressed by bilinear, piecewise-linear, or polynomial models, or when these curves fitted with tests are not smooth.

When the secant stiffness method using the present closed-form stiffness equations is applied to the analysis of flexibly jointed frames, two kinds of iteration are generally required. One is the calculation of nodal displacements from the structural stiffness equation and the other is for the calculation of axial force $\tilde{N}_1$. The latter iteration has to be carried out at every cycle of a structural iteration using the member stiffness equation if the stiffness equation is nonlinear in terms of member forces. The member stiffness equation becomes nonlinear when the stiffness equation (Eq. 3.3.2) derived from the nonlinear beam-column theory is used or a nonlinear flexible connection model is attached to the member under consideration. It should be noted, however, that member forces can be calculated without iteration for the stiffness equation (Eq. 3.3.3) ignoring the bowing effect (i.e., the equation is based on the linearized beam-column theory) and without nonlinear flexible connections.

The loading, unloading, and reloading behavior of connections is influenced by loading history and, hence, loads have to be divided into smaller increments to account for this behavior. In each load increment, the secant stiffness method can be employed to obtain convergent solutions.

Because the present method is primarily for design purposes, no special techniques are required in the iterative procedure. The secant stiffness procedure is no more than a repetition of the first-order analysis. The only difference here is to update the nonlinear terms in the stiffness equations at every cycle of iteration, using the latest information obtained from the previous cycle of iteration.

To obtain convergent solutions, the iteration is continued until unbalanced force becomes within the prescribed tolerance. In updating the nonlinear terms, connection moments are compared with those in the previous load step to determine whether the connection is in a state of loading or unloading. To begin an iteration for each load increment, the solutions in the previous loads steps are used as an initial estimation of the nonlinear terms. The present iterative procedures have yielded satisfactorily results regardless of the shape of the connection moment–rotation curves as commonly used in practical examples. When the bowing effect is ignored, the convergence of iteration comes quickly and solutions can be obtained even under a relatively large displacement.

On the equilibrium path in the vicinity of the limit point, it becomes difficult for the present load-control secant stiffness method to obtain convergent solutions. Furthermore, the present method cannot deal with critical behaviors of structural systems such as limit-load instability and bifurcation. For second-order elastic analysis in frame design, however, the limit-load instability is seldom encountered. This is in contrast to the design using the second-order inelastic analysis. As a result, the present method should be adequate for design purpose.

To investigate critical behaviors more accurately, special formulations are required in such an analysis. Goto et al. (1991a, 1991b) proposed an accurate, yet practical method for the analysis of critical and postcritical behaviors of semi-rigid frames with inelastic connections, where the incremental arc-length method was combined with the Newton–Raphson iterative procedure to deal specifically with the limit-load instability problem, whereas the inelastic bifurcation problems were analyzed by use of Hill's bifurcation theory for elastic-plastic solids (Hill, 1958). Some interesting numerical results based on the previous analysis are shown later in Section 3.7.

3.7 NUMERICAL EXAMPLES

Example 1: Cantilever Column with Small End Moment A cantilever column with the small end moment shown in Fig. 3.6 was analyzed numerically to demonstrate the accuracy and limitations of the second-order theories: the nonlinear beam-column theory and the linearized beam-column theory. Computed results of vertical load versus end displacement relations are summarized in Fig. 3.7. Closed-form elliptic integral solutions for the exact theory (Goto et al., 1987, 1990) are also shown in Fig. 3.7 for comparison. This exact theory referred here to as the *theory of finite displacements with finite strains*. In its derivation, no approximations are introduced except for the usual beam assumptions: no change of cross-sectional shapes and the Bernoulli–Euler hypothesis where the transverse plane is assumed to remain plane and normal to the beam axis. This theory coincides with the one shown by Reissner (1972). The earlier elliptic integral solutions are considered to be more accurate than the well known solutions of inextensional elastica (Timoshenko and Gere, 1961), which ignores the elongation of centroidal axis. To solve the nonlinear stiffness equations in the postbuckling range, the incremental arc-length method combined with the Newton–Raphson iterative procedures were used.

From Fig. 3.7, it can be seen that both the nonlinear and linearized beam-column theories yield accurate solutions at least up to the buckling load. So far as the vicinity of the buckling point is concerned, these theories can also be used to analyze the postbuckling behavior in terms of horizontal displacement. However, as the displacements increase in the postbuckling range, the

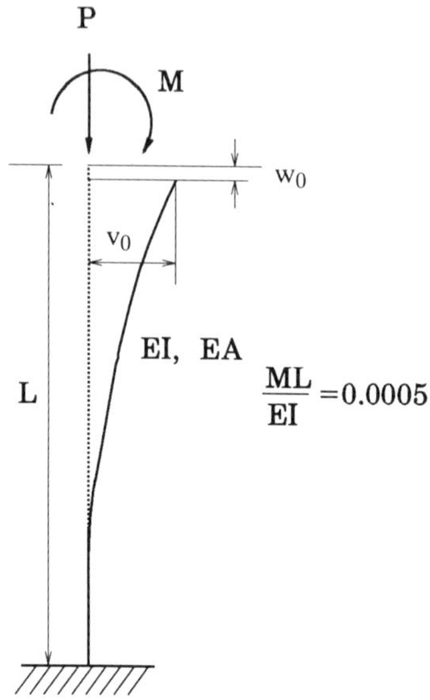

Fig. 3.6 Cantilever column with small end moment.

discrepancy between the exact theory and the beam-column theories becomes large. As can be seen from the linear constitutive equation given by Eq. (3.2.21), the linearized beam-column theory cannot simulate the buckling behavior with regard to the vertical displacement.

Example 2: Bowing Effect on the Design Analysis of Semi-Rigid Rectangular Frames (Goto and Chen, 1987b) As seen from the results of Example 1, it is more accurate to use the nonlinear beam-column theory, which considers the bowing effect in the second-order analysis. However, the stiffness equation including the bowing effect, Eq. (3.3.2), becomes much more complicated than that ignoring this effect, Eq. (3.3.3). Furthermore, the inclusion of the bowing effect results in a slower convergence in the numerical iterative procedures. Therefore, it is desirable to ignore the bowing effect in design analyses where the structures seldom exhibit postbuckling behavior. For the rigid frames, it has been shown by numerical examples that the bowing effect is negligibly small for design purposes (Goto and Chen, 1987a; Korn, 1981). However, for Type PR construction, this bowing effect resulting from the rotational displacements of structures may have a significant influence on the results of analysis, because the frames are more deformable and the nonlinear behavior of the connections is also more sensitive to the magnitude of applied moments. The bowing effect for the design analysis of Type PR construction is therefore investigated numerically.

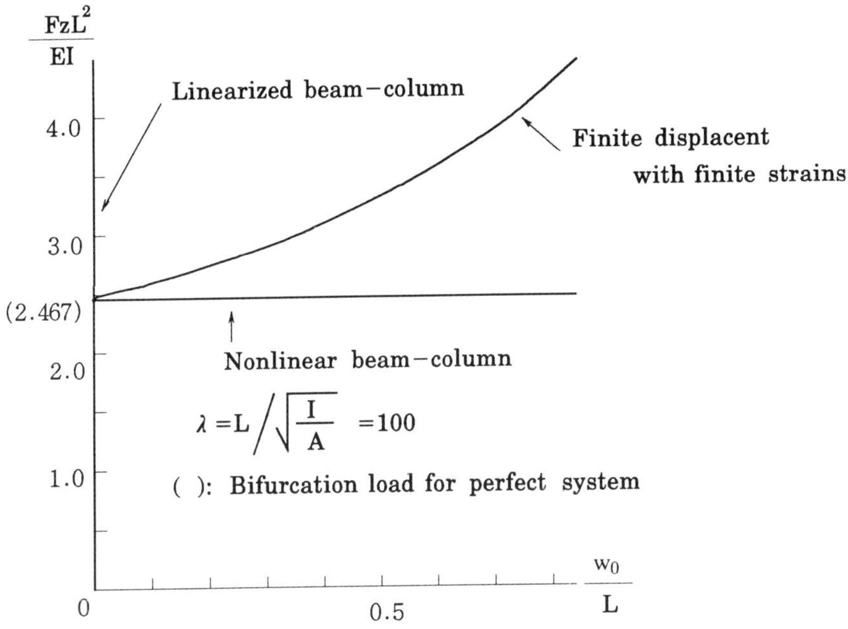

(a) Vertical displacement at column end

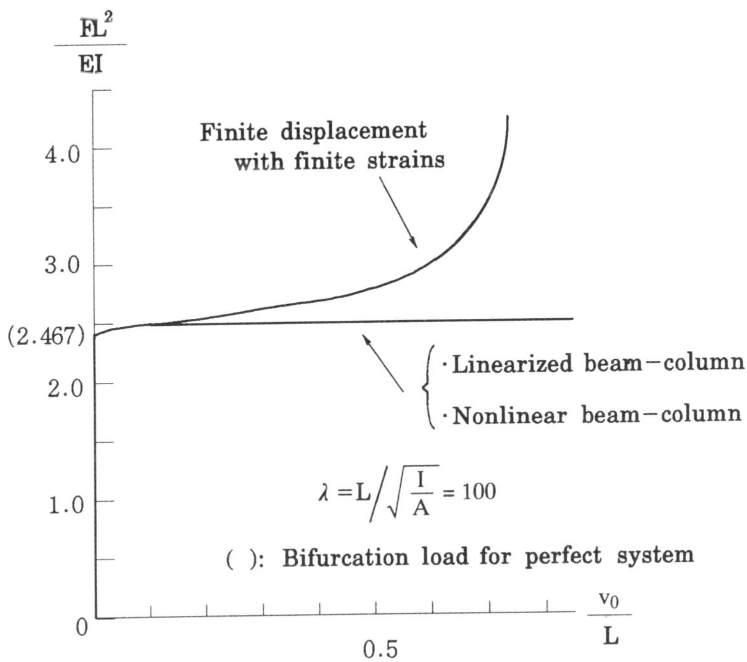

(b) Horizontal displacement at column end

Fig. 3.7 Load-displacement curves of cantilever column. (a) Vertical displacement at column end. (b) Horizontal displacement at column end.

Two typical rectangular semi-rigid frames with fixed base designed as Type PR construction (Moncarz and Gerstle, 1981; Lindsey, Ioannides, and Goverdhan, 1985) are used here. These rectangular frames are illustrated in Fig. 3.8 along with the loads considered in the analysis.

Model I is the two-story–one-bay frame used by Moncarz and Gerstle (1981). They chose the connections from among the top and seat angles tested by Hechtman and Johnston (1947); specimen 23 and 25 were selected for the upper and the lower beams, respectively. Here, due to the lack of information about specimen 25, specimen 23 is used for all the connections. Further, to examine the effect of connection flexibility on bowing, two other kinds of connections, that is, specimens 16 and 24 in the same experiment series, are added for comparison as the two extreme cases in terms of connection flexibility. The frame was designed based on the AISC ASD and so the load is multiplied by 1.4 to obtain the factored load shown in Fig. 3.8.

Model II is the two-story–four-bay frame used by Lindsey et al. (1985), where the double web angles (4 in. $\times$ 3.5 in. $\times$ 3/8 in.) were used for all connections. Because the angle heights are not indicated by them, the connections are selected from those shown by Rathbun (1936), considering the depth of beams; specimens A-4 and A-6 are used for the lower and the upper beams, respectively. In addition, following the same philosophy as explained for Model I, specimens A1-1 and E1-3 are selected from the newer experiment (Thompson, McKee, and Visintainer, 1970) as two extreme cases in terms of connection flexibility. In these two cases, the same connecton is used throughout the structure. The moment–rotation relationship of the connections used for Models I and II are shown in Figs. 3.9 and 3.10, respectively.

In the numerical analysis, the factored loads are divided equally into five steps to account for the unloading of connections and the loads are increased monotonically up to the factored loads. At each load step, the structural iteration is repeated until the ratio of an unbalanced force to the corresponding nodal force becomes less than 0.01%, while the member iteration required at every cycle of the structural iteration is also continued until the error of a member force comes within 0.01% of the corresponding member force.

The ratios of the physical quantities obtained by the second-order analysis to those by the first-order analysis are summarized for two types of frames in Tables 3.2 and 3.3, which are classified according to the method of analysis as well as the connection type. It should be noted that the first-order analysis used here only ignores the geometrical nonlinearity. For comparison, the results for a rigid frame are also added to Tables 3.2 and 3.3. Further, in Tables 3.4 and 3.5, it is shown how the difference of the methods has an effect on the loading/unloading characteristics of the connection in a steel frame.

It is recognized from Tables 3.2 and 3.3 that the bowing effect along with the geometrical nonlinearity generally have more influence on flexibility jointed frames than on rigid frames. The bowing also affects the loading/unloading behavior of connections, as shown in Table 3.5, although this effect is not so significant as that of the geometrical nonlinearity. Throughout Tables

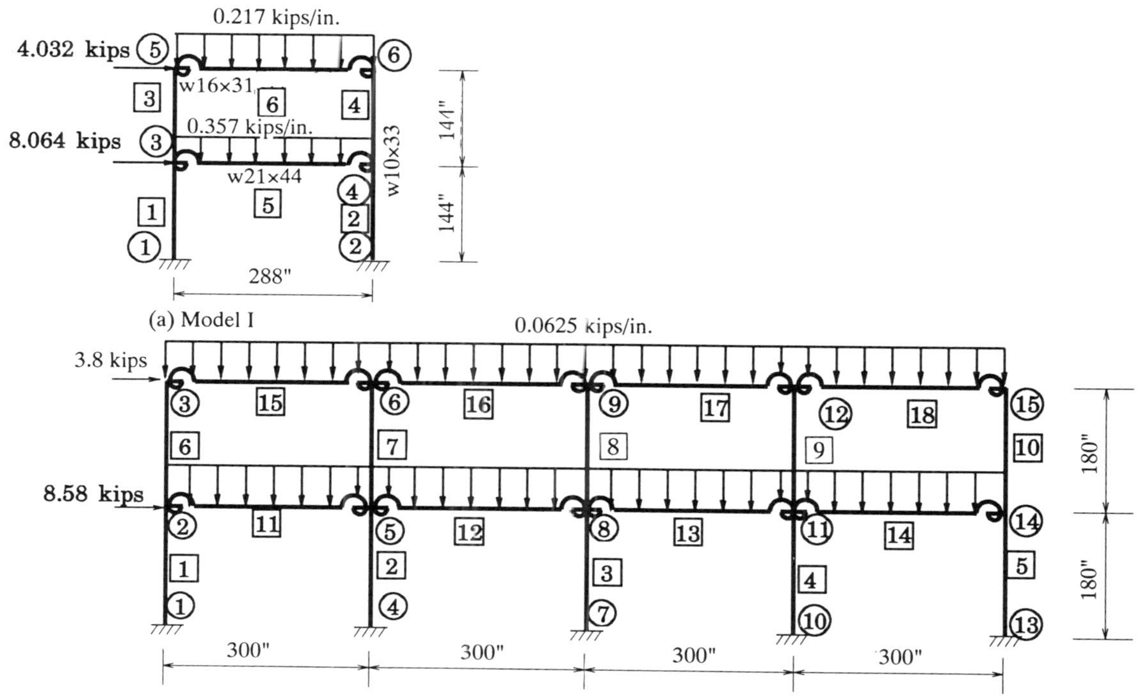

All Roof Girders = W14×22, All Floor Girders = W18×46
Exterior Columns = W8×24, Interior Columns = W8×31

(b) Model II

○: **Node No.** □: **Element No.**

Fig. 3.8 Semi-rigid rectangular frames with fixed base.

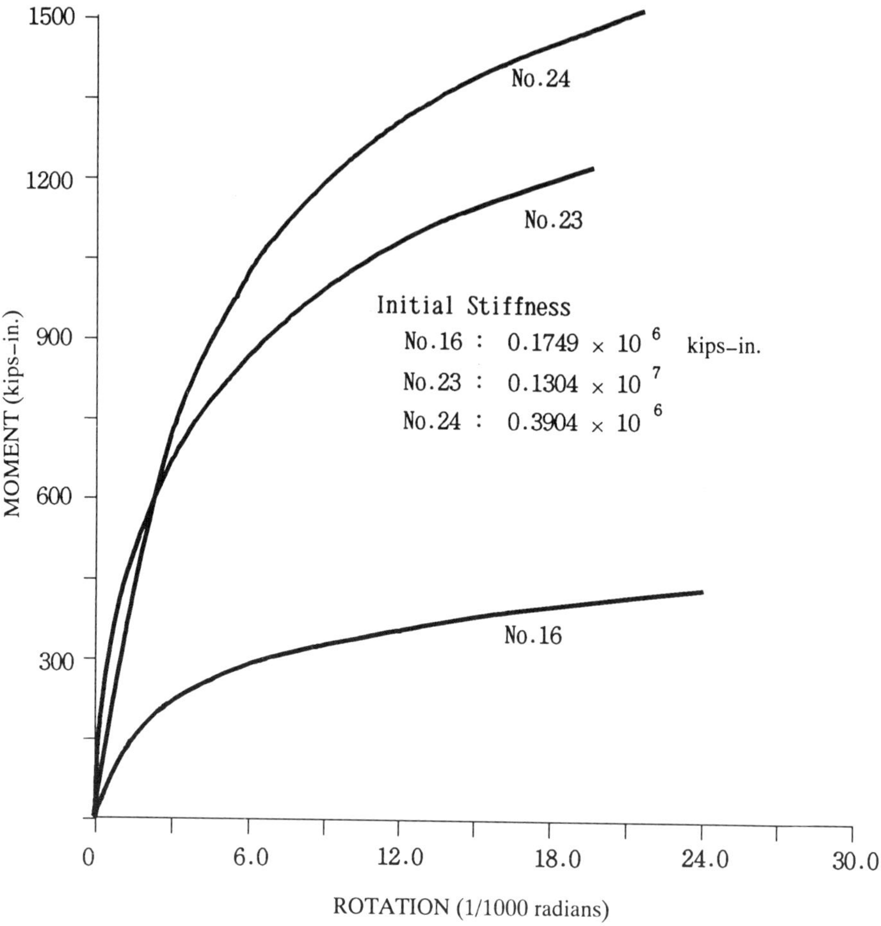

Fig. 3.9 Experimental moment–rotation relationship of connections for Model I.

3.2–3.5, the effect of the geometrical nonlinearity is significant in the drift as well as in the column moment of flexibly jointed frames. Hence, it can be said that the precise consideration on the geometrical nonlinearity may be more important in the design analysis of Type PR construction. Indeed, the bowing effect on the column moment is more evident in flexibly jointed frames than in rigid frames, but it seems that the bowing effect can still be ignored in the design analysis for flexibly jointed frames.

Example 3: Bifurcation and Postbifurcation Behavior of Semi-Rigid Frames (Goto et al., 1991a, 1993) Postbifurcation analysis was carried out to examine the imperfection sensitivity of semi-rigid sway frames. A semi-

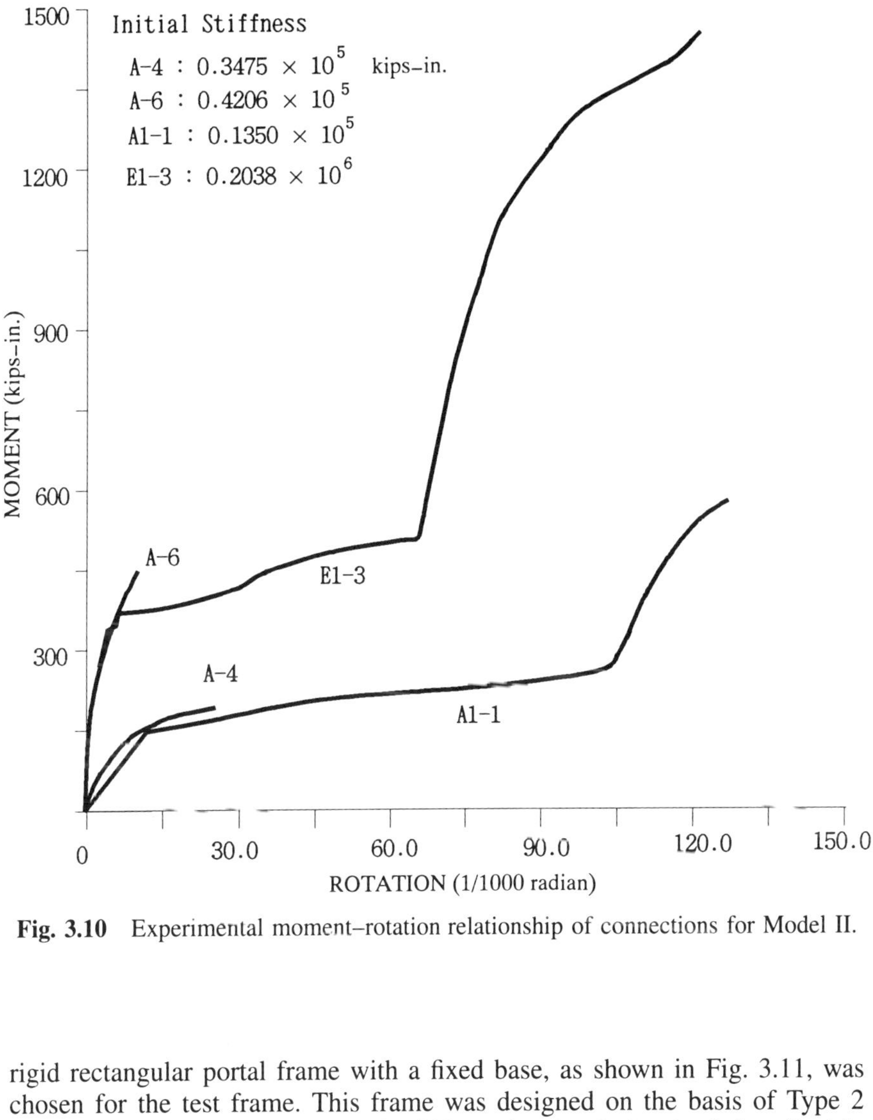

Fig. 3.10 Experimental moment–rotation relationship of connections for Model II.

rigid rectangular portal frame with a fixed base, as shown in Fig. 3.11, was chosen for the test frame. This frame was designed on the basis of Type 2 construction in AISC ASD (AISC 1989) under the design load described in Table 3.6. All the beam-to-column connections were assumed to have the same moment–rotation characteristics as the top- and seat-angle connection, specimen 23, tested by Hetchmann and Johnston (1947). The moment–rotation curve for this connection is illustrated in Fig. 3.12, where the curve expressed by the modified exponential model is also shown for comparison. The constants of the modified exponential model are summarized in Table 3.7. After a first unloading occurs in the connections, the bounding surface model is used. The hardening shape parameter h, the initial stiffness R_{ki}, and

TABLE 3.2 Bowing Effect in Two-Story–One-Bay Frame

	Drift				Maximum Column Moment in 1st Story				Maximum Column Moment in 2nd Story			
	Flexibly Jointed Frames			Rigid	Flexibly Joined Frames			Rigid	Flexibly Jointed Frames			Rigid
Connection	No. 16	No. 23	No. 24	Frame	No. 16	No. 23	No. 24	Frame	No. 16	No. 23	No. 24	Frame
Location of maximum value	Node 5	Node 5	Node 5	Node 5	Node 5	Node 5	Node 5	Node 5	Node 6	Node 6	Node 6	Node 6
Considering bowing effect	1.113	1.089	1.081	1.042	1.048	1.041	1.072	1.008	1.014	1.015	1.014	1.007
Ignoring bowing effect	1.112	1.087	1.079	1.039	1.058	1.047	1.082	1.009	1.014	1.016	1.015	1.007

TABLE 3.3 Bowing Effect in Two-Story–Four Bay Frame

	Drift				Maximum Column Moment in 1st Story				Maximum Column Moment in 2nd Story			
	Flexibly Jointed Frames			Rigid	Flexibly Joined Frames			Rigid	Flexibly Jointed Frames			Rigid
Connection	A-4/A-6	A1-1	E1-3	Frame	A-4/A-6	A1-1	E1-3	Frame	A-4/A-6	A1-1	E1-3	Frame
Location of maximum value	Node 3	Node 3	Node 3	Node 3	Node 4	Node 4	Node 4	Node 4	Node 14	Node 15	Node 15	Node 14
Considering bowing effect	1.149	1.268	1.125	1.074	1.133	1.211	1.134	1.065	0.991	1.033	1.020	1.001
Ignoring bowing effect	1.140	1.262	1.121	1.072	1.125	1.202	1.124	1.064	0.981	1.033	1.018	1.000

TABLE 3.4 Connection Behavior of Model I

Connection	Method of Analysis	Location of Unloading Connections		Load Step				
		Element No.	Node No.	1	2	3	4	5
No. 16	Considering/ ignoring bowing effect	5	3		←	—	—	→
		6	5			←	—	→
	First-order analysis	5	3			←	—	→
		6	5					

←——→: unloading.

the equation for the bounding line are determined, as shown in Table 3.8, by using the data under monotonic loading. The connection models adopted here are explained in Section 2.4.

In the analysis of nonlinear equilibrium paths, except for bifurcation points, the incremental arc-length method (Riks, 1979), combined with the Newton–Raphson iterative procedure, is employed. At the bifurcation point, solutions have an inherent singularity and this singularity of the tangent stiffness matrix cannot be eliminated even by the use of the arc-length method. Thus, bifurcation analysis is carried out by using Hill's theory of uniqueness of elastic-plastic solids (Hill, 1958). Although the present analysis considers the geometrical nonlinearity of structural systems based on the nonlinear beam-column theory, the nonlinearity due to member plastification is ignored.

By using the method explained before, the bifurcation and postbifurcation behaviors of the aforementioned semi-rigid test frames subject to two vertical load conditions (a uniformly distributed load on the beams or concentrated loads on the columns) shown in Fig. 3.13 were analyzed. In addition, the effect of the cyclic wind load on the critical behavior was also discussed.

Figures 3.14a and 3.14b show the vertical load vs. horizontal displacement relations of a rectangular semi-rigid portal frame with or without imperfections. These figures also include the postbifurcation curves of frames with pinned or rigid connections, together with those of the frames with linearly elastic connections. The elastic constant for these connections coincides with the initial stiffness of the connection curve shown in Fig. 3.12. In the imperfect systems, small horizontal force was applied at the top of the column as initial imperfections. The frame in Fig. 3.14a is subjected to uniformly distributed load applied on the beams, whereas the frame in Fig. 3.14b is subjected to concentrated loads applied on the columns. In Fig. 3.14b, the dashed-line curve corresponds to the perfect frame under distributed load and this curve is added to show the difference due to the loading condition. The magnitude of the concentrated loads is such that they give the same value of

TABLE 3.5 Connection Behavior of Model II

Connection	Method of Analysis	Location of Unloading Connection: Element No.	Location of Unloading Connection: Node No.	Load Step 1	2	3	4	5
Upper story: A-4 Lower story: A-6	Considering bowing effect	11	2				↔	↔
		12	5				↔	↔
		14	11				↔	↔
	Ignoring bowing effect	11	2				↔	↔
		12	5					
		14	11				↔	↔
	First-order analysis	11	2					
		12	5					
		14	11					
A1-1	Considering/ignoring bowing effect	11	2		↔	↔	↔	↔
		12	5		↔	↔	↔	↔
		13	8		↔	↔	↔	↔
		14	11		↔	↔	↔	↔
		15	3				↔	↔
		16	6				↔	↔
		17	9				↔	↔
		18	12				↔	↔
	First-order analysis	11	2				↔	↔
		12	5				↔	↔
		13	8				↔	↔
		14	11				↔	↔
		15	3				↔	↔
		16	6					
		17	9					
		18	12					
E1-3	Considering/ignoring bowing effect	11	2				↔	↔
		12	5		↔	↔	↔	↔
		13	8		↔	↔	↔	↔
		14	11		↔	↔	↔	↔
	First-order analysis	11	2					
		12	5					
		13	8					
		14	11					

⟷: unloading.

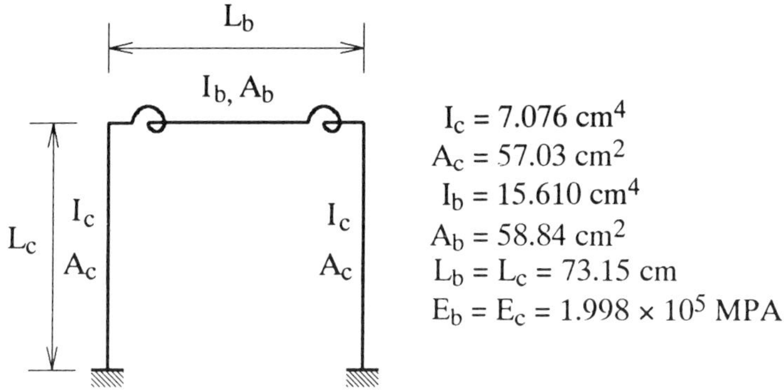

Fig. 3.11 Geometry of semi-rigid rectangular portal frame.

vertical load on a story as do the distributed loads. From the figures, it can be seen that the stability behavior of the frames is quite different, depending on the type of the vertical loads applied. For the perfect semi-rigid frames subjected to uniformly distributed loads, the postbifurcation load-deflection curve rises above the bifurcation load until the limit point is reached, being different from the elastic frames with rigid, pinned, or elastic connections. This implies that the semi-rigid frames under uniformly distributed load do not lose stability even after the bifurcation. On the contrary, for the perfect semi-rigid frames subjected to concentrated loads applied on the top of the columns, the postbifurcation load-deflection curve drops dramatically after bifurcation. This indicates that the semi-rigid frame becomes unstable just after bifurcation and no postbuckling strength is present. The reason for this behavior can be explained in terms of the loading and unloading behavior of the connections.

For the frame subjected to distributed loads, the connections start to load as soon as the loads are applied. As the applied loads increase, the stiffness of the connections decreases because the slope of the connection moment–rotation curve decreases at higher values of moment, as shown in Fig. 3.12. At bifurcation, some connections unload as a result of buckling. When a connection unloads, its stiffness increases because the unloading stiffness is

TABLE 3.6 Design Loads

Dead load (vertical)	$g = 0.335$ N-cm^{-2} or 0.245 kN-cm^{-1} of each girder
Wind load (horizontal)	$w = 0.0957$ kN-cm^{-2}, which results in the concentrated floor loads
	$W_1 = 25.6$ kN for the floors of the one-story frame
	$W_2 = 12.6$ kN for the upper floor of the two-story frame

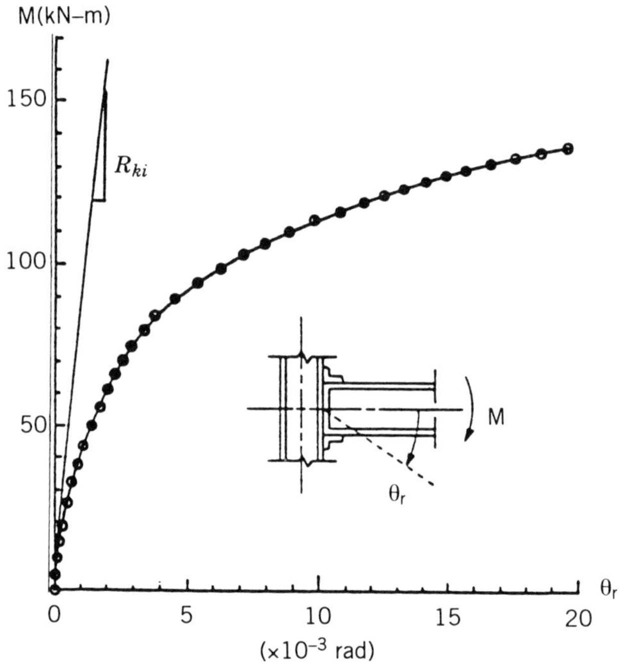

○ : Experimental value

— : Modified exponential model

R_{ki} : Initial stiffness

Fig. 3.12 Modeling of connection.

larger than the loading stiffness. This increase in connection stiffness also increases the stiffness of frames. Hence, additional vertical loads can be applied beyond bifurcation. On the other hand, if concentrated loads are applied to the top of the columns, none of the connections will load until bifurcation. Once bifurcation occurs, the connections will continue to load and their stiffness will continue to degrade. Thus, the frame becomes unstable just after bifurcation and no additional loadings can be applied beyond bifurcation.

TABLE 3.7 Constants of Modified Exponential Model (Table 3.1)

α = 0.17, 345, 750, M_0 = 0.		
A_i = 0.15, 077, 925 × 10^3,	−0.74, 457, 301 × 10^3,	0.13, 416, 943 × 10^4
0.68, 622, 477 × 10^2,	−0.22, 747, 253 × 10^4,	0.15, 788, 006 × 10^4
R_j = 0.18, 289, 890 × 10,	T_j = 9.78	

A_j (i = 1–6) (kN-m); R_j (j = 1) (kN-m); α, T_j (j = 1) (rad).

TABLE 3.8 Constants of Bounding Surface Model

Bounding line:	$M = 1.829\ \theta_r^p + 102.7$ (kN-m)
Initial stiffness:	$R_{ki} = 142.7$ (kN-m)
Hardening shape parameter:	$h = 21.34$ (kN-m)

For the imperfect frames, although a bifurcation does not occur, a limit-load instability of the same type can be found on the equilibrium path. The limit-point load of an imperfect frame subjected to uniformly distributed loads rises above the bifurcation loads of its perfect counterpart, thus indicating that the bifurcation load is a conservative estimate of the stability-limit state of the frames. In contrast, if the imperfect frame is subjected to a concentrated load applied to the top of the column, the limit-point load is always located below the bifurcation load of its perfect counterpart. This implies that the

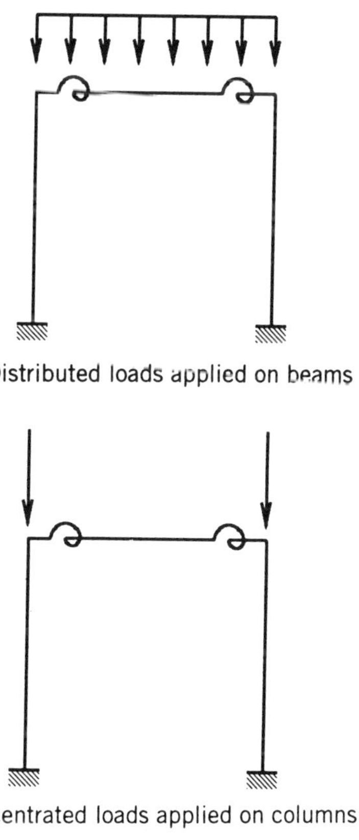

Fig. 3.13 Loading conditions. (a) Distributed loads applied on beams. (b) Concentrated loads applied on columns.

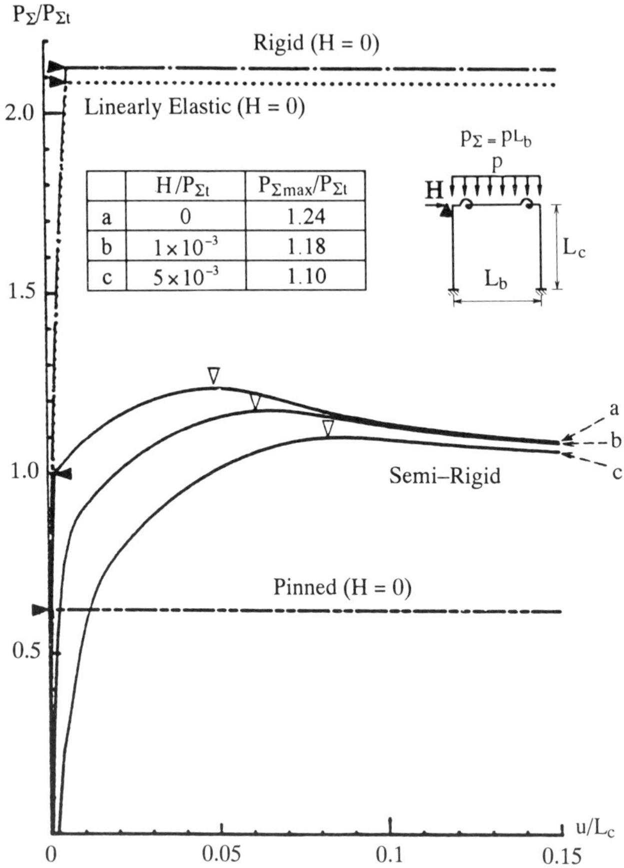

(a) Portal frame under distributed load

Fig. 3.14 Vertical-load vs. lateral-displacement behavior of a semi-rigid frame. (a) Portal frame under distributed load. (b) Portal frame under concentrated load. $P_{\Sigma_{max}}$ is the maximum load, P_{Σ_t} is the tangent modulus load, ▲ is the node whose lateral displacement is shown, ▽ is the limit point, and ▶ is the bifurcation point.

bifurcation load for this case is an unconservative estimate of the stability-limit state.

Figures 3.15a and 3.15b show the critical behavior of frames with semi-rigid connections that were subjected to a prior history of cyclic wind load. In the analysis, the load was applied in three stages. The vertical loads, either distributed or concentrated, were applied first and increased to their design load level. Second, a cycle of horizontal wind load was applied. This cyclic loading was given in the following manner. The wind load was increased monotonically up to the maximum value and then reduced to zero. Next, the

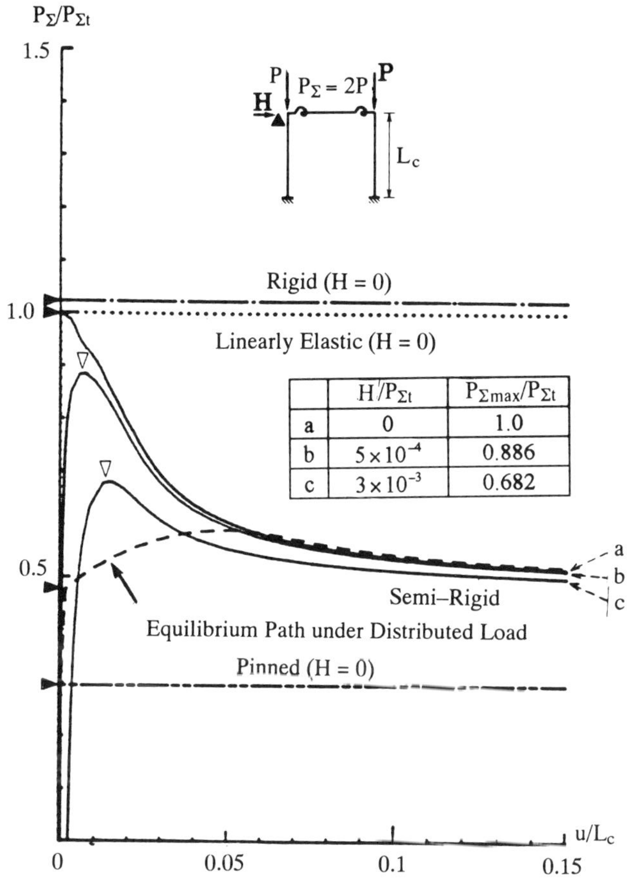

(b) Portal frame under concentrated load

Fig. 3.14 *(Continued)*

wind direction was reversed and this load was applied from the opposite side in a similar manner. This cycle of wind load was repeated until the hysteresis loops of connections became convergent. In the present wind cycles, two amplitudes were considered. One corresponds to the design wind load in Table 3.7 and the other is twice this value. Figures 3.16a and 3.16b show how the hysteresis loops of semi-rigid connections converge when twice the design wind load is applied. As can be seen from Figs. 3.15a and 3.15b, the load history caused by the cyclic wind load has a negligible influence on the critical behavior of the frame with distributed load, but it has a significant influence on the critical behavior of the frame with concentrated loads. Due to the initial imperfection resulting from the load history, the frame with

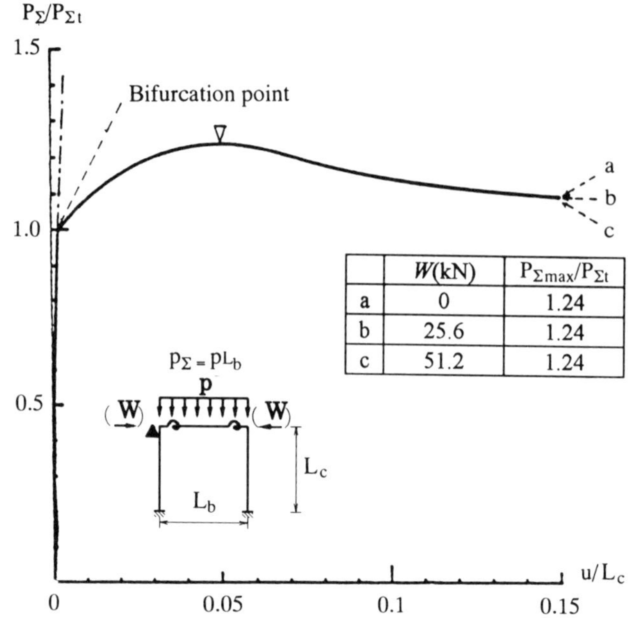

(a) Portal frame under distributed load

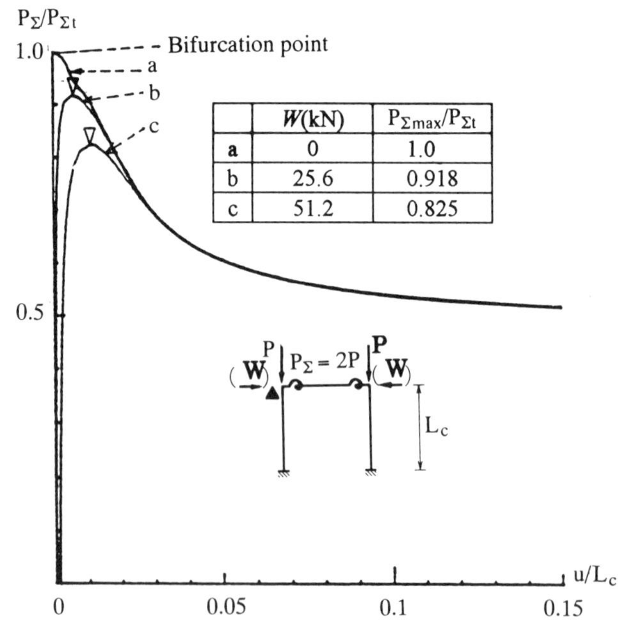

(b) Portal frame under concentrated load

Fig. 3.15 Vertical-load vs. lateral-displacement behavior of a semi-rigid frame with a history of cyclic wind load. (a) Portal frame under distributed load. (b) Portal frame under concentrated load. W is the cyclic wind load, $P_{\Sigma \max}$ is the maximum load, $P_{\Sigma t}$ is the tangent modulus load, ▲ is the node whose lateral displacement is shown, and ▽ is the limit point.

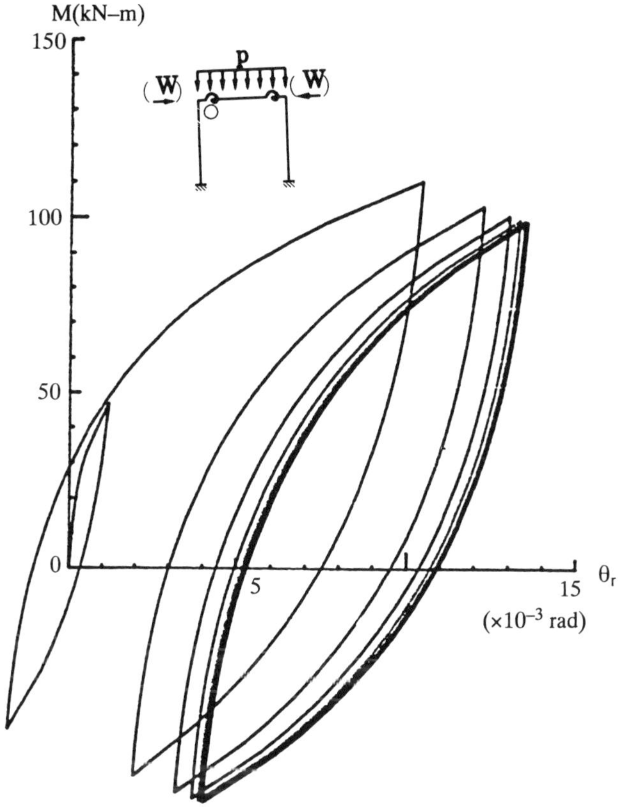

p = 0.24kN/cm, W = 51.2kN

(a) Portal frame under distributed load

Fig. 3.16 Hysteresis loops of connections under cyclic wind load. (a) Portal frame under distributed load. (b) Portal frame under concentrated load. ○ is the connection whose hysteresis loop is shown.

concentrated loads exhibits a limit-load instability similar to the imperfect frame shown previously in Fig. 3.14b. This limit load is located at a load level lower than the bifurcation load of a perfect frame. Based on this observation, it can be concluded that the bifurcation load of a perfect system represents a conservative estimate of the failure load of a semi-rigid frame subjected to uniformly distributed loads, but it represents an unconservative estimate of the failure load of the frame when it is subjected to concentrated loads.

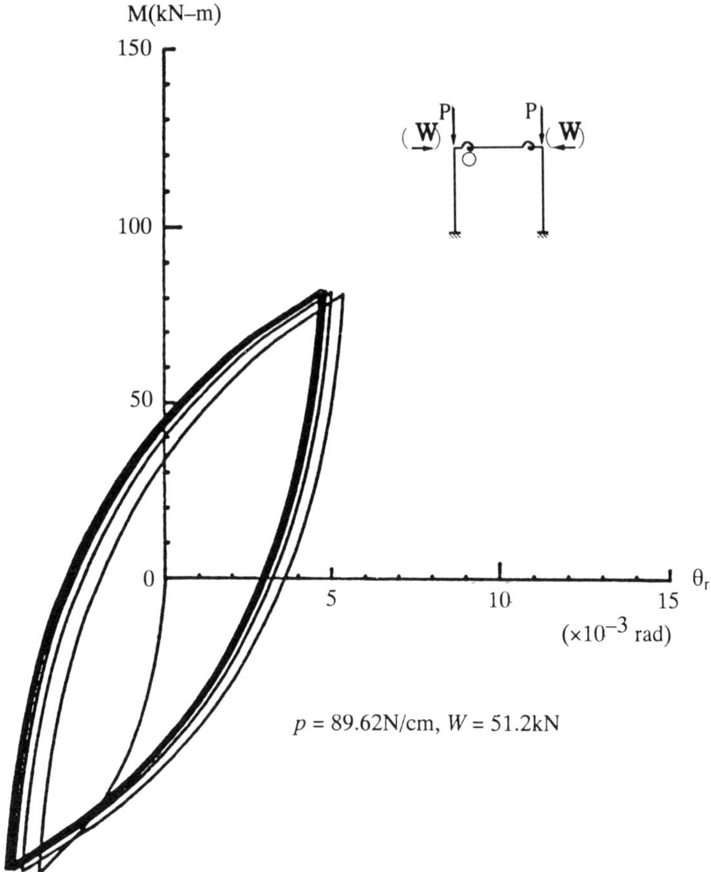

(b) Portal frame under concentrated load

Fig. 3.16 (*Continued*)

3.8 COMPUTER PROGRAM

This program (Kishi et al., 1993), a refined version of FLFRM (Goto and Chen, 1986), is based on the analysis method explained in the foregoing Sections 3.2–3.6 where the second-order frame analysis, including the effect of connection flexibility, is considered. It should be noted, however, that member plastification is not considered in this program.

Two types of theory shown in Section 3.2 are used here for the second-order frame analysis. These are (1) nonlinear beam-column theory and (2) linearized beam-column theory. The nonlinear beam-column theory considers the bowing effect, whereas the linearized beam-column theory ignores it. In addition to the second-order analysis, the first-order analysis is included in this program.

Three connection models described in Table 3.1 are considered in this program. These are (1) the modified exponential model, (2) the polynomial model, and (3) the three-parameter model. For the models mentioned before, either elastic or inelastic behavior can be taken into account. To express inelastic behavior, the independent hardening model explained in Section 3.4 is used.

As an iterative procedure to solve nonlinear stiffness equations, the load-control secant stiffness method explained in Section 3.6 is adopted. This procedure is found not only simple in programming, but also less influenced by the shape of the connection's moment–rotation curves. This iteration is continued until an unbalanced force comes within some prescribed tolerance. It should be noted that the preceding numerical method cannot be used to analyze the bifurcation and limit-load instability of structural systems, as shown in Example 3 (Section 3.7). However, in the design analysis for civil engineering structures, it seldom happens that we have to analyze the highly nonlinear behaviors such as bifurcation and limit-load instability. Therefore, in most cases, the load-control secant stiffness method yields satisfactory results.

3.9 USER'S MANUAL

The program FLFRM written in FORTRAN language is provided by the file named flfrm.f in the diskette. This program is tested by a Sun SPARCstation IPX using a Sun FORTRAN 77 compiler and a Sony RISC NEWS using RISC NEWS FORTRAN. makefile is also included in the diskette to show how an executable program can be made by the utility of the UNIX system. Users can compile the source program and make the executable file by themselves.

In order to execute this program, two sets of data have to be given according to the users' purpose. One set of data is called input.d. The other is toframe. input.d includes all the data except for the connection characteristics that are supplied by toframe. Both of these files should be in the same directory as the executable program. The computed results are written in output.d. If the connections in the data bank are used in the analysis, the values of the parameters that are to be put in toframe can be calculated by this data-bank program. The details of the semi-rigid connection data bank are explained in Chapter 2. For the users' convenience, input.d, toframe, and output.d corresponding to the semi-rigid frame shown later in "(3) Example" are included in the diskette.

(1) input.d

Line 1 (4I5)
IDSI: Flag to select units used in the analysis
- 0: U.S. customary units (pounds, inches)
- 1: MKS units (kg, m)

NANL: Flag to specify the theory used for the beam-column element
- 1: Small displacement theory (first-order theory)
- 2: Nonlinear beam-column theory (bowing effect is considered)
- 3: Linearized beam-column theory (bowing effect is ignored)

MDLC: Flag to specify the connection model used in the analysis
- 1: Elastic model including the nonlinear elastic model
- 2: Elastic-plastic model based on the independent hardening rule

ITLM: Flag to control the iterative procedure
- 0: Iteration is continued until the relative error for displacements comes within a prescribed tolerance given by Line 12.
- 1: Iteration is stopped after one cycle of iteration. (The accuracy of this iterative procedure is guaranteed only when linearized beam-column theory is applied to rigid frames.)

Line 2 (4I5)
NE: Number of elements $\leq$ 210
ND: Number of nodes $\leq$ 121
NMAT: Number of element materials (element materials are specified in Line 6)
NMATC: Number of different flexible connections $\leq$ 100

Line 3 (3I5, F10.0)
NLSQ: Number of load sequences $\leq$ 20
NL: Largest number of all the "NCL(I)" and "NDLI(I)" for $1 \leq I \leq NLSQ$ ("NCL" and "NDLI" are given in Line 7) $\leq$ 100
KS: Number of terms adopted for a power series to calculate the stiffness equation. "KS" = 20 is enough to obtain the convergent solution. If "KS" = -1, the number of the terms is determined so that the relative error comes within the tolerance specified by "DE".
DE: Relative tolerance in the calculation of the power series. Leave the space blank, unless "KS" = -1. Generally, "DE" = 10^{-5} is enough.

Line 4 (9I5)
N: Element number
IEB(1): Node number of the first end of an element
IEB(2): Node number of the second end of an element
IEB(3): Element material identifier that corresponds to "KK" given in Line 6.
IEB(4): Element type identifier
- 1: Frame element

2: Element with flexible connections at both ends
3: Frame element which a hinge attached to the first end
4: Frame element with a hinge attached to the second end
5: Truss element with hinges at both ends
6: Frame element with a flexible connection attached to the first end
7: Frame element with a flexible connection attached to the second end
8: Element with a flexible connection attached to the first end and a hinge to the second end
9: Element with a hinge attached to the first end and a flexible connection to the second end

IEB(5): Flexible connection identifier for the first end of an element. This identifier corresponds to the DATA NO. defined in toframe. Leave this data space blank, unless IEB(4) is 2, 6, or 8.

IEB(6): Flexible connection identifier for the second end of an element. This identifier corresponds to the DATA NO. defined in toframe. Leave this data space blank, unless IEB(4) is 2, 7, or 9.

IEB(7): Flag to use only the initial stiffness of the moment–rotation relation given by IEB(5) for the first connection. Leave this data space blank, unless IEB(4) is 2, 6, or 8.
0: Use only the initial stiffness
1: Use the full moment–rotation relation

IEB(8): Flag to use only the initial stiffness of the moment–rotation relation given by IEB(6) for the second connection. Leave this data space blank, unless IEB(4) is 2, 7, or 9.
0: Use only the initial stiffness
1: Use the full moment–rotation relation
Line 4 has to be supplied "NE" times

Line 5 (I5,2(F8.0,I2))
KK: Element number
DDX(N), IP: Horizontal projection of element "KK" before deformation DDX(N) and IP denote the mantissa and the exponent, respectively.
DDY(N), JP: Vertical projection of element "KK" before deformation DDY(N) and JP denote the mantissa and the exponent, respectively.
Coordinate systems are shown in Fig. 3.17.
Line 5 must be supplied "NE" times.

Line 6 (I5,3(F8.0,I2))
KK: Element material identifier
E(N), KP: Young's modulus. E(N) and KP denote the mantissa and the exponent, respectively.
BA(N), IP: Cross-sectional area. BA(N) and IP denote the mantissa and the exponent, respectively.
BI(N), JP: Second moment of inertia. BI(N) and JP denote the mantissa and the exponent, respectively.
Line 6 has to be supplied "NMAT" times.

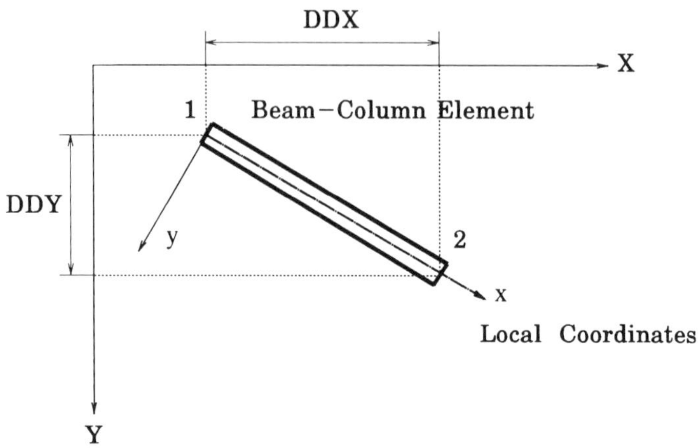

Fig. 3.17 Coordinate systems.

Line 7 (4I5)

KK: Load sequence number

NCL(I): Number of concentrated loads applied in the KKth load sequence (Fig. 3.18).

NDLI(I): Number of the types of distributed loads applied in the KKth load sequence. As explained in Line 9, only the uniformly distributed load acting transverse to the beam-element axis can be considered in this program (Fig. 3.19). Therefore, the types of distributed loads are classified just by the magnitude of their local y-component.

NDIN(I): Number of load increments in the KKth load sequence.

NDIN(I) has to be determined such that the convergent solution can be obtained in the nonlinear analysis.

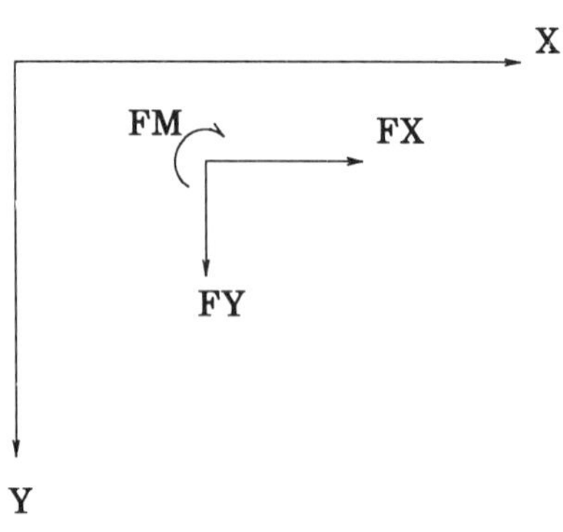

Fig. 3.18 Sign conventions for concentrated loads.

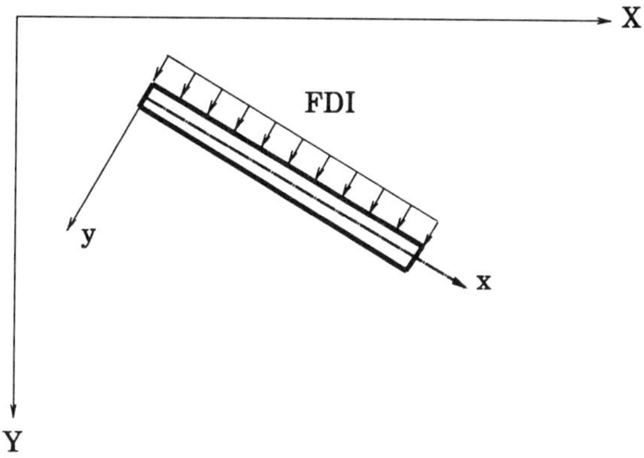

Fig. 3.19 Sign conventions for uniformly distributed loads.

When small displacement analysis is conducted on linear elastic frames, "NDIN(I)" = 1 is enough.

Line 8 (I5,3(F8.0,I2))

NNC(I): Node number where concentrated load is applied at the KKth load sequence

FX(I,II), IP: X-component of the concentrated load at the end of the KKth load sequence. FX(I,II) and IP denote the mantissa and the exponent, respectively.

FY(I,II), JP: Y-component of concentrated load at the end of the KKth load sequence. FY(I,II) and JP denote the mantissa and the exponent, respectively.

FM(I,II), KP: Concentrated moment at the end of the KKth load sequence. FM(I,II) and KP denote the mantissa and the exponent, respectively. These force components follow the sign convention defined in Fig. 3.18. This data must be supplied "NCL(I)" times. Without any concentrated load, this data should be omitted.

Line 9 (F8.0,I2,I5/10I5)

FDI(I,II), IP: Local y-component of the uniformly distributed load at the end of KKth load sequence (Fig. 3.19). FDI(I,II) and IP denote the mantissa and the exponent, respectively.

NEDN: The number of elements where the uniformly distributed load is applied

NEDI(J): Element number where the uniformly distributed load is applied. This data have to be supplied "NEDN" times, following the format specified as 10I5.

It should be noted that only the uniformly distributed load in the direction of local y-axis can be considered in this program. In the case of the variably distributed load or the load with a local x-component, a member has to be divided into smaller elements in order to approximate the load distribution by piecewise uniformly distributed loads or by concentrated loads applied at small intervals.

Line 9 has to be supplied "NDLI(I)" times. Without any distributed load, this data should be omitted.

Lines 7, 8, and 9 have to be supplied "NLSQ" times.

Line 10 (I5)
NBCSRT: Number of the types of restrained nodes. These types are explained in Line 11.

Line 11 (2I5/10I5)
NBTYPE: Identifier for restraining condition.
0: Fixed
1: Hinged
2: Hinged and free to translate in the direction of X-axis
3: Hinged and free to translate in the direction of Y-axis
NBCI: The number of nodes that have the restraining condition defined by "NBTYPE"
NNB(I): The node number where the preceding restraining condition is applied. These data have to be supplied "NBCI" times following the format specified as 10I5.
Line 11 has to be supplied "NBCSRT" times.

Line 12 (2F10.0)
ERITE: Tolerance for structural iteration. "ERITE" = 10^{-3} generally gives a satisfactory result.
ERFL: Tolerance for element iteration. "ERFL" = 10^{-4} generally gives a satisfactory result.
These data can be left blank in the case of the first-order elastic analysis. The following convergence criterion is used in this program.

- Structural iteration

$$\sum_{k=1}^{ndf} |\Delta d_k^i| \Big/ \sum_{k=1}^{ndf} |d_k^i| \geq ERITE, \ \Delta d_k^i = d_k^i - d_k^{i-1}$$

- Element iteration

$$\sum_{k=1}^{6} |\Delta f_k^i| \Big/ \sum_{k=1}^{6} |f_k^i| \geq ERFL, \ \Delta f_k^i = f_k^i - f_k^{i-1}$$

where d_k^i and f_k^i are the kth component of the displacement vector and the sectional force vector, respectively, obtained at the ith cycle of iteration.

(2) toframe This data set is necessary to define the connection characteristics.

Line 1 (A9, I3)
Input the title "DATA NO." in format A9 and then input the data number in format I3. The data number must start from 1.

Line 2 (A8, I3)
Input the title "MODEL = " in format A8 and then input the connection model number in format I3. The numbers are allocated as follows to the connection models shown in Table 3.1.

1: Modified exponential model
2: Polynomial model
3: Three-parameter model

The input format after Line 2 is different according to the connection models. Notations used herein are exactly the same as those shown in Table 3.1. In the case when the connection data are made by the data-bank program explained in Chapter 3, these data are given in U.S. customary units.

Modified Exponential Model

Line 4 (A8, I3)
This line specifies the number of linear terms adopted in this model. Input title "NLINEAR = " in format A8 and n in format I3 ($n \leq 3$).

Line 5 (6X, A5, E17.8, 2X, A4, E17.8, 2X, A5, I3)
This line specifies the values of constants α, M_0, and m. First, input "AL = " in format A5 and the value of α in format E17.8. Second, input "BMO = " in format A4 and the value of M_0 in format E17.8. Third, input "NEXP = " in format A5 and the value of m in format I3 ($m \leq 6$).

Line 6 (6X, A5, 3E17.8)
This line specifies the values of constants A_i ($1 \leq i \leq 3$). First, input "AI = " in format A5, followed by the values of A_1, A_2, and A_3 in format 3E17.8.

Line 7 (11X, 3E17.8)
This line specifies the values of constants A_i ($4 \leq i \leq 6$). Input the values of A_4, A_5, and A_6 in format 3E17.8.

Line 8 (6X, A5, 3E17.8)
This line specifies the values of T_j ($1 \leq i \leq n$). Input "TJ = " in format A5 and the values of T_1, T_2, and T_3 in format 3E17.8.

Line 9 (6X, A5, 3E17.8)
This line specifies the values of R_j $(1 \le i \le n)$. Input "RJ = " in format A5 and the values of R_1, R_2, and R_3 in format 3E17.8.
Lines 8 and 9 must be omitted when $n = 0$.

Polynomial Model

Line 4 (3(3X, A4, F10.6,) 3X, A4, F10.6)
Input "C1 = " in format A4 and the value of C_1 in format F10.6. Similarly, input "C2 = ", the value of C_2, "C3 = ", and the value of C_3. Then, input "K = " in format A4 and the value of K in format F10.6.

Line 5 (3(3X, A4, I10))
Input "P1 = " in format A4 and the value of p_1 in format I10. Similarly, input "P2 = ", the value of p_2, "P3 = ", and the value of p_3.

Line 6 (3(3X, A4, I10))
Input "N1 = " in format A4 and the value of n_1 in format I10. Similarly, input "N2 = ", the value of n_2, "N3 = ", and the value of n_3.

Three-Parameter Model

Line 4 (7X, A4, F6.3, 2(2X, A5, E17.8))
First, input "RN = " in format A4 and the value of n in format F6.3. Second, input "RKI = " in format A5 and the value of R_{ki} in format E17.8. Third, input "RMU = " in format A5 and the value of M_u in format E17.8.

(3) ***Example*** The two-story–one-bay frame illustrated in Fig. 3.20 is selected as an example to demonstrate how to use this program. This frame is a semi-rigid frame with fixed bases, where all the beams are connected to columns with the same type of semi-rigid connections. For these connections, the modified exponential model is used to represent the moment–rotation relationship shown in Fig. 3.21. The theory used for beam-column elements is the linearized beam-column theory. All the loads are applied at the same time with one load increment. Although one load increment is used in this example, more than one load increments are normally required for the analysis of semi-rigid frames with nonlinear connection characteristics.

Two sets of input data, input.d and toframe for this example, are given in Figs. 3.22 and 3.23, respectively. Output data output.d is given in Fig. 3.24. As computed results, Fig. 3.25 shows the displacement diagram, moment diagram, and shear-force diagram. The aforementioned two sets of input data and the output data are provided in the diskette, named inpex.d, tofrex and outex.d, respectively.

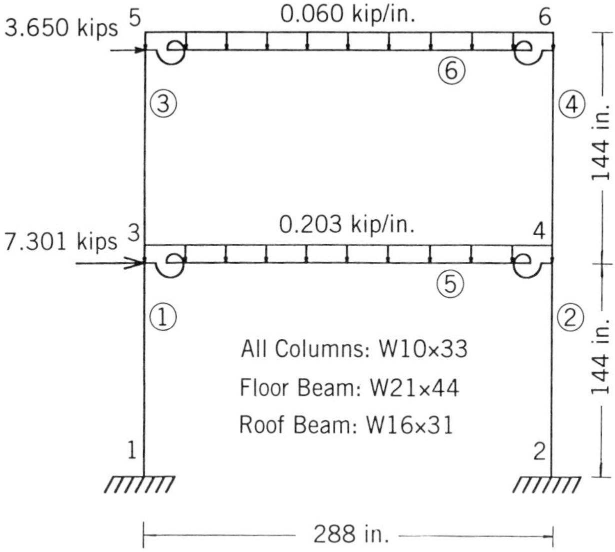

Fig. 3.20 Two-story–one-bay frame.

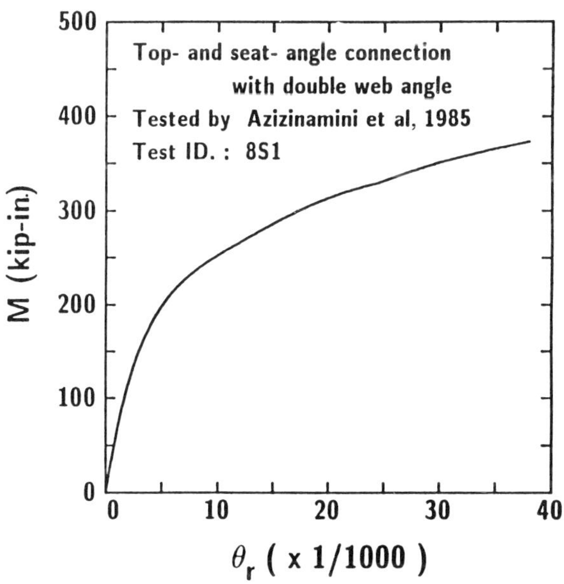

Fig. 3.21 Moment–rotation relationship of connections.

```
 0     3     2     0
 6     6     3     1
 1     2    23          0. 0
 1     1     3     1     1     0     0     0     0
 2     2     4     1     1     0     0     0     0
 3     3     5     1     1     0     0     0     0
 4     4     6     1     1     0     0     0     0
 5     3     4     2     2     1     1     1     1
 6     5     6     3     2     1     1     1     1
 1   0. 0000 0  -1. 4400 2
 2   0. 0000 0  -1. 4400 2
 3   0. 0000 0  -1. 4400 2
 4   0. 0000 0  -1. 4400 2
 5   2. 8800 2   0. 0000 0
 6   2. 8800 2   0. 0000 0
 1     2. 900 4  9. 7100 0   1. 700   2
 2     2. 900 4  1. 3000 1   8. 430   2
 3     2. 900 4  9. 1200 0   3. 750   2
 1     2     2     1
 3  7. 3008   0   . 00000 0   . 00000 0
 5  3. 6504   0   . 00000 0   . 00000 0
0. 2032  0     1
 5
0. 0600  0     1
 6
 1
 0     2
 1     2
  0. 001       0. 0001
```

Fig. 3.22 Input data input.d.

```
DATA NO.    1
MODEL =   1
NLINER=   1
       C =     0. 39499417E+00   BMO=     0. 00000000E+00   NEXP=   6
      AI =    -0. 30968749E+02    0. 62361139E+03   -0. 50361980E+04
               0. 16127627E+05   -0. 20637821E+05    0. 93013973E+04
     RJ0 =            24. 7000
     RKJ =     0. 20517930E+01
```

Fig. 3.23 Input data toframe.

```
************************************************
****                                        ****
****      SECOND ORDER FRAME ANALYSIS       ****
****        WITH/WITHOUT BOWING EFFECT      ****
****                                        ****
************************************************

US UNIT IS USED IN THIS CALCULATION
LENGTH........INCH
WEIGHT........KIPS

FLAG FOR METHOD OF ANALYSIS..........  3
FLAG FOR CONNECTION MODEL............  2
FLAG FOR ITERATIVE PROCEDURE (ITLM)..  0

NO. OF ELEMENT (NE)...........................    6
NO. OF NODE (ND)..............................    6
NO. OF ELEMENT MATERIAL (NMAT)................    3
NO. OF PREPARED CONNEC. DATA (NMATC)..........    1
NO. OF LOAD SEQUENCE (NLSQ)...................    1
NO. OF LOADING POINTS (NL)....................    2


********************CONNECTION PROPERTY********************
DATA NO. 1 FOR NONLINEAR FRAME ANALYSIS
MODIFIED EXPONENTIAL MODEL PARAMETERS
     C =   0. 39499417E+00              BMO=    0. 00000000E+00
  NEXP=  6        NLINER=  1
    AI =  -0. 30968749E+02   0. 62361139E+03  -0. 50361980E+04   0. 16127627E+05  -0. 20637821E+05   0. 93013973E+04
   RJ0 =           24. 7000
   RKJ =   0. 20517930E+01

********************ELEMENT PROPERTY********************

EL. NO.  ( KA, KB)     (      DX ,  DY          L        E              A              BI           EL. TP  CON. TYPE (KA, KB)  INDXLN
   1     (  1,  3)     (     0. 00,  -144. 00)  144. 00  ( 0. 29000E+05, 0. 97100E+01, 0. 17000E+03)    1          (  0,  0)      0  0
   2     (  2,  4)     (     0. 00,  -144. 00)  144. 00  ( 0. 29000E+05, 0. 97100E+01, 0. 17000E+03)    1          (  0,  0)      0  0
   3     (  3,  5)     (     0. 00,  -144. 00)  144. 00  ( 0. 29000E+05, 0. 97100E+01, 0. 17000E+03)    1          (  0,  0)      0  0
   4     (  4,  6)     (     0. 00,  -144. 00)  144. 00  ( 0. 29000E+05, 0. 97100E+01, 0. 17000E+03)    1          (  0,  0)      0  0
   5     (  3,  4)     (   288. 00,     0. 00)  288. 00  ( 0. 29000E+05, 0. 13000E+02, 0. 84300E+03)    2          (  1,  1)      1  1
   6     (  5,  6)     (   288. 00,     0. 00)  288. 00  ( 0. 29000E+05, 0. 91200E+01, 0. 37500E+03)    2          (  1,  1)      1  1


********************LOADING CONDITION********************

***** LOAD SEQUENCE =  1  ,  NO. OF LOAD STEPS =    1  *****
*** CONCENTRATED AT THE END OF LOAD SEQUENCE 1 ***
 NODE NO.         FX          FY          BM
      3         7. 3008     0. 0000     0. 0000
      5         3. 6504     0. 0000     0. 0000
*** DISTRIBUTED LOAD AT THE END OF LOAD SEQUENCE  1 ***
 DP : VERTICAL LOAD DEFINED IN TERMS OF LO. COORD
      DP     NEDI
   0. 20320    1
ELE. NO.
     5
      DP     NEDI
   0. 06000    1
ELE. NO.
     6

********************BOUNDARY CONDITION********************
 NBTY :  NODE NUMBERS
   0 :     1   2
**********************************************************
 ***** ALLOWABLE ERROR IN CONVERGENCE *****
ERR. IN STRUCTURAL ITERATION (ERITE) :    0. 10000E-02
ERR. IN ELEMENT ITERATION (ERFL)     :    0. 10000E-03
*************************************************************
```

Fig. 3.24 Output data output.d.

```
************************************************
NO. OF ITERATION (ITE) ......       8
ERROR IN ITERATION (ERR) ....   0.3674829E-03
             (SMRD) ..........  0.2379053E-02
             (SMD) ...........  0.6473914E+01

LOAD SEQUENCE (NSL)..................    1
PRESENT TOTAL LOAD RATIO (CNINC).....    1.00000
NO. OF STRCTURAL ITERATION (ITE).....    8
NO. OF TOTAL STEPS (NTST)............    1
************************************************


 *** NODAL DISPLACEMENTS ***      (  ( ) : VALUE FOR ONE CYCLE BEFORE COVERGENCE  )
NOD NO              DX                                  DY                                 ROTATION OF CONNECTION
     1   0.000000E+00 (  0.000000E+00)       0.000000E+00 (  0.000000E+00)       0.000000E+00 (  0.000000E+00)
     2   0.000000E+00 (  0.000000E+00)       0.000000E+00 (  0.000000E+00)       0.000000E+00 (  0.000000E+00)
     3   0.944687E+00 (  0.944438E+00)       0.217084E-01 (  0.217028E-01)       0.936764E-02 (  0.936446E-02)
     4   0.938088E+00 (  0.937841E+00)       0.243125E-01 (  0.243071E-01)       0.873637E-02 (  0.873295E-02)
     5   0.222641E+01 (  0.222550E+01)       0.311595E-01 (  0.311376E-01)       0.774969E-02 (  0.774479E-02)
     6   0.222181E+01 (  0.222090E+01)       0.351601E-01 (  0.351387E-01)       0.710918E-02 (  0.710316E-02)

*** MEMBER FORCE ***

ELM NO  KA  KB      AXIALA       SHEARA       MOMENTA         AXIALB       SHEARB      MOMENTB  ITER        ERROR
     1   1   3  0.35257E+02 -0.51091E+01 -0.70125E+03    -0.35257E+02  0.51091E+01 -0.67774E+02    2   0.19880E-05
     2   2   4  0.40537E+02 -0.58420E+01 -0.73447E+03    -0.40537E+02  0.58420E+01 -0.14480E+03    2   0.15542E-05
     3   3   5  0.72930E+01 -0.91080E+00 -0.15000E+02    -0.72930E+01  0.91080E+00 -0.12550E+03    2   0.74112E-05
     4   4   6  0.99642E+01 -0.27392E+01 -0.14811E+03    -0.99642E+01  0.27392E+01 -0.25913E+03    2   0.26450E-05
     5   3   4  0.31022E+01 -0.27957E+02  0.82747E+02    -0.31022E+01 -0.30565E+02  0.29286E+03    3   0.36682E-05
     6   5   6  0.27391E+01 -0.73049E+01  0.12545E+03    -0.27391E+01 -0.99751E+01  0.25904E+03    3   0.23439E-04

*** CONNECTION INFORMATION ***    ( LOC=1 : LOADING,  LOC=0 : UNLOADING PASS )
ELE NO  KA  LOC RELATIVE R.      MOMENT   SECANT STIF     KB  LOC RELATIVE R.       MOMENT   SECANT STIF
     5   3    1  0.13274E-02  0.82747E+02  0.62337E+05     4    1  0.16021E-01  0.29286E+03  0.18280E+05
     6   5    1  0.22701E-02  0.12545E+03  0.55261E+05     6    1  0.10863E-01  0.25904E+03  0.23846E+05

**** UNBALANCED FORCE ****      (  ( ) : PRESENT SURCHARGED NODAL FORCE (GRP)  )
NOD NO              FX                                  FY                                    BM
     1   0.000000E+00 (  0.000000E+00)       0.000000E+00 (  0.000000E+00)       0.000000E+00 (  0.000000E+00)
     2   0.000000E+00 (  0.000000E+00)       0.000000E+00 (  0.000000E+00)       0.000000E+00 (  0.000000E+00)
     3   0.280236E-03 (  0.128362E+02)       0.724769E-02 (  0.262185E+02)       0.274510E-01 (  0.402103E+03)
     4   0.550274E-03 ( -0.553542E+01)       0.776829E-02 (  0.240659E+02)       0.430558E-01 ( -0.127762E+03)
     5   0.553720E-03 (  0.513326E+01)       0.119023E-01 (  0.201664E+02)       0.554760E-01 (  0.189205E+03)
     6   0.188298E-03 ( -0.148286E+01)       0.108670E-01 (  0.195502E+02)       0.911275E-01 ( -0.918895E+02)
*********************************************************************************************
```

Fig. 3.24 (*Continued*)

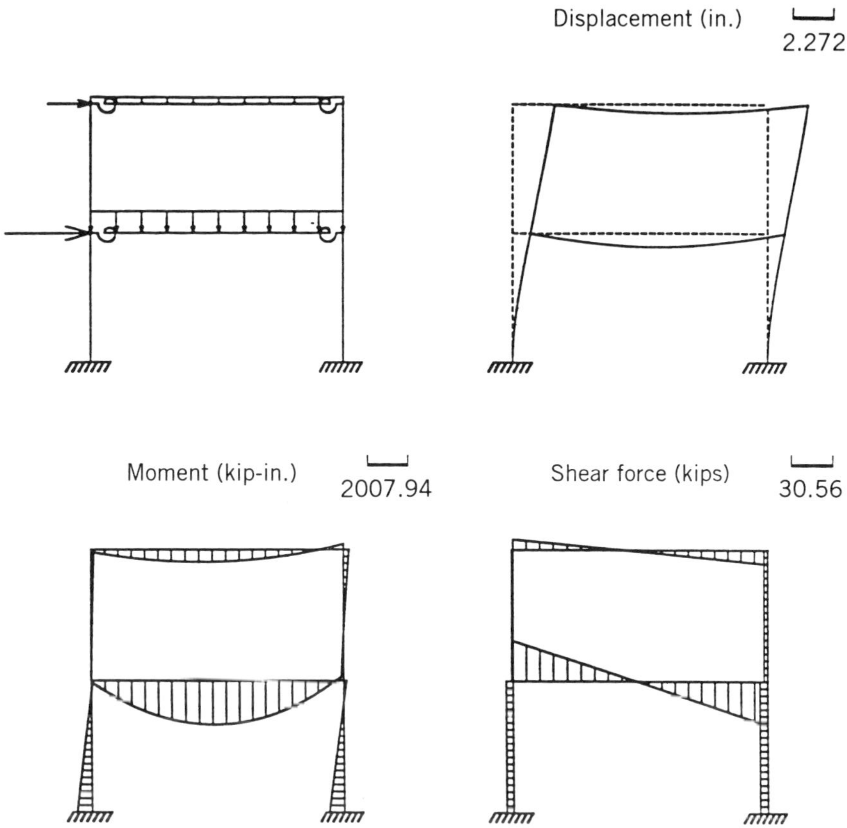

Fig. 3.25 Computed results.

REFERENCES

AISC (1989). *Allowable Stress Design and Plastic Design: Specifications for Structural Steel in Buildings,* 9th Ed., Chicago: American Institute of Steel Construction.

AISC (1993). *Load and Resistance Factor Design Specifications for Structural Steel Buildings,* 2nd Ed., Chicago: American Institute of Steel Construction.

Allen, H. G., and P. S. Bulson (1980). *Background to Buckling,* New York: McGraw-Hill.

Azizinamini, A., J. H. Bradburn, and J. B. Radziminski (1985). *Static and Cyclic Behavior of Semi-Rigid Steel Beam-Column Connections,* Columbia: Department of Civil Engineering, University of South Carolina.

Byskov, E. (1989). "Smooth Post-Buckling Stress by a Modified Finite Element Method," *International Journal for Numerical Methods in Engineering,* Vol. 28, pp. 2877–2888.

Chen, W. F., and A. F. Saleeb (1982). *Uniaxial Behavior and Modeling in Plasticity,* Structural Engineering Report No. CE-STR-82-35, West Lafayette, IN: School of Civil Engineering, Purdue University.

Connor, J. J., R. D. Logcher, and S. C. Chan (1968). "Nonlinear Analysis of Elastic Framed Structures," *Journal of Structural Division,* ASCE, Vol. 94, No. ST6, pp. 1525–1547.

Dafalias, Y. E., and E. P. Popov (1976). "Plastic Internal Variables Formalism of Cyclic Plasticity," *Journal of Applied Mechanics,* Vol. 43, pp. 645–651.

Frye, M. J., and G. A. Morris (1975). "Analysis of Flexibly Connected Steel Frames," *Canadian Journal of Civil Engineering,* Vol. 2, pp. 280–291.

Goto, Y., and W. F. Chen (1986). *Documentation of Program FLFRM for Design Analysis of Flexibly Jointed Frames,* Structural Engineering Report No. CE-STR-86-34, West Lafayette, IN: School of Civil Engineering, Purdue University.

Goto, Y., and W. F. Chen (1987a). "Second-Order Elastic Analysis for Frame Design," *Journal of Structural Engineering,* ASCE, Vol. 113, No. ST7, pp. 1501–1519.

Goto, Y., and W. F. Chen (1987b). "On the Computer-Based Design Analysis for the Flexibly Jointed Frames," *Journal of Constructional Steel Research,* Special Issue on Joint Flexibility in Steel Frames, W. F. Chen, ed., Vol. 8, pp. 203–231.

Goto, Y., S. Matsuura, A. Hasegawa, and F. Nishino (1985). "A New Formulation of Finite Displacement Theory of Curved and Twisted Rods," *Structural Engineering/Earthquake Engineering,* JSCE, Vol. 2, No. 2, pp. 151s–160s.

Goto, Y., and W. F. Chen (1989). Closure to the discussions by Wan T. Tsai and by M. McGuire and R. D. Ziemian, *Journal of Structural Engineering,* ASCE, Vol. 115, No. ST2, pp. 500–506.

Goto, Y., S. Suzuki, and W. F. Chen (1991a). "Analysis of Critical Behavior of Semi-Rigid Frames with or without Load History in Connections," *International Journal of Solids and Structures,* Vol. 27, No. 4, pp. 467–483.

Goto, Y., S. Suzuki, and W. F. Chen. (1991b). "Bowing Effect on Elastic Stability of Frames under Primary Bending Moments," *Journal of Structural Engineering,* ASCE, Vol. 117, No. ST1, pp. 111–127.

Goto, Y., S. Suzuki, and W. F. Chen (1993). "Stability Behavior of Semi-Rigid Sway Frames," *Engineering Structures,* Vol. 15, No. 3, pp. 209–219.

Goto, Y., T. Yamashita, and S. Matsuura (1987). "Elliptic Integral Solutions for Extensional Elastica with Constant Initial Curvature," *Structural Engineering/Earthquake Engineering,* JSCE, Vol. 4, No. 2, pp. 299s–309s.

Goto, Y., T. Yoshimitsu, and M. Obata (1990). Elliptic Integral Solutions of Plane Elastica with Axial and Shear Deformations, *International Journal of Solids and Structures,* Vol. 26, No. 4, pp. 375–390.

Hechtman, R. A., and B. G. Johnston (1947). *Riveted Semi-Rigid Beam-to-Column Building Connections,* Progress Report No. 1, Bethlehem, PA: AISC Research, Lehigh University.

Hill, R. (1958). "A General Theory of Uniqueness and Stability in Elastic-Plastic Solids," *Journal of Mechanics and Physics of Solids,* Vol. 6, pp. 236–249.

Kishi, N., and W. F. Chen (1986). *Data Base of Steel Beam-to-Column Connections,* Structural Engineering Report No. CE-STR-86-26, West Lafayette, IN: School of Civil Engineering, Purdue University.

Kishi, N., W. F. Chen, Y. Goto, and K. Matsuoka (1993). "Analysis Program for the Design of Flexibly Jointed Frames," *Computers and Structures,* Vol. 49, No. 4, pp. 705–713.

Kishi, N., and W. F. Chen (1990). "Moment–Rotation of Semi-Rigid Connections with Angles," *Journal of Structural Engineering,* ASCE, Vol. 116, No. ST7, pp. 1813–1834.

Korn, A. (1981). "Effect of Bowing on Rectangular Plane Frames," *Journal of Structural Division,* ASCE, Vol. 107, No. ST3, pp. 569–574.

Lindsey, S. D., S. A. Ioannides, and A. V. Goverdhan (1985). "The Effect of Connection Flexibility on Steel Members and Frame Stability," in *Connection Flexibility on Steel Members and Frame Stability,* W. F. Chen, ed., New York: ASCE, pp. 6–12.

Mallet, R. H., and P. V. Marcal (1968). "Finite Element Analysis of Nonlinear Structures," *Journal of Structural Division,* ASCE, Vol. 94, No. ST9, pp. 2081–2103.

Moncarz, P. D., and K. H. Gerstle (1981). "Steel Frames with Nonlinear Connections," *Journal of Structural Division,* ASCE, Vol. 107, No. ST8, pp. 1427–1441.

Nishino, F., and A. Hasegawa (1979). "Thin-Walled Beam Theory," *Journal of Faculty of Engineering,* University of Tokyo, Vol. B-35, pp. 109–190.

Nishino, F., C. Kasemset, and S. L. Lee (1973). "Variational Formulation of Stability Problem for Thin-Walled Members," *Ingenieur Archives,* Vol. 43, pp. 58–68.

Nishino, F., Y. Kurakata, and Y. Goto (1975). "Finite Displacement Beam Theory," *Proceedings of JSCE,* No. 237, pp. 11–25.

Oran, C. (1973). "Tangent Stiffness in Plane Frames," *Journal of Structural Division,* ASCE, Vol. 99, No. 6, pp. 973–985.

Rathbun, J. C. (1936). "Elastic Properties of Riveted Connections," *Transactions of the ASCE,* Paper No. 1933, Vol. 101, pp. 524–563.

Reissner, E. (1972). "On One-dimensional Finite-Strain Beam Theory," *Journal of Applied Mathematics and Physics (ZAMP),* Vol. 23, pp. 795–804.

Riks, E. (1979). "An Incremental Approach to the Solutions of Snapping and Buckling Problem," *International Journal of Solids and Structures,* Vol. 15, No. 7, pp. 529–551.

Saafan, S. A. (1963). "Nonlinear Behavior of Structural Plane Frames," *Journal of Structural Division,* ASCE, Vol. 89, No. ST4, pp. 557–579.

SSRC (1988). In chapter 21-. "Finite Element Analysis of Stability Problems," in *Guide to Structural Design Criteria for Metal Structures,* 4th Ed., T. V. Galambos, ed., New York: Wiley, pp. 673–686.

Timoshenko, S. P., and J. M. Gere (1961). *Theory of Elastic Stability,* 2nd Ed., New York: McGraw-Hill.

Thompson, L. E., R. J. McKee, and D. A. Visintainer (1970). *An Investigation of Rotation Characteristics of Web Shear Framed Connections Using A-36 and A-441 Steels,* Rolla: Department of Civil Engineering, University of Missouri.

Washizu, K. (1968). *Variational Method in Elasticity and Plasticity,* Oxford and New York: Pergammon.

White, D. W. (1985). "Material and Geometric Nonlinear Analysis of Local Planar Behavior in Steel Frames Using Iterative Computer Graphics," M.S. thesis, Cornell University, Ithaca, New York.

4

STABILITY DESIGN OF SEMI-RIGID FRAMES WITH LRFD

4.1 INTRODUCTION

In the 1993 AISC LRFD specification (AISC, 1993), design of semi-rigid framing systems is allowed under Type PR construction. Type PR construction includes both Type 2 and Type 3 constructions of the conventional ASD specification (AISC, 1989).

Under Type 2 construction in the ASD specification, the semi-rigid framing system can be designed by using simplifying assumptions that the connections are pinned for the design of beams and that the same connections are rigid for the design of columns. This design approach, however, results in an overdesign for the beams and an underdesign for the columns. Type 3 construction assumes that the connections of the beam and columns possess a predictable end restraint intermediate in degree between that of the fully rigid and ideally pinned conditions. A designer who engages in the design of a semi-rigid frame must have a sufficient knowledge of the behavior of the connections.

In the LRFD specification, a designer can design the semi-rigid frame by using the same simplifying assumptions that were adopted for Type 2 construction. Alternatively, the designer can carry out the design considering the actual behavior of connections, similar to the design based on Type 3 provisions. The latter design procedure is more rigorous and rational because the actual behavior of connections is fully taken into account.

In order to perform the design of semi-rigid frames based on the actual behavior of connections, the designer must have a sufficient knowledge of the moment–rotation characteristics of connections and carry out a second-order elastic frame analysis that also takes into account the connection behavior. In this book, we have already provided tools to evaluate the connection behavior and reflect it in the second-order elastic frame analysis in Chapter 2 and 3.

To design Type PR construction based on the AISC LRFD interaction equations for beam-columns, Eqs. (3.1.1a) and (3.1.1b), the designer further has to evaluate the column effective-length factor K by taking into account the behavior of semi-rigid connections. In this chapter, we first present the method to evaluate the column effective-length factor K of semi-rigid frames. Then, we proceed to show the design procedures and examples, by considering the actual behavior of connections.

4.2 DETERMINATION OF *K*-FACTOR FOR COLUMNS IN SEMI-RIGID FRAMES

4.2.1 Simplified Method Using Alignment Chart

The K-factor of the columns in semi-rigid frames has to be evaluated by considering the connection characteristics being different from the rigid frames. An analytically exact method is to carry out a buckling analysis of semi-rigid frames (Goto, Suzuki, and Chen, 1991, 1993). However, this method requires nonlinear structural analysis that takes into account the nonlinearity of connections as well as the geometrical nonlinearity of frames. Analysis of this kind is complicated and is not suitable for design use. Thus, we show an approximate method (Lui, 1985; Barakat, 1988; Barakat and Chen, 1990) based on the philosophy similar to that using the alignment chart (Galambos, 1988). This approximate method adopts the following assumptions.

1. All beams and columns are purely elastic.
2. All members have constant cross-section.
3. For braced frames, rotations at opposite ends of the restraining beams are equal in magnitude, producing reverse-curvature bending. This deformation pattern become possible only for infinite periodic rectangular frames with the same columns, beams, and connections.
4. For unbraced frames, rotations at opposite ends of the restraining beams are equal in magnitude, producing reverse-curvature bending.
5. The stiffness parameter $L\sqrt{P/EI}$ of all columns are equal.
6. All beam-to-column connections have the same moment–rotation characteristics.
7. All columns buckle simultaneously.
8. No significant axial force exists in the beams

Where the actual conditions differ considerably from these assumptions, the K-factor obtained from the alignment chart may be inaccurate.

The portion of the semi-rigid frames are illustrated in Fig. 4.1. Column *AB* is our concern. By following the same procedure as that for rigid frames, the stability equations for column *AB* can be obtained as follows according to whether the frame is sway-permitted or sway-inhibited (Lui, 1985; Chen and Lui, 1991).

Unbraced:

$$\frac{G_A G_B(\pi/K)^2 - 36}{6(G_A + G_B)} = \frac{\pi}{K}\cot\left(\frac{\pi}{K}\right) \tag{4.2.1}$$

Braced:

$$\frac{G_A G_B}{4}\left(\frac{\pi}{K}\right)^2 + \left(\frac{G_A + G_B}{2}\right)\left[1 - \frac{\pi}{K}\cot\left(\frac{\pi}{K}\right)\right] + 2\frac{\tan(\pi/2K)}{\pi/K} - 1 = 0 \tag{4.2.2}$$

where K is the effective length factor of column *AB*; G_A and G_B are the relative stiffnesses defined at ends A and B, respectively. In case of semi-rigid frames, these values must be evaluated as follows by taking into account the connection stiffness.

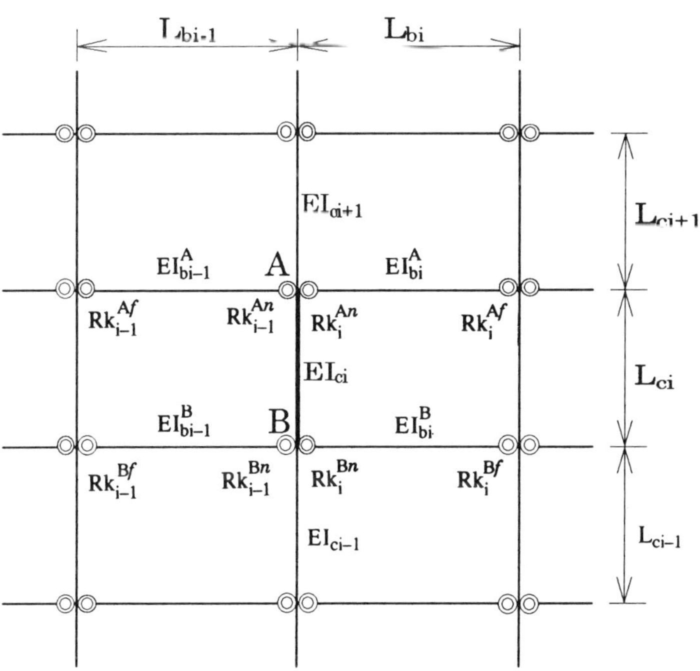

Fig. 4.1 Subassemblage model for frames.

$$G_A = \frac{EI_{ci+1}/L_{ci+1} + EI_{ci}/L_{ci}}{\eta_{i-1}^A EI_{bi-1}^A/L_{bi-1} + \eta_i^A EI_{bi}^A/L_{bi}} \tag{4.2.3}$$

$$G_B = \frac{EI_{ci}/L_{ci} + EI_{ci-1}/L_{ci-1}}{\eta_{i-1}^B EI_{bi-1}^B/L_{bi-1} + \eta_i^B EI_{bi}^B/L_{bi}} \tag{4.2.4}$$

where η_M^α ($\alpha = A, B$; $M = i - 1, i$) is a modification factor to consider the effect of connections. This factor is given by the following:

Unbraced:

$$\eta_M^\alpha = \frac{(2 + \overline{R}_{kM}^{\alpha f})\overline{R}_{kM}^{\alpha n}}{12 + 4(\overline{R}_{kM}^{\alpha n} + \overline{R}_{kM}^{\alpha f}) + \overline{R}_{kM}^{\alpha n}\overline{R}_{kM}^{\alpha f}} \tag{4.2.5}$$

Braced:

$$\eta_M^\alpha = \frac{(6 + \overline{R}_{kM}^{\alpha f})\overline{R}_{kM}^{\alpha n}}{12 + 4(\overline{R}_{kM}^{\alpha n} + \overline{R}_{kM}^{\alpha f}) + \overline{R}_{kM}^{\alpha n}\overline{R}_{kM}^{\alpha f}} \tag{4.2.6}$$

where

$$\overline{R}_{kM}^{\alpha f} = R_{kM}^{\alpha f}\frac{L_{bM}}{EI_{bM}^\alpha} \tag{4.2.7}$$

$$\overline{R}_{kM}^{\alpha n} = R_{kM}^{\alpha n}\frac{L_{bM}}{EI_{bM}^\alpha} \tag{4.2.8}$$

R_{kM} is the rotational stiffness of connections at node M. Once the relative stiffnesses G_A and G_B are calculated by Eqs. (4.2.3) and (4.2.4), the K-factor can be found from the alignment charts shown in Fig. 4.2. If the connection behavior is linearly elastic, the rotational stiffness R_k can be easily evaluated. However, the moment–rotation relations of semi-rigid connections usually exhibit nonlinear characteristics. Thus, a problem arises as to how to determine the value of R_k. Theoretically, it is exact to adopt the tangent stiffness of connections R_{kt} (Fig. 4.5) when buckling occurs (Goto et al., 1991, 1993). In case there is no prebuckling deformations in connections, the initial connection stiffness R_{ki} can be used for R_k. However, the use of the initial stiffness may underestimate the K-factor, because the load drop after buckling is significant for this case, as demonstrated by Goto et al. (1993). Some reduction in initial connection stiffness will be necessary. When prebuckling deformations exist in connections, a nonlinear structural analysis is required to find their tangent stiffness.

Herein, we show two methods to evaluate the rotational stiffness of connections to be used for the calculation of the K-factor. One is a method that

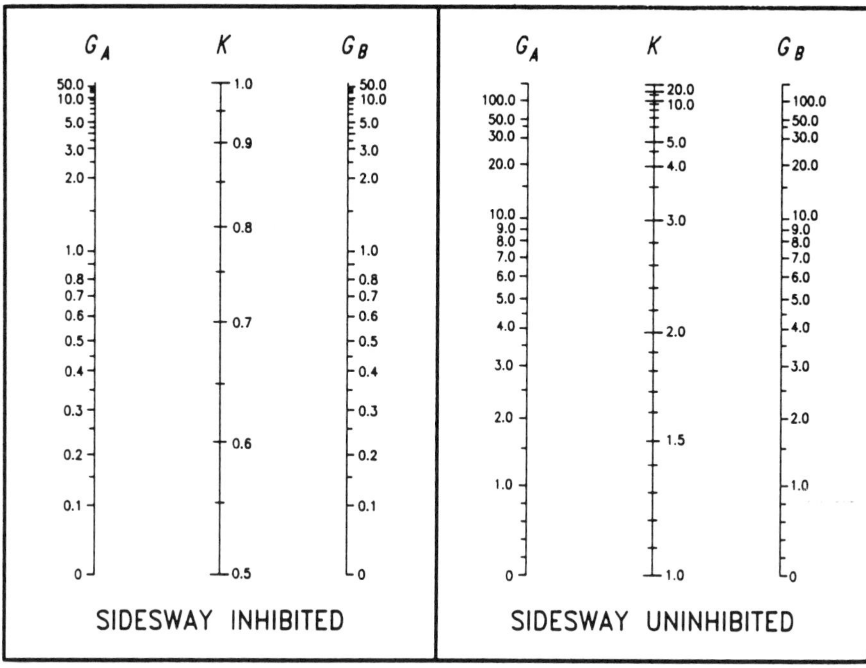

Fig. 4.2 Alignment charts.

utilizes the tangent stiffness of connections under a factored load. The other is a simplified method proposed by Barakat and Chen (1990).

4.2.2 Evaluation of Rotational Stiffness of Connections

4.2.2.1 Use of Tangent Stiffness of Connections Under Factored Load (Method 1) This method adopts the tangent stiffness of connections under the factored load. In order to evaluate the tangent stiffness of connections, the corresponding connection moment has to be calculated first. For this calculation, the second-order elastic analysis described in Chapter 3 can be used.

As an alternative simplified method, the connection moment may be computed by a model shown in Fig. 4.3, where columns are assumed to be rigid. This simplified method is referred to as the beam-line method. In what follows, we explain some details of the beam-line method.

By assuming elastic behavior and symmetric loading conditions, the relationship between the beam end moment M_E and end rotation ϕ_b can be determined as

$$M_E = M_F - \frac{2EI}{L}\phi_b \qquad (4.2.9)$$

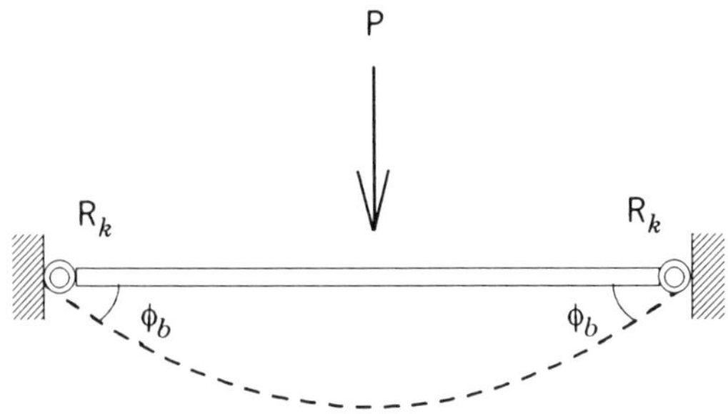

Fig. 4.3 Simplified beam model with semi-rigid connections.

where M_F is the fixed-end moment. For a uniformly loaded beam, Eq. (4.2.9) can be written as

$$M_E = \frac{wL^2}{12}\left(1 - \frac{\phi_b}{\phi_{b0}}\right) \tag{4.2.10}$$

where $\phi_{b0} = wL^3/24EI$ is the end rotation of the beam when its ends are pinned. For a given value of w, Eq. (4.2.10) represents a straight line (beam line) in an end moment vs. rotation plot. Compatibility of rotational deformation at a joint will be satisfied at the intersection of the beam line (curve 2) and the connection curve 1 (point A in Fig. 4.4). Connection moment M_E is obtained as a moment at the intersection point A.

The tangent stiffness of connections R_{kt} are then determined based on the moment–rotation curves, as illustrated in Fig. 4.5. To calculate the K-factor, R_{kt} is substituted into R_k in Eqs. (4.2.5)–(4.2.8).

The connection moment–rotation curves can be known either by experimental results in the data bank or by the prediction equations presented in Chapter 2.

4.2.2.2 Use of Barakat and Chen's Approach (1990) (Method 2) For braced frames, stiffness of connections are determined by the two connection parameters. They are the initial connection stiffness R_{ki} and the ultimate moment capacity M_u of the connection. In addition to the previous connection parameters, a connection M–θ_r curve is also required for determining the connection stiffness. Barakat and Chen adopted the three-parameter power model of Section 2.4.2.3. With R_{ki}, M_u, and a connection model, R_{k0} is determined graphically, as depicted in Fig. 4.6. Note that R_{k0} is the secant stiffness corresponding to a rotation of ϕ_0. ϕ_0 is obtained as the intersection of initial stiffness R_{ki} and the ultimate connection moment M_u. R_{k0} is substituted into R_k in Eqs. (4.2.7) and (4.2.8) to calculate the K-factor.

For unbraced frames, the connection is presumed to undergo noticeable deformation when the effect of lateral loads is added to that of gravity loads.

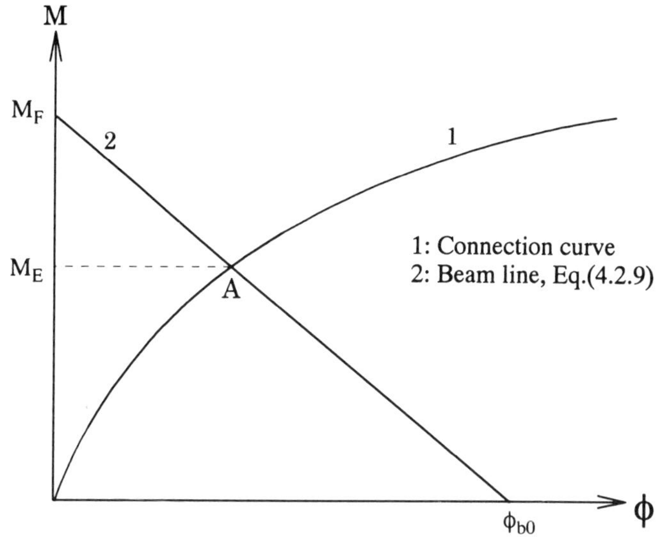

Fig. 4.4 Determination of beam end moment M_E.

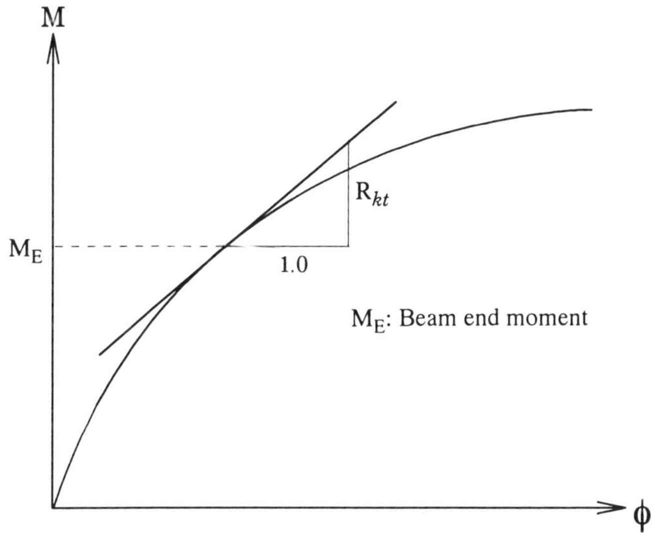

Fig. 4.5 Determination of R_{kt}.

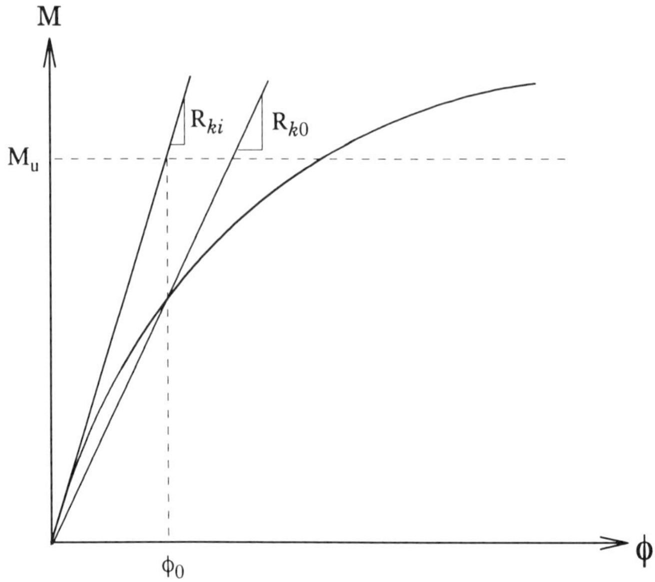

Fig. 4.6 Determination of R_{k0}.

Therefore, Barakat and Chen considered that the connection stiffness in this phase should be less than R_{k0} used for determining the K-factor of braced frames. Then, they proposed a stiffness value of R_{kb} to be used specifically for unbraced frames. The model to evaluate the connection stiffness is similar to the one illustrated in Fig. 4.3. The determination of R_{kb} is schematically shown in Fig. 4.7. Curve 1 represents the moment–rotation curve of connection. Curve 2 is the so-called beam line, which is given by Eq. (4.2.9). Compatibility of rotational deformation at a joint will be satisfied at the intersection of the beam line (curve 2) and curve 1 (point A in Fig. 4.7). R_{kb} is obtained as the secant stiffness corresponding to a rotation defined by the intersection point A. The value of R_{kb} so obtained is larger than that of the tangent stiffness R_{kt}, which is recommended for use in method 1. R_{kb} is substituted into Eqs. (4.2.7) and (4.2.8) to evaluate the K-factor.

4.3 DESIGN PROCEDURES

The stability design procedure with LRFD is as follows:

1. Select sections for columns, beams, and beam-to-column connections. In a preliminary design, some simplified design method such as the one based on conventional Type 2 construction provisions (AISC, 1989) may be used.

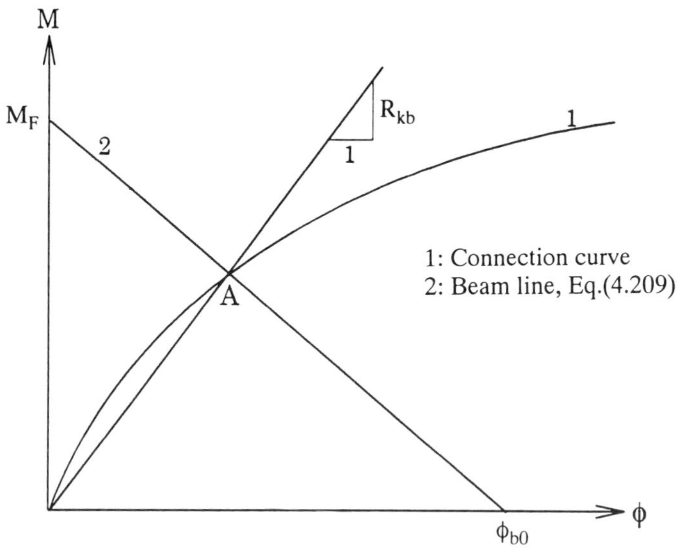

Fig. 4.7 Determination of beam end moment R_{kb}.

2. Predict moment–rotation relations for respective connections. For this purpose, the prediction equations (Section 2.4) or the connection curves stored in the data bank (Section 2.5) may be used.
3. Carry out the second-order elastic analysis (Chapter 3) under factored load, considering the connection behavior.
4. Calculate the effective-length factor for columns (Section 4.2) by taking into account the connection characteristics.
5. Check the stability of frames by the AISC LRFD interaction equations, Eqs. (3.1.1a) and (3.1.1b).
6. Check the strength and deformation capacity of connections.
7. Return to step 1 until the results become satisfactory.

4.4 DESIGN EXAMPLES

4.4.1 Nonsway Semi-Rigid Frame

The nonsway frame considered herein is shown in Fig. 4.8 along with the loading condition.

1. A top- and seat-angle connection without a double web-angle (see Section 2.3.4) is adopted for the semi-rigid beam-to-column connections. Select W12 × 45 for the beam and W8 × 40 for the columns. These sizes of the member sections are chosen such that the width of the beam flange is less

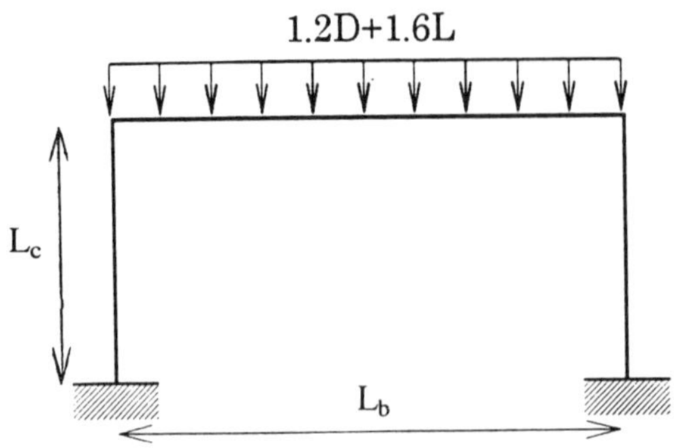

Fig. 4.8 Nonsway frame.

than that of the column flange. This is due to the details of the beam-to-column connection adopted in the present frame. Both top and seat angles of the connection have the same section of the L6-4-3/4 (6 in. width, 2 in. edge distance). A490 7/8-in. diameter bolts are used for this connection. The details of the connection are shown in Fig. 4.9. The specified minimum yield stress F_y of the steel is 36.0 ksi. All the member sections selected here are classified as compact.

2. Predict the moment–rotation relation of the top- and seat-angle connection without a double web angle, based on the three-parameter power model (see Section 2.4.2.3). The initial stiffness R_{ki} and the ultimate moment capacity M_u are respectively calculated by Eqs. (2.4.12) and (2.4.15). Shape

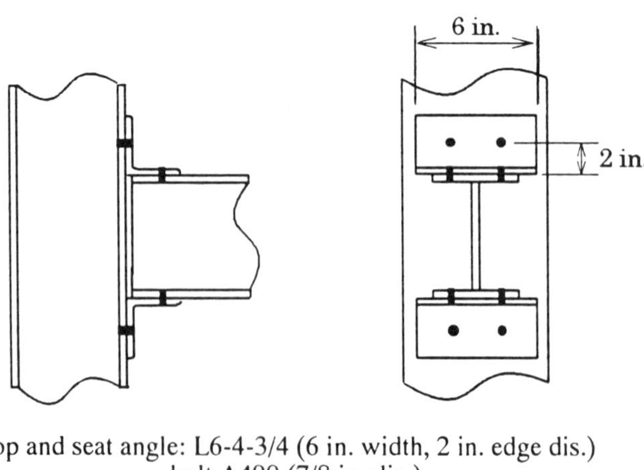

Top and seat angle: L6-4-3/4 (6 in. width, 2 in. edge dis.)
bolt A490 (7/8 in. dia.)

Fig. 4.9 Top- and seat-angle connection without double web angle (nonsway frame).

parameter n is obtained from Table 2.2. The values of the three parameters so obtained are as follows:

$$R_{ki} = 2.25 \times 10^6 \text{ (kips-in.)}, \qquad M_u = 1.22 \times 10^3 \text{ (kips-in.)}, \qquad n = 0.302$$

3. Carry out the second-order elastic analysis using the program FLFRM (Chapter 3) by considering the connection characteristics predicted in step 2. The maximum bending moments and axial forces computed for the respective members are as follows.

Column:

$$N_u = 36.1 \text{ kips}, \qquad M_{uy} = 393 \text{ kips-in.}$$

Beam:

$$N_u = 4.88 \text{ kips}, \qquad M_{uy} = 1897 \text{ kips-in.}$$

4. Calculate the effective-length factor of columns by taking into account the rotational stiffness of beam-to-column connections. Here, we adopt the simplified method using the alignment charts shown in Fig. 4.2. To evaluate the connection stiffness, two methods are proposed in Section 4.2.2. Method 1 adopts the tangent stiffness of connections under a factored load. Method 2 utilizes Barakat and Chen's approach. Both methods may be applicable to the present design. For comparison, we show the values of the effective-length factor calculated by the respective methods.

Method 1 First, we have to calculate the connection moment under the factored load in order to evaluate the tangent stiffness of connections. The connection moment can be directly obtained from the results of the second-order elastic analysis carried out in step 3. The connection moment M_E so obtained is 394 kips-in. Tangent stiffness R_{kt} is calculated by Eq. (2.4.6) as 1.09×10^4 kips-in. The modification factor η_M^A is used to consider the effect of the connection stiffness on the evaluation of the relative stiffness G_A at the top of the column. The values of η_M^A and G_A are calculated respectively by Eqs. (4.2.6) and (4.2.4) as $\eta_M^A = 0.254$ and $G_A = 3.285$. The column base is fixed, so the relative stiffness G_B at the column base is zero. With the values of the relative stiffnesses G_A and G_B calculated before, the effective-length factor K of the column is obtained from the alignment chart (Fig. 4.2) as $K = 0.682$.

A simplified method to calculate the connection moment is described in Section 4.2.2.1. This method adopts the so-called beam-line method. The calculation of connection moment based on this method is illustrated in Fig. 4.10. The connection moment is given as $M_E = 413$ kips-in. Thus, the tangent stiffness of the connection is calculated as $R_{kt} = 1.08 \times 10^4$ kips-in. By

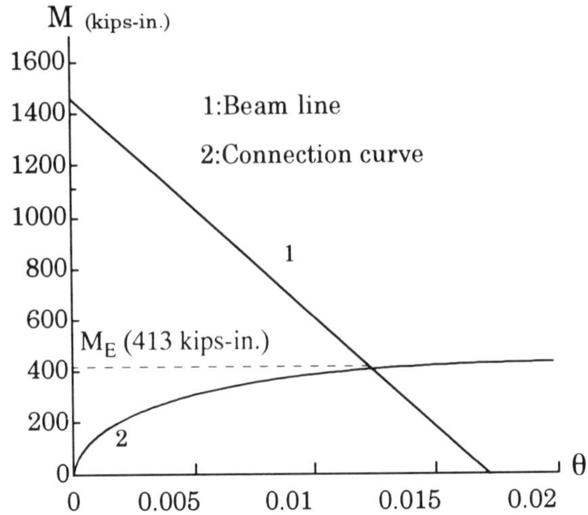

Fig. 4.10 Determination of M_E.

following the same procedure to calculate the modification factor and the relative stiffness as explained previously, the K-factor corresponding to this connection moment is obtained as $K = 0.683$. This value is not so much different from that obtained by the second-order elastic analysis.

Method 2 The procedure to calculate R_{k0} is illustrated in Fig. 4.11. R_{k0} so obtained is 2.28×10^5 kips-in. η_M^A is calculated from Eq. (4.2.6) as 0.729,

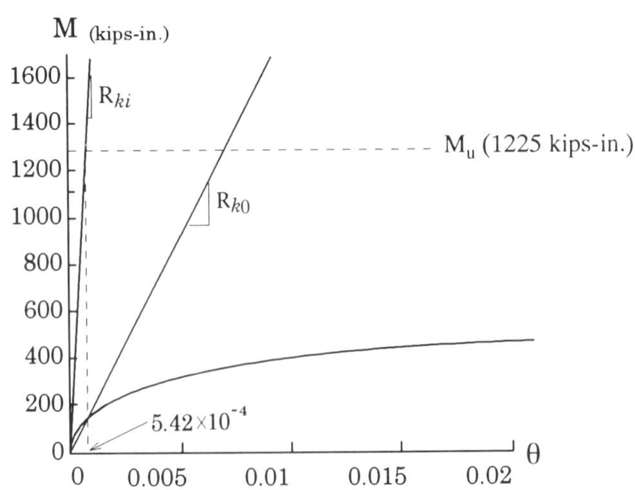

Fig. 4.11 Determination of R_{k0}.

by substituting R_{k0} into R_{kM}. Thus, the value of G_A becomes 1.14. Considering $G_B = 0$, the K-factor is obtained as 0.589. This value is smaller than that obtained by Method 1 because Method 2 does not explicitly consider the reduction of the connection stiffness caused by the vertical load applied on the beam.

5. The stability of the frame is checked by the AISC LRFD interaction equation of Eq. (3.1.1). The compression flange of the column is assumed to be braced against the lateral displacement such that only the limit state of yielding is applicable to the evaluation of M_{ny}. This implies that M_{ny} can be evaluated by the equation of $F_y Z$, where Z is the plastic section modulus. The values of the left side of Eq. (3.1.1) are summarized in Table 4.1, classified according to the methods used to calculate the K-factor of the column in step 4.

The axial force acting on the beam is small. Therefore, the beam may be checked by the following equation:

$$\frac{M_{uy}}{\phi_b M_{ny}} \leq 1.0 \tag{4.4.1}$$

where $\phi_b = 0.90$ and M_{ny} is the moment resisting capacity of the beam. In the calculation of M_{ny}, the compression flange of the beam is assumed to be braced against the lateral displacement such that only the limit state of yielding is applicable. The left side of Eq. (4.4.1) for the present case is calculated as 0.905.

As can be seen from Table 4.1, there is a large margin in the strength of the column compared to that of the beam. However, the cross-sectional size of the column cannot be reduced because of the connection details that were explained in step 1.

TABLE 4.1 Stability Check of Column Based on LRFD (Nonsway Frames)

Evaluation of K-Factor	Method 1 a	Method 1 b	Method 2
K	0.682	0.683	0.589
P_n (kips)	409.	409.	412.
M_{ny} (kips-in.)	1432.	1432.	1432.
P_u (kips)	36.1	36.1	36.1
M_{uy} (kips-in.)	393.	393.	393.
Value of the left side of Eq. (3.1.1)	0.356	0.356	0.356

a: connection moment is calculated by the second-order elastic analysis. b: connection moment is calculated by the beam-line method.

6. The nominal flexural strength of the connection M_{En} is assumed to coincide with the ultimate moment capacity M_u calculated in step 2. Thus, connection moment M_E shall be limited by the following equation:

$$\frac{M_E}{\phi M_{En}} \leq 1.0 \tag{4.4.2}$$

where ϕ is assumed to take the same value of 0.75 that is used for fasteners. For the present case, the left side of Eq. (4.4.2) is calculated as 0.428.

Because there is no information regarding the deformation capacity of the connection, the limiting value for rotation is assumed here to be calculated by the three-parameter model. That is, the limiting rotational displacement of the connection is given by substituting ϕM_{En} into M in Eq. (2.4.7). This implies that the deformation of the connection is always within the limiting value when Eq. (4.4.2) is satisfied.

4.4.2 Sway Semi-Rigid Frame

The sway frame with a fixed base to be designed here is illustrated in Fig. 4.12. As with the nonsway frame, the compression flanges of beams and columns are assumed to be braced against lateral displacements such that only the limit state of yielding is applicable to the evaluation of their moment resisting capacity.

1. A top- and seat-angle connection without a double web angle is used for the beam-to-column connection. Select W12 × 22 for the beam and W6 × 20 for the columns. The width of the beam flange is chosen to be less than that of the column flange. Both top and seat angles of the connection have the same section of the L4-4-3/4 (5 in. width, 2.5 in. edge distance). A325 7/8-in.-diameter bolts are used for this connection. The details of the connection are shown in Fig. 4.13. The specified minimum yield stress F_y of the steel is 36.0 ksi. All the member sections are compact.

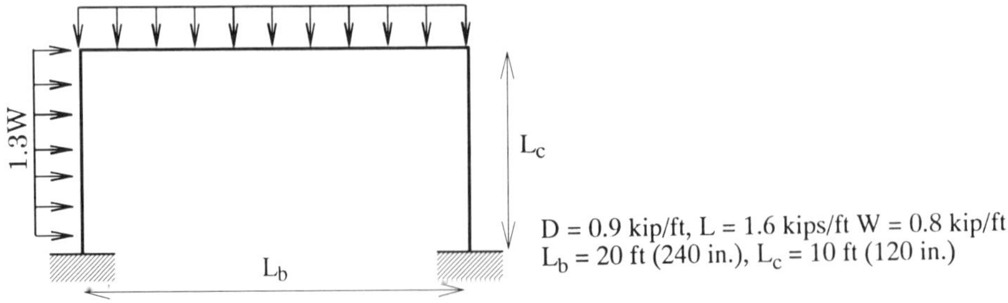

Fig. 4.12 Sway frame.

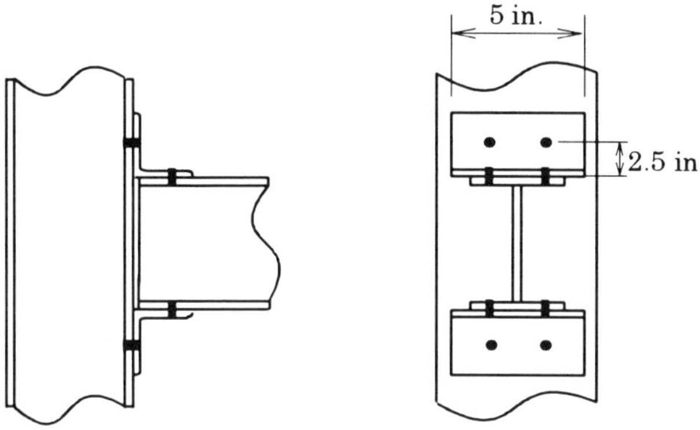

Top and seat angle: L4-4-3/4 (5 in. width, 2.5 in. edge dis.)
bolt A490 (7/8 in. dia.)

Fig. 4.13 Top- and seat-angle connection without double web angle (sway frame).

2. The moment–rotation curve of the connection is predicted by the three-parameter power model (See section 2.4.2.3). The initial stiffness R_{ki}, the ultimate moment capacity M_u, and the shape parameter n are calculated as follows:

$$R_{ki} = 6.84 \times 10^5 \text{ (kips-in.)}, \qquad M_u = 8.54 \times 10^2 \text{ (kips-in)}, \qquad n = 0.302$$

3. Carry out the second-order elastic analysis of the frame. The connection curve predicted in step 2 is considered in this analysis. The maximum bending moments and axial forces for the respective members are calculated as follows.

Right column:

$$N_u = 19.3 \text{ kips}, \qquad M_{uy} = 306 \text{ kips-in}$$

Left column:

$$N_u = 18.2 \text{ kips}, \qquad M_{uy} = 281 \text{ kips-in.}$$

Beam:

$$N_u = 4.47 \text{ kips}, \qquad M_{uy} = 896 \text{ kips-in.}$$

4. Calculate the K-factor of the columns by the simplified method using alignment charts (Fig. 4.2). The two methods, that is, Method 1 (Section 4.2.2.1) and Method 2 (Section 4.2.2.2), are used to evaluate the connection stiffness.

Method 1 The beam-to-column connection moments calculated from the second-order elastic analysis in step 3 are given by the following:

Left connection: $M_E = 335$ kips-in.
Right connection: $M_E = 541$ kips-in.

The tangent stiffness R_{kt} is computed for respective connections from Eq. (2.4.6) as follows:

Left connection: $R_{kt} = 2.09 \times 10^4$ kips-in.
Right connection: $R_{kt} = 4.30 \times 10^3$ kips-in.

The modification factors to evaluate the relative stiffness are obtained from Eq. (4.2.5) as follows:

Left column: $\eta_M^A = 0.141$
Right column: $\eta_M^A = 0.040$

The relative stiffness is calculated from Eq. (4.2.3) for each column as follows:

Left column: $G_A = 3.77$
Right column: $G_A = 13.1$

Considering that the relative stiffness G_B at the fixed column base is zero, the effective-length factors are finally obtained from the alignment charts as follows:

Left column: $K = 1.43$
Right column: $K = 1.79$

If a simplified method based on the beam-line method (Figs. 4.3 and 4.4) is used to calculate the connection moment, this value becomes the same for both connections. That is, $M_E = 223$ kips-in. Following the similar procedure explained previously, the K-factor is calculated as 1.71 for both columns.

Method 2 The connection stiffness R_{kb} is determined as 1.49×10^4 kips-in. by the procedure illustrated in Fig. 4.14. Because the beam-line method shown in Figs. 4.3 and 4.4 is used for the evaluation of R_{kb}, the values of R_{kb} for both connections are the same. The modification factor η_M^A is calculated as 0.116 by substituting the value of R_{kb} into R_k in Eq. (4.2.6). Thus, the relative stiffness G_A is evaluated from Eq. (4.2.3) as 4.55. Because $G_B = 0$, the K-factor is finally obtained as $K = 1.47$ for both columns.

5. The stability of the frame is checked by the interaction equation, Eq. (3.1.1). The values of the left side of Eq. (3.1.1) are summarized in Table 4.2, classified according to the methods adopted to calculate the K-factor.

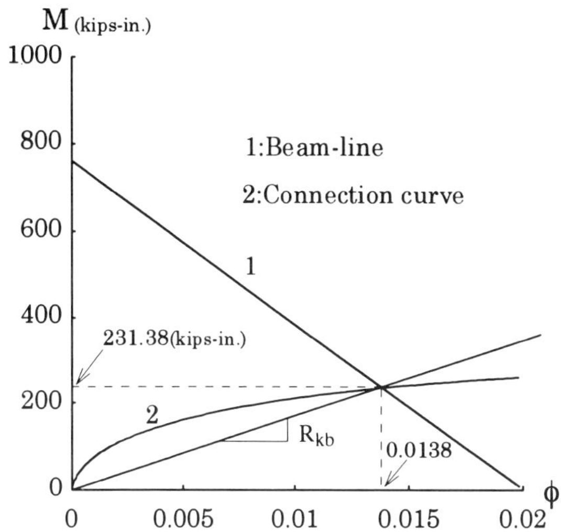

Fig. 4.14 Determination of R_{kb}.

The beam is checked here by Eq. (4.4.1) in which the axial force is ignored due to the same reasoning as was explained for the nonsway frame. The left side of Eq. (4.4.1) is calculated as 0.944.

Table 4.2 shows that there is some margin in the strength of the columns compared to that of the beam. However, we cannot use a smaller section size for columns due to the connection details.

6. The nominal flexural strength of the connection M_{En} in Eq. (4.4.2) is assumed to coincide with the ultimate moment capacity M_u obtained in step 2. The left side of Eq. (4.4.2) is calculated for the right and left connections

TABLE 4.2 Stability Check of Column Based on LRFD (Sway Frames)

Evaluation of K-Factor	Method 1			
	a			
	Left	Right	b	Method 2
K	1.43	1.72	1.71	1.47
P_n (kips)	169.	153.	153.	167.
M_{ny} (kips-in.)	536.	536.	536.	536.
P_u (kips)	18.2	19.3	19.3	19.3
M_{uy} (kips-in.)	281.	306.	306.	306.
Value of the left side of Eq. (3.1.1)	0.645	0.708	0.708	0.702

a: connection moment is calculated by the second-order elastic analysis. b: connection moment is calculated by the beam-line method.

respectively as 0.522 and 0.844. The limiting rotational capacity is assumed to correspond to the rotation when the moment of ϕM_{En} is applied to the connection. Thus, the deformation of the connection under a factored load is less than its limiting value.

REFERENCES

AISC (1989). *Specification for Structural Steel Buildings—Allowable Stress Design and Plastic Design,* 9th Ed., Chicago: American Institute of Steel Construction.

AISC (1993). *Load and Resistance Factor Design Specifications for Structural Steel Buildings,* 2nd Ed., Chicago: American Institute of Steel Construction.

Barakat, M. A. (1988). "Simplified Design Analysis of Frames with Semi-Rigid Connections," Ph.D. dissertation, School of Civil Engineering, Purdue University, West Lafayette, IN.

Barakat, M. A., and W. F. Chen (1990). "Practical Analysis of Semi-Rigid Frames," *Engineering Journal,* AISC, Vol. 27, No. 2, pp. 54–68.

Chen, W. F., and E. M. Lui (1991). *Stability Design of Steel Frames,* Boca Raton, FL: CRC Press.

Galambos, T. V., ed. (1988). *Guide to Stability Design Criteria for Metal Structures,* 4th Ed., New York: Wiley.

Goto, Y., S. Suzuki, and W. F. Chen (1991). "Analysis of Critical Behavior of Semi-Rigid Frames with or without Load History in Connections," *International Journal of Solids and Structures,* Vol. 27, No. 4, pp. 467–483.

Goto, Y., S. Suzuki, and W. F. Chen (1993). "Stability Behavior of Semi-Rigid Sway Frames," *Engineering Structures,* Vol. 15, No. 3, pp. 209–219.

Lui, E. M. (1985). "Effects of Connection Flexibility and Panel Zone Deformation on the Behavior of Plane Steel Frames," Ph.D. dissertation, School of Civil Engineering, Purdue University, West Lafayette, IN.

5

AN OVERVIEW OF ADVANCED ANALYSIS FOR FRAME DESIGN

5.1 INTRODUCTION

This chapter provides an overall view of the current impetus behind the research in inelastic advanced analysis and design. The current state of the art with respect to the methods of analysis of the ultimate strength performance of steel frames is discussed. Advanced analysis, by definition, represents the methods that capture the behavior associated with beam, column, and beam-column strength limit states with sufficient accuracy such that the method alone suffices as a verification of the adequacy with respect to these limit states. By the nature of this definition, advanced analysis must account for the interrelated effects of material inelasticity and geometric nonlinearity on the response of structural systems and their component elements.

Modern steel design codes around the world are mostly based on limit states design philosophy. The second edition of the AISC LRFD specification (AISC, 1993) gained information in implementing limit states design since 1986; thus, it represents a more rational approach to the design of steel structures. Limit states specifications such as that of the AISC LRFD are based directly on factored loads and limits of resistance; it is expected that structural systems and their members will behave nonlinearly before their respective limit states of resistance are reached. There is an increasing awareness of using second-order analysis for the consideration of strength and stability effects in the design of the overall structural system. The most direct approach for structural design is to model all the significant nonlinear effects in the analysis. However, not until recently, rigorous consideration of system as well as member stability and inelastic strength in the analysis of a large-scale structural system was not feasible and practical. As a result, contemporary specification provisions have been based primarily on simpler methods of analysis and member interaction equations that account approximately for the interaction of strength and stability between the member and structural system.

One major draw back of elastic analysis/design is that overall structural strength and stability are treated only in an indirect manner. SSRC Technical Memorandum No. 5 reflects this problem by stating (Galambos, 1988, Appendix B.5):

> Although the maximum strength of frames and maximum strength of the component members are interdependent (but not necessarily coexistent), it is recognized that in many structures it is not practical to take this interdependence into account rigorously. At the same time, it is known that difficulties are encountered in complex frameworks when attempting to compensate automatically in column design for the instability of the entire frames (for example by adjustment of column effective length). Therefore it is recommended that, in design practice, the two aspects, stability of individual members and elements of the structure and stability of the structure as a whole, be considered independently.

The conventional design procedure based on elastic analysis and member-capacity checks have been shown to produce safe and serviceable design (Liew, White, and Chen, 1991). However, recent advancements in computer hardware and software, particularly in computing and graphics performance of personal computers and their software, are making advanced methods of analysis more and more feasible for design use. It is now possible for the engineer to adopt the limit states design philosophy in a wider perspective. Advanced inelastic analyses are now permitted, with certain restrictions on application, by several steel building codes around the world (Eurocode 3, 1992; Standards Australia, 1990). Two committees within the structural design profession of the United States are now working toward the implementation of advanced analysis for practical design. These committee task groups are (1) the AISC Technical Committee 117 on Inelastic Analysis and Design and (2) the SSRC Task Group 29 on Second-Order Inelastic Analysis for Frame Design. The efforts from these two task groups have led to a workshop being organized to discuss the prospect of developing computer-based advanced analysis methods that are suitable for design implementation. A state-of-the-art report (White, Liew, and Chen, 1991) and book (White and Chen, 1993) have been published. Educational programs for advanced analyses are also available (Chen and Toma, 1994; Chen and Sohal, 1995).

The main goal of this chapter is to provide an overall view of current and ongoing efforts directed toward providing computer-based analysis methodologies for inelastic analysis and design of steel frame structures. An encapsulation of various analysis approaches for frame design ranging from elastic analysis and plastic analysis to the more sophisticated computer-based advanced analysis approach is presented. Methods that are essential for representing initial imperfections and distributed plasticity effects on the side-sway instability of frames and local instability of members are discussed. Behavioral effects essential for advanced analyses are highlighted. Finally, design issues related to advanced analysis in steel frame design are discussed, and future research needs are identified.

5.2 CATEGORIZATION OF ANALYSIS METHODS

It is necessary to define the definitions of general types of analysis that will be referred to throughout the book. This section highlights the unique characteristic of several analysis methods that are utilized in research and practice. Figure 5.1 shows schematic representations of the load-displacement curves of a rigid sway frame generated for each type of analysis to be considered.

5.2.1 First-Order Elastic Analysis

The main distinguishing characteristics of a first-order elastic analysis is that the deformations are assumed to be small, so that equilibrium and compatibility relationships can be formulated with respect to the undeformed (or

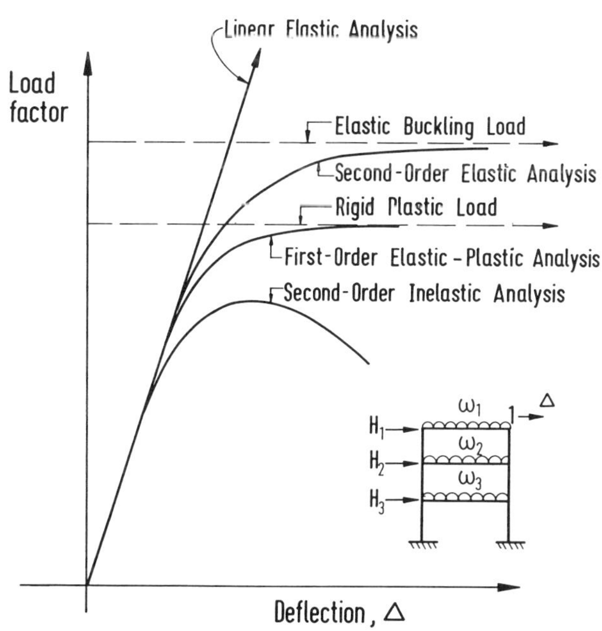

Fig. 5.1 Load-displacement characteristics of general analysis types.

original) geometry, ignoring the yielding of material. The analysis is the current staple of the profession because the principle of superposition remains valid. The analysis provides a simple means of estimating the force distributions within the structural system. However, it does not provide any information about the influence of plasticity and stability on the behavior of the structure. These influences are normally provided indirectly in member-capacity checks through the use of column effective-length and moment-amplification factors. The load-displacement curve obtained from the first-order elastic analysis is a linear line, as shown in Fig. 5.1.

5.2.2 Linear Buckling Analysis

The elastic buckling load is calculated by linear buckling or bifurcation (or eigenvalue) analysis. The buckling loads are obtained from the solutions of idealized elastic frames subjected to idealized loads that do not produce direct bending in the structure. The only displacements that occur before buckling are those in the directions of the applied loads. When buckling (bifurcation) occurs, the displacements begin and increase without bound, assuming the linearized theory of elastic stability, as shown as the horizontal straight line in Fig. 5.1. The load at which these displacements occur is known as the buckling load, commonly referred to as bifurcation load.

Buckling analysis is a common tool for calculations of column effective lengths. The effective length factor K of a column member can be calculated using the following expression:

$$K = \sqrt{\frac{P_e}{P_{cr}}} = \sqrt{\frac{\pi^2 EI/L^2}{P_{cr}}} \tag{5.2.1}$$

where P_e is the Euler buckling load of the column, and P_{cr} is the axial force in the column at the incipient buckling of the frame. The buckling analysis provides useful indices of the stability behavior of structures; however, it does not predict actual behavior of all but idealized structures, with gravity loads applied only at the joints.

5.2.3 Second-Order Elastic Analysis

The analysis is formulated based on the deformed configuration of the structure. When derived rigorously, a second-order analysis includes both the member curvature (P–δ) and the side-sway (P–Δ) stability effects. The P–δ effect is associated with the influence of the axial force acting through the member displacement with respect to the rotated chord, whereas the P–Δ effect is the influence of axial force acting through the relative side-sway displacements of the member ends. It is interesting to note that a structural system becomes stiffer when its members are subjected to tension. Conversely, the structure becomes softer when its members are in compression. Such behavior is illustrated for a simple model shown in Fig. 5.2. There is a clear advantage for a designer to take advantage of the stiffer behavior for tension structures. However, the detrimental effects associated with second-order deformations due to the compression forces must be considered in designing structures subjected to predominant gravity loads.

Unlike the first-order analysis in which solutions can be obtained in a rather simple and direct manner, a second-order analysis often requires an iterative procedure to obtain solutions. Although second-order analysis can account for all the stability effects, it does not provide information on the actual inelastic strength of the structure. Therefore, in design, these combined stability and inelasticity effects must be accounted for in the specification member-capacity checks. The load-displacement curve generated from a second-order analysis gradually approaches the horizontal straight line, which is the buckling load obtained from the elastic buckling analysis, as shown in Fig. 5.1.

Recent works on second-order elastic analysis have been reported (Chen and Lui, 1991; Liew, White, and Chen, 1991; White and Hajjar, 1991). Second-order analysis programs that can take connection flexibility into consideration are also available (Ackroyd, 1985 Dhillon and Abdel-Majid, 1990; Frye and Morris, 1975; Goto and Chen, 1987; Heish, 1990; Liew, White, and Chen, 1993a, 1993b).

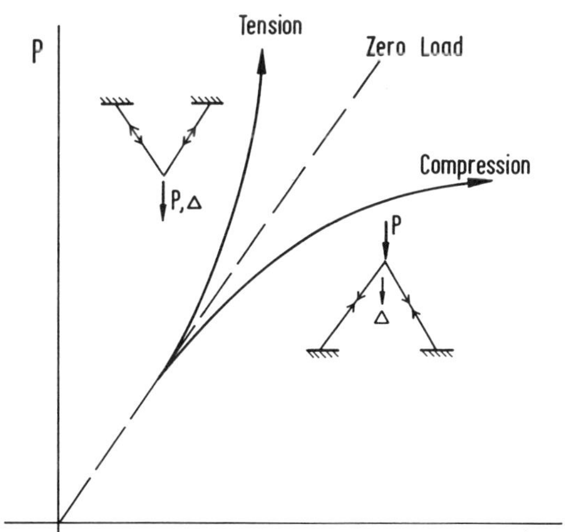

Fig. 5.2 Stiffness enhancement and stiffness degradation from tension and compression structure.

5.2.4 First-Order Plastic-Hinge Analysis

The principles of plastic-hinge analysis are based on the assumption that as a structure is loaded and sections within that structure reach their plastic moment resistances, plastic hinges form at these points and redistribution of forces within the structure is allowed to occur. Plastic-hinge analysis assumes that the member cross-section behaves elastically until full plasticity is achieved at which a plastic hinge is said to have formed. However, full yielding is rather difficult to achieve in reality because it requires a large change in slope to occur over a small length of member.

Most of the plastic analysis theories assume that plastic hinges are concentrated at zero-length plasticity. In fact, the yield zone is developed over a certain length, normally called the *hinge length,* depending upon the loading, boundary conditions, and geometry of the section. Figure 5.3 shows the various plastic-hinge lengths (ΔL) developed in beams with different support and loading conditions. Plastic hinges are normally developed first at the sections subjected to the greatest curvature deformation. The possible locations for plastic hinges to develop are at the points of concentrated loads, at the intersections of members involving a change in geometry, and at the point of zero shear for a member under uniform distributed load, as shown in Fig. 5.3. When a cross-section develops a plastic hinge, any addition of moment causes the beam to rotate with little increase in moment at the plastic-hinge location. The fully yielded zone acts as if it was a hinge that can undergo indefinite rotation at a constant restraining moment M_p.

Two plastic-hinge methods are now commonly used in practice for first-order analysis, namely: (1) rigid-plastic and (2) elastic-plastic methods. Rigid-plastic analysis is based on the determination of the final collapse mechanism for a structure. A collapse mechanism forms when the structure has developed sufficient number of plastic hinges to allow the frame to collapse under the applied loading. Essentially, the structure has reached the point where no further redistribution of moment is possible, and if more loads were to be applied, the frame would collapse. A collapse mechanism is established by comparing a number of predetermined mechanisms to evaluate which one gives the lowest load factor and hence represents the resistance that could be carried by the structure before the formation of a collapse mechanism.

The elastic-plastic-hinge analysis is an alternate plastic analysis method that determines the collapse load of a structure and gives additional information about the redistribution process prior to achieving the collapse mechanism. The method essentially determines the order in which the hinges form, the load factor associated with each hinge formation, and how the bending moments in the structure vary between each hinge formation. The main difference between rigid-plastic- and elastic-plastic-hinge analyses is that the former is based on direct determination of the resistance of the final collapse mechanism of a structure, whereas the latter gives, in addition to the collapse

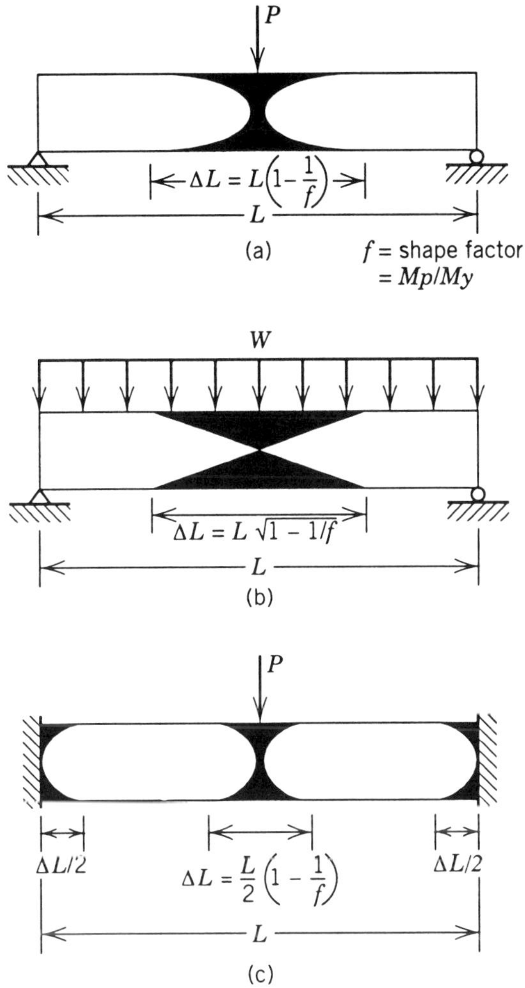

Fig. 5.3 Plastic-hinge lengths in beams with different boundary conditions and loading.

load of the frame, additional information about the redistribution process prior to collapse. In comparison, the elastic-plastic analysis allows a much more accurate determination of the bending moment diagram at the design load level.

In rigid-plastic analysis, it is always assumed that the bending moments in the frame are always increasing as the loading is increased. No account is taken of the possibility of bending moments to increase to a point and then decrease after one or more plastic hinges have been set. On the other hand, bending moment reversal can be implemented in the elastic-plastic-hinge analysis.

Figure 5.1 shows the curve obtained from the first-order elastic-plastic analysis. The collapse load obtained from the elastic-plastic analysis corresponds to that obtained by the rigid-plastic analysis. Plastic-hinge analysis generally provides a more economical design. It is commonly used to assess the strength of narrow range of structures, mostly continuous beams, lightly loaded single-story rigid frames and braced frames that are not "stability-critical."

5.2.5 Second-Order Inelastic Analyses

Inelastic analysis refers to any method of analysis in which the effects of material yielding are accounted for in the analysis. Second-order inelastic analysis may be generalized into two main types: (1) plastic-zone, or distributed plasticity, approach, and (2) plastic-hinge-based approach. This generalization is based on the degree of refinement in the representation of yielding effects. The plastic-hinge-based method is the simplest approach, whereas the plastic-zone model exhibits the greatest refinement. The plastic-hinge approach requires less computational effort and is less costly than the plastic-zone analysis.

5.2.5.1 Second-Order Plastic-Zone Analysis This type of analysis is generally called distributed-plasticity analysis, spread-of-plasticity analysis, fiber analysis, and is referred to here as plastic-zone analysis. There are generally two types of plastic-zone analyses. The first involves the use of three-dimensional finite-shell elements in which the elastic constitutive matrix, in the usual incremental stress–strain relations, is replaced by an elastic-plastic constitutive matrix once yielding is detected. Based on a deformation theory of plasticity, the effects of the combined effects of normal and shear stresses may be accounted for in the analysis. This analysis approach typically requires modeling of structures using a large number of finite three-dimensional shell elements, and numerical integration for the evaluation of the elastic-plastic stiffness matrix. The three-dimensional spread-of-plasticity analysis when combined with second-order theory, which deals with frame stability, is computationally intensive and, therefore, best suited for analyzing small-scale structures or if the detailed solutions for member local instability and yielding behavior are required. Because detailed analysis of local effects in realistic building frames is not the common practice in engineering design, this approach is considered too costly for practical use.

The second-approach for second-order plastic-zone analysis is based on the use of beam-column theory, in which the member is discretized into many line segments, and the cross-section of each segment is further subdivided into a number of finite elements. Inelasticity is typically modeled by the consideration of normal stress only. When the computed stresses at the centroid of any fibers reach the uniaxial normal strength of the material, the fiber is considered as yielded. Also, compatibility is treated by assuming that full continuity is retained throughout the volume of the structure in the same manner as for elastic range calculations. Although there may exist a quite sharp curvature in the vicinity of inelastic portions of the structure, the so-called "plastic hinges" can never develop. In plastic-zone analysis, the calculation of forces and deformations in the structure after yielding requires iterative trial-and-error processes because of the nonlinearity of the load-deformation response, and the change in cross-section effective stiffness at inelastic regions associated with the increase in the applied loads and the change in structural geometry. Although most of the plastic-zone analysis methods have been developed for planar analysis (Alvarez and Birnstiel, 1957; Clarke et al., 1992; El-Zanaty, Murray, and Bjorhovde, 1980; Vogel, 1985; White, 1985), three-dimensional plastic-zone techniques are also available involving various degrees of refinements (Chen and Atsuta, 1977; Clarke, 1994; Wang, 1988; White, 1988).

A plastic-zone analysis that includes the spread of plasticity, residual stresses, initial geometric imperfections, and any other significant second-order behavioral effects would certainly eliminate the need for checking individual member capacities in the frame. Therefore, this type of method is generally classified as advanced inelastic analysis. In fact, the member interaction equations in the modern limit states specifications were developed, in part, by curve fitting to the results from this type of analysis. However, plastic-zone analysis is generally too computationally intensive for routine design use. Its applications are limited only for research studies, generation of design charts, and very special design problems (Clarke, 1994). In reality, some significant behavioral effects such as joint and connection performance tend to defy precise numerical and analytical modeling. In such cases, a simpler method of analysis that adequately represents the significant behavior would be sufficient for engineering applications.

5.2.5.2 Second-Order Plastic-Hinge-Based Analysis The simplest and most direct second-order inelastic analysis is to adopt the elastic-plastic-hinge formulation. This analysis approach typically involves the use of one beam-column element for each frame member and assumes that the element remains elastic except at its ends where zero-length plastic hinges are allowed to form (Orbison, 1982; Liew, 1992; Ziemian, McGuire, and Deierlien, 1992a, 1992b; White et al., 1993; White, 1993; Chen and Toma, 1994; Chen and Sohal, 1995). The major benefit of using second-order elastic-plastic-hinge analysis for design is a more reliable estimate of structural stiffness, strength, and performance than the elastic-analysis procedure can attain. Secondary benefits are primarily due because elastic analysis does not explicitly account for any degradation of stiffness and the corresponding redistribution of forces within the structural system prior to achievement of the design load. This degradation in stiffness is in general due to second-order effects associated with change in geometry (i.e., P–Δ and P–δ effects), as well as plastification within the structure.

Second-order plastic-hinge analysis allows efficient analysis of large-scale building frames. This is particularly true for structures in which the axial forces in the component members are small and the predominant behavior is associated with bending actions. In such cases, second-order elastic-plastic-hinge analysis may be applied to describe the inelastic behavior with sufficient accuracy, assuming that lateral–torsional and local buckling modes of failure are prevented (Liew, 1992). However, research by Liew et al. (1993a, 1993b, 1994) has shown that second-order elastic-plastic-hinge analysis is only an approximate method. When used to analyze a single beam-column element subjected to combined axial load and moment, this method often overestimates the strength and stiffness of the element when it is loaded into the inelastic range. Although elastic-plastic-hinge approaches can provide essentially the same load-displacement predictions as second-order plastic-zone methods for many frame problems, they cannot be classified as advanced analysis for use in frame design. Some modifications to the elastic-plastic hinge are required to qualify the methods as advanced analysis; they are discussed in Section 5.5.

Figure 5.1 shows the load-displacement curve (a smooth curve with a descending branch) obtained from the second-order inelastic analysis. The computed limit load should be close to that obtained from the plastic-zone analysis.

5.3 CATEGORIZATION OF DESIGN METHODS

The steel building design specifications in the United States have undergone a metamorphosis from the *Allowable Stress Design* (ASD) approach through the *Plastic Design* (PD) approach to the present stage of the *Load and Resistance Factor Design* (LRFD) approach. The two major causes that brought about the evolution of a new generation of design codes are (1) the emergence of limit state design and (2) the use of probabilistic methods in the development of load and resistance factors (Galambos, 1991).

The move from ASD design to LRFD is because the specification recognizes that near failure, structural materials do not behave elastically and that a reasonable calculation of strength must account for inelasticity. One fundamental feature that demarcates the LRFD approach to steel design from ASD and PD approaches is that it is a probability-based design methodology that uses statistical evidence and reliability theory to prove the logistic and rationale of the design. Design considerations are guided by the identification of a set of limit states. A limit state is a condition that represents the limit of structural usefulness. Limit states could relate to many different behavior but generally they can be related to serviceability and strength. A design is considered acceptable if none of the limit states is violated. Because a probabilistic mathematical model was used in the development of the load and resistance factors, it is possible to give proper weight to the degree of importance for each load effect and resistance. The design based on this approach often yields greater economy and more uniform margin of safety.

Recently, an emerging design methodology called advanced analysis has been proposed (Liew et al., 1993a, 1993b, 1994; White and Chen, 1993; Chen and Toma, 1994). This analysis approach has the potential to unlock the power of the computer for direct assessment of system and member strength and stability without the need of specification member-capacity checks. Advanced analysis holds the promise of more realistic prediction of load effects and overall structural performance, and thus it has the potential to provide more innovative and cost-effective design.

5.3.1 Allowable Stress Design

Under the Allowable Stress Design (ASD) philosophy, a design is said to be satisfactory if the stresses computed under service loads (or working loads) do not exceed some predesignated allowable values. It has the format of

$$\frac{R_n}{\text{F.S.}} \geq \Sigma Q_i \tag{5.3.1}$$

where

F.S. = factor of safety
i = type of load (i.e., dead load, live load, and wind load)
R_n = nominal resistance of the structural member expressed in units of stress
Q_i = nominal working stresses computed under working-load conditions

The right side of Eq. (5.3.1) is the required strength, which is the summation of the load effects, Q_i (i.e., forces and moments). The left side, the design strength, is the nominal strength or resistance divided by a factor of safety. The two sides of the inequality are often divided by the appropriate section property (e.g., corss-sectional area), thus becoming the actual stress and allowable stress, respectively. The left-hand side of Eq. (5.3.1) represents the allowable stress of the structural member or component under various loading conditions (e.g., tension, compression, bending, and shear). The right-hand side of the equation represents the computed combined stresses produced by various load combinations (e.g., dead load, live load, and wind load).

ASD is characterized by the use of "working" (or unfactored) loads in conjunction with a single factor of safety applied to the resistance. Because of the greater availability and unpredictability of live load and other loads in comparison with the dead load, a uniform reliability is not possible. The stress computation is carried out using first-order elastic analysis. Because safety is evaluated in the elastic range, the actual ultimate capacity of the structure is not known. However, this design approach has been set based on experience and proven to be acceptable for many years.

The ASD has its limitations in adopting to new structural concepts such as the use of composite systems, partially restrained connections, and eccentrically braced frames. Such structural concepts are evolved based on the ultimate strength design concept, which is not suitable for implementation in the ASD domain. Some of the new seismic codes provide loading in the form of ultimate loads. These loads need to be scaled back for use in the ASD procedures.

5.3.2 Plastic Design

As discussed in Section 5.2, plastic analysis refers to a concentrated plasticity, or zero-length plastic-hinge approach, which may be rigid plastic analysis, first-order elastic-plastic-hinge analysis, or second-order elastic-plastic-hinge analysis. In plastic design, it must be demonstrated that the structures must satisfy both the ultimate strength and serviceability limit states. At the strength limit state, the structure must then satisfy two design criteria: (1) the strength criterion, in which no collapse mechanism should exist before the design load, and (2) the stability criterion, in which members, frames, and plastic hinges must be adequately braced or restrained to provide adequate resistance against buckling. Of course, if one performes a second-order analysis, the second criterion mentioned can be somewhat relaxed.

In plastic design, a structure is said to satisfy the strength limit state if the load effects computed based on factored-load combinations do not exceed the plastic resistance of the structural member or component. It has the following format:

$$R_n \geq \gamma \Sigma Q_i \tag{5.3.2}$$

where R_n is the nominal plastic resistance of the structure, and γ is a load factor.

The load factor γ is taken as 1.7 by the AISC (1978, 1989) if the nominal-load effect Q_i consists of only dead and live gravity loads, or as 1.3 if Q_i consists of dead and live gravity loads plus wind or earthquake loads.

In most of the plastic design methods, the structural resistance is evaluated based on the lowest collapse load factor from the rigid-plastic mechanism analysis or the failure-load factor from first-order elastic-plastic-hinge analysis. Although material nonlinearity has been considered in these analysis methods, the effects of spread of yielding and member/frame stability are neglected. Thus, like ASD, any significant nonlinear effects have to be accounted for approximately in the member strength equations. In other words, member-capacity checks are required when these plastic analyses are employed.

The plastic design method allows the engineer to design a more economical structure by allowing inelastic force redistribution to occur in the structure. However, the analysis relies on the ability of steel to accept considerable deformation in excess of yield. Therefore, the regions of the structure in which significant plasticity (plastic hinges) develops must have adequate deformation capacity to carry a load. To achieve this, members containing plastic hinges must satisfy the limitations on the flange and web proportions for plastic design sections. In addition, sufficient lateral and torsional restraints must be provided to prevent the member from buckling laterally.

Plastic analysis involves the calculation of the collapse load of the structure. The working load was derived by dividing the collapse load by a load factor. To carry out member-capacity checks at the design load factor, the bending moment diagram (BMD) at collapse is defactored by the ratio α, where α = design load/collapse load. This is to achieve an approximation to the true BMD at the design-load factor. In the hinge-by-hinge analysis, a much more accurate determination of the bending moment diagram at the design-load factor can be obtained directly.

The effects of residual stresses and member and frame geometrical imperfections generally are not included in the analysis, but are accounted for separately in the section capacity and member design rules corresponding to plastic design. A second-order plastic-hinge analysis including both material and geometrical nonlinearities will certainly lead to an improvement in the estimates of the moment and force distributions in the frame under the factored loads.

5.3.3 Load and Resistance Factor Design (LRFD)

Under the Load and Resistance Factor Design (LRFD) format, a design is said to be satisfactory if it can be demonstrated that the structure satisfies both the ultimate strength and serviceability limit states. The LRFD method may be summarized by the following format:

$$\phi R_n \geq \Sigma \gamma_i Q_i \tag{5.3.3}$$

where Q_i and γ_i are the nominal-load effects and their corresponding load factors, respectively, R_n is the nominal resistance, and ϕ is the resistance factor corresponding to R_n. The required strength on the left side of the inequality is the summation of the various load effects Q_i multiplied by their respective load factor γ_i. The design strength, on the left side, is the nominal strength or resistance R_n multiplied by the resistance factor ϕ.

The load factors γ_i are derived based on statistical analyses, and the resistance factor ϕ is developed based on probability theory and calibration. These factors and the load combinations used in LRFD are summarized in Tables 5.1 and 5.2, respectively.

LRFD is a method for proportioning a structure so that no applicable limit state is exceeded when the structure is subjected to all appropriate load combinations. Strength limit states are related to safety and load-carrying capacity. Serviceability limit states relate to performance of the structure under normal service-load conditions. In general, a structure member will have several limit states. Because AISC specification is concerned primarily with safety, strength limit states are emphasized.

In the design of steel frames using LRFD, the analysis must account for second-order effects that arise from the change in geometry of the structure under the design loads. AISC LRFD (AISC, 1993) provides simplified methods of including second-order effects by increasing the moments calculated from a first-order elastic analysis using B_1 and B_2 moment-amplification factors. In addition, column effective lengths are required to evaluate the column resistance term in the member-interaction equations. The column effective lengths are commonly evaluated based on elastic buckling analysis of the individual members, the frame subassemblage, or the frame as a whole.

In LRFD, a direct second-order elastic analysis is the preferred method to obtain member forces for member-capacity checks. A direct second-order analysis, which employs either a matrix displacement or finite-element approach, will include both P–Δ and P–δ effects, thereby avoiding the need to compute moment-amplification factors. However, it is still required to compute the column effective length (Liew et al., 1991, 1992). Although the member-interaction equations implicitly account for the ultimate strength behavior of a beam-column member, it does not satisfy compatibility between

TABLE 5.1 Load Factors and Load Combinations

$1.4D$
$1.2D + 1.5L + 0.5(L_r \text{ or } S \text{ or } R)$
$1.2D + 1.5(L_r \text{ or } S \text{ or } R) + (0.5L \text{ or } 0.8W)$
$1.2D + 1.3W + 0.5L + 0.5(L_r \text{ or } S \text{ or } R)$
$1.2D \pm 1.0E + 0.5L + 0.2S$
$0.9D \pm (1.3W \text{ or } 1.0E)$

where D = dead load due to weight of structural elements and permanent features on structure.
L = live load due to occupancy and movable equipment
L_r = roof live load
W = wind load
S = snow load
E = earthquake load (determined in accordance with Part 1 of AISC seismic provisions for structural steel buildings)
R = load due to initial rainwater on ice exclusive of ponding contribution

TABLE 5.2 Resistance Factors

Member Type and Limit State	ϕ
Tension member, limit state: yielding	0.90
Tension member, limit state: fracture	0.75
Pin-connected member, limit state: tension	0.75
Pin-connected member, limit state: shear	0.75
Pin-connected member, limit state: bearing	1.00
Column, all limit states	0.85
Beams, all limit states	0.90
High-strength bolts, limit state: tension	0.75
High-strength bolts, limit state: shear	
A307 bolts	0.50
others	0.55

the inelastic forces, as assumed by the ultimate strength member-capacity checks, and the elastic forces from the global frame analysis. Nevertheless, the introduction of the LRFD concept, which utilizes the maximum strength resistances of members, has laid the groundwork for a more rational approach for structural steel design.

Recent research by Liew et al. (1993a, 1993b) has shown that elastic-analysis procedures based on specification member-capacity checks are limited in their ability to provide a true assessment of the maximum strength behavior of highly redundant structures. The design procedure employed by the member-capacity equations does not represent the true behavior of a frame after inelastic redistribution of forces occur. Therefore, its application is truly valid for frames that are essentially elastic. The research also found that the LRFD elastic-analysis/design approach is conservative for most conventional frames. In general, the conservativeness of the LRFD elastic-analysis/design approach increases with the degree of structural redundancy. The elastic-analysis/design method should be conservative because the reserved capacities due to inelastic force redistribution prior to reaching the design-load level have been ignored, and the system's strength is essentially taken as the load at which the most critically loaded member reaches its individual strength.

Member and frame imperfections such as initial geometric imperfections and residual stresses affect force distribution and their strengths. These effects have been accounted for in the AISC LRFD member-capacity equations, but they are not included in the global analysis. In Eurocode 3 (1992) approach, frame and member imperfection effects are explicitly accounted for in the global frame analysis using either enlarged geometric imperfections or equivalent notional forces.

In summary, any specification checks based on elastic analysis are limited in their ability to estimate the true strength of redundant structures. The more rational way of estimating the interactive effects of member and system strength and stability—which also avoids complex issues such as effective length concept—is through second-order advanced inelastic analysis.

5.3.4 Advanced Analysis/Design

The development of computer programs suitable for advanced analysis has followed three main directions. The first approach is based on second-order elastic-plastic-hinge analysis without any modification to the basic theory. In this approach, the effects of residual stresses, gradual yielding, and geometrical imperfections are accounted for approximately using either "enlarged geometrical imperfections" or "fictitious loads," which are referred herein as notional loads.

The second approach to advanced analysis is based on modification or refinement of the basic plastic-hinge theory by allowing for smooth degradation of stiffness due to the effects of spread of plasticity. The effects of residual stresses, gradual yielding, and geometrical imperfections can be accommodated implicitly in the element formulation by proper calibration of phenomenological or behavioral models (Deierlein, Zhao, and McGuire, 1991; King, White, and Chen, 1992; Liew et al., 1993a, 1993b). This approach is commonly called the modified plastic-hinge or refined plastic-hinge method.

The third approach to advanced analysis, as discussed in Section 5.2, is based on second-order plastic-zone analysis. A detailed treatment of plastic-zone analysis is given by Chen and Toma (1994). Compared to the methods of advanced inelastic analysis based on the plastic-hinge concept, plastic-zone analysis is more generally applicable to arbitrary cross-sections and geometry, and the effects of specialized stress–strain characteristics and the distribution of residual stresses can be more readily accommodated. However, plastic-zone analysis suffers from the disadvantage of high computational expense, and it requires more resources to prepare input data for analysis and to manage output data for postprocessing.

The advanced analysis concept can be implemented in the LRFD format by moving a number of resistance effects from the left- to right-hand side of Eq. (5.3.3) and, thus, eliminating a number of equation checks from the design. In particular, member-capacity checks can be waived when an advanced analysis is performed. The reason for this is that because advanced analysis can directly assess the strength and stability of the overall structural system as well as interdependence of member and system strength and stability, separate member-capacity checks are therefore not required.

It is expected that use of advanced analysis techniques will simplify the design process considerably. However, this task cannot be accomplished fully in the present PD and LRFD formats, even if all the significant nonlinear effects are considered directly in the analysis. The current LRFD specifications require the engineer to perform beam-column capacity checks even though an inelastic analysis is used. In contrast, the provisions for advanced analysis in Eurocode 3 (1992) and Australian limit states specifications (Standards Australia, 1990) postulate that if all the significant planar behavioral effects are modeled properly in the analysis, the checking of conventional beam-column equations is not required. Thus, for advanced analysis to provide the most benefit in frame design, specification provisions of this type are required. The main distinction between advanced analysis and other simplified analysis methods is that advanced analysis combines the theories of plasticity and stability, and the effects of imperfections in the analysis/design of structural steel frameworks. Other analysis and design methods treat stability, plasticity, and imperfections separately—usually through the use of beam-column interaction equations and member effective-length factors.

5.4 DESIRABLE ATTRIBUTES FOR AN ADVANCED ANALYSIS MODEL

Table 5.3 provides a summary of nonlinear behavioral phenomena and physical attributes that may be considered in an advanced analysis of frame members, either by explicit or implicit means. Other useful categorizations of members as well as system behavior are provided by Galambos (1988), Ackroyd (1991), and White and Chen (1993). For an analysis method to be classified as advanced, not all of the items shown in Table 5.3 need to be addressed or represented in the design/analysis situation. When certain attributions cannot be considered in the analysis, they may be viewed as the limitations of the analysis model, and these limitations must be considered in the final design.

The following outlines the desirable attributes for the development of an advanced inelastic analysis model:

1. The formulation of the model should be mathematically rigorous and based on sound and well-understood theories of engineering mechanics.

2. The strengths, internal deformation, and internal force distribution predicted by the model for various types of member and systems should be close to those predicted by benchmark solutions. The benchmark solutions used for calibrating advanced analyses should be established either by rigorous analysis (such as second-order plastic-zone analysis) or by full-scale tests. Other possible benchmark solutions are those that have been defined in the context of the design specifications under which the advanced analysis model might be used.

3. The analysis should be capable of capturing the characteristic behavior of a wide range of structural system and member types. For instance, the model should represent accurately (a) strong- and weak-axis bending action of rolled and welded beam-column sections, (b) behavior of braced and unbraced beam-columns, (c) failure of members by predominantly elastic or

TABLE 5.3 Behavioral Effects for Frame Analysis/Design

1. Physical Attributes

- Frame topology: member center-to-center lines or member clear-span length with finite joint size
- Two- or three-dimensional structures with orthogonal or inclined members
- Initial geometric imperfections due to camber, sweep, and twist of member, frame and column out of plumbness, member out of straightness, cross-section distortion
- Residual stresses due to manufacturing and fabrication processes, erection out of fit, and construction sequencing
- Internal and external restraints due to bracing, supports, foundations, walls, etc.
- Connection types: pinned, rigid, semi-rigid
- Cross-section shapes: symmetry/nonsymmetry, closed/open profile
- Prismatic/nonprismatic member profiles
- Composite actions from slabs, infilling, etc.
- Construction sequences
- Interaction with foundations

2. Nonlinear Response Phenomena

Geometric Nonlinearity

- P–Δ moment: second-order moments due to axial force acting through displacements associated with member chord rotation
- P–δ moment: second-order moments due to axial force acting through displacements associated with member curvature bending
- Axial deformation due to axial strain and bowing effects
- Shear deformation of members
- Local buckling and distortions
- Interaction of local and overall buckling
- Geometric interaction due to axial force, moments, torsion, and shear deformation
- Connection deformation
- Panel-zone deformation

Material Nonlinearity

- Zero-length plastic hinge: elastic-plastic or elasto-plastic
- Spread of plasticity in members
- Strain hardening
- Unloading due to plastic deformation
- Inelastic interaction of axial force, moments, torsion, and shear
- Cyclic plasticity effects
- Strain aging

3. Loading Effects

- Proportional or nonproportional loading
- Conservative or nonconservative loading
- Load position from shear center
- Distributed loads implemented as equivalent point loads or as "pressure" loads
- Variable repeated loading
- Dynamic loading
- Construction loading in stages

4. Uncertainty

- Variability of loading
- Variability of joint, member, and structural resistances
- Variability of material resistance

inelastic instability, and (d) load-deflection and strength behavior of beam-columns subjected to axial and transverse forces.

4. The model should be capable of representing P–Δ and P–δ effects, including the instability effects from unequally loaded columns, unequal-length columns, leaning columns, and inclined columns, as shown in Fig. 5.4. It should predict the response due to distributed plasticity associated with residual stresses and geometric imperfections at a wide range of axial force and bending moment combinations.

5. The model should capture, accurately and conservatively, the stability and strength of individual frame components in resisting load actions. The analysis should not be more than 5% unconservative when compared to a

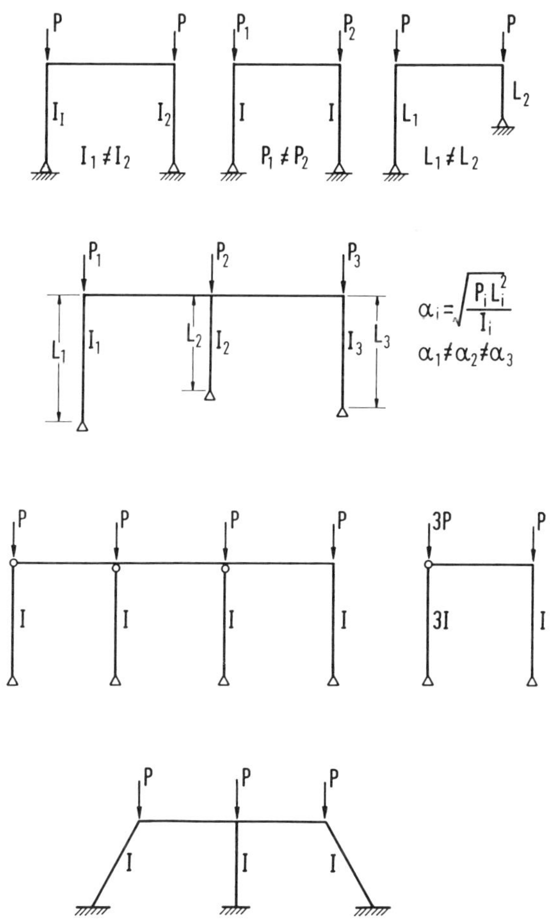

Fig. 5.4 Basic instability effects on frames.

wide range of benchmark solutions for beam, column, and beam-column strength.

6. The element force-displacement relationships should be derived analytically and implemented in a convenient form for analysis. Initial geometrical imperfections and residual stress effects should be included in the analysis either implicitly or explicitly.

7. The model should reduce to a well-recognized behavioral model in the limits of pure beam and pure column action. That is, when the axial force approaches zero, the element ultimate strength behavior should approach that of the elastic-plastic-hinge model. For the case of a member loaded by pure axial load, the column strength should be close to that associated with the column equations of the design specification used by the engineer.

8. The member forces calculated by the model should not violate the cross-section strength associated with the condition of full plasticity.

9. The model should accommodate force point movement due to continued loading after full plasticity is attained. If the axial force supported by a beam-column increases after a plastic hinge has formed, the moment at the plastic hinge must decrease.

10. The possible benefits of strain hardening should not be relied upon. This clause is, however, debatable. The ability of a beam-column to develop significant strain hardening is dependent on many factors, such as the moment gradient and interaction of local and lateral–torsional buckling effects, and distributed yielding along the member length. At the present time (1996), the precise effects of yielding, strain hardening, and possible local and lateral torsional buckling on the full moment–rotation characteristics of beam-column members are still not well enough quantified for direct implementation in a practical inelastic analysis.

11. The model should represent the effects of distributed yielding on both the axial and flexural deformations. Framing members subjected to high axial forces and small bending moments, such as members in a truss-type system as well as in most tall building frames, may exhibit significant distributed yielding along their lengths prior to failure. The member effective axial stiffness provides a significant contribution to the response of these types of structures.

12. The model should be readily extendable to three-dimensional analysis. The framework of the model should accommodate the formulation of three-dimensional elements.

The next section introduces two practical plastic-hinge-based advanced analysis models that will satisfy most of the preceding attributes. These plastic-hinge-based analyses involve idealization of the inelastic behavior by zero-length plastic hinges, but the behavior of the plastic hinges can be elastic-perfectly plastic or elasto-plastic.

5.5 PRACTICAL ADVANCED ANALYSIS TECHNIQUES

Research by Liew et al. (1993a, 1993b) shows that the conventional elastic-plastic-hinge approach is limited because it overpredicts the strength of beam-columns that fails by inelastic flexural buckling. Furthermore, conventional plastic-hinge solutions often do not include geometric imperfections and residual stresses in the analysis. Thus, the effects of distributed plasticity as well as imperfections must be represented in a proper advanced analysis. Two possible plastic-hinge-based approaches for advanced analysis are outlined in what follows.

5.5.1 Notional-Load Plastic-Hinge Approach

One possible approach to advance the use of second-order elastic-plastic-hinge analysis for frame design is to specify artificially large values of frame imperfections (i.e., initial out of plumbness) in the analysis. This is the general approach adopted by Eurocode 3 (1992) for frame design. In addition to accounting for standard erection tolerance for out of plumbness, these artificial large imperfections are intended to account for the effects of residual stresses, frame imperfections, such as lack of verticality, lack of straightness, lack of fit, and the unavoidable minor eccentricities present in practical connections. In Eurocode 3, the effects of imperfections are allowed for in frame analysis by means of an equivalent geometric imperfection in the form of an initial sway imperfection Ψ_0, as shown at the top of Fig. 5.5. The Eurocode 3 also allows the initial sway imperfection to be replaced by a closed system of equivalent horizontal forces.

Liew et al. (1994) have proposed an approach that satisfies the planar strength requirements for advanced analysis. This approach is called the *notional-load plastic-hinge analysis,* which utilizes a set of equivalent notional lateral loads to account approximately for the influence of member imperfections and distributed plasticity effects on the overall system strength. These notional-lateral loads are expressed as a fraction of the story gravity loads or member axial force, and they should be applied in addition to the design loads in the global frame analysis.

The notional-load plastic-hinge approach is similar to the concept of "enlarged" geometric imperfection approach adopted by Eurocode 3. The similarities and differences between these two methods are summarized as shown in Fig. 5.5. Basically, the specification of frame imperfections is the same in both approaches, with the maximum out of plumbness defined as $\Psi_0 = L/200$, where L is the column length. Furthermore, the modeling of member initial out-of-straightness imperfections is not required when the column axial load P is less than $0.25\pi^2EIL^2$ in both approaches. However, the modeling of member imperfection effects for cases in which this limit is exceeded is different in the two procedures.

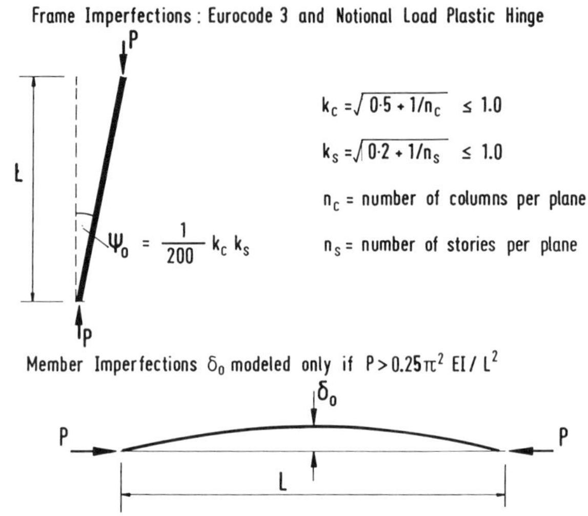

Eurocode 3	
δ_0	Column Curve
L/400	a
L/250	b
L/200	c
L/150	d

Notional-Load Plastic Hinge
$\delta_0 = L/400$

Fig. 5.5 Comparison between the notional-load plastic-hinge method with Eurocode 3 procedures.

Eurocode 3 specifies different equivalent initial out-of-straightness imperfections based on the relevant column curve for the member under consideration. The maximum out-of-straightness magnitudes range from $L/400$, or rolled I-sections under strong-axis bending, to $L/150$, for heavy I-sections under strong- or weak-axis bending. The equivalent maximum out-of-straightness magnitude assumed in the notional-load plastic-hinge approach is $L/400$. This value is consistent with the use of a single column curve in AISC LRFD and also it is equivalent to application of the maximum equivalent out-of-plumbness rule to an individual member with a maximum midspan deflection of $(L/2)/200$. Of course, the equivalent notional load corresponding to this member imperfection is $0.01P$, and the equivalent notional load corresponding to the maximum out of plumbness of $1/200$ is $0.005P$. The application of these notional loads to a building frame is illustrated in Fig. 5.6.

The general application of the notional-load analysis to beam-column and frame cases has been studied in Liew et al. (1994). Figure 5.7 illustrates the interaction strength curve obtained by application of this approach. The results are compared with the adjusted plastic-zone solutions generated by Kanchanalai (1977). The "exact" plastic-zone solutions are obtained by subtracting a

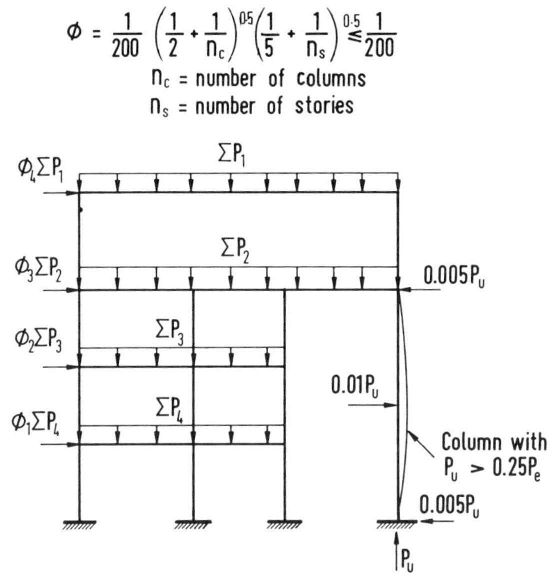

Fig. 5.6 Examples of the application of notational loads for second-order elastic-plastic-hinge analysis.

"used-up" P–Δ moment from the results for perfect frame geometry, as described in Liew et al. (1991). This procedure results in the adjusted "exact" strength that is consistent with the AISC LRFD column strength for the case of pure column action. It is observed that the notional-load plastic-hinge approach compares well with the plastic-zone solutions for all range of axial force vs. moment. If the notional load is not employed, the maximum unconservative error can be as high as 40%. The frame example shown in Fig. 5.7 is only one of the worst cases from the more comprehensive set of problems considered in Liew (1992). A detailed study of the notional-load plastic-hinge method is given in Chapter 7.

5.5.2 Refined Plastic-Hinge Analysis

This type of analysis model is sometimes called the force-based distributed plasticity analysis. The analysis seeks to represent the effects of distributed plasticity through the cross-section, assuming that the plastic-hinge stiffness degradation is smooth. The inelastic behavior in the member is modeled in terms of member forces instead of the detailed level of stresses and strains as used in the plastic-zone analysis model. Like elastic-plastic-hinge analysis, this approach can obtain efficient solution for inelastic analysis by modeling each frame member as one element. Refined plastic-hinge models have been

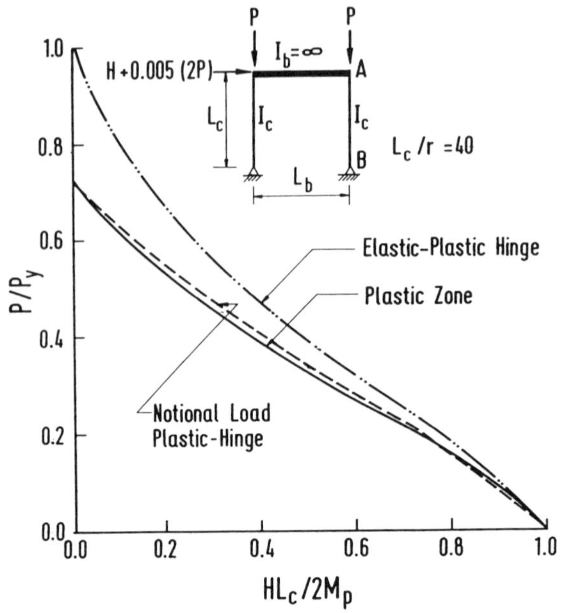

Fig. 5.7 Comparison of strength curves from notational-load plastic-hinge analysis with plastic-zone strength curves (Kanchanalai, 1977).

proposed by Abdel-Ghaffar, White, and Chen (1991); Al-Mashary and Chen (1991); King et al. (1991); Deierlien et al. (1991); Liew et al. (1993a, 1993b); Attalla, Deierlien, and McGuire (1994); King and Chen (1994); Liew and Chen (1994); and Chan and Chen (1995), among others. Currently, there is a large amount of work in this area of research, and there are a number of detailed behavioral effects that can be represented by refinements to the existing plastic-hinge model.

The refined plastic-hinge approach proposed by Liew et al. (1993a, 1993b) and White et al. (1993) is based on two fundamental improvements to the elastic-plastic-hinge model. First, the member elastic core is represented by an effective stiffness associated with tangent modulus E_t. The column tangent modulus E_t can be evaluated based on the specification equations for column design. Thus, it includes the effects of residual stresses and initial out of straightness in modeling the member effective stiffness (Liew et al., 1993a, 1993b). The second improvement is to capture the gradual degradation in stiffness associated with inelasticity that occurs at highly yielded regions under combined axial force and bending. This is accomplished by using a two-surface stiffness-degradation model, as shown in Fig. 5.8. When the force state in the cross-section crosses the initial yield-surface function, yielding is said to occur and the stiffness of the member starts to degrade.

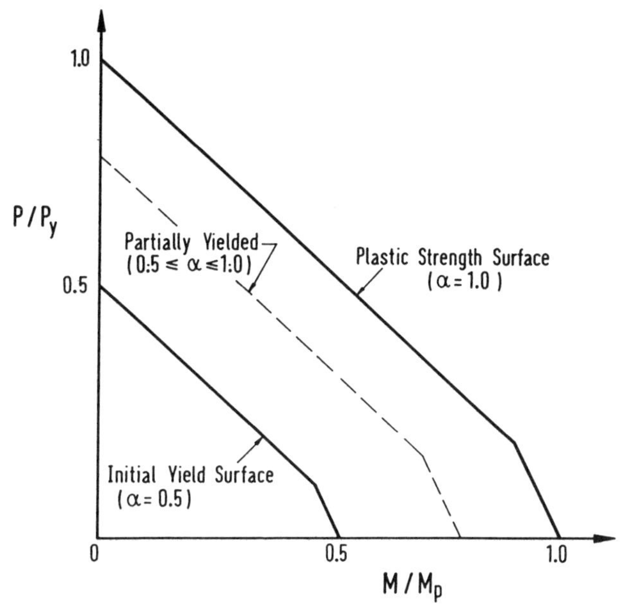

Fig. 5.8 Plastic strength surface and initial and yield surface for refined plastic-hinge analysis.

The preceding refinements lead to better representation of inelastic strength and stability behavior of frame structures without significant computational effort. The effects of connection flexibility can also be incorporated in a second-order refined plastic-hinge analysis. The connections can be modeled as rotational springs that are physically tied to the ends of the element by enforcing equilibrium and compatibility between the connection element and the beam-column element.

The principal merits of the refined plastic-hinge model are that the method is as simple and as efficient as the elastic-plastic-hinge analysis approach, and it is sufficiently accurate for the assessment of members and system strength and stability. The investigations carried out by Liew et al. (1993a, 1993b) further suggest that the refined plastic-hinge method has the most promise for use in analyzing a large-scale frame structure without risk of overestimating the strength and stiffness of the individual member in the frame.

5.6 DESIGN ISSUES AND RULES RELATED TO ADVANCED ANALYSIS

Design issues and rules pertinent to accomplishing the advanced analysis are discussed in what follows.

5.6.1 Cross-Section Compactness

For members with significant inelastic rotation upon yielding, their cross-section compactness may be checked based on the ASCE plastic design guide (ASCE-WRC, 1971) to ensure that local buckling does not limit the member's rotation capacity. For I-sections, wide-flange beams, and welded H-shapes, the limitation on flange and web proportions for plastic design sections (or Class 1 section) are as follows (ASCE-WRC, 1971):

1. Flanges in compression due to bending, axial force, or both:

$$\text{A36 Steel} \qquad \frac{b}{T} = 17 \tag{5.6.1}$$

$$\text{A441 Steel} \qquad \frac{b}{T} = 14 \tag{5.6.2}$$

$$\text{A572 Steel} \qquad \frac{b}{T} = 12 \tag{5.6.3}$$

2. Web under compression:

$$\frac{d}{t} = 43\sqrt{\frac{36}{F_y}} \tag{5.6.4}$$

3. Web under bending and compression:

$$\frac{d}{t} = 43\sqrt{\frac{36}{F_y}}, \qquad \text{for } \frac{P}{P_y} > 0.27 \tag{5.6.5}$$

$$\frac{d}{t} = \left(70 - 100\frac{P}{P_y}\right)\sqrt{\frac{36}{F_y}}, \qquad \text{for } \frac{P}{P_y} \leq 0.27 \tag{5.6.6}$$

where

b = half width of flange plate
d = section depth
F_y = design strength in ksi
t = thickness of web
T = thickness of flange

The provision given by (ASCE-WRC, 1971) is more restrictive than those given in the AISC LRFD specification (AISC, 1986). The cross-section compactness provisions in AISC LRFD (AISC, 1986) essentially correspond to Class 2 sections (compact sections) of many other specifications, especially in a higher grade of steel. Dawe and Kulak (1986) have shown that beam-columns of Class 2 sections may not be able to develop sufficient rotation capacity for force redistribution due to the formation of plastic hinges. Hence, the provisions for compactness requirements in the AISC LRFD specifications are not recommended in the present design.

5.6.2 Cross-Section Plastic Strength

It is important to define the equations for cross-section plastic strength in advanced inelastic analysis using the plastic-hinge model. Figures 5.9 and 5.10 show the "exact" analytical strength surfaces for various hot-rolled wide-flange sections subjected to strong- and weak-axis bending. The dashed-line curves are the AISC LRFD beam-column interaction curves for a member of compact cross-section and of zero length. These equations are as follows:

$$\frac{P}{P_y} + \frac{8M}{9M_p} = 1.0, \qquad \text{for } \frac{P}{P_y} \geq 0.2 \tag{5.6.7}$$

$$\frac{P}{2P_y} + \frac{M}{M_p} = 1.0, \qquad \text{for } \frac{P}{P_y} < 0.2 \tag{5.6.8}$$

where

M = second-order moment at cross-section under consideration
M_p = plastic moment capacity for member under pure bending action
P = second-order axial force at cross-section under consideration
P_y = squash load

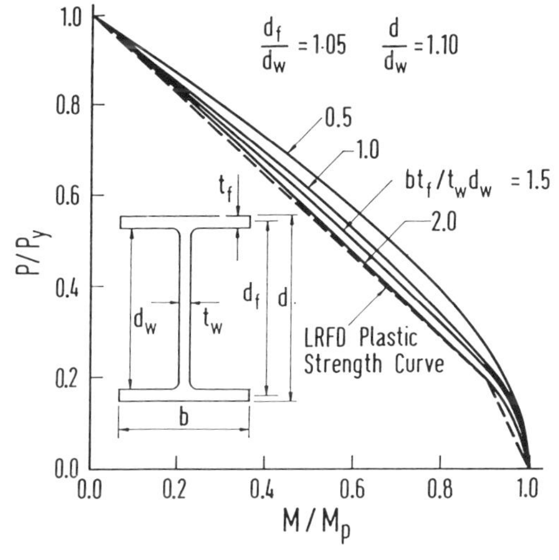

Fig. 5.9 Strength interaction curves for wide-flange sections bending about the strong axis.

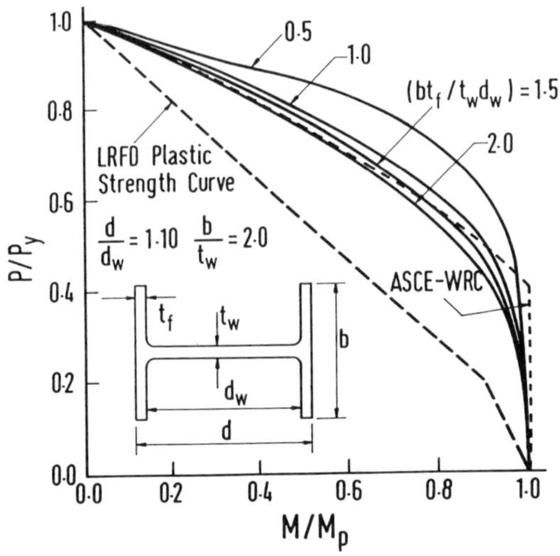

Fig. 5.10 Strength interaction curves for wide-flange sections bending about the weak axis.

Equations (5.6.7) and (5.6.8) are plotted in Figs. 5.9 and 5.10 for comparison with the more "exact" analytical solutions. It can be observed that Eqs. (5.6.7) and (5.6.8), which assume the same functional relationships for both strong- and weak-axis strengths, provide a good lower-bound fit to most of the strong-axis strengths, but they are rather conservative for the weak-axis strength when the bending moment and axial force are dominant. A simplified weak-axis cross-section strength equation proposed by ASCE-WRC (1971), as shown in Fig. 5.10, gives a better fit to the weak-axis analytical strength results.

Duan and Chen (1990) have proposed accurate strength functions for axial load and biaxial bending of various types of doubly symmetric cross-sections. For wide-flange sections bending about the strong axis, the function is

$$\left(\frac{P}{P_y}\right)^{1.3} + \frac{M}{M_p} = 1.0 \tag{5.6.9}$$

For wide-flange sections bending about the weak axis, the function is

$$\left(\frac{P}{P_y}\right)^{2.7} + \frac{M}{M_p} = 1.0 \tag{5.6.10}$$

The actual Duan and Chen plastic strength surface is a three-dimensional function with no sharp "corner" except at $P = P_y$. Compared to the AISC

LRFD plastic strength function, the Duan and Chen curve is a single equation curve that is easy to use in general computation. The functions also give a good lower-bound fit to the exact strong-axis strengths for rolled wide-flange shapes. The curve is similar in accuracy to the AISC LRFD strong-axis strength curve.

The Duan and Chen function fits the weak cross-section strength much more accurately than the AISC LRFD curves, particularly in the range of $0 \leq P/P_y \leq 0.7$. The function produces smooth curves in contrast to the AISC LRFD and the ASCE-WRC (1971) curves.

5.6.3 Resistance Factors

In analysis for limit state design, the design resistance factors would be used in an advanced analysis that does not require member-capacity checks. These resistance factors may be applied to the plastic strength surface following the recommendation given in AISC LRFD for column and beam design, which is $\phi_c P_y = 0.85P_y$ and $\phi_b M_p = 0.9M_p$. The final design plastic strength equations posted in the AISC LRFD format are

$$\frac{P}{0.85P_y} + \frac{8}{9}\left(\frac{M}{0.9M_p}\right) = 1.0, \qquad \text{for } \frac{P}{0.85P_y} \geq 0.2 \tag{5.6.11}$$

$$\frac{1}{2}\left(\frac{P}{0.85P_y}\right) + \frac{M}{0.9M_p} = 1.0 \qquad \text{for } \frac{P}{0.85P_y} < 0.2 \tag{5.6.12}$$

It should be noted that the AISC LRFD specifications' resistance factors are obtained from member-based probability models. Further research is required to address the appropriateness and limitations of these approaches in overall frame reliability.

If a plastic-zone analysis is employed, the resistance factors may be assigned to at least the material yield strength, and other parameters such as the residual stresses, the magnitudes of initial geometrical imperfections, and the elastic modulus. The Australian approach for advanced analysis uses a uniform resistance factor of 0.9 for the material for full system analysis. Recent research (Galambos, 1988) suggests that the conventional resistance factors may be multiplied by an additional system factor associated with system reliability in design. Hence, it is logical to prescribe a system resistance factor to the strength parameters utilized for advanced inelastic analysis.

Adopting the concept of system resistance factors and load factors in the design of structural frameworks is a natural progression in structural analysis and design. When the computational methods available are restricted to linear analysis, the codes will be based on methods that depend on linear analysis and member-capacity checks. With the advance in computational methods, nonlinear analyses are available as a design tool, so eventually the codes should reflect this.

5.6.4 Formation of a Plastic Hinge within the Member

Chen and Atsuta (1976) show that the insertion of a plastic hinge at the "exact" maximum moment point within the member is not required for an accurate estimate of the maximum strength of the member. As long as the "exact" location of the plastic hinge in a member is not more than a distance $L/5$ away from the assumed position (where L is the length of the member), the difference in strength prediction is not more than 5%. The preceding observation is true for beam-columns subjected to various types of transverse loading.

Beam-column elements may be checked by the following analytical expression to ensure that peak second-order moments occur at the member end (Liew et al., 1991):

$$\frac{x}{L} = -\frac{1}{kL}\tan^{-1}\left[\frac{(M_1/M_2)\cos(kL) + 1}{(M_1/M_2)\sin(kL)}\right] > 1.0 \text{ or } <0 \qquad (5.6.13)$$

where x is the location of the maximum moment along the length of the member, L is the member length, $k = \sqrt{P/E_t I}$, P is the member axial load, and M_1 and M_2 are the smaller and larger of the beam-column end moments, respectively, obtained from the advanced analysis, as shown in Fig. 5.11.

M_1/M_2 is positive when the member is bent in double curvature, and negative when it is bent in single curvature. If the peak moment occurs in between the member ends, there is a possibility that a plastic hinge may form within the member. In such cases, the structural frame needs to be analyzed again by using two elements per member with a node inserted at the peak moment point.

Recently, an elasto-plastic large-deflection analysis based on a beam-column element that allows the formation of plastic hinges at its midlength

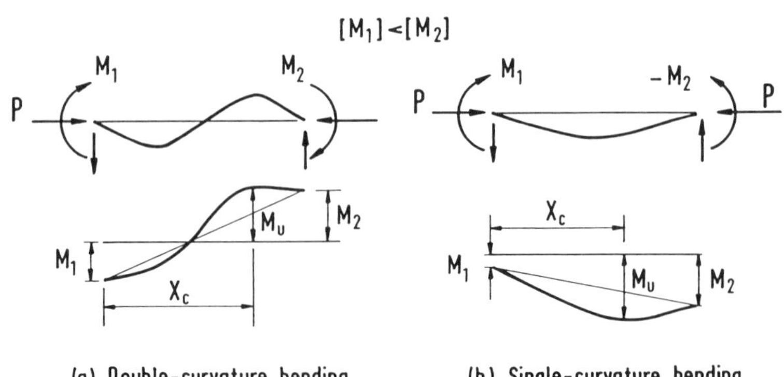

Fig. 5.11 Maximum moment location in a beam-column member subjected to end loading.

and two ends was proposed by Chen and Chan (1995). The analysis can model distributed loads and point loads acting along the element length without the need to divide the member into several elements for these loads and for capturing the plastic-hinge formation.

5.6.5 Initial Imperfections

The notional-load approach for second-order elastic-plastic hinge approximates the effects of initial residual stresses and geometric imperfections in a different manner than the refined plastic-hinge model mentioned in the previous section. The notional-load plastic-hinge approach requires the application of horizontal fictitious loads in the analysis and the use of the elastic-plastic-hinge model. The additional notional loads are applied to achieve "equivalent geometric imperfections," which approximate the effects of initial residual stresses and geometrical imperfections that are neglected in the elastic-plastic-hinge analysis. Initial column out-of-straightness effects are not considered in the analysis unless the column axial force is significant. Directions on how the designer should propose these effects in frame analysis are discussed in Chapter 7.

The notional-load plastic-hinge approach and the refined plastic-hinge approach may be compared by stating that the notional approach approximates the effects by applying additional loads, whereas the refined plastic-hinge analysis seeks to approximate the effects of both residual stresses and initial geometrical imperfections through the use of a suitable tangent modulus. This approach is, of course, a behavioral model that cannot compute exactly the effects of initial imperfections on the distribution of forces in a system. If the engineer wishes to compute this effect accurately, the initial imperfections need to be modeled explicitly in the analysis.

5.6.6 Lateral Bracing Requirements

Beam-columns containing plastic hinges are checked for adequate rotation capacity by limiting the lateral brace spacing according to the AISC LRFD plastic design provision, which is

$$L_b \leq \left(0.124 + 0.0759\,\frac{M}{M_p}\right)\frac{Er_y}{F_y} \tag{5.6.14}$$

where

E = modulus of elasticity
F_y = material design strength
L_b = maximum distance between points of restraint
M/M_p = ratio of end moments in which one end reaches full plastic capacity
r_y = radius of gyration about weak axis

The ratio of end moments is positive for double-curvature bending and negative for single-curvature. This requirement corresponds to a rotation capacity of at least 3 in a flexural member.

For a beam under single-curvature bending of A36 steel, Eq. (5.6.14) gives a limit of $38r_y$, which is similar to the values specified in several other limit state specifications. Adequate bracing should be provided in the immediate vicinity of a plastic hinge. If the plastic-hinge position cannot be braced due to some practical constraints, then the brace should be positioned at a distance of not more than half the depth of the member from the plastic-hinge location.

Most modern specifications give guidance on the minimum force and stiffness requirements for the bracing in order for the member to be considered fully braced and achieve its required strength for the effective length between the braces. The brace may be designed to withstand 2.5% of the member compression force as given by Standards Australia (1990) and BSI (1990). Clarke et al. (1993) have found that the bracing stiffness is within the range of $5P_u/L$ to $10P_u/L$ in order to achieve within 5% the ultimate strength of the column (P_u) with infinite brace stiffness. Several standards such as BSI (1990) and CSA (1989) require web stiffeners to be placed at plastic hinges, or within half the depth of the member from the hinge point. These stiffeners are to be designed as load-carrying stiffeners with the additional requirements that their proportions be within the limits for plastic design sections. The provisions of web stiffeners and lateral bracing at the plastic-hinge locations are to prevent instability by bracing the member against lateral deflections and twist.

5.6.7 Serviceability

In practice, many parameters affect the design serviceability of a steel building frame. The two common parameters are building deflections and accelerations.

Building deflections are usually measured in terms of a nondimensional parameter called drift. Interstory drift is the ratio of interstory deflection to story height. Building drift is the ratio of the lateral deflection at the top of the building to the building height. For steel building frames, commonly used values for first-order drift ranges from 1/450 to 1/600 under a 50-year wind. If the drift is calculated for a different return period, the drift limit should be adjusted appropriately (Iyengar, Baker, and Sinn, 1992). Maximum interstory drift from first-order analysis is normally limited to the range of 1/350 to 1/500. Service beam live-load deflections may be limited to $L/360$ when the beam carries brittle finishers such as plaster ceilings. Otherwise, the beam-deflection limit may be relaxed to $L/240$, where L is the span length of the beam. A maximum absolute deflection limit must be employed depending on the constructional materials and the manufacturers that supply those materials. This absolute limit should be in the range of 3/4 to 1 inch (Fisher and West, 1990).

Another parameter that is closely related to building drift is the second-order P–Δ effects. Second-order deflection occurs due to the presence of gravity loads. A steel building of typical dimensions and designed to a first-order drift of 1/500 may experience additional second-order deflections in the range of 20 to 30% of the first-order deflection. Therefore, if a second-order analysis is used directly to assess the serviceability performance of a building frame, the limit for lateral drift should be relaxed to between the range of 1/350 to 1/480. Similarly, interstory drifts can be relaxed to the range of 1/250 to 1/400.

Plastic-hinge formation under service loads should be prohibited to prevent the possibility of excessive localized damage before the full service load is reached. In Eurocode 3, it is suggested that the possibility of plastic redistribution of forces occurring at serviceability limit states should be investigated. This should be permitted only where it can be shown that the loading will not be repeated, and it should be taken into account in calculating deformations.

5.6.8 Live-Load Reduction

Subject to certain limitations, ASCE 7-88 (ASCE, 1990) suggests that members having an influence area of 400 ft^2 or more may be designed for a reduced live load determined by applying the following equation:

$$L = L_0\left(0.25 + \frac{15}{\sqrt{A_I}}\right) \tag{5.6.15}$$

where

A_I = influence area in square feet
L_0 = unreduced design live load per square foot of area supported by the member
L = reduced design live load per square foot of area supported by the member

The influence area is four times the tributary area for a column, two times the tributary area for a beam, and equal to the panel area for a two-way slab. The reduced designed live load shall not be less than 50% of the unit live load L_0 for members supporting one floor nor less than 40% of the unit live load L_0 otherwise.

Ziemian (1990) proposed a live-load-reduction procedure for nonlinear analysis. The method is based on estimating the difference in column axial forces, calculating by applying live-load-reduction factors to (1) only the beam and (2) only the columns. The difference in column axial forces is called the "compensating force." The structural analyses are then performed by (1) distributing live loads to the beams and using beam reduction factors and (2) applying the "compensating forces" in an upward direction at the top of the column. The procedure to obtain the "compensating forces" is as follows (Ziemian, 1990):

1. Based on the tributary area, estimate the axial force in each column without applying any reduction factors. For the frame example shown in Figs. 5.12a and 5.12b, the unreduced axial force in column C_3 is

$$P_{c3} = \sum_{i=1}^{9} 0.5W_{bi}L_{bi} \tag{5.6.16}$$

   where W_{bi} is the unreduced live load along beam B_i, and L_{bi} is the length of beam B_i.

2. Based on the influence area of each column, reduce the preceding axial force by Eq. (5.6.15). For column C_3, the reduced axial force is

$$P^1_{C3} = P_{C3}\left(0.25 + \frac{15}{\Sigma_{i=1}^{6} A_i}\right) > 0.4P_{C3} \tag{5.6.17}$$

   where $\Sigma_{i=1}^{6} A_i$ is the total influence area for column C_3.

3. Based on the tributary area, calculate the axial force in each column applying only the beam reduction factors. The reduced axial force for column C_3 is

$$P^2_{C3} = \sum_{i=1}^{9} 0.5W_{rBi}L_{Bi} \tag{5.6.18}$$

   where W_{rBi} is the reduced beam live load for beam B_i based on beam reduction factor, and L_{Bi} is the length of beam B_i.

4. Determine the difference in axial force calculated in steps 2 and 3. For column C_3, this force is

$$P^3_{C3} = P^1_{C3} - P^2_{C3} \tag{5.6.19}$$

5. Determine the "compensating force" to be applied at the top of each column. For column C_3, this force is

$$F_{C3} = P^3_{C3} - \sum_{i=1}^{2} F_{Ci} \tag{5.6.20}$$

   where F_{Ci} is the compensating force at column C_i.

A summary of the applied live load for the frame example is illustrated in Fig. 5.12b. It should be noted that both the beam live loads and the compen-

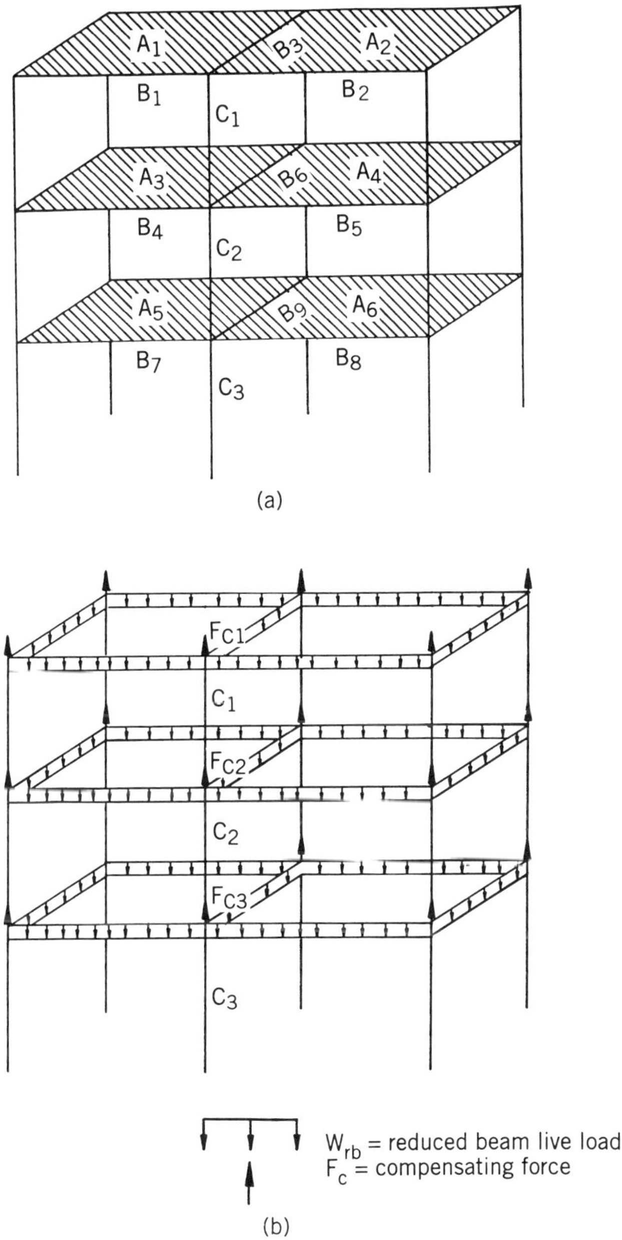

Fig. 5.12 (a) Live-load-reduction procedure. (b) Calculation of compensating forces in columns.

sating forces should be multiplied by the appropriate load factor, as shown in Table 5.1.

The preceding live-load-reduction procedure is oriented primarily toward specification member design rules and linear elastic analysis. The live-load-reduction procedure in ASCE 7-88 (ASCE, 1990) is applied on a member-by-member basis, and the amount of live-load reduction permitted for a beam member is generally lesser than an adjacent column member. In other words, the reduced live-load forces permitted by ASCE 7-88 (ASCE, 1990) do not occur in all the members of a frame at the same point in time, whereas in a nonlinear analysis, it is assumed that all the loads are applied to the structure at the same point in time. For future research, an appropriate live-load-reduction model is required based on a system analysis that considers both the effects of strength and stability.

5.6.9 Proportional/Nonproportional Loading

Eurocode 3 states that when elastic-plastic analysis is carried out, it is sufficient, in the case of building structures, to apply the loads in a series of increments, stopping when the full design load is reached, and to use the resulting internal forces and moments to check the resistances of the cross-sections and the buckling resistances of the members. It further states that when the global analysis is carried out by applying the loads in a series of increments, it may be assumed to be sufficient, in the case of building frame structures, to adopt simultaneous proportional increases of all loads.

This approach appears to be overconservative particularly for flexibly connected frames in which the connections on the windward side of a frame tend to elastically unload during wind loading. The elastic stiffness in the unloading path is much higher than the tangent stiffness of the connections in the loading path.

In many cases, the design loads may be applied proportionally from zero until the factored load is reached. Proportional loading is usually the most severe case for regular framing systems compared to the more realistic load case, which is to apply the gravity load first followed by the wind load. For proportional loading, the effects of repeated loading (i.e., alternating plasticity) need not be considered, as these effects are not as significant as in the nonproportional-load case. However, in some cases, the engineer might wish to consider nonproportional-load sequence effects. This is because of the elastic unloading that tends to occur at plastic hinges and connections on the windward side of the frame when the gravity load is applied first followed by the wind load. However, the analysis algorithm must correctly and efficiently follow the loading and unloading behavior. The alternate response of the load path can sometimes be numerically sensitive, especially in large frames, where a plastic hinge or connection may be caught in a numerical cycle of unloading and loading.

5.7 CONCLUSIONS

Various analysis and design concepts related to first-order elastic, first-order plastic, second-order inelastic, and advanced analyses have been presented. An overview of the current impetus behind the research in advanced inelastic analysis/design has been provided. An important concept of advanced analysis is that if the engineer employs an analysis method that adequately represents a limit state, then the checks of the corresponding specification rules would not be required. For complicated frameworks, a second-order inelastic analysis can provide a better estimate of structural responses than linear elastic analysis, especially when it is properly formulated to include initial imperfection effects, such as residual stresses and geometrical imperfections. However, for a second-order inelastic analysis to be used in this way in actual design practice, the capability of different analysis approaches for representing beam-column and connection performance must be clearly identified. Also, any limit states that are not properly modeled in the analysis must be well-understood such that appropriate specification checks can be performed.

It has been demonstrated that many of the behavioral phenomena and structural attributes can be considered directly in the analysis without much difficulty. The consideration of member and system in-plane strength and stability by second-order analysis is entirely achievable. However, an engineer who wishes to perform nonlinear analysis is typically faced with numerical problems such as solution strategies, convergence criteria and tolerances, load incremental size, yield-surface drift error, and so on. These problems often have subtle implications and often are the subject for misinterpretation, especially for engineers who do not often use the nonlinear program.

Advanced analysis techniques require substantial numerical expertise by the user. Today, there are robust analysis algorithms and user-friendly interfaces to assist the engineer in making decisions so that problems associated with nonlinear analysis techniques can be minimized. However, the engineer still needs to know the analysis process, but the procedural aspects of the analysis can be handled by the software.

The remaining chapters present detailed formulation related to advanced analyses that are feasible for practical implementation. These chapters represent a compendium of research documentation that provides engineers and researchers the necessary background and tools to use advanced analysis techniques for frame design. The chapters are arranged in such a way as to provide a smooth transition from the current state-of-the-art knowledge on computer-based frame design using second-order elastic-plastic analysis to a new methodology based on second-order refined plastic-hinge techniques.

Chapter 6 contains essential background information, basic theories, and numerical formulations for two-dimensional elastic-plastic-hinge analysis. Chapter 7 presents a practical method for advanced analysis using the notional-load concept. The implementation of this method in the context of

AISC LRFD is discussed. Chapter 8 focuses on the development of a refined plastic-hinge analysis and its verification for use in design practice. Chapter 9 demonstrates how the advanced analysis can be extended to include semi-rigid connection behavior in frame design. Advanced analysis holds many answers to real behavior of steel structures and, thus, these topics should be left for exploration by engineers seeking to improve their knowledge in the analysis and design of steel frame structures. Some educational programs with their user's manuals are given in Chapter11. The programs are helpful to those who wish to perform advanced planar analysis of frames using the techniques developed in Chapters 7–9.

REFERENCES

Abdel-Ghaffar, M., D. W. White, and W. F. Chen (1991). "Simplified Second-Order Inelastic Analysis for Steel Frame Design," *Special Volume of Session on Approximate Methods and Verification Procedures of Structural Analysis and Design, Proceedings at Structures Congress 91,* New York: ASCE, pp. 47–52.

Ackroyd, M. H. (1985). *Design of Flexibly Connected Steel Frames,* Final Research Report to the American Iron and Steel Institute for Project No. 333, Troy, NY: Rensselaer Polytechnic Institute.

Ackroyd, M. H. (1991). "Design Needs from Inelastic Analysis," in *Annual Technical Session Proceedings,* Bethlehem, PA: Structural Stability Research Council, Lehigh University, pp. 41–51.

AISC (1978). *Specification for the Design, Fabrication and Erection of Structural Steel for Buildings,* Chicago: American Institute of Steel Construction.

AISC (1986). *Load and Resistance Factor Design Specifications for Structural Steel Buildings,* Chicago: American Institute of Steel Construction.

AISC (1989). *Allowable Stress Design Specifications for Structural Steel Buildings,* Chicago: American Institute of Steel Construction.

AISC (1993). *Load and Resistance Factor Design Specifications for Structural Steel Buildings,* 2nd Ed., Chicago: American Institute of Steel Construction.

Al-Mashary, F. and W. F. Chen (1991). "Simplified Second-Order Inelastic Analysis for Steel Frames," *Journal of the Institution of Structural Engineers,* Vol. 59, No. 23, pp. 395–399.

Alvarez, R. J., and C. Birnstiel (1957). *Elasto-Plastic Analysis of Plane Rigid Frames,* New York: School of Engineering and Science, Department of Civil Engineering, New York University.

ASCE-WRC (1971). *Plastic Design in Steel—A Guide and Commentary,* 2nd Ed., ASCE Manuals and Reports on Engineering Practice No. 41, New York: Joint Committee of the Welding Research Council and ASCE.

ASCE (1990). *Minimum Design Loads for Buildings and Other Structures, ASCE Standard,* Revision of ANSI A58.1-1982, New York: American Society of Civil Engineers.

Attalla, M. R., G. G. Deierlein, and W. McGuire (1994). "Spread of Plasticity: Quasi-Plastic-Hinge Approach," *Journal of Structural Engineering,* ASCE, Vol. 120, No. 8, pp. 2451–2473.

BSI (1990). *BS5950: Structural Use of Steelwork in Building. Part 1: Code of Practice for Design in Simple and Continuous Construction: Hot Rolled Section,* London: British Standards Institution.

Chen, W. F., and T. Atsuta (1976). *Theory of Beam-Columns, Vol. 1: In-Plane Behavior and Design,* New York: McGraw-Hill.

Chen, W. F., and T. Atsuta (1977). *Theory of Beam-Columns: Vol. 2: Space Behavior and Design,* New York: McGraw-Hill.

Chen, W. F., and S. L. Chan (1995). "Second Order Inelastic Analysis of Steel Frames using Element with Mid-Span and End Springs," *Journal of Structural Engineering,* ASCE, Vol. 121, No. 3, pp. 530–541.

Chen, W. F., and E. M. Lui (1991). *Stability Design of Steel Frames,* Boca Raton, FL: CRC Press.

Chen, W.F., and I. S. Sohal (1995). *Plastic Design and Second-Order Analysis of Steel Frames,* New York: Springer-Verlag.

Chen, W. F., and S. Toma, eds. (1994). *Advanced Analysis of Steel Frames: Theory, Software, and Applications,* Boca Raton, FL: CRC Press.

Clarke, M. J. (1994) "Plastic Zone Analysis of Frames," in *Advanced Analysis of Steel Frames: Theory, Software, and Applications,* W. F. Chen and S. Toma, eds., Boca Raton, FL: CRC Press, pp. 259–319.

Clarke, M. J., R. Q. Bridge, G. J. Hancock, and N. S. Trahair (1993). "Benchmarking and Verification of Second-Order Elastic and Inelastic Frame Analysis Programs," in *Plastic Hinge Based Methods for Advanced Analysis and Design of Steel Frames—An Assessment of the State-of-the-Art,* Bethlehem, PA: Structural Stability Research Council, Lehigh University, pp. 245–274.

CSA (1989). *Limit States Design of Steel Structures,* CAN/CSA-S16.1-M89, Canadian Standards Association, Ontario, Canada.

Dawe, J. L., and G. L. Kulak (1986). "Local Buckling Behavior of Beam-Columns," *Journal of the Structural Division,* ASCE, Vol. 112, No. 11, pp. 2447–2461.

Deierlein, G. G., Y. Zhao, and W. McGuire (1991). "A Discrete Model for Gradual Plastification and Nonlinear Connection Response in Three-Dimensional Framed Structures," in *Annual Technical Session Proceedings,* Bethlehem, PA: Structural Stability Research Council, Lehigh University, pp. 423–432.

Dhillon, B. S., and S. Abdel-Majid (1990). "Interactive Analysis and Design of Flexibly Connected Frames," *Computer and Structures,* Vol. 36, No. 2, pp. 189–202.

Duan, L., and W. F. Chen (1990). "A Yield Surface Equation for Doubly Symmetrical Sections," *Engineering Structures,* Vol. 12, April, pp. 114–119.

ECCS (1984). *Ultimate Limit State Calculation of Sway Frames with Rigid Joints,* Technical Working Group 8.2, Systems, Publication No. 33, Brussels: European Convention for Constructional Steelwork.

El-Zanaty, M., D. Murray, and R. Bjorhovde (1980). *Inelastic Behavior of Multistory Steel Frames,* Structural Engineering Report No. 83, University of Alberta, Edmonton, Alberta.

Eurocode 3 (1992). "Design of Steel Structures: Part I—General Rules and Rules for Buildings," Brussels: European Committee for Standardization.

Fisher, J. M., and M. A. West (1990). *Serviceability Design Considerations for Low-Rise Buildings,* Chicago: American Institute of Steel Construction.

Frye, M. J., and G. A. Morris (1975). "Analysis of Flexibly Connected Steel Frames," *Canadian Journal of Civil Engineering,* Vol. 2, pp. 280–291.

Galambos, T. V. (1988). *Guide to Stability Design Criteria for Metal Structures,* 4th Ed., New York: Wiley.

Galambos, T. V. (1990). "Developments in Modern Steel Design Standards," *Journal of Constructional Steel Research,* Vol. 17, Nos. 1 & 2, pp. 141–162.

Goto, Y., and W. F. Chen (1987). "On the Computer-Based Design Analysis for the Flexibly Jointed Frames," *Journal of Constructional Steel Research,* Special Issue on Joint Flexibility in Steel Frames, W. F. Chen, ed., Vol. 8, pp. 203–231.

Heish, S. H. (1990). *Analysis of Three-Dimensional Steel Frames with Semi-Rigid Connections,* Structural Engineering Report 90-1, Ithaca, NY: School of Civil and Environmental Engineering, Cornell University.

Iyengar, S. H., W. F. Baker, and R. Sinn (1992). "Multi-Story Buildings," in *Constructional Steel Design, An International Guide,* P. J. Dowling et al., eds., London: Elsevier, pp. 645–670.

Kanchanalai, T. (1977). *The Design and Behavior of Beam-Columns in Unbraced Steel Frames,* AISI Project No. 189, Report No. 2, Austin: Civil Engineering/Structures Research Laboratory, University of Texas.

King, W. S., and W. F. Chen (1994). "Practical Second-Order Inelastic Analysis of Semi-Rigid Frames," *Journal of Structural Engineering,* ASCE, Vol. 120, No. 7, pp. 2156–2175.

King, W. S., D. W. White, and W. F. Chen (1992). "On Second-Order Inelastic Methods for Steel Frame Design," *Journal of Structural Engineering,* ASCE, Vol. 118, No. 2, pp. 408–428.

Liew, J. Y. R. (1992). "Advanced Analysis for Frame Design," Ph.D. dissertation, School of Civil Engineering, Purdue University, West Lafayette, IN.

Liew, J. Y. R., and W. F. Chen (1994). "Trends Toward Advanced Analysis," in *Advanced Analysis of Steel Frames: Theory, Software, and Applications,* W. F. Chen and S. Toma, eds., Boca Raton, FL: CRC Press, pp. 1–45.

Liew, J. Y. R., D. W. White, and W. F. Chen (1991). "Beam-Column Design in Steel Frameworks—Insights on Current Methods and Trends," *Journal of Constructional Steel Research,* Vol. 18, No. 4, pp. 259–308.

Liew, J. Y. R., D. W. White, and W. F. Chen (1992). "Beam-Columns," in *Constructional Steel Design, An International Guide,* P. J. Dowling et al., eds., London: Elsevier, pp. 105–132.

Liew, J. Y. R., D. W. White, and W. F. Chen (1993a). "Second-Order Refined Plastic Hinge Analysis for Frame Design: Part 1," *Journal of Structural Engineering,* ASCE, Vol. 119, No. 11, pp. 3196–3216.

Liew, J. Y. R., D. W. White, and W. F. Chen (1993b). "Second-Order Refined Plastic Hinge Analysis for Frame Design: Part 2," *Journal of Structural Engineering,* ASCE, Vol. 119, No. 11, pp. 3217–3237.

Liew, J. Y. R., D. W. White, and W. F. Chen (1994). "Notional Load Plastic Hinge Method for Frame Design," *Journal of Structural Engineering,* ASCE, Vol. 120, No. 5, pp. 1434–1454.

Lui, E. M., and W. F. Chen (1986). "Analysis and Behavior of Flexibly-Jointed Frames," *Engineering Structures,* Vol. 8, No. 2, pp. 107–118.

Orbison, J. G. (1982). *Nonlinear Static Analysis of Three-Dimensional Steel Frames,* Report No. 82-6, Ithaca, NY: Department of Structural Engineering, Cornell University.

Standards Australia (1990). *Steel Structures,* AS4100, Sydney.

Vogel, U. (1985). "Calibrating Frames," Stahlbau, Vol. 10, pp. 295–301.

Wang, Y. C. (1988). "Ultimate Strength Analysis of 3-D Beam Columns and Column Subassemblages with Flexible Connections," Ph.D. thesis, University of Sheffield, Sheffield, England.

White, D. W. (1985). "Material and Geometric Nonlinear Analysis of Local Planar Behavior in Steel Frames Using Iterative Computer Graphics," M.S. thesis, Cornell University, Ithaca, NY.

White, D. W. (1988). "Analysis of Monotonic and Cyclic Stability of Steel Frame Subassemblages," Ph.D. dissertation, Cornell University, Ithaca, NY.

White, D. W. (1993). "Plastic Hinge Methods for Advanced Analysis of Steel Frames," *Journal of Constructional Steel Research,* Vol. 24, No. 2, pp. 121–152.

White, D. W., and W. F. Chen, eds. (1993). *Plastic Hinge Based Methods for Advanced Analysis and Design of Steel Frames—An Assessment of the State-of-the-Art,* Bethlehem, PA: Structural Stability Research Council, Lehigh University.

White, D. W., and J. F. Hajjar (1991). "Application of Second-Order Elastic Analysis in LRFD: Research to Practice," *Engineering Journal,* AISC, Vol. 28, No. 4, pp. 133–148.

White, D.W., J. Y. R. Liew, and W. F. Chen (1991). *Second-Order Inelastic Analysis for Frame Design: A Report to SSRC Task Group 29 on Recent Research and the Perceived State-of-the-Art,* Structural Engineering Report No. CE-STR-91-12, West Lafayette, IN: Purdue University.

White, D.W., J. Y. R. Liew, and W. F. Chen (1993). "Toward Advanced Analysis in LRFD," in *Plastic Hinge Based Methods for Advanced Analysis and Design of Steel Frames—An Assessment of the State-of-the-Art,* Bethlehem, PA: Structural Stability Research Council, Lehigh University, pp. 95–173.

Ziemian, R. D. (1990). "Advanced Methods of Inelastic Analysis in the Limit States Design of Steel Structures," Ph.D. dissertation, School of Civil and Environmental Engineering, Cornell University, Ithaca, NY.

Ziemian, R. D., W. McGuire, and G. G. Deierlien (1992a). "Inelastic Limit States Design: Part I—Plannar Frame Studies," *Journal of Structural Engineering,* ASCE, Vol. 118, No. 9, pp. 2532–2549.

Ziemian, R. D., W. McGuire, and G. G. Deierlien (1992b). "Inelastic Limit States Design: Part II—Three-Dimensional Frame Study," *Journal of Structural Engineering,* ASCE, Vol. 118, No. 9, pp. 2550–2568.

6

SECOND-ORDER ELASTIC-PLASTIC-HINGE ANALYSIS

6.1 INTRODUCTION

Recent research has focused on the development of second-order elastic-plastic-hinge methods based on large-displacement small-strain theory and the assumption that inelasticity is concentrated at "zero-length" plastic hinges (Orbison, 1982; Powell and Chen, 1986; Chen and Lui, 1991; Ziemian, McGuire, and Deierlein, 1992a, 1992b; White, Liew, and Chen, 1993; Liew and Chen, 1994a, 1994b; Chen and Sohal, 1995). The research has demonstrated that the elastic-plastic-hinge analysis is capable of predicting the two-dimensional strength and load-displacement characteristics of a large-scale framework, and it is the preferred method compared to the more exact plastic-zone analysis approach (White et al., 1991, 1993). Although the elastic-plastic-hinge analysis is reasonably accurate for predicting the global behavior of a large framework, it is otherwise for local effects such as member stability and strength (see the discussion in Section 6.6).

The main advantages of the plastic-hinge approach over the plastic-zone method are as follows:

1. The model is capable of achieving the requisite accuracy using only one or few elements per member. This attribute is necessary for analysis of moderate and large-size structural systems.
2. Elastic-plastic-hinge formulation does not require numerical integration for calculation of element stiffness matrices and internal force vectors. Therefore, it gains significant economy over the plastic-zone method in terms of computational efficiency.

2. The element formulation can incorporate chord-rotational displacement (P–Δ) and member curvature (P–δ) effects quite efficiently. Geometric initial imperfections and force and displacement boundary conditions can be specified explicitly.
3. The basis of plastic-hinge formulation is cross-section strength. Cross-section strength curves developed by different design specifications may be adopted to approximate the exact strength of various cross-section types. Resistance factors (or strength-reduction factors) may be applied to the cross-section strength curve for the calculation of the design strength. This approach enables member capacity checks to be integrated directly in the analysis of the overall system.
4. The potential for extending to three-dimensional analysis is more straightforward compared to plastic-zone analysis in which three-dimensional shell-type elements are generally required. The discretization of a structural system into many small "fibers" in the plastic-zone analysis places a higher demand on hardware performance and it is comparatively more costly to implement.

Plastic-hinge analyses can be classified into two categories; first-order and second-order, depending on whether geometric second-order effects are accounted for in the analysis (see Section 5.2). For the first-order elastic-plastic-hinge analysis, equilibrium is formulated based on undeformed structural geometry. The analysis essentially predicts the same maximum load as the rigid-plastic analysis approaches. The advantage of the hinge-by-hinge analysis is that in addition to finding the collapse load of the frame, the analysis gives information about the distribution and redistribution of forces prior to reaching the collapse load. The hinge-by-hinge analysis determines the order of plastic-hinge formation, the load factor associated with each plastic-hinge formation, and member forces in the frame between each hinge formation. Thus, the state of the frame can be defined at any load factor rather than only at the state of collapse. This allows a more accurate determination of member forces at the design load level (see Chapter 5, Section 5.3.2).

If equilibrium is formulated based on deformed structural geometry, the analysis is called second-order. The need for a second-order analysis of steel frame is increasing in view of the AISC LRFD (AISC, 1986) specifications, which give explicit permission for the engineer to compute load effects from a direct second-order analysis. Second-order elastic analysis is also the "preferred" method in the AISC LRFD (AISC, 1993), Canadian limit states design specifications (CSA, 1989), and in many other limit states code provisions. Several approximated methods, which are based on the effective-length and moment-amplification concepts, have been proposed (Stevens, 1967; Yura, 1971; Cheong, 1977; LeMessurier, 1977). However, these methods do not always give satisfactory results because they are derived based on simplified assumptions and are applicable to regular frameworks with rigid connections and small displacements. Furthermore, it is rather cumbersome and tedious to calculate the moment-amplification factors applied to a first-order analysis. In view of this, it is more convenient and straightforward to use a computer-based method to calculate member forces directly as long as the method can be implemented easily.

Second-order elastic-plastic-hinge analysis considers inelasticity and stability effects at a certain level of approximation. The method often employs one element per member, and it is sufficiently accurate to capture the global behavior of the structures. At present, only Eurocode 3 (1992) and Standards Australia (1990) give some guidelines for the use of second-order elastic-plastic-hinge analysis for frame design. However, these design provisions do require engineers to perform individual member-capacity checks even if second-order analysis is employed.

This chapter presents the fundamental theory of concentrated plasticity models and investigates several implementation details of the conventional elastic-plastic-hinge approach. The incremental tangent stiffness equations for beam-columns and truss elements are derived. They are then modified to account for the formation of plastic hinges at the element ends. By comparing the plastic-hinge analysis results with the established benchmark solutions based on plastic-zone analysis, the qualities and limitations of the analysis method can be quantified.

The present scope of work is limited to two-dimensional steel frames only, and member cross-sections are assumed to be plastic sections. Information regarding spatial behavior of beam-columns and frames, including topics such as local and lateral-torsional buckling, will be considered in future endeavors. Also, the present work is limited to frames subjected to static loads only. Frames under dynamic and cyclic loading are not considered in the study.

6.2 MODELING ELASTIC BEAM-COLUMN ELEMENTS

This section presents a stiffness-matrix approach relating the element end forces and end displacements of a beam-column element. Attention is focused on elastic formulation based on large-displacement small-strain theory. The extension of the theory to include plastic-hinge formation is given in Section 6.4.

6.2.1 Assumptions

The general assumptions used in the modeling of a beam-column element are as follows:

1. All elements are initially straight and prismatic. Plane cross-sections remain plane after deformation.

2. All member cross-sections are fully compact such that local buckling effects are insignificant.
3. All members are sufficiently braced such that out-of-plane flexural or lateral-torsional buckling does not influence the member response prior to failure.
4. Cross-section distortions and strains are small, and thus can be ignored.
5. The member shear forces are small enough such that the effects of shear deformation can be neglected.
6. Axial shortening due to member curvature bending (bowling effects) is neglected.

The assumptions that the member is prismatic and member distortions are small are reasonable for ordinary steel frame structures. Although the steel frame may undergo large rigid-body displacements at collapse, the distortion of each member with respect to its chord length in the displaced configuration remains small. This assumption is normally valid because steel members with compact cross-sections generally exhibit high bending rigidity.

Large deformation theory is useful to model the full postcollapse behavior of members in the structure. However, for many types of steel structures, large strains usually do not occur until the members are loaded far into the postcollapse region.

Member curvature effects (bowling effects) are not considered in the present formulation because many practical frame members usually have small to intermediate slenderness ratios for which the axial shortening is predominated by inelastic axial deformation. For very slender beam-column members, the bowing effects may need to be considered in the stiffness formulation (Goto et al., 1991).

6.2.2 Beam-Column Stability Functions

Figure 6.1a shows a beam-column member subjected to end moments M_A and M_B (negative for clockwise moment) and acted on by axial force P. By using the free-body diagram of a segment of beam-column of length x, as shown in Fig. 6.1b, the following equilibrium equation can be derived (Chen and Lui, 1986):

$$\frac{d^2y}{dx^2} + \rho^2 y = -\frac{M_A + M_B}{LEI}x + \frac{M_A}{EI} \tag{6.2.1}$$

where

E = modulus of elasticity
I = moment of inertia
L = undeformed length of element
P = axial force, taken as positive in tension
$\rho = \sqrt{P/EI}$

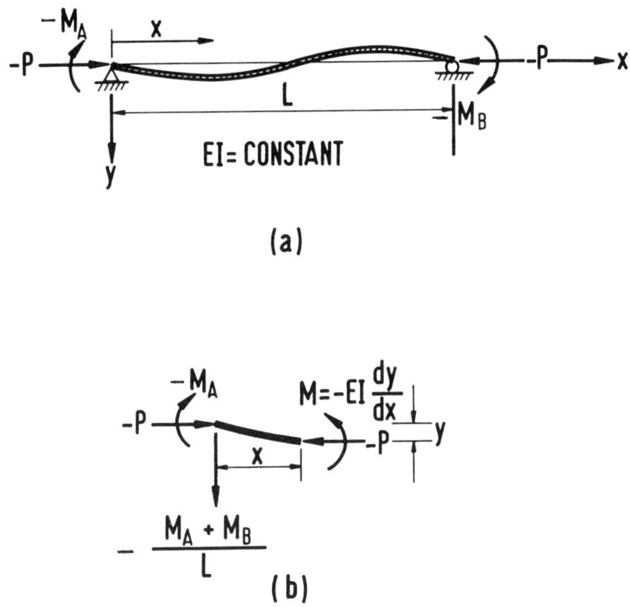

Fig. 6.1 Beam-column subject to end moments and axial force.

The general solution of this ordinary equation is

$$y = A\sin(\rho x) + B\cos(\rho x) + \frac{M_A + M_B}{LEI\rho^2}x - \frac{M_A}{EI\rho^2} \tag{6.2.2}$$

Constants A and B can be evaluated by using the boundary conditions

$$y(0) = 0, \qquad y(L) = 0 \tag{6.2.3}$$

The results may be written as

$$A = -\frac{1}{EI\rho^2\sin(\rho L)}[M_A\cos(\rho L) + M_B] \tag{6.2.4}$$

$$B = \frac{M_A}{EI\rho^2} \tag{6.2.5}$$

By substituting A and B into Eq. (6.2.2) and rearranging terms:

$$y = \frac{1}{EI\rho^2}\left[\frac{\cos(\rho L)}{\sin(\rho L)}\sin(\rho x) - \cos(\rho x) - \frac{x}{L} + 1\right]M_A + \frac{1}{EI\rho^2}\left[\frac{1}{\sin(\rho L)}\sin(\rho x) - \frac{x}{L}\right]M_B \quad (6.2.6)$$

from which the slope at any distance x can be derived:

$$\frac{dy}{dx} = \frac{1}{EI\rho}\left[\frac{\cos(\rho L)}{\sin(\rho L)}\cos(\rho x) + \sin(\rho x) - \frac{1}{\rho L}\right]M_A + \frac{1}{EI\rho}\left[\frac{1}{\sin(\rho L)}\cos(\rho x) - \frac{1}{\rho L}\right]M_B \quad (6.2.7)$$

The end rotation at end A, θ_A, is obtained by letting $x = 0$ in Eq. (6.2.7):

$$\theta_A = \left.\frac{dy}{dx}\right|_{x=0} = -\frac{L}{EI}\left[\frac{\sin(\rho L) - \rho L\cos(\rho L)}{(\rho L)^2\sin(\rho L)}\right]M_A - \frac{L}{EI}\left[\frac{\sin(\rho L) - \rho L}{(\rho L)^2\sin(\rho L)}\right]M_B \quad (6.2.8)$$

and the end rotation at end B, θ_B, can be obtained from Eq. (6.27) for $x = L$:

$$\theta_B = \left.\frac{dy}{dx}\right|_{x=L} = -\frac{L}{EI}\left[\frac{\sin(\rho L) - \rho L}{(\rho L)^2\sin(\rho L)}\right]M_A - \frac{L}{EI}\left[\frac{\sin(\rho L) - \rho L\cos(\rho L)}{(\rho L)^2\sin(\rho L)}\right]M_B \quad (6.2.9)$$

The axial force-displacement relationship ignoring the effect of curvature shortening may be expressed as

$$P = \frac{EA}{L}e \quad (6.2.10)$$

where

A = effective cross-section area
e = axial displacement of member
E = modulus of elasticity
L = member length

Equations (6.2.8)–(6.2.10) can be expressed in matrix form as

$$\begin{Bmatrix} M_A \\ M_B \\ P \end{Bmatrix} = \frac{EI}{L}\begin{bmatrix} S_1 & S_2 & 0 \\ S_2 & S_1 & 0 \\ 0 & 0 & A/I \end{bmatrix}\begin{Bmatrix} \theta_A \\ \theta_B \\ e \end{Bmatrix} \tag{6.2.11}$$

S_1 and S_2 are called *stability functions,* which may be written as follows:

$$S_1 = \begin{cases} \dfrac{\rho L\sin(\rho L) - (\rho L)^2 \cos \rho L}{2 - 2\cos(\rho L) - \rho L \sin(\rho L)}, & \text{for } P < 0 \\ \dfrac{(\rho L)^2 \cosh(\rho L) - \rho L \sinh(\rho L)}{2 - 2\cosh(\rho L) + \rho L \sinh(\rho L)} & \text{for } P > 0 \end{cases} \tag{6.2.12}$$

$$S_2 = \begin{cases} \dfrac{(\rho L)^2 - \rho L \sin(\rho L)}{2 - 2\cos(\rho L) - \rho L \sin(\rho L)} & \text{for } P < 0 \\ \dfrac{\rho L \sinh(\rho L) - (\rho L)^2}{2 - 2\cosh(\rho L) + \rho L \sinh(\rho L)} & \text{for } P > 0 \end{cases} \tag{6.2.13}$$

where parameter ρL is defined as

$$\rho L = \pi\sqrt{\frac{|P|}{P_e}} = \pi\sqrt{\frac{|P|}{\pi^2 EI/L^2}} = L\sqrt{\frac{|P|}{EI}}$$

Stability functions S_1 and S_2 account for the effect of the axial force on the bending stiffness of the member. However, Eqs. (6.2.12) and (6.2.13) are indeterminate when the axial force is zero. To circumvent this problem, the following simplified equations proposed by Lui (1985) are used to approximate the stability function when the axial force in the member falls within the range of $-1.0 \le \xi \le 1.0$ (Chen and Lui, 1986):

$$S_1 = 4 + \frac{2\pi^2\xi}{15} - \frac{(0.01\xi + 0.543)\xi^2}{4 + \xi} - \frac{(0.004\xi + 0.285)\xi^2}{8.183 + \xi} \tag{6.2.14}$$

$$S_2 = 2 - \frac{\pi^2\xi}{30} + \frac{(0.01\xi + 0.543)\xi^2}{4 + \xi} - \frac{(0.004\xi + 0.285)\xi^2}{8.183 + \xi} \tag{6.2.15}$$

where $\xi = \rho^2/\pi^2 = P/(\pi^2 EI/L)$. Equations (6.2.14) and (6.2.15) are applicable for members in tension and compression. When axial force P equals zero, S_1 and S_2 reduce to the values of 4 and 2, respectively. The comparison of stability functions is given in Table 6.1. Equations (6.2.14) and (6.2.15) give excellent predictions of S_1 and S_2 compared to the "exact" solutions obtained from Eqs. (6.2.12) and (6.2.13) when ξ falls within the range of $-1.0 \le \xi \le 1.0$. When ξ falls outside this range, a power-series expansion of the exact solution, Eqs. (6.2.12) and (6.2.13), proposed by Goto and Chen (1987), may be used.

TABLE 6.1 Comparison of Stability Functions

	Stability Function S_1		Stability Function S_2	
$\rho = L\sqrt{P/EI}$	Exact Eq. (6.2.12)	Approximate Eq. (6.2.14)	Exact Eq. (6.2.13)	Approximate Eq. (6.2.15)
−1.00	2.467	2.467	2.467	2.468
−0.80	2.816	2.816	2.346	2.346
−0.60	3.140	3.140	2.241	2.241
−0.40	3.444	3.444	2.150	2.150
−0.20	3.730	3.730	2.070	2.070
0.00	4.000	4.000	2.000	2.000
0.20	4.257	4.257	1.938	1.938
0.40	4.501	4.501	1.883	1.883
0.60	4.735	4.735	1.834	1.834
0.80	4.959	4.959	1.789	1.789
1.00	5.175	5.174	1.749	1.750

6.2.3 Tangent Stiffness Relationship

It is useful to express the element force-displacement equations in an increment form for nonlinear analysis. By denoting $\dot{M}_A$, $\dot{M}_B$, $\dot{\theta}_A$, and $\dot{\theta}_B$ as the incremental end moments and joint rotations at element ends A and B, and $\dot{P}$ and $\dot{e}$ as the incremental axial force and displacement in the longitudinal direction of the element, an incremental form of Eq. (6.2.11) may be written as follows (Chen and Lui, 1991):

$$\begin{Bmatrix} \dot{M}_A \\ \dot{M}_B \\ \dot{P} \end{Bmatrix} = \frac{EI}{L} \begin{bmatrix} S_1 & S_2 & 0 \\ S_2 & S_1 & 0 \\ 0 & 0 & A/I \end{bmatrix} \begin{Bmatrix} \dot{\theta}_A \\ \dot{\theta}_B \\ \dot{e} \end{Bmatrix} \tag{6.2.16}$$

The preceding equation ignores the incremental axial force in stability functions S_1 and S_2. The full expansion of the element tangent stiffness equation involves higher-order functions with respect to the axial-force term.

Equation (6.2.16) may be expressed symbolically as

$$\mathbf{f}_c = \mathbf{k}_c \dot{\mathbf{d}}_c \tag{6.2.17}$$

where $\mathbf{f}_c$ and $\dot{\mathbf{d}}_c$ are the incremental force and displacement vectors, respectively, and $\mathbf{k}_c$ is the element basic tangent stiffness matrix. For a plane frame member, three additional degrees of freedom are required to describe the total displacements of the member. If d_{g1}, d_{g2}, . . . , and d_{g6} are denoted as the global translational and rotational degrees of freedom of a frame member (see Fig. 6.2), it can be shown that the local displacements are related to the global displacements by the following:

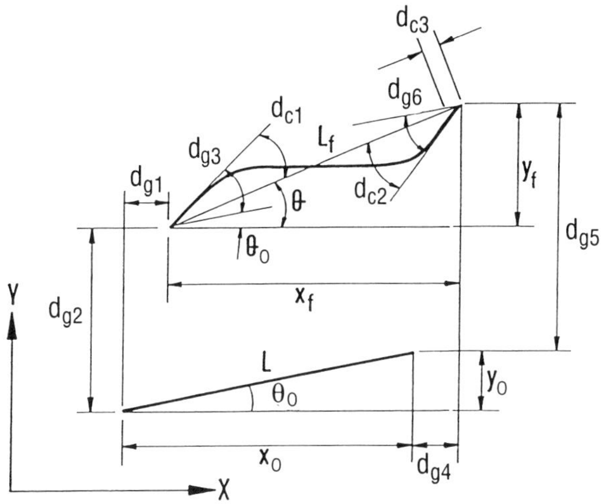

Fig. 6.2 Global and local displacements of a beam-column element.

$$d_{c1} = \theta_A = \theta_0 + d_{g3} - \tan^{-1}\frac{y_0 + d_{g5} - d_{g2}}{x_0 + d_{g4} - d_{g1}} \tag{6.2.18a}$$

$$d_{c2} = \theta_B = \theta_0 + d_{g6} - \tan^{-1}\frac{y_0 + d_{g5} - d_{g2}}{x_0 + d_{g4} - d_{g1}} \tag{6.2.18b}$$

$$d_{c3} = L_f - L = \sqrt{(x_0 + d_{g4} - d_{g1})^2 + (y_0 + d_{g5} - d_{g2})^2} - L \tag{6.2.18c}$$

Alternatively, d_{c3} may be written as

$$d_{c3} = \frac{(2x_0 + d_{g4} - d_{g1})(d_{g4} - d_{g1}) + (2y_0 + d_{g5} - d_{g2})(d_{g5} - d_{g2})}{L_f + L} \tag{6.2.18d}$$

where L and L_f are the undeformed and deformed chord lengths, respectively.

The expression for d_{c3} in Eq. (6.2.18d) is more accurate than the value calculated from Eq. (6.2.18c) based on $L_f - L$. This is because Eq. (6.2.18d) avoids finding the small difference between large member lengths (Cook, Malkus, and Plesha, 1989). This equation is obtained by writing $d_{c3} = (L_f^2 - L^2)/(L_f + L)$ and then solving for d_{c3}. In the denominator, $L_f + L \approx 2L$ may be assumed, because small-displacement theory is presumed for the corotational chord element (Belytschko and Hsieh, 1973).

Upon differentiation of Eqs. (6.2.18a), (6.2.18b), and (6.2.18d) with respect to each member end-displacement variable d_{gi}, where $i = 1, 2, \ldots, 6$, the incremental kinematics relationship relating the two sets of displacement vectors may be written in matrix form as

$$\begin{pmatrix} \dot{\theta}_A \\ \dot{\theta}_B \\ \dot{e} \end{pmatrix} = \begin{bmatrix} -(\sin\theta)/L & (\cos\theta)/L & 1 & (\sin\theta)/L & -(\cos\theta)/L & 0 \\ -(\sin\theta)/L & (\cos\theta)/L & 0 & (\sin\theta)/L & -(\cos\theta)/L & 1 \\ -\cos\theta & -\sin\theta & 0 & \cos\theta & \sin\theta & 0 \end{bmatrix} \begin{Bmatrix} \dot{d}_{g1} \\ \dot{d}_{g2} \\ \dot{d}_{g3} \\ \dot{d}_{g4} \\ \dot{d}_{g5} \\ \dot{d}_{g6} \end{Bmatrix} \tag{6.2.19}$$

where θ is the angle of inclination of the chord of the deformed member. Symbolically, the kinematics relationship of Eq. (6.2.19) may be written as

$$\dot{\mathbf{d}}_c = \mathbf{T}_{cg}\dot{\mathbf{d}}_g \tag{6.2.20}$$

Based on the principle of equilibrium, the forces in the two systems as shown in Fig. 6.3 are related by

$$\begin{Bmatrix} f_{g1} \\ f_{g2} \\ f_{g3} \\ f_{g4} \\ f_{g5} \\ f_{g6} \end{Bmatrix} = \begin{bmatrix} -(\sin\theta)/L & -(\sin\theta)/L & -\cos\theta \\ (\cos\theta)/L & (\cos\theta)/L & -\sin\theta \\ 1 & 0 & 0 \\ (\sin\theta)/L & (\sin\theta)/L & \cos\theta \\ -(\cos\theta)/L & -(\cos\theta)/L & \sin\theta \\ 0 & 1 & 0 \end{bmatrix} \begin{pmatrix} M_A \\ M_B \\ P \end{pmatrix} \tag{6.2.21}$$

Symbolically,

$$\mathbf{f}_g = \mathbf{T}_{cg}^T\mathbf{f}_c \tag{6.2.22}$$

Taking derivatives on both sides of Eq. (6.2.22) gives

$$\dot{\mathbf{f}}_g = \mathbf{T}_{cg}^T\dot{\mathbf{f}}_c + \dot{\mathbf{T}}_{cb}^T\mathbf{f}_c \tag{6.2.23}$$

In view of Eqs. (6.2.17) and (6.2.20), Eq. (6.2.23) may be further written as

$$\dot{\mathbf{f}}_g = \mathbf{T}_{cg}^T\mathbf{K}_c\mathbf{T}_{cg}\dot{\mathbf{d}}_g + \dot{\mathbf{T}}_{cg}^T\mathbf{f}_c \tag{6.2.24}$$

The transformation matrix $\dot{\mathbf{T}}_{cg}^T$ can be evaluated by taking the derivative of $\dot{\mathbf{T}}_{cg}$ with respect to each global degree of freedom, d_{gk}, as

$$\dot{\mathbf{T}}_{cg}^T = \left[\frac{\partial \mathbf{T}_{cg}^T}{\partial \mathbf{d}_{gk}}\right]\dot{\mathbf{d}}_{gk}, \qquad k = 1, 2, \ldots, 6 \tag{6.2.25}$$

The preceding equation may be written, in view of Eq. (6.2.20), as

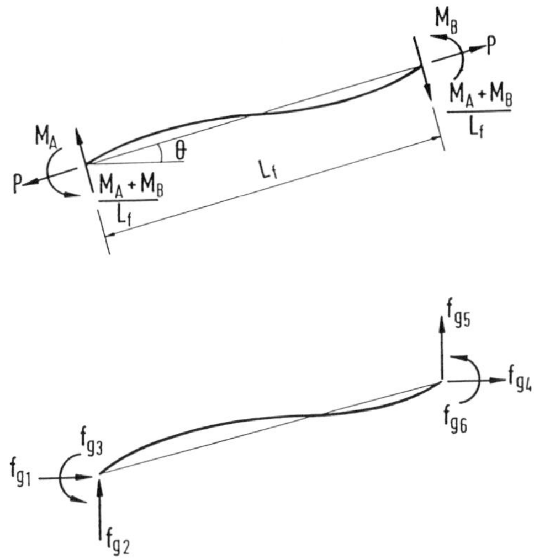

Fig. 6.3 Equivalent force system.

$$\dot{\mathbf{T}}_{cg}^{T} = \left[\frac{\partial^2 \mathbf{d}_{ci}}{\partial \mathbf{d}_{gj}\partial \mathbf{d}_{gk}}\right]^T \dot{\mathbf{d}}_{gk}, \qquad i = 1, 2, 3; j = 1, 2, \ldots, 6; k = 1, 2, \ldots, 6 \tag{6.2.26}$$

or

$$\dot{\mathbf{T}}_{cg}^{T} = \left[\frac{\partial^2 d_{c1}}{\partial d_{gj}\partial d_{gk}}\middle|\frac{\partial^2 d_{c2}}{\partial d_{gj}\partial d_{gk}}\middle|\frac{\partial^2 d_{c3}}{\partial d_{gj}\mathrm{j}\partial d_{gk}}\right]^T \dot{\mathbf{d}}_{gk} = [\mathbf{T}_1|\mathbf{T}_2|\mathbf{T}_3]\ \dot{\mathbf{d}}_{gk} \tag{6.2.27}$$

By carrying out the appropriate derivatives, matrices $\mathbf{T}_1$, $\mathbf{T}_2$, and $\mathbf{T}_3$ are given as follows (Powell, 1969):

$$\mathbf{T}_1 = \mathbf{T}_2 = \frac{1}{L^2}\begin{bmatrix} -2sc & c^2 - s_2 & 0 & 2sc & -(c^2 - s^2) & 0 \\ & 2sc & 0 & -(c^2 - s^2) & -2sc & 0 \\ & & 0 & 0 & 0 & 0 \\ & & & -2sc & c^2 - s^2 & 0 \\ \text{sym.} & & & & 2sc & 0 \\ & & & & & 0 \end{bmatrix} \tag{6.2.28}$$

$$\mathbf{T}_3 = \frac{1}{L}\begin{bmatrix} s^2 & -sc & 0 & -s^2 & sc & 0 \\ & c^2 & 0 & sc & -c^2 & 0 \\ & & 0 & 0 & 0 & 0 \\ & & 0 & s^2 & -sc & 0 \\ \text{sym.} & & 0 & & c^2 & 0 \\ & & 0 & & & 0 \end{bmatrix} \tag{6.2.29}$$

The complete derivation of an incremental force-displacement relationship is then obtained by substituting $\dot{\mathbf{T}}_{cg}^T$ from Eq. (6.2.27) into Eq. (6.2.24). The resulting equation is a large-displacement small-strain incremental force-displacement relationship of a beam-column element in the global coordinate system.

$$\dot{\mathbf{f}}_g = (\mathbf{T}_{cg}^T\mathbf{k}_c\mathbf{T}_{cg} + \mathbf{T}_1\mathbf{M}_A + \mathbf{T}_2\mathbf{M}_B + \mathbf{T}_3P)\dot{\mathbf{d}}_g \tag{6.2.30}$$

Equation (6.2.30) may be written symbolically as

$$\dot{\mathbf{f}}_g = \mathbf{k}_g\dot{\mathbf{d}}_g \tag{6.2.31}$$

where $\mathbf{k}_g$ represents the tangent stiffness matrix of the beam-column element in the global coordinate. It should be noted that in the derivation of the tangent stiffness matrix $\mathbf{k}_g$, the joints are assumed rigid. If plastic hinges or connections are presented at the element ends, the tangent stiffness matrix needs to be modified. These modifications are discussed in Section 9.4.2.

6.3 MODELING TRUSS ELEMENTS

For structures subjected to severe wind or lateral earthquake loading, truss diagonal bracing may be introduced to reduce the story drifts and to enhance the lateral-load resistance of the structure. In design, these braces are usually assumed to carry only axial force. Therefore, it is justifiable to use truss elements to model the bracing members.

The truss elements also may be used for modeling gravity columns that do not participate in the lateral-force-resisting system. These gravity columns (leaner columns), which are commonly used in low-rise industrial buildings and tall office building frames, are normally designed to carry only gravity loads. Therefore, they can be modeled by the truss element described in this section.

6.3.1 Assumptions

The following assumptions are made in the formulation of a truss element:

1. The element can undergo large rigid-body displacements, but member deformation remains small.
2. The axial-force displacement relationship in the local convected coordinates is written as

$$P = \frac{EA}{L}e \tag{6.3.1}$$

where

A = effective cross-section area
e = axial displacement
L = length of the truss member
P = axial force

3. Failure of the truss element may be assumed to occur when the applied axial force exceeds the specification column strength defined by the code. For a truss member designed based on AISC LRFD, the maximum compressive load capacity is (AISC, 1986)

$$P = (0.685^{\lambda_c^2})P_y, \qquad \text{for } \lambda_c \leq 1.5 \tag{6.3.2a}$$

$$P = \frac{0.877}{\lambda_c^2}P_y \qquad \text{for } \lambda_c > 1.5 \tag{6.3.2b}$$

For a truss member subjected to tension,

$$P = P_y = A_e F_y \tag{6.3.3}$$

where

A_e = effective cross-section area
E = modulus of elasticity
F_y = material yield stress
r = radius of gyration of cross-section of member
λ_c = $(L/r\pi)\sqrt{F_y/E}$, the non-dimensional member slenderness parameter

6.3.2 Tangent Stiffness Formulation

Figure 6.4 shows the degrees of freedom associated with the corotational and global coordinate systems of the truss element. The axial displacement of the truss element along the chord length may be expressed in terms of the global degrees of freedom as

$$\begin{aligned} e &= d_c \\ &= \frac{(2x_0 + d_{g3} - d_{g1})(d_{g3} - d_{g1}) + (2y_0 + d_{g4} - d_{g2})(d_{g4} - d_{g2})}{2L} \end{aligned} \tag{6.3.4}$$

The tangent stiffness relationship for a truss element can be obtained from the tangent stiffness relationship of a frame element by deleting the appropriate rows and columns in Eq. (6.2.30) that correspond to the rotational degrees of freedom of the element. The resulting tangent stiffness relationship of a truss element is (Lui and Chen, 1988)

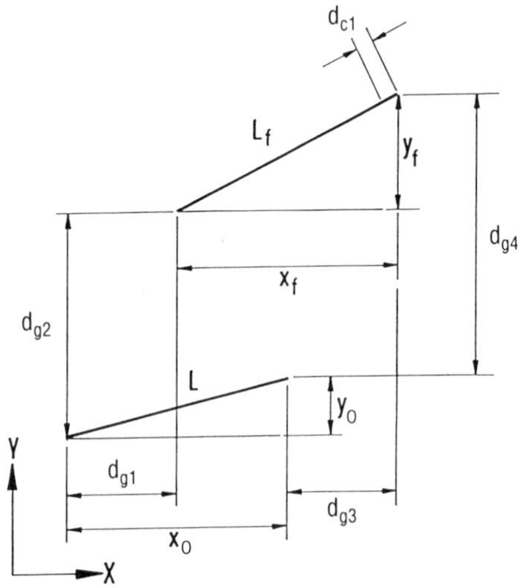

Fig. 6.4 Global and local displacements of a truss element.

$$\dot{\mathbf{f}}_g = (\mathbf{T}_{cg}^T \mathbf{K}_c \mathbf{T}_{cg} + \mathbf{T}P)\dot{\mathbf{d}}_g = \mathbf{k}_g \dot{\mathbf{d}}_g \tag{6.3.5}$$

where, referred to Fig. 6.4,

$$\dot{\mathbf{f}}_g = [\dot{f}_{g1}, \dot{f}_{g2}, \dot{f}_{g3}, \dot{f}_{g4}]^T \tag{6.3.6}$$

$$\dot{\mathbf{d}}_g = [\dot{d}_{g1}, \dot{d}_{g2}, \dot{d}_{g3}, \dot{d}_{g4}]^T \tag{6.3.7}$$

$$\mathbf{T}_{cg} = [-\cos\theta \quad -\sin\theta \quad \cos\theta \quad \sin\theta]^T \tag{6.3.8}$$

$$\mathbf{k}_c = \frac{EA}{L} \tag{6.3.9}$$

and

$$\mathbf{T} = \frac{1}{L}\begin{bmatrix} \sin^2\theta & -\sin\theta\cos\theta & -\sin^2\theta & \sin\theta\cos\theta \\ & \cos^2\theta & \sin\theta\cos\theta & -\cos\theta^2 \\ \text{sym.} & & \sin^2\theta & -\sin\theta\cos\theta \\ & & & \cos^2\theta \end{bmatrix} \tag{6.3.10}$$

where θ is the angle of inclination of the deformed element chord.

6.4 MODELING PLASTIC HINGES

6.4.1 General Approaches

If a plastic hinge forms at the element end, slope continuity is destroyed at that location. Therefore, the element force-displacement relationships need to be derived to reflect such change in the member behavior. There are two common approaches for modeling elastic-plastic hinges in a beam-column member (White et al., 1993; Liew and Chen, 1994b). The first approach is called the beam-column stability-function approach in which the plastic hinge is modeled as a "real hinge" that can undergo plastic rotation only. In this approach, the change in the axial force in a frame element is based solely on the element axial-force-displacement relationship with no effect from the inelasticity at the plastic hinges. When the cross-section forces in a new load increment violate the section plastic strength, a force-recovery step is required to bring the force point back to the plastic strength surface, as indicated in Fig. 6.5. The force-point movement on the plastic strength surface based on this approach can be viewed as a nonassociated flow-rule approach, because the plastic-hinge extension and contraction are not considered in the plasticity formulation. A detailed description of this procedure is given by Lui (1985), Lee and Basu (1989), and Liew (1992).

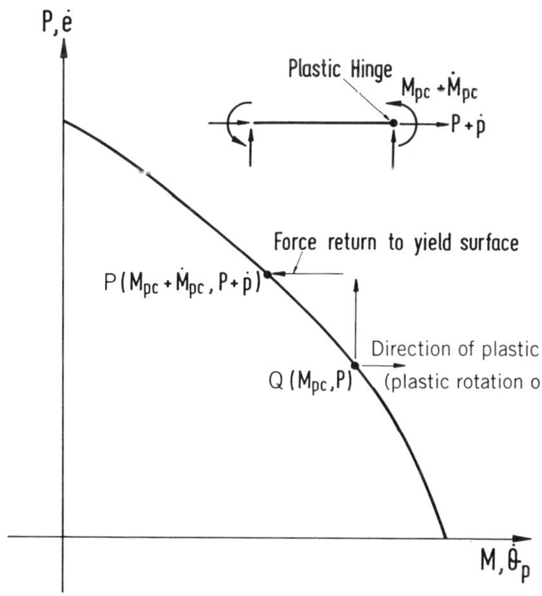

Fig. 6.5 Element force-recovery procedure of the "stability-function" plastic-hinge approach.

The second approach for the plastic-hinge formulation is based on a force-space plasticity approach in which an associated flow rule is used to describe the relationship between the axial and rotational plastic deformations at a fully plastified cross-section. For a state of force that satisfies the cross-section strength condition, the plastic hinge rotates and extends under continued plastic loading and axial tension, and it rotates and contracts for plastic loading and axial compression. This behavior leads to the condition of normality of the plastic flow relative to the cross-section surface, as shown in Fig. 6.6. The element incremental forces at a plastic hinge move tangentially to the yield surface, and these forces may fall outside the yield surface, as shown in Fig. 6.6. This force point has to be returned to the surface in order that the forces satisfy the cross-section strength equation.

The advantages of using the force-space plasticity formulation have been discussed in White et al. (1993). Derivations of beam-column force-displacement relationships based on this concept are given by Porter and Powell (1971), Gross (1980), Orbison (1982), and Hsieh (1990).

For most practical cases, the differences in results predicted by the two approaches are small, especially when the axial force in the member is small. Also, both approaches can account for force-point movement on the plastic strength surface. In other words, if the axial force is increased on a fully plastified cross-section, the bending moment would need to be decreased so that the cross-section plastic capacity is not violated.

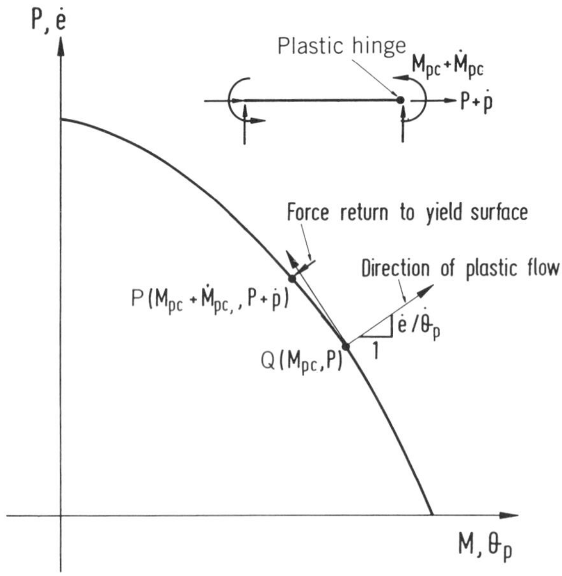

Fig. 6.6 Force-recovery procedure of the "force-space" plastic-hinge approach.

It is important to recognize that the stability-function and force-space plasticity approaches are simplified idealizations of the actual response. The actual cross-section response involves initial yielding, which occurs prior to reaching the plastic strength surface. Moreover, any partial yielding or strain-hardening in the cross-section would affect the physical plastic flow behavior, which is different from that predicted based on the elastic-plastic-hinge concept. In general, the beam-column stability-function approach tends to be more accurate in representing the element force-displacement relationship when the axial force is large because of the direct consideration of P–δ effects in the member formulation. Therefore, the use of one element per member in this approach for modeling structural members is justifiable for a wide range of axial-load and bending-moment combinations.

Several key advantages of the force-space plasticity approach have been reported in the recent work by White et al. (1993). The main advantage of adopting the force-space plasticity approach is that the direct extension to three-dimensional analysis is more straightforward in numerical formulation and implementation. However, accurate representation of inelastic torsion and warping of beam-columns in space still requires further research. Formulation based on the one-element-per-member approach in this extension is not feasible in the current state of the art. The two-dimensional formulation presented hereafter is based on the beam-column stability approach.

6.4.2 Assumptions

1. Inelastic behavior in the member is assumed to be contained within a zero-length plastic hinge.
2. Plastic hinges are allowed to form only at the element ends, and plastic deformations at plastic hinges involve inelastic rotation only.
3. Once a plastic hinge is formed, the cross-section forces are assumed to move on the plastic strength surface.
4. The AISC LRFD cross-section plastic strength equations (AISC, 1993) are adopted in the elastic-plastic-hinge formulation:

$$\frac{P}{P_y} + \frac{8M}{9M_p} = 1.0, \qquad \text{for } \frac{P}{P_y} \geq 0.2 \tag{6.4.1}$$

$$\frac{P}{2P_y} + \frac{M}{M_p} = 1.0, \qquad \text{for } \frac{P}{P_y} < 0.2 \tag{6.4.2}$$

where

M = second-order bending moment at cross-section under consideration

M_p = plastic moment capacity for member under pure bending action

P = second-order axial force at cross-section under consideration
P_y = squash load

The plastic strength equations are plotted in Fig. 6.7, and they are derived from the AISC LRFD interaction equations for a beam-column of zero length, that is, $L = 0$.

Although the present formulation is limited to modeling plastic hinges only at the element ends, the formation of a plastic hinge between both ends of the element can be detected based on a simple procedure discussed in Chapter 5, Section 5.6.4. Normally, the analysis of frame members with maximum moment within their span length and the analysis of beam-columns with in-span loading and inclined members subjected to gravity loads would require more than one beam-column element per member to capture the plastic-hinge formation in the member. Abdel-Ghaffar (1992) extends the conventional method and permits plastic-hinge formation at any location within the element length based on an assumed shape function for a beam with an intermediate hinge. This new element formulation, which can be implemented based on one element per member, avoids the need for dividing transversely loaded

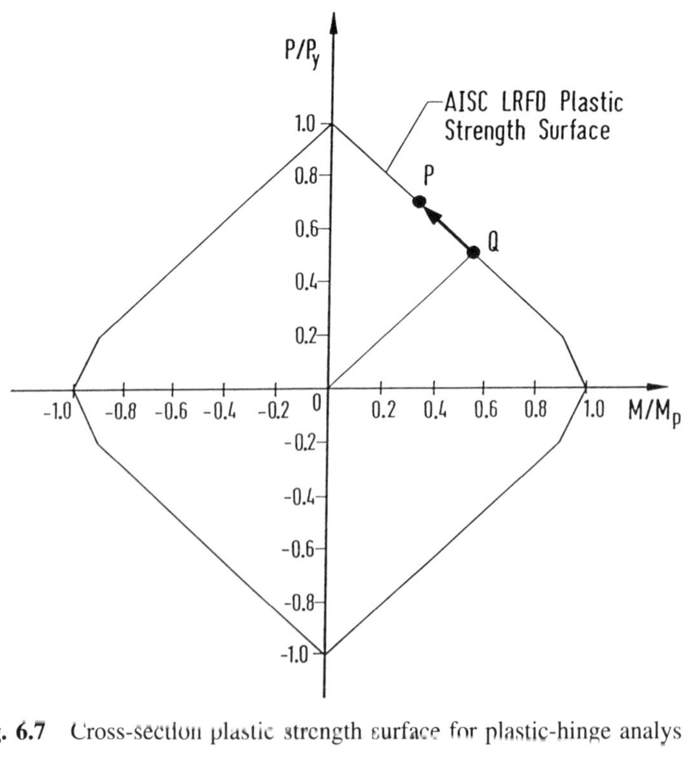

Fig. 6.7 Cross-section plastic strength surface for plastic-hinge analysis.

members into several elements in order to detect the plastic-hinge location within the member.

For structures subjected to static loading only, elastic unloading at the locations of plastic hinges may be neglected. The analytical results obtained based on this assumption are generally conservative. The numerical implementation following this assumption is also straightforward in concept, and it does not have the problem of being trapped into the recurring process of loading, unloading, and reloading of plastic hinges.

6.4.3 Modified Tangent Stiffness Relationship for the Presence of Plastic Hinges

As discussed in Section 6.2.2, if the state of forces at any cross-section equals or exceeds the plastic strength, a plastic hinge is formed. The element force-displacement relationships need to be modified to reflect the change in element behavior due to formation of plastic hinge(s) at the element end(s).

If a plastic hinge is formed at element end A, the incremental force-displacement relationships from Eq. (6.2.16) may be written as follows (Chen and Lui, 1986):

$$\begin{Bmatrix} \Delta M_{pcA} \\ \dot{M}_B \\ \dot{P} \end{Bmatrix} = \frac{EI}{L}\begin{bmatrix} S_1 & S_2 & 0 \\ S_2 & S_1 & 0 \\ 0 & 0 & A/I \end{bmatrix}\begin{Bmatrix} \dot{\theta}_A \\ \dot{\theta}_B \\ \dot{e} \end{Bmatrix} \tag{6.4.3}$$

where ΔM_{pcA} is the change in plastic moment capacity at end A as P increases to $P + \dot{P}$. From the first row of Eq. (6.4.3), $\dot{\theta}_A$ can be written as

$$\dot{\theta}_A = \frac{L\,\Delta M_{pcA}}{EIS_1} - \frac{S_2}{S_1}\dot{\theta}_B \tag{6.4.4}$$

By substituting Eq. (6.4.4) into the second and third row of Eq. (6.4.3), the modified element force-displacement relationship can be obtained as

$$\begin{Bmatrix} \dot{M}_A \\ \dot{M}_B \\ \dot{P} \end{Bmatrix} = \frac{EI}{L}\begin{bmatrix} 0 & 0 & 0 \\ 0 & (S_1 - S_2^2)/S_1 & 0 \\ 0 & 0 & A/I \end{bmatrix}\begin{Bmatrix} \dot{\theta}_A \\ \dot{\theta}_B \\ \dot{e} \end{Bmatrix} + \begin{Bmatrix} 1 \\ S_2/S_1 \\ 0 \end{Bmatrix}\Delta M_{pcA} \tag{6.4.5}$$

Equation (6.4.5) represents the modified incremental basic force-displacement relationship of a beam-column with a plastic hinge formed at end A. A similar approach can be followed if a plastic hinge is formed at end B. The resulting incremental force-displacement equation is

$$\begin{Bmatrix} \dot{M}_A \\ \dot{M}_B \\ \dot{P} \end{Bmatrix} = \frac{EI}{L}\begin{bmatrix} (S_1 - S_2^2)/S_1 & 0 & 0 \\ 0 & 0 & 0 \\ 0 & 0 & A/I \end{bmatrix}\begin{Bmatrix} \dot{\theta}_A \\ \dot{\theta}_B \\ \dot{e} \end{Bmatrix} + \begin{Bmatrix} S_2/S_1 \\ 1 \\ 0 \end{Bmatrix}\Delta M_{pcB} \tag{6.4.6}$$

If plastic hinges are formed at both ends of the element, θ_A and θ_B can be written in terms of the change in moment at the respective end of the element. The resulting modified force-displacement relationship is

$$\begin{Bmatrix} \dot{M}_A \\ \dot{M}_B \\ \dot{P} \end{Bmatrix} = \frac{EI}{L} \begin{bmatrix} 0 & 0 & 0 \\ 0 & 0 & 0 \\ 0 & 0 & A/I \end{bmatrix} \begin{Bmatrix} \dot{\theta}_A \\ \dot{\theta}_B \\ \dot{e} \end{Bmatrix} + \begin{bmatrix} \Delta M_{pcA} \\ \Delta M_{pcB} \\ 0 \end{bmatrix} \tag{6.4.7}$$

where ΔM_{pcA} and ΔM_{pcB} are the change in the plastic moment capacity at the respective ends of the member as P changes.

Equations (6.4.5)–(6.4.7) account for the presence of plastic hinge(s) at the element end(s). They may be written symbolically as

$$\dot{\mathbf{f}}_c = \mathbf{k}_{ch}\dot{\mathbf{d}}_c + \dot{\mathbf{f}}_{cp} \tag{6.4.8}$$

where $\mathbf{k}_{ch}$ is the modified tangent stiffness matrix due to the presence of plastic hinge(s), and $\dot{\mathbf{f}}_{cp}$ is an equilibrium force-correction vector that results from the change in moment capacity as P changes.

If Eq. (6.2.17) is replaced by Eq. (6.4.8) and the procedure in Section 6.2 is followed, the modified element force-displacement relationship in global coordinates may be written as

$$\dot{\mathbf{f}}_g = (\mathbf{T}_{cg}^T\mathbf{k}_{ch}\mathbf{T}_{cg} + \mathbf{T}_1 M_A + \mathbf{T}_2 M_B + \mathbf{T}_3 P)\dot{\mathbf{d}}_g + \mathbf{T}_{cg}^T\dot{\mathbf{f}}_{cp} \tag{6.4.9}$$

or

$$\dot{\mathbf{f}}_g = \mathbf{k}_{gh}\dot{\mathbf{d}}_g + \dot{\mathbf{f}}_{gp} \tag{6.4.10}$$

where

$$\mathbf{k}_{gh} = \mathbf{T}_{cg}^T\mathbf{k}_{ch}\mathbf{T}_{cg} + \mathbf{T}_1 M_A + \mathbf{T}_2 M_B + \mathbf{T}_3 P \tag{6.4.11}$$

is the modified tangent stiffness matrix, and

$$\dot{\mathbf{f}}_{gp} = \mathbf{T}_{cg}^T\dot{\mathbf{f}}_{cp} \tag{6.4.12}$$

is the global equilibrium force-correction vector.

Once a plastic hinge is formed in a member, the subsequent change in the plastic moment capacity due to the change in axial force affects the force-displacement relationships of the beam-column element. In other words, referring to Fig. 6.7, once the plastic strength is reached at point Q, the state of the moment is stationary at Q. However, because of the presence of axial force, the state of the moment changes. If the axial force is increased, the force point should move from Q to P. The force-point movement on the plastic strength surface must be considered in the tangent-stiffness formulation (White et al., 1991, 1993).

6.5 SOLUTION PROCEDURES

In the analysis of nonlinear behavior of structures, the element stiffness matrices have to be updated constantly in order to capture the state of equilibrium due to the change in structural geometry and inelastic yielding. It is therefore necessary to carry out the analysis in load incremental form. In the following two sections, two commonly used solution schemes are discussed: the simple-load incremental and the Newton–Raphson methods. The step-by-step procedure for implementing the two solution schemes are described in what follows.

6.5.1 Simple-Load Incremental Method

The simple-load incremental method is the simplest and most direct nonlinear global solution technique. This numerical procedure is straightforward in concept and implementation, and the numerical algorithm is generally well-behaved and often exhibits good computational efficiency. This is especially true when the structure is loaded into the inelastic region by which a trace of hinge-by-hinge formation is required in the element stiffness formulations. The procedure for implementing the load incremental scheme is as follows:

1. Calculate trial load increment $\dot{\mathbf{F}}_g$, equal to total loads multiplied by an incremental load factor.
2. Based on the latest available values of all geometric and static quantities, form the tangent stiffness matrix in global coordinates $\mathbf{k}_g$ for every element (frame or truss element) of the structure. If necessary, modify the tangent stiffness matrix of the beam-column element for the presence of plastic hinges.
3. Assemble the structure tangent stiffness matrix $\mathbf{K}_g$.
4. Solve for the structure displacement increment vector $\dot{\mathbf{D}}_g$ from

$$\dot{\mathbf{D}}_g = \mathbf{K}_g^{-1}\dot{\mathbf{F}}_g \tag{6.5.1}$$

5. Extract element incremental displacements $\dot{\mathbf{d}}_g$ from the structural incremental displacements $\dot{\mathbf{D}}_g$ and transform $\dot{\mathbf{d}}_g$ to the convected coordinate system by

$$\dot{\mathbf{d}}_c = \mathbf{T}_{cg}\dot{\mathbf{d}}_g \tag{6.5.2}$$

6. Determine incremental basic end forces $\dot{\mathbf{f}}_c$ for each type of element based on the procedures given in Section 6.5.3.
7. Update the element internal forces and the total load parameter.
8. Update the structure nodal geometry by

$$\mathbf{D}_{\mathbf{gi+1}} = \mathbf{D}_{\mathbf{gi}} + \dot{\mathbf{D}}_{\mathbf{g}}$$

9. The list of active plastic hinges is updated to include any newly formed plastic hinges.
10. Results are stored and the analysis is ready for the next load increment.

It should be noted that the applied load increment must not be excessive such that the element nodal forces violate the plastic cross-section strength. This aspect is discussed more in Section 6.5.5.

The major advantage of the simple-load incremental method is its simplicity in application. Also, this method does not encounter numerical instability problems associated with snap-through and snap-back buckling because iterations are not required. The drawback of this method is that errors tend to accumulate as the calculated equilibrium path is slowly diverted from the true path as the load increases, as shown in Fig. 6.8. To reduce the so called *drift-off error,* is it necessary to use a small load increment, especially in the region of high nonlinearity. When a nonpositive definite stiffness matrix is detected, an unloading from the maximum load point is suggested. In this case, the normal procedure of matrix decomposition and back substitution should be performed, and the descending branch of the load-displacement curve can be traced by simply reversing the sign of the load increment and displacement vectors (Orbison, 1982; King, 1990).

6.5.2 Newton–Raphson Iterative Method

In this approach, an iterative procedure is adopted to eliminate the unbalanced forces that exist at each load increment. The presence of unbalanced forces

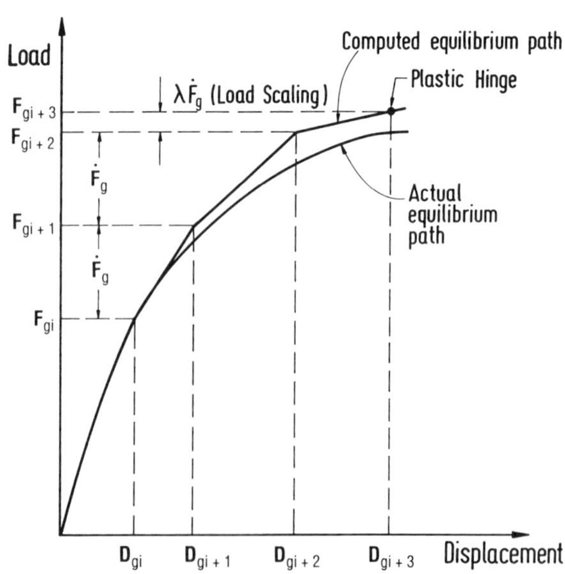

Fig. 6.8 Simple-load incremental scheme.

indicates that equilibrium between the external applied forces and internal forces is violated. The unbalanced forces arise because of the linearization process in which the updated stiffness matrix is formulated based on the previous configuration and the internal force distribution of the structure.

The procedure for implementing the Newton–Raphson technique follows. Unless otherwise stated, a subscript refers to the load step and a superscript refers to the cycle of calculation within each load step.

1. Calculate trial-load increment $\dot{\mathbf{F}}_g$, which is equal to total loads multiplied by an incremental load factor.
2. Form the tangent stiffness matrix in global coordinates $\mathbf{k}_g$ for an element. If necessary, modify the tangent stiffness matrix of the beam-column element for the presence of plastic hinges.
3. Assemble the element tangent stiffness matrices to form the structure tangent stiffness matrix $\mathbf{K}^1_{gi}$ based on the displacement configuration at the end of the previous load step.
4. Solve for the structure incremental displacement vector using

$$\dot{\mathbf{F}}_{gi} = \mathbf{K}^1_{gi}\dot{\mathbf{D}}^1_{gi} \tag{6.5.3}$$

from which

$$\dot{\mathbf{D}}^1_{gi} = [\mathbf{K}^1_{gi}]^{-1}\dot{\mathbf{F}}_{gi} \tag{6.5.4}$$

where

$\dot{\mathbf{D}}^1_{gi}$ = incremental structure nodal displacement vector at ith load step
$\dot{\mathbf{F}}_{gi}$ = prescribed incremental load vector of ith load step
$\mathbf{K}^1_{gi}$ = first estimate of the structure tangent stiffness matrix at beginning of ith load step

5. Update the structure nodal displacement vector from

$$\mathbf{D}^1_{gi} = \mathbf{D}_{gi} + \dot{\mathbf{D}}^1_{gi} \tag{6.5.5}$$

where

$\mathbf{D}_{gi}$ = structure nodal displacement vector at beginning of ith load step
$\mathbf{D}^1_{gi}$ = structure displacement vector after first cycle of calculation at load step i
$\dot{\mathbf{D}}^1_{gi}$ = incremental structure nodal displacement vector evaluated at step 4.

6. Extract the element end-displacement vector $\mathbf{d}_{gi}$ from $\mathbf{D}_{gi}$ for every element in the structure. More specifically, for frame elements, deter-

mine end rotations θ_A and θ_B and axial displacement e from Eqs. (6.2.18a), (6.2.18b), and (6.2.18c). For a truss element, determine the element axial displacement from Eq. (6.3.4).

7. For each element, evaluate the element end forces based on the newest geometric configuration of the structure obtained from step 5. The procedure for determining element end displacements and forces is given in Section 6.5.4.
8. Form the structure internal-force vector $\mathbf{F}^1_{gi}$ after the first cycle of calculation by assembling the element end forces evaluated in step 7 for all elements.
9. Calculate the external-force vector from

$$\mathbf{F}_{gi+1} = \mathbf{F}_{gi} + \dot{\mathbf{F}}_{gi} \tag{6.5.6}$$

10. Calculate the unbalanced force $\dot{\mathbf{Q}}^1_{gi}$ at the end of the cycle from

$$\dot{\mathbf{Q}}^1_{gi} = \mathbf{F}_{gi+1} - \mathbf{F}^1_{gi} \tag{6.5.7}$$

11. Using the current value of axial force P, update the element stiffness matrix $\mathbf{k}$ for every element. Assemble $\mathbf{k}$ for all elements to form an updated tangent structure stiffness matrix $\mathbf{K}^2_{gi}$. Evaluate the incremental displacement vector $\dot{\mathbf{D}}^2_{gi}$ from

$$\dot{\mathbf{D}}^2_{gi} = [\mathbf{K}^2_{gi}]^{-1}\dot{\mathbf{Q}}^1 \tag{6.5.8}$$

where $\dot{\mathbf{Q}}^1$ is the unbalanced force vector calculated in the previous cycle of calculation.

12. Update the structure nodal-displacement vector from

$$\mathbf{D}^2_{gi} = \mathbf{D}_{gi} + \sum_{\mathbf{k}=1}^{2} \dot{\mathbf{D}}^{\mathbf{k}}_{gi} \tag{6.5.9}$$

13. Extract the element end-displacement vector $\mathbf{d}_i$ from $\mathbf{D}^2_i$ calculated in Eq. (6.5.15) for every element. Update e, θ_A, and θ_B, and, hence, P, M_A, and M_B as done in step 7 for all elements in the frame.
14. Update the element end forces for all the elements and form the new structure internal-force vector $\mathbf{P}^2_{gi}$.
15. Evaluate the new unbalanced force $\dot{\mathbf{Q}}^2$ from

$$\dot{\mathbf{Q}}^2 = \mathbf{F}_{gi+1} - \mathbf{F}^2_{gi} \tag{6.5.10}$$

16. Repeat steps 11 to 15 until a convergence criterion is satisfied. Convergence is said to have been attained if both the unbalanced force

norm (FNORM) and the displacement norm (DNORM) fall within a prescribed tolerance.

The convergence criteria used in the Newton–Raphson scheme may be written as

$$\text{FNORM} = \left[\frac{|\dot{\mathbf{D}}^j|}{|\mathbf{F}^j|}\right]_{\max} \leq \varepsilon_0 \tag{6.5.11}$$

and

$$\text{DNORM} = \left[\frac{|\dot{\mathbf{D}}^j|}{|\mathbf{D}^j|}\right]_{\max} \leq \varepsilon_1 \tag{6.5.12}$$

where ε_0 and ε_1 are small numbers between 1×10^{-2} and 1×10^{-6}, depending on the accuracy required. Subscript j refers to the jth cycle of calculation at the specific load step.

17. After convergence is achieved, the structure nodal displacement at the end of load step i is obtained by

$$\mathbf{D}_{gi+1} = \mathbf{D}_{gi}^{n} = \mathbf{D}_{gi}^{n-1} + \dot{\mathbf{D}}_{gi} \tag{6.5.13}$$

18. Prescribe another load increment and repeat steps 2 to 17.

A schematic representation of the preceding procedure is shown in Fig. 6.9 for a structure with only one degree of freedom. Iterative methods, such as the Newton–Raphson method, satisfy the equilibrium equations at a specific external-load magnitude. In this method, the out-of balance forces that are present due to an initial linear load step are eliminated within a given tolerance by applying additional corrective steps. The Newton–Raphson method possesses the advantage of providing results that lie upon the "exact" load-displacement trace.

In performing the previous procedure, the complete load-deflection response of the frame can be traced, including the sequence of plastic-hinge formation. The stability-limit point is obtained as the peak point of the load-deflection trace. It is worth mentioning that the structural internal forces obtained from steps 7 and 8 of the preceding procedure are based on the total equilibrium condition and not on the incremental equilibrium condition as assumed in the load incremental scheme. Also, in Eqs. (6.2.18a), (6.2.18b), and (6.2.18d), the displacement variables θ_A, θ_B, and e are the total displacement quantities calculated without restoring any small-displacement assumption. Small-displacement and small-strain assumptions are introduced only in the element force-displacement relationship described by Eq. (6.2.11).

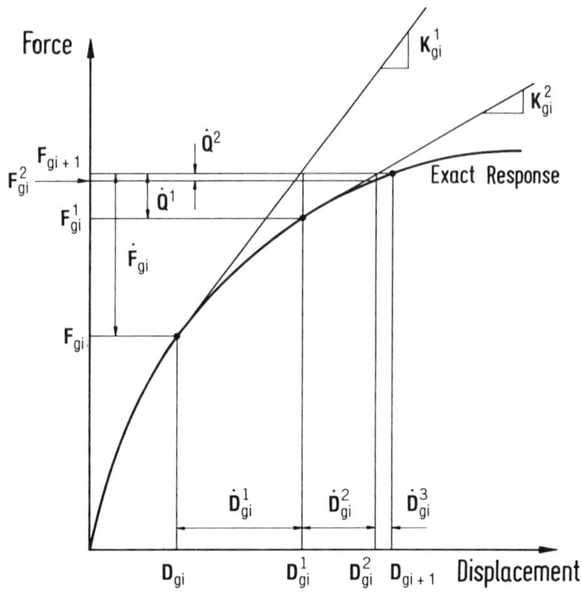

Fig. 6.9 Newton–Raphson incremental scheme.

6.5.3 Determination of Incremental End Forces $\dot{\mathbf{f}}_c$ in Simple-Load Incremental Procedure

The following procedure is used to determine the element incremental end forces for a *beam-column* element:

1. The total element end forces at the end of each increment are checked against the plastic strength surface to determine if a plastic hinge would form during the increment. If the element end forces are within the plastic strength surface, then no plastic hinge is formed at the element ends. The element incremental forces are determined from Eq. (6.2.17).
2. If the element end forces are lying on the plastic strength surface, a plastic hinge is formed at the element end. Equation (6.4.8) should be used for calculating $\dot{\mathbf{f}}_c$ instead of Eq. (6.2.17).
3. An appropriate modified stiffness matrix $\mathbf{k}_{ch}$ can be obtained from Eqs. (6.4.5)–(6.4.7). The choice of the $\mathbf{k}_{ch}$ matrix in Eq. (6.4.8) depends on the positions and number of plastic hinges formed in the element.
4. If the element end forces exceed the plastic strength surface, the attempted load increment is reduced by a fraction so that the forces would lie on the plastic strength surface within a given yield tolerance.

For a *truss element,* the element incremental end force is determined as

$$\dot{P} = \frac{EA}{L}\dot{e} \tag{6.5.14}$$

6.5.4 Determination of Element End Forces in Newton–Raphson Iterative Procedure

For Beam-Column Element

1. If no plastic hinge is formed at the element ends, determine the element axial displacement e and end rotations θ_A and θ_B from Eqs. (6.2.18a), (6.2.18b), and (6.2.18d). However, if a plastic hinge is formed at end A of the element, let $M_A = M_{pcA}$ and solve for θ_A from the first row of Eq. (6.2.11):

$$\theta_A = \frac{EIM_{pcA}/L - S_2\theta_B}{S_1} \tag{6.5.15}$$

where θ_B is given by Eq. (6.2.18b).

If a plastic hinge is formed at end B of the element, let $M_B = M_{pcB}$ and solve for θ_B from the second row of Eq. (6.2.11):

$$\theta_B = \frac{EIM_{pcB}/L - S_2\theta_A}{S_1} \tag{6.5.16}$$

θ_A in Eq. (6.5.6) can be determined from Eq. (6.2.18a).

If a plastic hinge is formed at both ends of the element, let $M_A = M_{pcA}$, $M_B = M_{pcB}$, and solve for θ_A and θ_B from Eq. (6.2.11), upon which we have

$$\theta_A = \frac{L(S_1M_{pcA} - S_2M_{pcB})}{EI(S_1^2 - S_2^2)}. \tag{6.5.17}$$

and

$$\theta_B = \frac{L(S_1M_{pcB} - S_2M_{pcA})}{EI(S_1^2 - S_2^2)}. \tag{6.5.18}$$

2. Determine element basic end forces from Eq. (6.2.11) based on the element end displacements calculated from step 1.
3. Determine element forces in global coordinates from Eq. (6.2.22).

For Truss Element

1. Determine the element end displacement e from Eq. (6.3.4).
2. Determine the element axial force from

$$P = \frac{EAe}{L} \tag{6.5.19}$$

3. Determine the element force in global coordinates from

$$\mathbf{f}_g = \mathbf{T}_{cg}^T \mathbf{f}_c \tag{6.5.20}$$

where $\mathbf{T}_{cg}^T$ is the transformation matrix given by Eq. (6.3.8).

6.5.5 Prediction of Limit Load

To prevent plastic hinges from forming within a constant-stiffness load increment, load step sizes less than or equal to the specified increment magnitude are internally computed such that plastic hinges form only after the load increment. Thus, the subsequent element stiffness formulation will account for the stiffness reduction due to the presence of plastic hinges.

In addition, the applied-load increment may be reduced for any of the following reasons:

1. The total load parameter is reached.
2. The increment in the element nodal forces at plastic hinges is excessive.
3. The nonpositive definiteness of the structural stiffness matrix.

As the stability-limit point is approached in the analysis, large step increments may overstep a limit point. Therefore, a smaller step size is automatically computed and used near the limit point to obtain accurate collapse displacements and second-order forces. Unloading from the maximum load point may be detected by checking the definiteness of the stiffness matrix; that is, by checking the appearance of one or more negative coefficients on the main diagonal of the decomposed stiffness matrix. The maximum load point is detected by noting a change from a positive definite matrix to a nonpositive definite matrix occurring in a load increment. If a negative diagonal coefficient appears during an increment and the previous step exhibited a positive definite stiffness matrix, then the load increment being attempted is automatically scaled back. The analysis is terminated when the load increment near the limit point becomes very small compared to the total load applied.

However, the number of load increments required to reach the collapse load is typically unknown at the start of the analysis; therefore, a large number of increments may be prescribed to assure that the analysis will continue to the limit point. However, by some modifications, the number of load increments can be controlled, and small-load increments are prescribed only at the range of severe nonlinearity. Such modification is based on an error estimate and step-size control method, which is described by Abdel-Ghaffar, White, and Chen (1993).

Detailed implementations of both the simple-load incremental and Newton–Raphson methods have been implemented for second-order plastic-hinge-based analysis (Liew, 1992). Computer program and source codes for second-order elastic-plastic-hinge analysis are given in Chapter 11. By using the computer to perform a second-order analysis, the forces obtained from the analysis can be used directly for member proportioning in the member-interaction equations.

6.6 VERIFICATION STUDIES

Although the elastic-plastic-hinge approach is capable of predicting the overall behavior of many types of frames, it cannot be used without modification to predict accurately or conservatively the capacity and load behavior of general beam-column members. Its accuracy is suspected for frames in which failure is influenced to a great extent by the strength and stability of individual members.

This section aims to explore the limitations of the elastic-plastic-hinge analysis for assessment of maximum strength behavior of planar frames. The performance of the analysis method is evaluated by comparing the results with simple portal frame and beam-column benchmark solutions. The benchmark solutions used for the verification studies are the results obtained from second-order plastic-zone analysis. Other benchmark solutions for columns are those that have been defined in the context of the design such as AISC LRFD. The design equations in AISC LRFD is derived based on extensive studies involving both experimental and theoretical investigations.

All the primary benchmark problems utilized in the present studies involve small subassemblies and isolated members with imperfections. These benchmark problems cover a wide range of loading, ranging from pure bending to pure axial load. Studies of the performance of large framing systems are not as critical as for subframe systems and individual members for full verification of the inelastic analysis model. The reason is if the performance of an isolated member is predicted properly for general loading cases, the proper prediction of the system's response should follow from the principles of equilibrium and compatibility.

6.6.1 Axially Loaded Columns

Figure 6.10 compares the AISC LRFD column strength curve with the strength curve generated by second-order elastic-plastic-hinge analysis. For

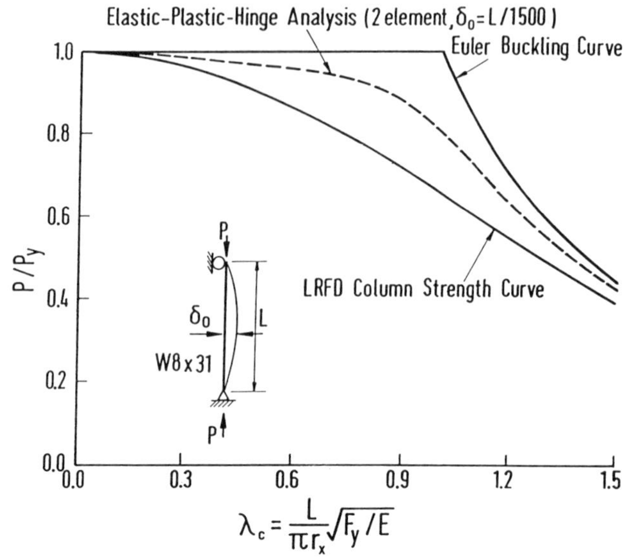

Fig. 6.10 Comparison of strength curves for axially loaded pinned-end column.

comparison with the AISC LRFD column strength curve, the elastic-plastic-hinge analysis is carried out using a W8 $\times$ 31 section, bent about its strong axis. The column is discretized into two elements with an initial deflection δ_0 at midlength equal to $L/1500$. This initial out-of-straightness magnitude is implicitly assumed in the development of the AISC LRFD column equations. Residual stresses are, however, not modeled in the plastic-hinge analysis.

It is observed from Fig. 6.10 that the elastic-plastic-hinge analysis over-predicts the column capacity implied by the AISC LRFD column equations. The maximum error is about 23%, which corresponds to the column with a slenderness parameter of unity, at which value residual stress and member initial out-of-straightness effects interact to produce the greatest reduction in strength from the theoretical value for perfect column. However, for columns whose slenderness parameter is less than or equal to 0.4, the maximum error from the elastic-plastic-hinge analysis is not more than 5% unconservative when compared to the AISC LRFD column strength curve. These results suggest that for stocky column members with a slenderness parameter λ_c less than 0.4, the column maximum strength can be predicted within 5% error using the second-order elastic-plastic-hinge analysis.

6.6.2 Beam-Columns

Zhou, Duan, and Chen (1990) analyzed the beam-columns shown in Figs. 6.11 and 6.12 using the plastic-zone method. The results are presented for

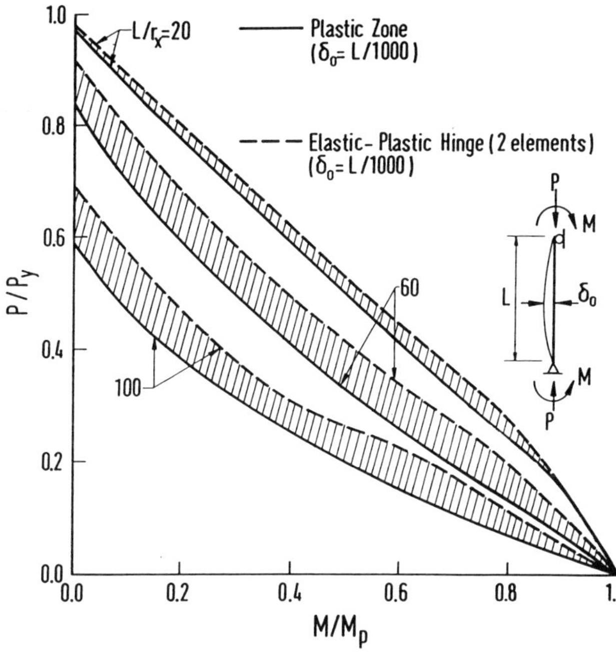

Fig. 6.11 Comparison of beam-column strengths from elastic-plastic hinge analysis with "exact" strength curves–single-curvature bending.

beam-columns subjected to equal end moments producing single- and reversed-curvature bending. The analyses were carried out using a W8 × 31 section of length L bending about the strong axis. The members in the plastic-zone analysis were assumed to have the residual stress pattern described by Galambos and Ketter (1957) with a maximum compressive residual stress of $0.3F_y$. Also, the member's initial out of straightness was assumed to vary sinusoidally with a maximum midlength deflection of $\delta_0 = L/1000$. The member cross-sections are fully compact, satisfying the compactness requirements for plastic design sections. All members are adequately braced against lateral-torsional buckling.

In Fig. 6.11, the plastic-zone strength curves for a column under single-curvature bending are compared with the second-order elastic-plastic-hinge solutions. The plastic-hinge results were generated using two elements per member. The analyses explicitly model the member initial out of straightness, which has a maximum magnitude of $\delta_0 = L/1000$ at midlength. It can be observed from Fig. 6.11 that for beam-columns with L/r less than 20 and P less than $0.2P_y$, the elastic-plastic-hinge analysis is sufficiently accurate for the prediction of the ultimate strengths of these beam-columns. Otherwise,

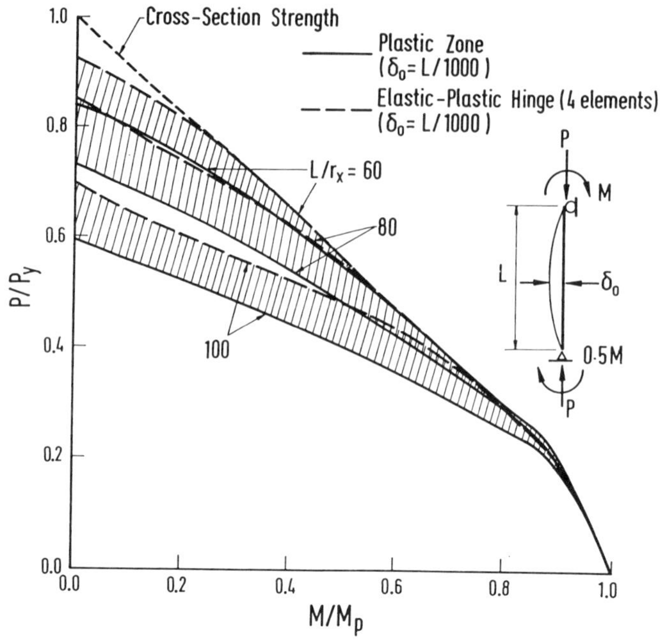

Fig. 6.12 Comparison of beam-column strengths from elastic-plastic hinge analysis with "exact" strength curves–reversed-curvature bending.

the elastic-plastic-hinge analysis always overpredict the capacities of the beam-columns by more than 5%.

For members in reversed-curvature bending (Fig. 6.12), the elastic-plastic-hinge method does not yield significant error (less than 5%) when the member slenderness ratio L/r) is less than or equal to 40. In addition, when the axial force in the member is less than $P = 0.3P_y$, the plastic-hinge results are no more than 5% unconservative compared to the plastic-zone results. In fact, for beam-column members that fall within the preceding categories, member failure is controlled more by the cross-section strength at the end of the member rather than by inelastic instability. However, for members subjected to high axial load ($P > 0.3P_y$), member failure is generally controlled by inelastic instability. In these cases, the plastic-hinge analyses overestimate the capacities of beam-columns, particularly when L/r is greater than 60. For the beam-columns with $L/r = 80$ and 100, the strong-axis strengths from the plastic-hinge analyses are in error by a maximum of 15 and 17%, respectively. For all cases, the axial force in each member should not be more than 0.85 times its axial yield load, because the plastic-hinge analysis performs rather poorly when the axial load in the member is high.

6.6.3 Sway Frames

Figures 6.13 through 6.15 compare the in-plane strength curves obtained by the second-order elastic-plastic hinge with the plastic-zone results from Kanchanalai (1977). Kanchanalai's plastic-zone solutions were used as benchmarks for the development of the AISC LRFD beam-column equations (Liew, White, and Chen, 1991, 1992), any analysis methods that can match these benchmarks are expected to satisfy the requirements for two-dimensional advanced inelastic analysis, for which separate specification member-capacity checks are not required.

In Kanchanalai's plastic-zone analyses, initial geometric imperfections as well as strain hardening of steel are neglected. The plastic-zone analyses assume a W8 × 31 section and a residual stress distribution with a peak compressive stress at the flange tips of $0.3F_y$, and a constant tensile stress in the web. These inelastic strength curves were generated for both strong- and weak-axis bending and for columns having slenderness ratios of L/r from 20 to 60.

In Figs. 6.13 through 6.15, both the strong- and weak-axis strength curves are presented for the plastic-zone analyses. However, only the strong-axis

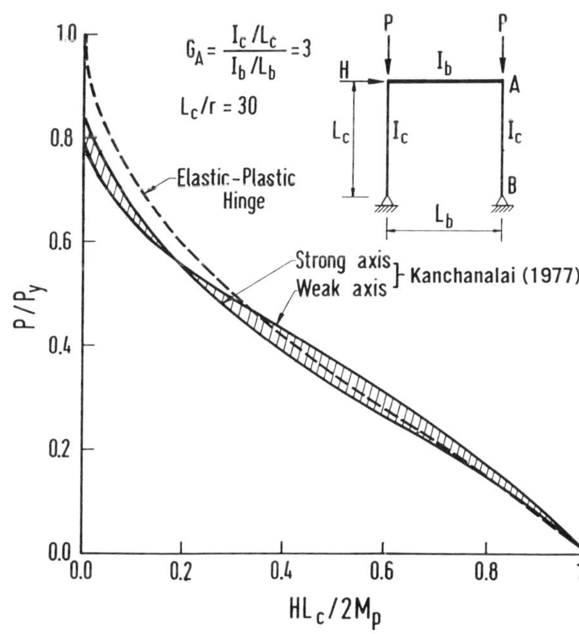

Fig. 6.13 Comparison of strength curves for a portal frame with $G_A = 3$ and $L_C/r = 30$.

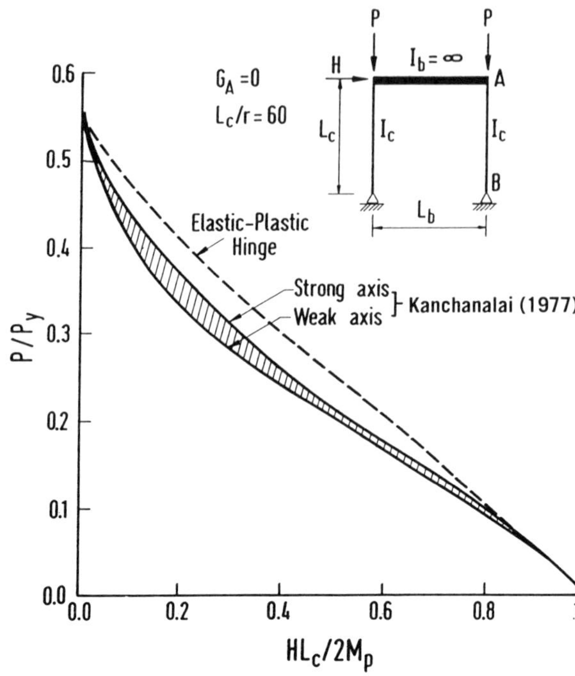

Fig. 6.14 Comparison of strength curves for a portal frame with $G_A = 0$ and $L_C/r = 60$.

strength curves are shown for the plastic-hinge analyses. The weak-axis curves from the plastic-hinge analysis are identical with the strong-axis strength curves because the results are presented in a nondimensional form, and only one plastic strength curve (i.e., the LRFD beam-column interaction equations for a member of zero length) was used for both the strong- and weak-axis section strengths.

Comparisons of results show that the elastic-plastic-hinge analyses overpredict the maximum strengths of the frames. For the portal frame shown in Figs. 6.13 and 6.14, the strong-axis strengths from the plastic-hinge model are approximately in error by a maximum of 20 and 9%, respectively. The errors are measured radially from the origin of the plots. For the frame with a leaning column, as shown in Fig. 6.15, the strong axis strengths are overpredicted with a maximum error of 20%.

The weak-axis strengths are also significantly overpredicted by the elastic-plastic-hinge model for the frames shown in Figs. 6.13 through 6.15. In these cases, the maximum unconservative error in the weak-axis strength predictions based on the elastic-plastic-hinge analyses ranges from about 10 to 30% unconservative.

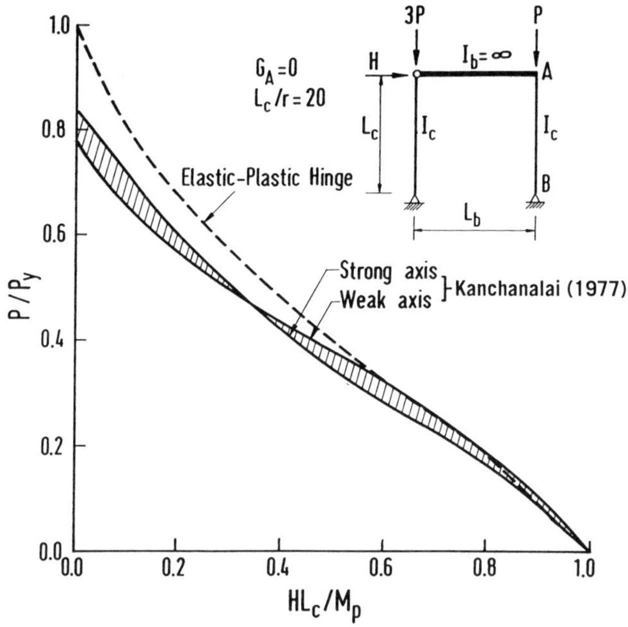

Fig. 6.15 Comparison of strength curves for a leaning-column frame with $G_A = 0$ and $L_C/r = 20$.

Research (Liew et al., 1993a, 1993b) shows that for very stocky or very flexible frames, the elastic plastic-hinge results are quite comparable to the strong-axis plastic-zone results for almost all ranges of axial force vs. moment. This means that second-order elastic-plastic-hinge analysis is sufficiently accurate in predicting the failure load either at elastic buckling or at the collapse load associated with the formation of a plastic mechanism. However, for frames such as those shown in Figs. 6.13 through 6.15, both the strong- and weak-axis strengths are significantly overpredicted by plastic-hinge analyses for a wide range of moment vs. axial force. Therefore, it is necessary to refine the conventional elastic-plastic-hinge model to generalize its application for the analysis of a wider range of structural systems.

6.7 SUMMARY AND CONCLUSIONS

Several important summaries can be drawn from the numerical examples with regard to the accuracy of the conventional elastic-plastic-hinge method for advanced design of frame structures.

1. For lightly loaded frame members, particularly beam members, the second-order elastic-plastic-hinge analysis should provide reliable results. In addition, this method is sufficiently accurate in predicting the failure load of a frame either at the elastic buckling load or at the collapse load associated with the formation of a plastic mechanism. For beam-column members that fall within the following categories, the elastic-plastic-hinge analysis may be applied with an unconservative error less than 5% when compared to plastic-zone solutions.
   - Beam-columns in single-curvature bending: $P/P_y < 0.2$ and/or $L/r \le 20$.
   - Beam-columns in double-curvature bending: $P/P_y < 0.3$ and/or $L/r \le 40$.
   - Columns subjected to axial load only: $\lambda_c \le 0.4$.
   - For all cases, it is recommended that the axial force in the member should not be more than $0.85P_y$.
2. The conventional plastic-hinge analysis does not perform adequately in several numerical examples cited in the previous sections. By using the plastic-zone solutions as a standard for comparison, the maximum unconservative errors associated with the use of second-order elastic-plastic-hinge analyses are summarized in what follows (Liew, 1992):
   - Single-story portal frames bent about the strong axis: 4 to 21% error for L/r from 20 to 60.
   - Single-story portal frames bent about the weak axis: 0 to 30% error for L/r from 20 to 60.
   - Isolated beam-columns subjected to single-curvature bending: 1 to 18% error for L/r from 20 to 100.
   - Isolated beam-columns subjected to double-curvature bending: 0 to 17% error for L/r from 20 to 100.
   - Axially loaded columns: 23% maximum error at $\lambda_c = 1.0$.
3. The elastic-plastic-hinge analysis always predicts column failure either at the elastic buckling load P_{cr} or at the member squash load P_y. This is because the columns are modeled as perfectly straight elements that are perfectly elastic between the plastic-hinge locations without the consideration of initial residual stresses and column out of straightness. Likewise, the beam-column strength and stiffness are overestimated in cases where there is significant distributed yielding through the cross-section along the member length.
4. The conventional elastic-plastic-hinge analysis cannot be classified as an advanced inelastic analysis in general. Also, the elastic-plastic-hinge model performs poorly for structural systems subjected to a high gravity load and that have little reserve capacity for inelastic redistribution of forces.

The conclusions from the studies show that although the elastic-plastic-hinge approach can provide good predictions of the ultimate strength behavior of many practical frames as the plastic-zone methods, they are not adequate as advanced inelastic analysis methods in general. This is mainly due because members are modeled as fully elastic elements between the plastic-hinge locations, and the analysis fails to consider the effects of residual stresses and geometrical imperfections.

REFERENCES

Abdel-Ghaffar, M. (1992). "Post Failure Analysis of Braced and Unbraced Steel Structures," Ph.D. dissertation, School of Civil Engineering, Purdue University, West Lafayette, IN.

Abdel-Ghaffar, M., D. W. White, and W. F. Chen (1993). "An Error Estimate and Step Size Control Method for Nonlinear Solution Techniques," *Journal of Finite Element in Analysis and Design,* Vol. 13, pp. 137–148.

AISC (1986). *Load and Resistance Factor Design Specifications for Structural Steel Buildings,* Chicago: American Institute of Steel Construction.

AISC (1993). *Load and Resistance Factor Design Specifications for Structural Steel Buildings,* 2nd Ed., Chicago: American Institute of Steel Construction.

ASCE (1990). *Specification for the Design of Cold-Formed Stainless Steel Structural Members,* ANSI/ASCE 8-90, New York: American Society of Civil Engineers.

Belytschko, T., and B. J. Hsieh (1973). "Non-Linear Transient Finite Element Analysis with Convected Coordinates," *Journal of Numerical Methods in Engineering,* Vol. 7, No. 3, pp. 255–271.

CSA (1989). *Limit States Design of Steel Structures,* CAN/CSA-S15.1-M89, Canadian Standards Association. Ontario, Canada.

Chen, W. F., and E. M. Lui (1986). *Structural Stability—Theory and Implementation,* New York: Elsevier.

Chen, W. F., and E. M. Lui (1991). *Stability Design of Steel Frames,* Boca Raton, FL: CRC Press.

Chen, W. F., and I. S. Sohal (1995). *Plastic Design and Second-Order Analysis of Steel Frames,* New York: Springer-Verlag.

Chen, W. F., and S. Toma (1994). *Advanced Analysis of Steel Frames: Theory, Software, and Applications,* Boca Raton, FL: CRC Press.

Cheong, S. M. (1977). "Consideration of Secondary Effects in Frame Design," *Journal of the Structural Division,* ASCE, Vol. 103, No. ST10, pp. 2005–2019.

Cheong, S. M. (1991). "Column Design in Gravity-Loaded Frames," *Journal of Structural Engineering,* ASCE, Vol. 117, No. 5, pp. 1448–1461.

Cook, R. D., D. S. Malkus, and M. E. Plesha (1989). *Concepts and Applications of Finite Element Analysis,* 3rd Ed., New York: Wiley.

Galambos, T. V., and R. L. Ketter (1957). *Further Studies of Columns under Combined Bending and Thrust,* Fritz Engineering Laboratory Report No. 205A.19, Lehigh University, Bethlehem, PA.

Eurocode 3 (1992). *Design of Steel Structures: Part 1.1—General Rules and Rules for Buildings,* National Application Document for use in the UK with ENV1993-1-1: 1991, Draft for Development.

Goto, Y., and W. F. Chen (1987). "Second-Order Elastic Analysis for Frame Design," *Journal of Structural Engineering,* ASCE, Vol. 113, No. ST7, pp. 1505–1519.

Goto, Y., S. Suzuki, and W. F. Chen (1991). "Bowing Effect on Elastic Stability of Frames under Primary Bending Moments," *Journal of Structural Engineering,* ASCE, Vol. 117, No. 1, pp. 111–127.

Gross, J. L. (1980). "Design for the Prevention of Progressive Collapse using Iterative Computer Graphics," Ph.D. dissertation, Department of Structural Engineering, Cornell University, Ithaca, NY.

Hsieh, S. (1990). *Analysis of Three-Dimensional Steel Frames with Semi-Rigid Connections,* Structural Engineering Report 90-1, Cornell University, Ithaca, NY.

Kanchanalai, T. (1977). *The Design and Behavior of Beam-Columns in Unbraced Steel Frames,* AISI Project No. 189, Report No. 2, Austin: Civil Engineering/Structures Research Laboratory, University of Texas.

King, W. S. (1990). "Second-Order Inelastic Analysis of Semi-Rigid Frames," Ph.D. dissertation, School of Civil Engineering, Purdue University, West Lafayette, IN.

Lee, S. L., and P. K. Basu (1989). "Secant Method for Nonlinear Semi-Rigid Frames," *Journal of Constructional Steel Research,* Vol. 14, No. 2, pp. 49–67.

LeMessurier, W. J. (1977). "A Practical Method of Second Order Analysis, Part 2—Rigid Frames," *Engineering Journal,* Vol. 14, No. 2, pp. 49–67.

Liew, J. Y. R. (1992). "Advanced Analysis for Frame Design," Ph.D. dissertation, School of Civil Engineering, Purdue University, West Lafayette, IN.

Liew, J. Y. R., and W. F. Chen (1994a). "Trends Toward Advanced Analysis," in *Advanced Analysis of Steel Frames: Theory, Software, and Applications,* W. F. Chen and S. Toma, eds., Boca Raton, FL: CRC Press, pp. 1–45.

Liew, J. Y. R., and W. F. Chen (1994b). "Second-Order Plastic Hinge Analysis of Frames," in *Advanced Analysis in Steel Frames: Theory, Software, and Applications,* W. F. Chen and S. Toma, eds., Boca Raton, FL: CRC Press, pp. 139–194.

Liew, J. Y. R., D. W. White, and W. F. Chen (1991). "Beam-Column Design in Steel Frameworks—Insights on Current Methods and Trends," *Journal of Constructional Steel Research,* Vol. 18, No. 4, pp. 269–308.

Liew, J. Y. R., D. W. White, and W. F. Chen (1992). "Beam-Columns," in *Constructional Steel Design, An International Guide,* P. J. Dowling et al., eds., London: Elsevier, pp. 105–132.

Liew, J. Y. R., D. W. White, and W. F. Chen (1993a). "Second-Order Refined Plastic Hinge Analysis for Frame Design: Part 1," *Journal of Structural Engineering,* ASCE, Vol. 119, No. 11, pp. 3196–3216.

Liew, J. Y. R., D. W. White, and W. F. Chen (1993b). "Second-Order Refined Plastic Hinge Analysis for Frame Design: Part 2," *Journal of Structural Engineering,* ASCE, Vol. 119, No. 11, pp. 3217–3237.

Lui, E. M. (1985). "Effects of Connection Flexibility and Panel Zone Deformation on the Behavior of Plane Steel Frames," Ph.D. dissertation, School of Civil Engineering, Purdue University, West Lafayette, IN.

Lui, E. M., and W. F. Chen (1988). "Behavior of Braced and Unbraced Semi-Rigid Frames," *Journal of Solids and Structures,* Vol. 24, No. 9, pp. 893–913.

Orbison, J. G. (1982). *Nonlinear Static Analysis of Three-Dimensional Steel Frames,* Report No. 82-6, Ithaca, NY: Department of Structural Engineering, Cornell University.

Porter, F. L., and G. M. Powell (1971). *Static and Dynamic Analysis of Inelastic Frame Structures,* Report No. EERC 71-3, Berkeley: Earthquake Engineering Research Center, University of California.

Powell, G. H. (1969). "Theory of Nonlinear Elastic Structures," *Journal of the Structural Division,* ASCE, Vol. 95, No. ST12, pp. 2687–2701.

Powell, G. H., and P. E. S. Chen (1986). "3D Beam-Column Element with Generalized Plastic Hinges," *Journal of Engineering Mechanics,* ASCE, Vol. 112, No. 7, pp. 627–641.

Standards Australia (1990). *Steel Structures,* AS4100, Sydney.

Stevens, L. K. (1967). "Elastic Stability of Practical Multistory Frames," *Proceedings of the Institution of Civil Engineers,* Vol. 36, London: ICE.

White, D. W., J. Y. R. Liew, and W. F. Chen (1991). *Second-Order Inelastic Analysis for Frame Design: A Report to SSRC Task Group 29 on Recent Research and the Perceived State-of-the-Art,* Structural Engineering Report No. CE-STR-91-12, West Lafayette, IN: Purdue University.

White, D. W., J. Y. R. Liew, and W. F. Chen (1993). "Toward Advanced Analysis in LRFD," in *Plastic Hinge Based Methods for Advanced Analysis and Design of Steel Frames—An Assessment of the State-of-the-Art,* D. W. White and W. F. Chen, eds., Bethlehem, PA: Structrual Stability Research Council, Lehigh University, pp. 95–173.

Yura, J. A. (1971). "The Effective Length of Columns in Unbraced Frames," *Engineering Journal,* AISC, Vol. 8, No. 2, pp. 37–42.

Zhou, S. P., L. Duan, and W. F. Chen (1990). "Comparison of Design Equations for Steel Beam-Columns," *Structural Engineering Review,* Vol. 2, No. 1, pp. 45–53.

Ziemian, R. D., McGuire, W., and G. G. Deierlien (1992a). "Inelastic Limit States Design: Part I—Planar Frame Studies," *Journal of Structural Engineering,* ASCE, Vol. 118, No. 9, pp. 2532–2549.

Ziemian, R. D., W. McGuire, and G. G. Deierlien (1992b). "Inelastic Limit States Design: Part II—Three-Dimensional Frame Study," *Journal of Structural Engineering,* ASCE, Vol. 118, No. 9, pp. 2550–2586.

7

NOTIONAL-LOAD PLASTIC-HINGE ANALYSIS FOR FRAME DESIGN

7.1 BACKGROUND

The specification beam-column interaction equations (e.g., AISC, 1993) are general-purpose formulas that are intended to provide a simple, but approximate means for designing a wide range of beam-column problems such as beam-columns with different end-restraint conditions, loading, and slenderness ratios. These equations can accommodate a confluence of many effects, such as residual stresses, member initial out of straightness and out of plumbness, distributed plasticity, local and lateral-torsional buckling, and other member behavioral effects. However, the specification beam-column design approach does introduce several difficulties in general frame design; one of these problems lies in addressing the stability of an individual member as part of an overall structural system. In the AISC LRFD approach, this interaction effect is accounted for by using a column effective-length procedure. In evaluating the column effective length, a distinction has to be made between braced and unbraced frames or sway and nonsway frames.

In view of recent trends toward elimination of column effective length in Canadian and other specifications, recent research has investigated the accuracy of the AISC LRFD beam-column equations with and without using effective-length factors (K-factors) for predicting system performance (CSA, 1989; Cheong, 1991; Liew, White, and Chen, 1991). The studies by Liew et al. (1991) conclude that the effective-length factor concept is an important tool for elastic analysis and design of steel frames. For nonredundant frames in which the limit load of the frame is governed by the failure of the most critical member, the elastic analysis/design procedure provides a good estimate of the actual system strength. For redundant systems, the elastic analysis/design procedure is generally conservative provided that the column effective length and second-order design moments are evaluated properly.

Strictly speaking, the beam-column equation approach is valid only for elastic frames. That is, the elastic forces obtained from the analysis and the column effective length used for frame design do not represent the actual behavior of a frame after inelastic redistribution of forces occurs. Procedures that attempt to compute a more precise effective-length factor to account for the formation of plastic hinges in a frame are rather cumbersome. Although the K-factor approach has been shown to give reliable predictions for subassemblies subjected to gravity load only, there is no theoretical reason to suppose that the capacity determined through interaction equations for the most critical member in a frame represents the limit load of the frame. Although a direct second-order elastic analysis eliminates the need to compute moment-amplification factors, an effective-length factor is still required in the axial-force-resistance term to ensure general validity of the beam-column equations for design using the current AISC LRFD provisions (Liew et al., 1991, 1992). The only way to avoid the use of K-factors completely and to eliminate separate member capacity checks is by using advanced analysis.

A plastic-zone analysis that includes spread of plasticity, residual stresses, initial geometric imperfections, and other significant behavioral effects would certainly eliminate the need for checking individual member capacities in the frame. Therefore, this type of method is generally classified as an advanced inelastic analysis. Although the current computational technology allows use of plastic-zone analysis of steel frames, the cost and effort of such a procedure are so great that analysis of a complete framework often would be prohibitive. Fortunately, it is possible to accommodate the dominant behavioral effects of members and joints through simple direct analysis approaches that are feasible for routine design use. One common approach for representing inelasticity effects in frames is to employ the elastic-plastic-hinge analysis.

As discussed in Chapter 6, the elastic-plastic-hinge approach is limited by its ability to provide strength assessment of columns that fail by inelastic buckling. The key reason for this limitation is that the analysis assumes the element to be fully between the plastic-hinge locations. Furthermore, plastic-hinge analysis often does not include the effects of initial imperfections in the modeling of a practical beam-column. Conversely, the physical columns always have locked-in residual stresses due to the manufacturing process, and they are not perfectly straight in nature.

In order to capture the physical behavior of a beam-column member, the effects of distributed plasticity and member initial imperfections must be represented in a properly advanced analysis. The frame-imperfection effects also need to be taken into consideration in analyzing the global behavior of the structural system. Although the AISC LRFD (AISC, 1986), Standards Australia (1990), and other specifications permit use of second-order plastic-hinge analysis to determine the magnitude and distribution of member forces in a system, the member forces so obtained must be checked against code-specified interaction equations.

This chapter presents a plastic-hinge-based procedure, which does not involve any modifications to the basic elastic-plastic-hinge theory, and which promises to satisfy the in-plane strength requirements of many frame members, for advanced analysis of moment-resisting frames. This method is called the *notional-load plastic-hinge approach.* It involves the use of equivalent lateral loads to account approximately for the influence of member imperfections and distributed plasticity effects on the overall system strength. The determination of the notional-lateral-load factor for frame design is discussed, and the accuracy of this method is verified by comparison with the benchmark solutions generated by Kanchanalai (1977) and Zhou, Duan, and Chen (1990). All the primary benchmark problems utilized in the present studies involve small subassemblies and isolated members with imperfections. As discussed in Chapter 6, studies of the performance of a large framing system are not necessary for verification of the inelastic analysis model, because if the performance of an isolated member is predicted properly for general loading cases, the proper prediction of the system's response should follow from the principles of equilibrium and compatibility.

7.2 CONSIDERATION OF INITIAL IMPERFECTIONS FOR FRAME DESIGN

7.2.1 ECCS Approach

The European Convention for Construction Steelwork (ECCS) has published several reports (ECCS, 1984, 1991) that address issues pertinent to the limit state design, and recommend simplified techniques for elastic-plastic-hinge analysis of structural steel frames. The ECCS (1991) also provides recommendations for the use ofsecond-order elastic-plastic-hinge analysis in frame design, but it requires the analysis to model explicitly the frame initial geometric imperfections. Because the effects of residual stresses and distributed plasticity are not considered in the plastic-hinge analysis, the frame's geometric imperfections are "enlarged" to simulate these deleterious effects. An "equivalent" initial out of plumbness, as shown in Fig. 7.1, is required for elastic-plastic-hinge analysis using the ECCS approach.

Explicit modeling of member initial out of straightness is not required if the column axial force P is less than $0.260\pi^2EI/L^2$ (or $L/\sqrt{P/EI} > 1.6$). If the column axial load is higher than $0.260\pi^2EI/L^2$, an exaggerated member initial out of straightness must be modeled simultaneously with the specified initial out of plumbness to compensate for possible member instability and residual stress effects.

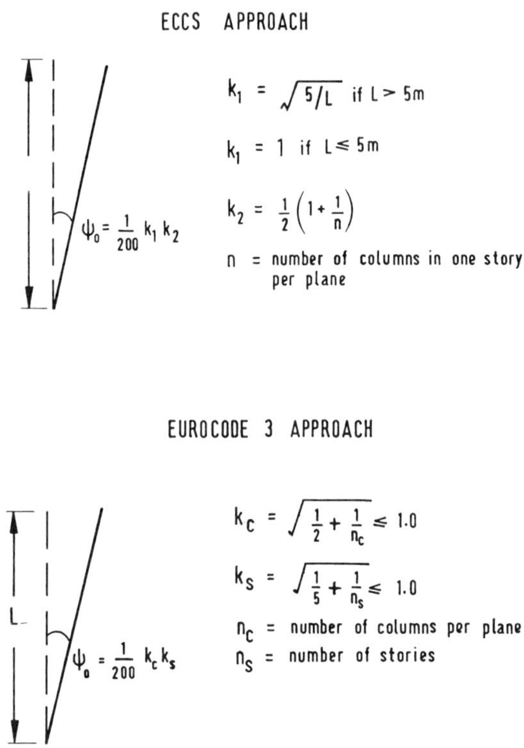

Fig. 7.1 Initial sway imperfections according to ECCS and Eurocode 3 approaches.

7.2.2 Eurocode 3 Approach

Similar to the ECCS (1991) approach, Eurocode 3 (1992) also requires explicit modeling of initial sway imperfections, as shown in Fig. 7.1, for elastic-plastic-hinge analysis. These initial sway imperfections apply in all horizontal directions, but need only be considered in one direction at a time. Columns that carry a vertical load of less than 50% of the mean value of the vertical load per column in the plane considered shall not be included in n_c for the calculation of the sway imperfection. Also, columns that do not extend through all the stories included in n_s shall not be included in n_c. Those floor levels and roof levels that are not connected to all the columns included in n_c shall not be included when determining n_s. The possible torsional effects on the structure of antisymmetric sways, on two opposite faces, should be considered. Eurocode 3 also permits the initial sway imperfection to be replaced by a closed system of equivalent horizontal forces.

If the column axial force is less than $0.250\pi^2 EI/L^2$, modeling of initial out of straightness is not required in the analysis. Otherwise, an enlarged member initial out of straightness of magnitude ranging from $L/400$ to $L/150$ must be modeled explicitly for elastic-plastic-hinge analysis. The precise magnitude of initial out of straightness is dependent on the applicable column strength given in Eurocode 3, as shown in Fig. 5.5.

7.3 THE CONCEPT OF NOTIONAL LOADS

Instead of specifying the "enlarged" initial geometric imperfections for every column in a story, the sway imperfections (column out of plumbness) in a frame may be replaced by an equivalent lateral load that is expressed as a fraction of the gravity loads acting on a story. The equivalent lateral loads, which are called *notional loads,* are chosen to adequately allow for frame instability when they are combined with the applied lateral loads used for second-order frame analysis. The distributed plasticity and residual stress effects that are neglected in the conventional elastic-plastic-hinge analysis are approximated by the notional lateral loads such that part of the frame capacity is used up by the second-order moment associated with the notional loads. This approach shows promise as an effective mean of accounting for geometrical imperfections and distributed plasticity without physically altering the geometry of the frame model. The resulting approach usually yields conservtive and accurate strength predictions for a wide range of frame problems (Liew et al., 1994).

7.3.1 Illustrative Examples

The basic concept of the notional-load procedure is illustrated by studying the cantilever column in Fig. 7.2. The purpose is to obtain a practical value of the notional horizontal load to be applied at the top of the cantilever column by calibrating the results obtained from the second-order plastic-hinge analysis against the AISC LRFD column strength. The axial-load capacity of a cantilever column evaluated based on the AISC LRFD column equations is

$$P_n = P_y(0.658^{\lambda_c^2}), \qquad \lambda_c \leq 1.5 \tag{7.3.1}$$

where $\lambda_c^2 = (2L)^2/(\pi^2 r^2)(F_y/E)$, because the effective length of the cantilever column is twice the actual height L of the column.

A notional lateral load αP_u is now applied at the top of the column, as shown in Fig. 7.2, in which α is the ratio of the notional lateral load to the total gravity load P_u in the column. In the present study, the maximum second-order end moment, which is at the fixed end moment of the cantilever column, is computed based on LeMessurier's amplification factor (LeMessurier, 1977):

$$M_{\max} = A_F M \tag{7.3.2}$$

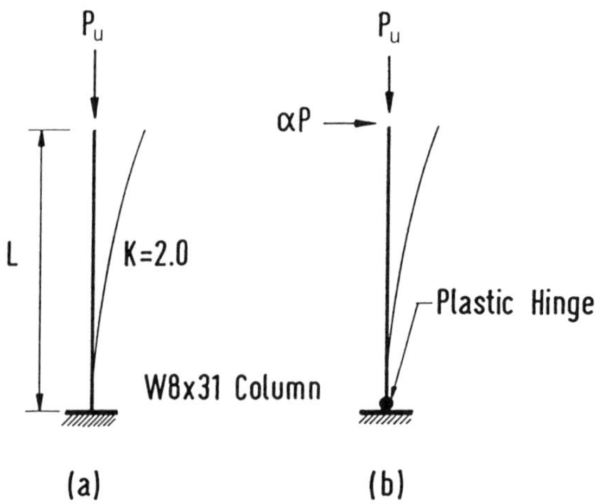

Fig. 7.2 Cantilever column for calibration of α.

where $M = \alpha P_u L$ is the first-order moment due to the notional lateral load, and the amplification factor is written as (LeMessurier, 1977):

$$A_F = \frac{1}{1 - \left(\frac{\Sigma P_u}{\Sigma P_L - \Sigma C_L P_u}\right)} \tag{7.3.3}$$

where

$C_L = \beta \frac{K_n^2}{\pi^2} - 1$
I_i = moment of inertia of column i
K_n = effective length factor; $K_n = 2$ for cantilever column
L_i = height of column i
$P_L = \frac{\beta EI}{L^2}$
ΣP_u = sum of all vertical forces acting on the story at factored load level
P_{ui} = axial force on column i
β = column end-restraint coefficient; $\beta = 3$ for cantilever column with one end fixed and the other end free

For a cantilever column subjected to an axial load and notional force αP_u, the second-order fixed end moment from Eq. (7.3.2) becomes

$$M_{\max} = \frac{1}{1 - \dfrac{P_u}{3EI/L^2 - (12/\pi^2 - 1)P_u}} \alpha P_u L \tag{7.3.4}$$

LeMessurier's procedure provides essentially the "exact" second-order elastic moment for this problem. The column is assumed to have reached its maximum strength when the second-order elastic forces at the fixed end reach the cross-section strength. By adopting the AISC LRFD cross-section strength equation for a member with zero length as the plastic strength surface for the elastic-plastic-hinge analysis, the second-order moment and axial-force interactions that produce a plastic hinge at the base of the cantilever may be written as

$$\frac{P_u}{P_y} + \frac{8}{9}\frac{\alpha P_u L}{M_p}\left[\frac{1}{1 - \dfrac{P_u/P_y}{1.216/\lambda_c^2 - 0.216\, P_u/P_y}}\right] = 1.0, \qquad \text{for } \frac{P_u}{P_y} \geq 0.2 \tag{7.3.5}$$

The bracket term in Eq. (7.3.5) is LeMessurier's moment-amplification factor after normalization. In the second order plastic-hinge analysis, the overturning moment at the column base, $\alpha P_u L$, produced by the notional lateral load (αP_u) is amplified by this factor.

The notional-load factor α can be obtained by solving Eq. (7.3.5) as

$$\alpha = \frac{9}{8}\left(1 - \frac{P_u}{P_y}\right)\frac{Z/rA}{(L/r)(P_u/P_y)}\left[1 - \frac{P_u/P_y}{12/\pi^2\lambda_c^2 - 0.216 P_u/P_y}\right], \qquad \text{for } \frac{P_u}{P_y} \geq 0.2 \tag{7.3.6}$$

in which L is the member length, Z is the cross-section plastic modulus, r is the radius of gyration, and A is the cross-section area.

When a W8 × 31 section is used, the term Z/rA in Eq. (7.3.6) is equal to 0.960 and 0.765 for strong- and weak-axis bending, respectively. It should be noted that the α parameter calculated from Eq. (7.3.6), for an assumed W8 × 31 section bending about the strong axis, is not more than 5% different from the result obtained for other wide-flange sections available in the AISC LRFD manual. However, for sections bending about the weak axis, the variation of α is larger and the maximum difference is about 25%—the α values calculated for other sections may be 25% lower than the α value calculated based on a W8 × 31 section.

The W8 × 31 section is used here for the calculation of the parameter α that results in the formation of a plastic hinge at a load equal to the AISC LRFD column capacity implied by Eq. (7.3.1). The cantilever column is bent about the strong axis. The results are shown in Table 7.1 for L/r_x ranging from 10 to 80. It is observed that the parameter α in the table is not a constant value. To simplify the design process, a simplified expression for α is required.

By adopting the ECCS geometric imperfection approach, the "enlarged" initial out of plumbness of the cantilever column is equal to 1/200 of the column height. For the column subjected to an axial force P_u, the axial force acting on the column out of plumbness produces an overturning moment of $P_u L/200$. In using the proposed notional-load method, a notional lateral load of $0.005P_u$ has to be applied at the top of the cantilever column to obtain an overturning moment of $P_u L/200$, which is equivalent to the moment generated by the gravity load acting on the "enlarged" column out of plumbness of $L/200$. To summarize, an initial lean of 1 to 200 is equivalent to a notional horizontal-load factor of 0.5%. This is greater than the erection tolerance in order to allow for the effects of other sorts of imperfection.

Following the preceding idea, an investigation is undertaken to study the discrepancy involved in the maximum-strength predictions when the cantilever column in Fig. 7.3 is analyzed using a notional-load factor of α equal to 0.005.

The strong-axis column capacities computed by the proposed notional-load method with $\alpha = 0.005$ are shown in Table 7.1. The predicted results are compared with the AISC LRFD column-strength equation (Eq. 7.3.1). The A_F factors in Table 7.1 are the LeMessurier's moment-amplification factors computed based on the bracket term in Eq. (7.3.2). These results indicate that the second-order effects become larger as the height of the cantilever column increases from $L/r = 10$ to 80.

It is interesting to observe that the errors in the predicted P_n/P_y values are not more than 5% for the cantilever columns with L/r_x ranging from 10 to 80. The results in Table 7.1 show that a large error may be made in selecting the α value without much sacrifice in the accuracy of the maximum strength. A design parameter of $\alpha = 0.005$ is therefore proposed. Although more complicated expressions for α could be developed and proposed (α may be normalized by the Z/rA parameter as given in Eq. (7.3.6) for wide-flange sections other than W8 × 31), the value of $\alpha = 0.005$ is chosen to keep the design process simple. Also, the notional-load factor of 0.005 is consistent with the ECCS column initial out-of-plumbness deflection of $L/200$, which is the upper-bound value recommended for use with the second-order elastic-plastic-hinge analysis.

In the next two sections, several portal frames with and without "leaning" columns are analyzed by the proposed notional-load plastic-hinge method. The predicted results are compared with the more "exact" plastic-zone solutions.

TABLE 7.1 Calibration of α—Cantilever Columns Subjected to Gravity Load Only

$\frac{L}{r}$ (1)	$\frac{KL}{r}$ (2)	λ_c (3)	$\frac{P_{cr}}{P_y}$ (4)	$\frac{P_n}{P_y}$ (5)	A_F (6)	Required α to Obtain LRFD Column Strength (7)	Predicted P_n/P_y Based on $\alpha = 0.005$ (8)	Ratio (8)/(5)
10	20	0.224	19.93	0.979	1.043	0.0022	0.954	0.974
20	40	0.449	4.960	0.919	1.187	0.0040	0.901	0.980
40	80	0.897	1.243	0.714	2.111	0.0051	0.717	1.004
60	120	1.346	0.552	0.469	5.618	0.0036	0.445	0.950
80	160	1.794	0.311	0.272	6.857	0.0053	0.273	1.007

A_F = moment amplification factors (Eq. 7.3.3)
K = effective-length factor
P_n = nominal axial capacity
P_y = axial capacity at full-yield condition or squash load
α = notional-load factor
$P_{cr} = \pi^2 EA/(KL/r)^2$

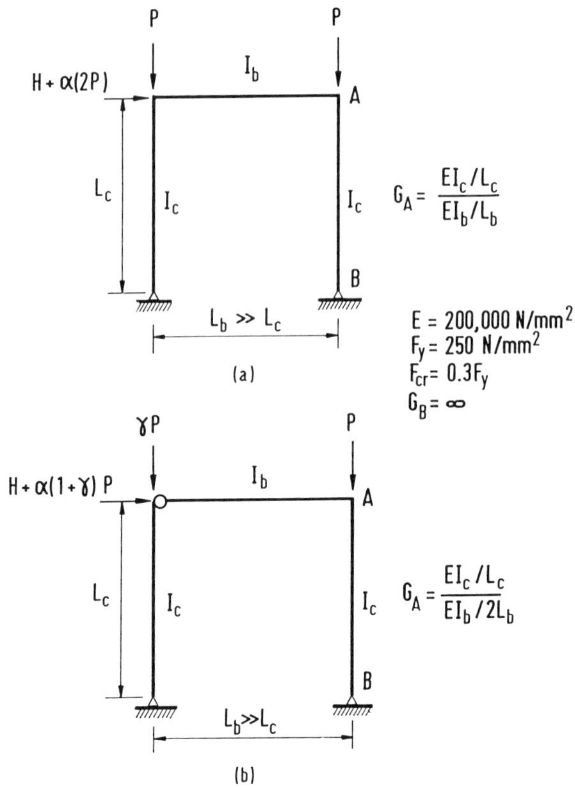

Fig. 7.3 (a) Portal frames (Kanchanalai, 1977). (b) Leaning-column frames (Kanchanalai, 1977).

7.4 SWAY FRAMES SUBJECT TO GRAVITY LOAD

The frames under consideration are shown in Figs. 7.3a and 7.3b. In the notional-load plastic-hinge analysis, the second-order forces in the beam-column members are computed with a notional lateral load equal to $0.005\Sigma P_u$ applied at the top of the left-hand column, where ΣP_u is the sum of all gravity loads in the story. The maximum strength computed by this approach is compared with the column capacity predicted by AISC LRFD column equations. The present AISC LRFD column formulas are based on the ultimate strength of an initially crooked member, including the effects of initial residual stresses. Also, distributed yielding effects in the columns are considered in the effective-length factor, which is computed based on LeMessurier's K-factor procedure (LeMessurier, 1977). Inelastic stiffness-reduction factors are employed with LeMessurier's procedure for these calculations (Liew et al., 1991).

The nominal column capacities calculated from the preceding procedures are presented in Tables 7.2 and 7.3. Strong-axis bending is assumed for determination of P_n/P_y based on $\alpha = 0.005$. Several parameters relevant to the behavior of the frames are also shown in these tables. The results cover a wide range of planar frames with beam-to-column stiffness ratios G_A ranging from 0 to 4, and slenderness ratios L/r_x ranging from 10 to 80.

The results in Tables 7.2 and 7.3 show that the notional-load parameter α required for the plastic-hinge analysis to achieve the nominal strength implied by the LRFD column equations is somewhat smaller than the recommended value of $\alpha = 0.005$. However, the proposed plastic-hinge analysis with a notional load of $0.005\Sigma P_u$ applied at the tops of the columns provides excellent results for all portal frames, and the results are largely conservative for all frames with a leaning column. These observations are true for a wide range of simple portal frames with G_A values ranging from 0 to 4 and with L/r_x from 10 to 80.

7.5 SWAY FRAMES SUBJECT TO COMBINED GRAVITY AND LATERAL LOADS

For sway frames subject to combined gravity and lateral loads, an additional notional lateral load of 0.005 times the total story gravity loads must be applied with the lateral load to achieve an overall improvement in strength prediction. In the present studies, the second-order plastic-hinge analyses are performed for the benchmark frames shown in Figs. 7.4–7.10 with a notional lateral load of $0.005\Sigma P_u$ applied at the top of the frame. Plastic-zone solutions for these benchmark frames have been generated and reported by Kanchanalai (1977), as discussed in Chapter 6.

Because the notional-load plastic-hinge analysis implicitly considers the column initial out-of-plumbness and residual stress effects through the use of notional lateral loads, the comparisons are valid only for frames with initial geometric imperfections. For comparison purposes, the Kanchanalai plastic-zone strength curves (considering only residual stress effects) are adjusted based on an effective P–Δ moment that would reduce the plastic-zone strength for the pure column case to the capacity predicted by the AISC LRFD column curve. The details of the procedures are reported in Liew et al. (1991, 1992). In fact, these adjusted strength curves were employed for the calibration of the present AISC LRFD beam-column equations (LeMessurier, 1977); therefore, they are used as benchmarks for comparisons.

Also included in Figs. 7.4–7.10 are the strength curves generated by the conventional elastic-plastic-hinge analysis *without* the application of notional lateral load. It should be noted that both the strong- and weak-axis strength curves obtained by the conventional plastic-hinge analyses are identical because the analyses are based on the use of the LRFD beam-column equations with $L = 0$, and the results are presented in a nondimensional form.

TABLE 7.2 Calibration of α—Frames with Leaning Columns Subjected to Gravity Load Only

Case	G	γ	$\frac{L}{r}$	K	$\frac{P_{cr}}{P_y}$	$\frac{P_n}{P_y}$	Required α to Obtain P_n/P_y	Predicted P_n/P_y Based on α = 0.005	Ratio
	(1)	(2)	(3)	(4)	(5)	(6)	(7)	(8)	(8)/(6)
1	0	2	10	3.255	7.505	0.946	0.0018	0.865	0.915
2	0	2	20	3.255	1.876	0.800	0.0027	0.699	0.873
3	0	2	30	3.255	0.834	0.605	0.0023	0.500	0.826
4	0	2	40	3.255	0.469	0.410	0.0017	0.341	0.832
5	0	2	50	3.255	0.300	0.263	0.0026	0.239	0.906
6	0	2	60	3.255	0.200	0.183	0.0035	0.174	0.949
7	0	3	10	3.726	5.727	0.930	0.0017	0.823	0.886
8	0	3	20	3.726	1.432	0.747	0.0023	0.612	0.820
9	0	3	30	3.726	0.636	0.518	0.0016	0.404	0.780
10	0	3	40	3.726	0.358	0.314	0.0019	0.266	0.846
11	0	3	50	3.726	0.229	0.201	0.0028	0.183	0.911
12	4	2	10	3.752	5.648	0.929	0.0018	0.832	0.896
13	4	2	20	5.253	0.720	0.559	0.0023	0.490	0.875
14	4	2	30	5.464	0.296	0.260	0.0042	0.254	0.979
15	4	2	40	5.464	0.166	0.146	0.0063	0.150	1.026
16	4	3	10	4.534	3.867	0.897	0.0017	0.769	0.857
17	4	3	20	6.272	0.505	0.437	0.0022	0.385	0.881
18	4	3	30	6.302	0.222	0.195	0.0046	0.193	0.989

G = ratio of the bending stiffness of the column to the beam $(I_c/L_c)/(I_b/2L_b)$.

TABLE 7.3 Calibration of α—Simple Portal Frames Subjected to Gravity Load Only

Case	G (1)	$\frac{L}{r}$ (2)	K (3)	$\frac{P_{cr}}{P_y}$ (4)	$\frac{P_n}{P_y}$ (5)	Required α to Obtain P_n/P_y (6)	Predicted P_n/P_y Based on $\alpha = 0.005$ (7)	Ratio (7)/(5)
1	0	10	2.000	19.809	0.979	0.0022	0.954	0.974
2	0	20	2.000	4.952	0.919	0.0040	0.901	0.981
3	0	30	2.000	2.201	0.827	0.0050	0.828	1.002
4	0	40	2.000	1.238	0.713	0.0051	0.717	1.006
5	0	50	2.000	0.792	0.590	0.0044	0.576	0.977
6	0	60	2.000	0.550	0.467	0.0036	0.445	0.953
7	0	70	2.000	0.404	0.355	0.0041	0.346	0.975
8	0	80	2.000	0.310	0.271	0.0053	0.273	1.007
9	3	10	2.061	18.72	0.978	0.0022	0.951	0.973
10	3	20	2.258	3.900	0.898	0.0038	0.873	0.972
11	3	30	2.607	1.300	0.725	0.0042	0.704	0.971
12	3	40	2.872	0.602	0.499	0.0040	0.485	0.971
13	3	50	2.918	0.374	0.328	0.0056	0.332	1.013
14	3	60	2.918	0.259	0.227	0.0077	0.238	1.045

G = ratio of the bending stiffness of the column to the beam, $(I_c/L_c)/(I_b/L_b)$.

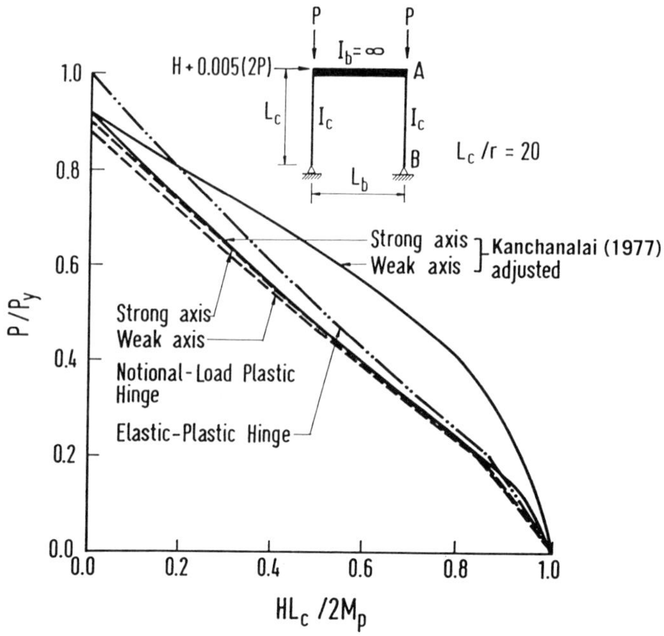

Fig. 7.4 Comparison of strength curves from notional-load plastic-hinge analysis to adjusted strength curves (Kanchanalai, 1977): portal frame with $G_A = 0$ and $L_c / r = 20$.

The following observations are made from the strength comparisons:

1. For the simple portal frames in Figs. 7.4–7.7, the notional-load plastic-hinge analyses compare well with Kanchanalai's strong-axis results for all ranges of axial force vs. moment. These observations reinforce the point that a notional lateral load must be applied with the design lateral load to obtain an adequate prediction of the system strength. Because the conventional elastic-plastic-hinge analysis does not consider gradual stiffness degradation due to distributed plasticity and residual-stress effects, it is suggested that these effects on the strength be simulated by applying the notional lateral load in the analyses. If the notional lateral load is not applied in the second-order plastic-hinge analysis, the maximum unconservative error can be as high as 40% (see Figs. 7.5 and 7.6).
2. For the frames with a leaning column bent about the strong- and weak-axis (Figs. 7.8–7.10), the results from the notional-load plastic-hinge analyses are generally conservative in comparison with Kanchanalai's strength curves. In most cases, the maximum *conservative* error occurs when the frame is subjected to gravity load only. The errors are expected

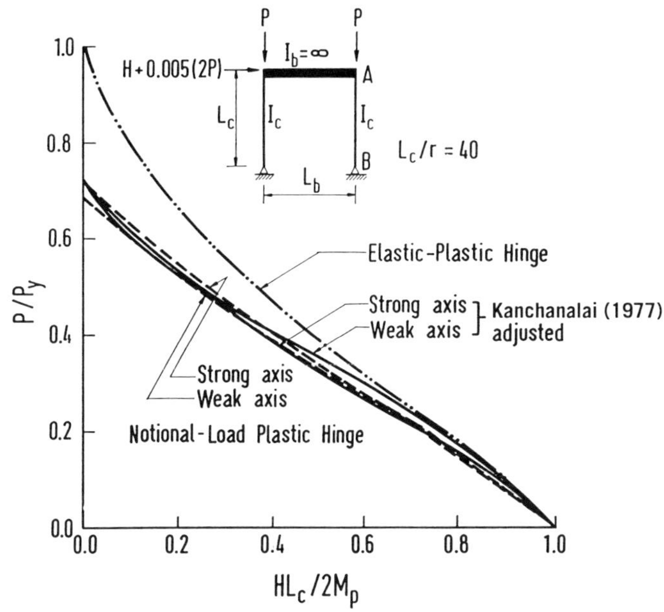

Fig. 7.5 Comparison of strength curves from notional-load plastic-hinge analysis to adjusted strength curves (Kanchanalai, 1977): portal frame with $G_A = 0$ and $L_c/r = 40$.

to be smaller for the frames with a leaning column when the axial loads are decreased.

3. For all the benchmark portal frames, the weak-axis strengths are conservatively predicted by the notional-load plastic-hinge analysis. This is because the LRFD cross-section strength curve used in the analyses (i.e., the LRFD interaction equations based on $L = 0$) does not distinguish between strong- and weak-axis bending. In Figs. 7.5–7.10, the "exact" strong-axis strength curves computed by Kanchanalai (1977) do not differ very much from the weak-axis strength curves. The weak-axis strengths of these frames tend to be dominated more by the member instability than by the cross-section strength. In these cases, the use of the LRFD cross-section strength curve for the analysis to predict both the strong- and weak-axis strengths is adequate. But for the frame in Fig. 7.4, the beam-columns are relatively stocky compared to those shown in Figs. 7.5–7.10. In these cases, the weak-axis strengths are controlled by the formation of a plastic hinge at the top of the columns. Therefore, the more "exact" weak-axis cross-section strength curve, such as that proposed by ASCE-WRC (1971) or Duan and Chen (1990) for plastic design, is more representative of the actual behavior (see Chapter 5, Section 5.6.2).

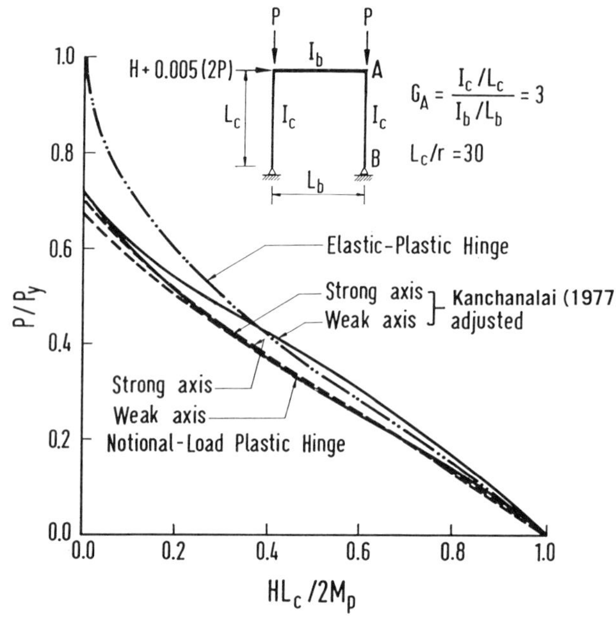

Fig. 7.6 Comparison of strength curves from notional-load plastic-hinge analysis to adjusted strength curves (Kanchanalai, 1977): portal frame with $G_A = 3$ and $L_c/r = 30$.

7.6 APPLICATION TO A LARGE FRAMEWORK

For multistory and multibay steel frames, the notional lateral load may be reduced because not all the columns in a building are arranged with out of plumbness leaning in the same direction. By adopting the expressions for out-of-plumbness imperfections recommended by Eurocode 3 (1992), the expression for story notional lateral load H^* may be expressed as

$$H^* = \frac{1}{200} K_c K_s \Sigma P_u \tag{7.6.1}$$

where

$$K_c = \sqrt{\frac{1}{2} + \frac{1}{n_c}} \leq 1.0 \tag{7.6.2}$$

$$K_s = \sqrt{\frac{1}{5} + \frac{1}{n_s}} \leq 1.0 \tag{7.6.3}$$

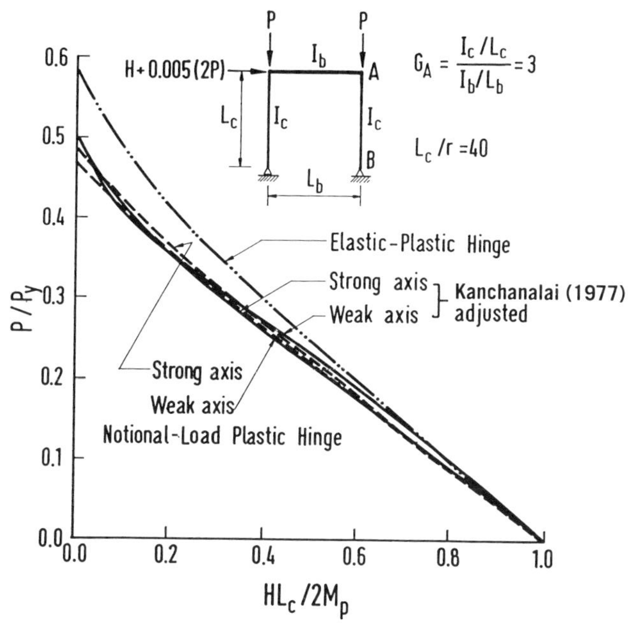

Fig. 7.7 Comparison of strength curves from notional-load plastic-hinge analysis to adjusted strength curves (Kanchanalai, 1977): portal frame with $G_A = 3$ and $L_c/r = 40$.

where n_c is the number of columns in the plane of the story, n_s is the number of stories in the plane of the frame, and ΣP_u represents the sum of gravity loads acting on the story.

Coefficients K_c and K_s in Eqs. (7.6.2) and (7.6.3) account for the fact that a large number of columns in a story and a large number of stories in a frame would probabilistically reduce the influence of initial imperfection effects on frame or story stability; therefore, the notional lateral load may be reduced accordingly. Notional load factors α calculated based on the preceding procedure are shown in Table 7.4 for structural systems with n_c from 2 to 12 and n_s from 1 to 80. These α factors are compared with the AISC erection tolerance for columns, which is equivalent to $\alpha = 0.002$ (or 1/500). It can be observed from Table 7.4 that for multibay and multistory systems, the α values are not significantly smaller than the AISC erection tolerance for columns. Therefore, the effects produced by the notional lateral loads cannot be neglected in the analysis of a large structural system.

Eurocode 3 requires the explicit modeling of geometric imperfections in analysis using the second-order elastic-plastic-hinge method. The code specifies the use of artificially large values of frame imperfections to account for actual out of plumbness of frame members within erection tolerances, and

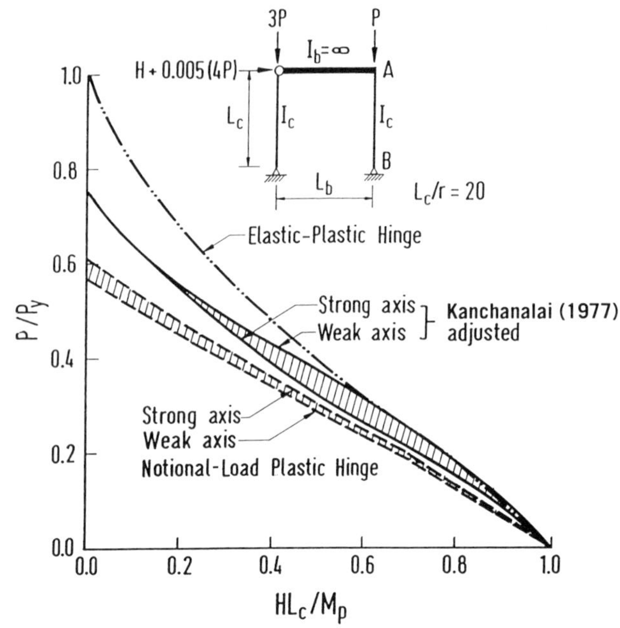

Fig. 7.8 Comparison of strength curves from notional-load plastic-hinge analysis to adjusted strength curves (Kanchanalai, 1977): leaning-column frame with $G_A = 0$ and $L_c/r = 20$.

the effects of residual stresses, load eccentricities, and member imperfections that are not considered in the analysis. The notional-load approach utilizes an equivalent horizontal shear-force system for representation of the imperfection effects. The similarities and differences between the notional-load plastic-hinge method and the Eurocode 3 procedures are outlined in Fig. 7.11. The specification of frame imperfections is the same for both approaches, with the initial out of plumbness being defined as $\varphi_0 = (1/200)K_c K_s$.

7.7 NOTIONAL LATERAL LOADS FOR BRACED SYSTEMS AND MEMBERS

For braced frames where lateral stability is provided by adequate attachment to shear walls, or other systems that provide lateral-force resistance, the member initial imperfection effects often have a greater influence on the stability of individual members and the overall frame than the sway imperfection effects. The member imperfection effects are particularly important for beam-column members subjected to high axial forces, for example, $P > 0.25\pi^2 EI/L^2$. In this case, the effects of member imperfections and the associated P–δ effects must be considered in the analysis of the entire system.

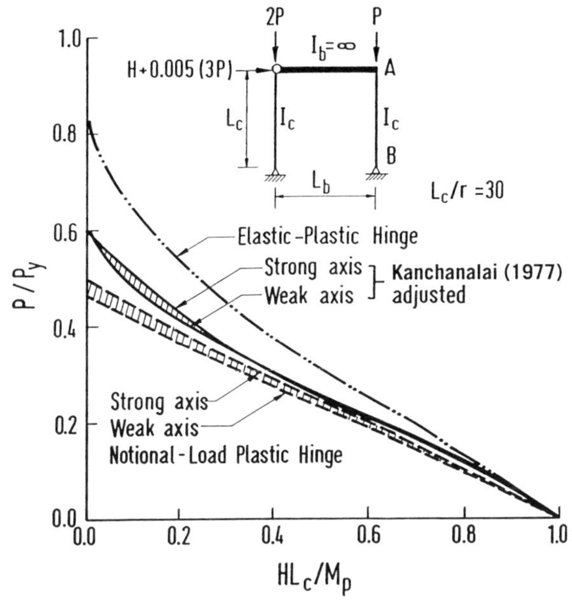

Fig. 7.9 Comparison of strength curves from notional-load plastic-hinge analysis to adjusted strength curves (Kanchanalai, 1977): leaning-column frame with $G_A = 0$ and $L_c/r = 30$.

One way to account for these local P–δ effects is to apply a notional force at the maximum moment point of the member. This notional force should be counterbalanced by the two end-reaction forces to maintain equilibrium of forces at the individual member level. However, research by Chen and Atsuta (1976) indicates that the insertion of a plastic hinge at the "exact" maximum moment point in a frame member is not required for an accurate estimate of the maximum strength of the member. As long as the "exact" location of the plastic hinge in a member is not more than $L/6$ away from the assumed position (where L is the length of the member), the difference in strength prediction is not more than 5%. The foregoing observation is true for beam-columns subjected to various types of transverse loading.

The case studies shown in what follows demonstrate how the initial member imperfections of a column or beam-column can be compensated for with sufficient accuracy by using the member notional lateral-load concept. Based on these studies, a suitable notional-load factor is recommended for practical design use.

7.7.1 Axially Loaded Columns

A suitable notional lateral load to be applied at the column's midlength can be obtained by calibrating the results from the second-order plastic-hinge

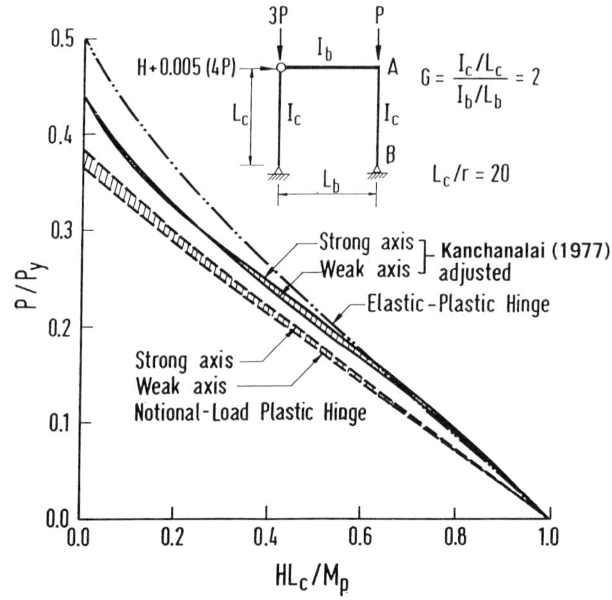

Fig. 7.10 Comparison of strength curves from notional-load plastic-hinge analysis to adjusted strength curves (Kanchanalai, 1977): leaning-column frame with $G = 2$ and $L_c/r = 20$.

analysis against the results from the AISC LRFD column-strength equations, as shown in Fig. 7.12. The calibration is carried out on a W8 × 31 column pinned at both ends, subjected to a compressive axial force. The column is assumed to have reached its maximum strength when the second-order elastic forces at midlength reach the cross-section plastic strength. The required α

TABLE 7.4 Notional-Load Factor α for a Gravity-Loaded System

No. of Stories,	No. of Columns in the Plane, n_c					
n_s	2	4	6	8	10	12
1	0.0050	0.0043	0.0041	0.0040	0.0039	0.0038
2	0.0042	0.0038	0.0036	0.0034	0.0032	0.0032
3	0.0036	0.0032	0.0030	0.0029	0.0028	0.0028
4	0.0034	0.0031	0.0029	0.0027	0.0026	0.0026
5	0.0032	0.0027	0.0026	0.0025	0.0025	0.0024
10	0.0027	0.0024	0.0022	0.0022	0.0022	0.0021
20	0.0025	0.0022	0.0020	0.0020	0.0019	0.0019
40	0.0024	0.0021	0.0019	0.0019	0.0018	0.0018
80	0.0023	0.0020	0.0019	0.0018	0.0018	0.0018

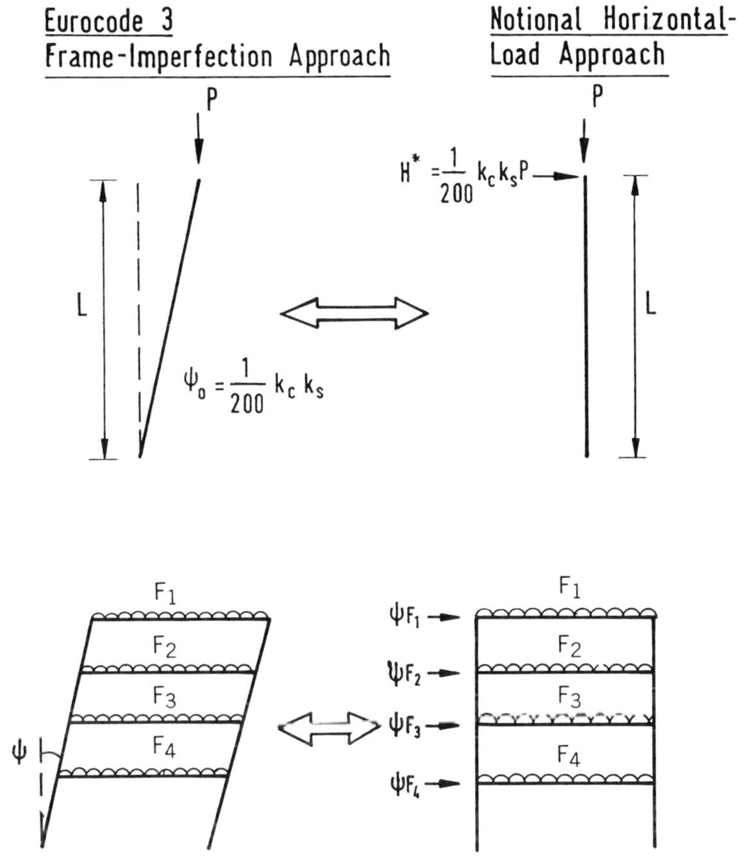

Fig. 7.11 Modeling of frame imperfections using Eurocode 3 and notional-load procedures.

value for the second-order plastic-hinge analysis to attain the same column capacity implied by the LRFD column-strength equation is computed for the column bent about the strong axis. The results are shown in Fig. 7.12 for column parameter λ_c ranging from 0.1 to 1.5. It is observed that the notional-load factor α is not a constant value for columns with different λ_c values. For practical design purposes, an upper-bound value of $\alpha = 0.01$ is proposed. This is equivalent to the use of $\alpha = 0.005$ to account for out of plumbness and other detrimental effects for the cantilever column shown in Section 7.3.1.

Figure 7.13 shows the strong-axis column-strength curves generated by the proposed notional-load plastic-hinge method with $\alpha = 0.01$, and by the conventional plastic-hinge analysis without notional load. In the conventional elastic-plastic-hinge analysis, the column is discretized into two elements with an initial deflection $\delta_0 = L/1500$ at midlength. The predicted results are

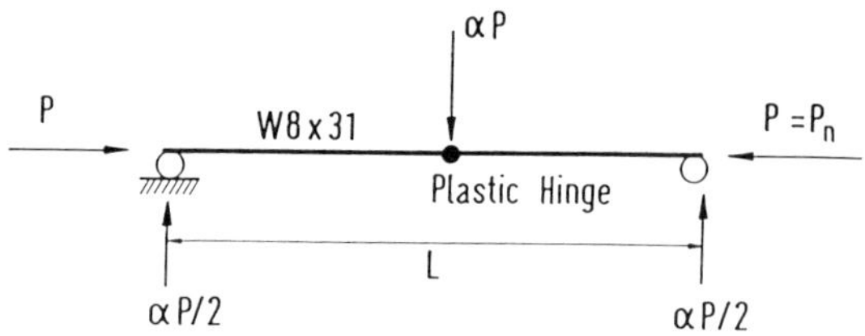

$$M_{max} = \frac{\alpha P_n L}{4}\left[\frac{\tan u}{u}\right] = \left(1 - \frac{P_n}{P_y}\right)\frac{9M_p}{8}$$

$$P_n = 0.658^{\lambda_c^2} \qquad u = (\pi/2)\sqrt{P/P_e}$$

λ_c	Required α	Percentage Error Based on $\alpha = 0.01$ $(1 - P/P_n) \times 100\%$
0.10	0.0020	1.6
0.20	0.0039	2.5
0.30	0.0058	2.6
0.40	0.0073	2.3
0.50	0.0086	1.5
0.60	0.0096	0.5
0.80	0.0104	-0.8
0.90	0.0103	-0.6
1.10	0.0091	1.9
1.20	0.0083	3.4
1.30	0.0076	4.4
1.40	0.0072	4.6
1.50	0.0074	3.8

Fig. 7.12 Notional lateral load for an axially loaded column.

compared with the LRFD column-strength curve. The errors in the predicted maximum strengths associated with the use of the notional-load plastic-hinge method are within 5% for column parameter λ_c ranging from 0.1 to 1.50. The elastic-plastic-hinge analysis without applying notional load overpredicts the column strength up to a maximum error of 22.5% (see Fig. 7.13). This maximum error corresponds to the column with a slenderness parameter of unity, at which value residual stress and initial out-of-straightness effects interact to produce the greatest reduction in strength from the theoretical value for a perfect column.

7.7.2 Isolated Beam-Columns

The plastic-zone solutions for isolated beam-columns generated by Zhou et al. (1990) are included to assess the performance of the notional-load plastic-hinge approach for which the P–δ effects within the member can be signifi-

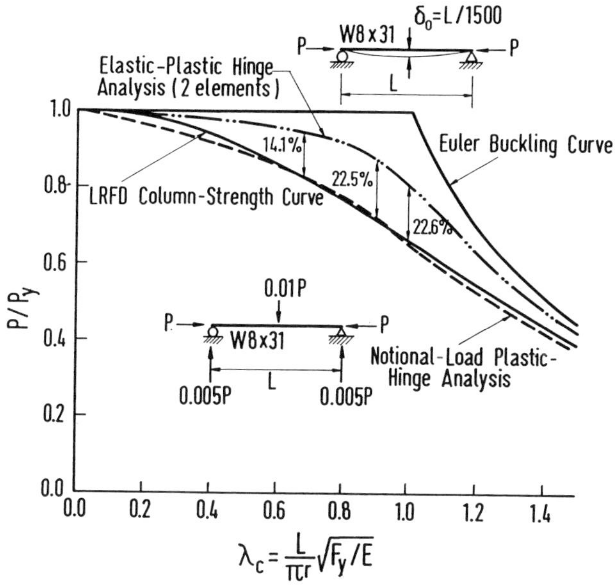

Fig. 7.13 Comparison of column-strength curves from plastic-hinge analyses with the LRFD column-strength curve.

cant. In the notional-load plastic-hinge approach, the member initial imperfection effects are simulated by applying a notional load, which is equal to 1% of the axial load, at midlength of the beam column shown in Fig. 7.14. The plastic-hinge analysis was carried out using two elements per member. Also plotted in Fig. 7.14 is a set of beam-column strength curves obtained based on the AISC LRFD interaction equations. The maximum moment in the AISC LRFD interaction equations is computed based on direct second-order elastic analysis.

Figure 7.14 shows that the results predicted by the notional-load plastic-hinge analysis are generally greater than the plastic-zone results. However, these results compare well with the strength curves obtained using the AISC LRFD beam-column interaction equations. In summary, the maximum error from the notional-load plastic-hinge analysis is about 8% compared to the plastic-zone solutions. The maximum error is, however, less than 5% in comparison with the AISC LRFD beam-column strength curves, as shown in Fig. 7.14.

Additional benchmark checks involving multistory and gable frame systems subjected to combined gravity and lateral loads are provided in Chapter 8. The reader is referred to this chapter to further confirm the validity of the proposed method.

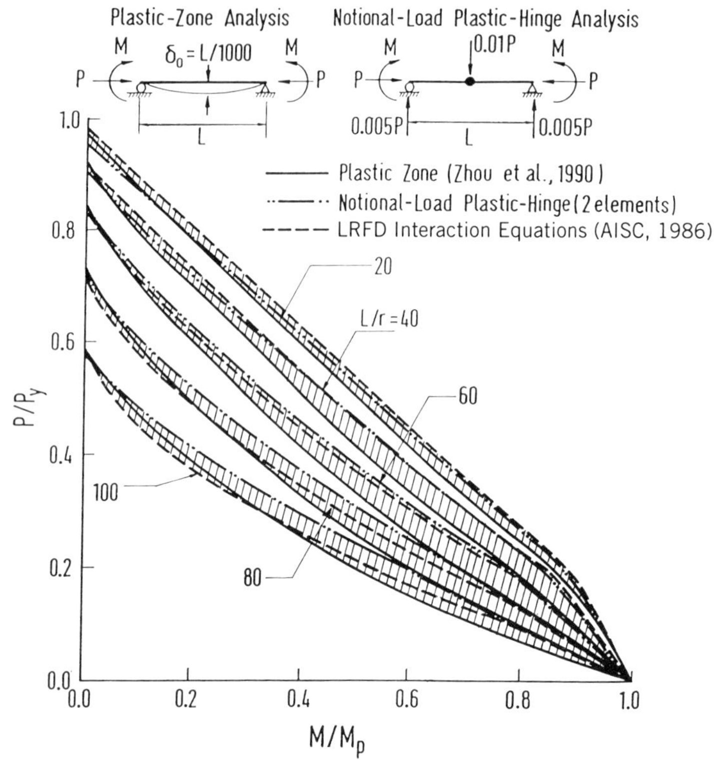

Fig. 7.14 Comparison of beam column-strength curves from notional-load plastic-hinge analysis with LRFD beam-column interaction strength curves and "exact" strength curves from Zhou et al. (1990).

7.8 DESIGN IMPLEMENTATIONS

In local subassemblies where columns are slender ($\lambda_c > 0.4$) and subjected to high axial loads (say, $P > 0.25\,\pi^2 EI/L^2$), the predictions of the distribution of loads and the magnitude of deformations by the conventional elastic-plastic-hinge method are not sufficiently accurate. However, if the P–δ effects in the members are significant, the error in the estimation of the local distributions of forces and deformations would have some influence on the prediction of the overall system strength. But not all the members in a large structural system are loaded to their maximum strengths prior to reaching the system's maximum strength. Therefore, errors in the estimation of the local effects often have very little influence on the overall system strength and stability, particularly for a highly redundant structure.

Yet, there are cases where possible local failures of members may have significant effects on the overall system stability. For these cases, the direct assessment of system and member strengths using conventional second-order elastic-plastic-hinge analysis may yield results on the unconservative side. On the other hand, the proposed notional-load plastic-hinge analysis appears to be sufficiently accurate and conservative, and, thus, it is suitable for use in the analysis of a wider range of structural systems.

7.8.1 Second-Order Elastic Analysis for Frame Design

It is important to note that the second-order elastic-plastic-hinge approach is a direct extension of the elastic-analysis approach. Therefore, the notional lateral-load concept is applicable also for second-order elastic analysis. The present AISC LRFD specification does not require the modeling of notional loads (or geometric imperfections) for second-order elastic analysis, but it requires beam-column capacity checks to be carried out for all members. If notional loads are specified in a second-order elastic analysis, member stability checks are also not required; instead, it would be sufficient to check the maximum moment and axial force within the member against the member's cross-section strength.

7.8.2 Noncompact Section

If the member cross-sections are fully compact, the elastic-plastic-hinge analysis may be used advantageously to account for the additional reserve strength in the system subsequent to reaching the capacity of the most critically loaded member. However, if the cross-sections are noncompact, the cross-section strength may be derived based on the effective-width concept given in ASCE (1990), and frame design can proceed using a second-order elastic analysis.

7.8.3 Combining Member and System Notional Loads

The member notional load is expressed as 1% of the axial load in the beam-column. It is therefore affected by the tributary area of the beam-column under consideration. It is important to note that both the member notional lateral load as well as the self-equilibrating end reactions should be included in the analysis of frame members that are subjected to high axial force. However, for stocky members and for members subjected to low axial load (e.g., $P < 0.25P_e$), the member notional loads tend to have negligible effects on the member strength and stability. Thus, they can be ignored from a practical point of view.

The similarities and differences between the member notional-load approach and Eurocode 3 procedures are outlined in Fig. 7.15. Both approaches do not require the modeling of member imperfections if the axial force P in the column is less than $0.25\,\pi^2 EI/L^2$. However, the modeling of member-imperfection effects for cases in which the axial force is high ($P > 0.25\,\pi^2 EI/L^2$) is different in the two procedures. Eurocode 3 specifies different

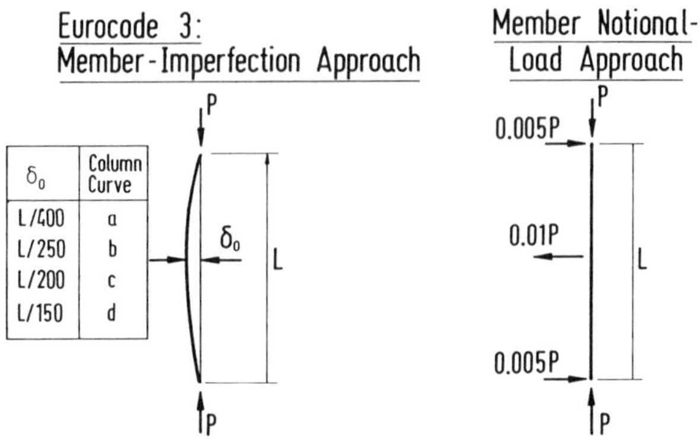

Fig. 7.15 Modeling of member imperfections using Eurocode 3 and notional-load procedures.

equivalent initial out-of-straightness imperfections based on the relevant column curve for the member under consideration. The initial out-of-straightness magnitude ranges from $L/400$ for column curve a to $L/150$ for column curve d. The equivalent initial out-of-straightness magnitude assumed in the notional-load approach is $L/400$. This value is consistent with the use of one column curve in AISC LRFD.

The story notional load is derived based on the factored gravity load on the story, and it is not cumulative from story to story. The story notional loads should be assumed to act in any one direction at a time and should be applied at each roof and floor level or their equivalent. They should be considered to act simultaneously with the factored dead, imposed, and wind loads.

The inclusion of member notional loads for advanced analysis of simple portal frames should be a straightforward task. However, for a complex structural system, the directions on how these notional loads should be applied together with the story notional loads to arrive at a critical mode of failure are not at all obvious. One way to apply the member notional loads is to observe the deflected shape of the columns. The member notional load should be applied in the direction that causes further transverse deflection of the member, as shown in Fig. 7.16. The story notional loads should be applied in conjunction with the design lateral load to cause additional frame drift. Schematic diagrams shown in Fig. 7.16 demonstrate how the notional-load approach may be applied to irregular and nonrectangular frameworks. Other more direct methods to accommodate member imperfections and distributed plasticity effects in advanced analysis of frames, which are based on simple modifications to the elastic-plastic-hinge theory, are discussed in Chapter 8.

7.8.4 Adopting the Notional-Load Concept in Codes

The notional-load concept has also been used in CSA (1989) and BSI (1990) specifications for the design of gravity-loaded system. These notional forces

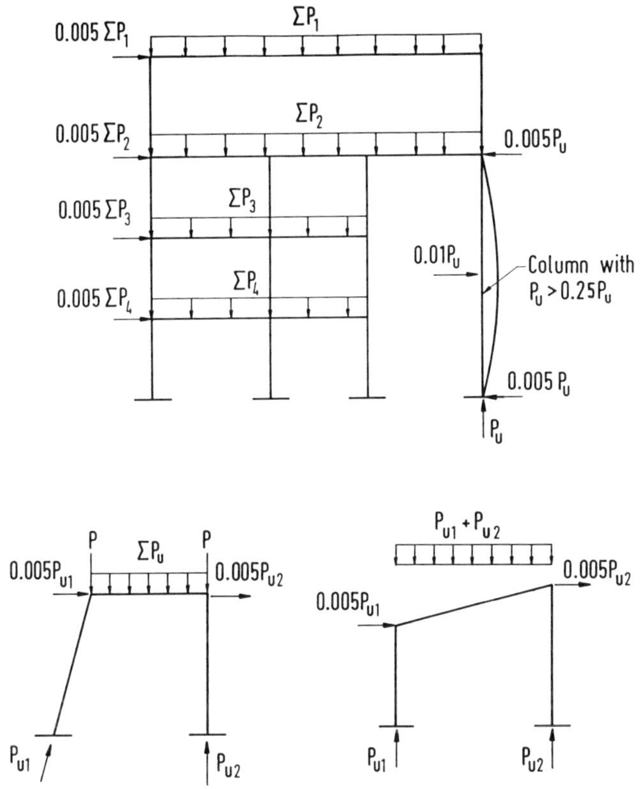

Fig. 7.16 Notional lateral loads for irregular and nonrectangular frameworks.

are applied to account for practical imperfections such as lack of verticality. In BSI (1990), the notional forces are taken as the greater of 1% of the factored dead load or 0.5% of the factored load (dead plus imposed) from that lever, applied horizontally. As specified in both the CSA (1989) and BSI (1990) specifications, the notional forces should not be combined with (a) the applied horizontal loads, (b) with temperature effects, and (c) should not be taken to contribute to the net reactions at the foundations. However, in the work published by Liew et al. (1994), the authors demonstrate that for extreme situations in which the beam-columns are loaded by "substantial" axial loads and the factored lateral loads required for design are small, the use of notional loads only for the gravity-load check could be unsafe. Even though the preceding concerns should not present any problem for most practical cases, however, from the point of view of the accuracy of predicting beam-column strength, it appears that a notional horizontal force equal approximately to $\Sigma P_u/200$ is required to match advanced analysis strengths consistently over the entire range axial force and bending moment. For most practical frames, notional horizontal forces often will be a small portion of the total factored horizontal loads that a frame must be designed for in load combinations involving wind forces.

In the design of beam-columns using CSA specifications, a limit of $B_2 <$ 1.4 is placed on the moment amplification associated with the sidesway effects. This can be interpreted as a limit on the maximum gravity load that the frame would be allowed to be designed for without the use of elasto-plastic analysis. One important point to observe is that if the notional force is applied together with the factored horizontal loads, then the column strength curve will match the "exact" beam-column strength quite well. This seems to justify that the $B_2 < 1.4$ limit would not be necessary to control error associated with the cases where the beam-columns are subjected to high axial load (Liew et al., 1991).

7.8.5 Design Checks for Out-of-Plane and Spatial Beam-Column Strength

It is important to note that separate equation checks for lateral-torsional buckling such as those of the LRFD specifications are still required for two-dimensional notional-load plastic-hinge analysis and design. Although the notional lateral-load method eliminates the need for separate in-plane member-capacity checks, the specification equation checks for out-of-plane lateral-torsional buckling are required to ensure that the inherit assumptions of the analysis are not violated. Alternatively, if the member is adequately braced against out-of-plane deformation, such checks become unnecessary. Rules for lateral bracing spacing for beams designed based on plastic analysis are given in Chapter 5, Section 5.6. Additional discussions regarding consideration of out-of-plane strength and rotation capacity in frames designed based on advanced analysis are discussed by White et al. (1991) and White (1993).

Accurate consideration of the effects of detailed aspects such as warping end restraints, inelastic torsion, and general inelastic torsional-flexural coupling on member three-dimensional response will likely require a three-dimensional plastic-zone type of analysis model. However, it is doubtful whether three-dimensional plastic-zone analysis ever will be feasible for design practice. The proposed notional-load concept may be extended for three-dimensional frame design by applying the notional lateral loads in addition to the design lateral loads (wind and/or equivalent earthquake loads) in two orthogonal directions at each story in the frame. Currently, elastic-plastic-hinge-analysis capabilities for three-dimensional frame analysis are available (e.g., Mondkar and Powell, 1975; Orbison, 1982; Ziemian, 1990) and may be used for analyzing moderate large-scale three-dimensional frames. However, these models cannot accommodate the effects of warping, inelastic torsion, and general torsional-flexural coupling on member three-dimensional response. Therefore, until an advanced analysis is developed that considers these effects on member instability, design checks such as the presently used semiempirical equations for the prevention of lateral-torsional instability must be carried out. The development of advanced analysis for three-dimensional frame design is likely to be the next priority of research in the coming years.

7.9 CONCLUSIONS

This chapter provides a method that permits the use of second-order elastic-plastic-hinge analysis for design of planar frames without the risk of overestimating the maximum strength of the component members in the framework. The proposed notional lateral-load approach removes, to a large extent, the limitation of the conventional elastic-plastic-hinge method for representing individual member strength. Furthermore, this analysis approach allows the accommodation of inelastic force redistribution and, therefore, provides a more rational assessment of the system strength and stability. If this method is used, the need for separate member equation checks for in-plane member strength can be waived. The validity and reliability of this method have been demonstrated through use of a set of benchmark frames, which are subjected to pure gravity loads and combined gravity and lateral loads.

For gravity-loaded frames, the proposed method is similar to the approach proposed by the Canadian standard (CSA, 1989), which requires a notional-load factor of $\alpha = 0.005$ for member and cross-section capacity checks. The only difference between the Canadian standard and the proposed approach is that the former requires member stability checks based on an effective-length factor of 1, and the latter requires *only* cross-section capacity checks. Also, the Canadian standard does not require the use of notional lateral loads in the analysis if the design lateral load at a story level is greater than 0.5% of the total gravity load in the story. In the proposed approach, however, the notional loads must be combined with the designed lateral and gravity loads in order to produce solutions that are safe and reasonably accurate.

The use of a story notional-load factor $\alpha = 0.005$ is shown to be adequate for all practical purposes, and the results obtained from the analyses using this load factor are never more than 5% unconservative compared to the benchmark solutions for frames with a wide range of L/r ratios. Although the notional-load factor is derived based on a W8 × 81 section, the results are applicable for other wide-flange sections, bending about the strong axis, with an expected error of not more than 5%. For weak-axis bending cases, the proposed notional load factor of $\alpha = 0.005$ is mostly conservative with a maximum error of 25%. It is therefore suggested that for weak-axis bending cases, the notional factor should be expressed as the cross-section properties, as discussed in Section 7.3.

For a frame system in which the members are subjected to high axial forces and for members in which P–δ effects are significant ($P > 0.25P_e$), a notional lateral load equal to 1% of the column axial force and the counterbalanced end-reaction forces must be applied in the analysis to simulate imperfection and distributed plasticity effects on P–δ instability of the member. It should be noted that the notional-load concept is applicable to both the second-order elastic as well as elastic-plastic-hinge analyses. In fact, it can be argued that the notional lateral-load method is also applicable for frame members made of noncompact cross-sections provided that the cross-section strength is not violated. However, inelastic force redistribution should not be relied upon once local buckling occurs in the frame member.

The notional-load approach shows promise as an effective means of accounting for member initial imperfections without physically altering frame geometry. The studies conducted in this chapter are applicable to both sway and nonsway frames. For sway frames, the story notional loads usually play a more important role in degrading frame stiffness and strength than member notional loads. In contrast, the strengths of many braced frames are affected more by member notional loads and are relatively unaffected by story notional loads. Although the inclusion of both story and member notional lateral loads is necessary to obtain a realistic lower-bound estimate of the frame's maximum strength, it is important to recognize that frames that have lightly loaded beam-columns, and for which plastic collapse is the predominant mode of failure, are insensitive to member notional-load effects.

REFERENCES

AISC (1986). *Load and Resistance Factor Design Specifications for Structural Steel Buildings,* Chicago: American Institute of Steel Construction.

AISC (1993). *Load and Resistance Factor Design Specifications for Structural Steel Buildings,* 2nd Ed., Chicago: American Institute of Steel Construction.

ASCE (1990). *Specification for the Design of Cold-Formed Stainless Steel Structural Members,* ANSI/ASCE-8-90, New York: American Society of Civil Engineers.

ASCE-WRC (1971). *Plastic Design in Steel—A Guide and Commentary,* 2nd Ed., ASCE Manuals and Reports on Engineering Practice No. 41, New York: Joint Committee of the Welding Research Council and ASCE.

BSI (1990). *BS5950: Structural Use of Steelwork in Building. Part 1: Code of Practice for Design in Simple and Continuous Construction: Hot-Rolled Sections,* London: British Standards Institution.

Chen, W. F., and T. Atsuta (1976). *Theory of Beam-Columns, Vol. 1: In-Plane Behavior and Design,* New York: McGraw-Hill.

Cheong, S. M. (1991). "Column Design in Gravity-Loaded Frames," *Journal of Structural Engineering,* ASCE, Vol. 117, No. 5, pp. 1448–1461.

CSA (1989). *Limit States Design of Steel Structures,* CAN/CSA-S16.1-M89. Ontario, Canada.

Duan, L., and W. F. Chen (1990). "A Yield Surface Equation for Doubly Symmetrical Sections," *Engineering Structures,* Vol. 12, April, pp. 114–119.

ECCS (1984). *Ultimate Limit State Calculation of Sway Frames with Rigid Joints,* Technical Working Group 8.2, Systems, Publication No. 33, Brussels: European Convention for Constructional Steelwork.

ECCS (1991). *Essentials of Eurocode 3 Design Manual for Steel Structures in Buildings,* ECCS Advisory Committee 5, No. 65, Brussels: European Convention for Constructional Steelwork.

Eurocode 3 (1992). *Design of Steel Structures: Part 1.1—General Rules and Rules for Buildings,* National Application Document for use in the UK with ENV1993-1-1: 1991, Draft for Development.

Kanchanalai, T. (1977). *The Design and Behavior of Beam-Columns in Unbraced Steel Frames,* AISI Project No. 189, Report No. 2, Austin: Civil Engineering/Structures Research Laboratory, University of Texas.

LeMessurier, W. J. (1977). "A Practical Method of Second Order Analysis, Part 2—Rigid Frames," *Engineering Journal,* Vol. 14, No. 2, pp. 49–67.

Liew, J. Y. R., D. W. White, and W. F. Chen (1991). "Beam-Column Design in Steel Frameworks—Insights on Current Methods and Trends," *Journal of Constructional Steel Research,* Vol. 18, No. 4, pp. 269–308.

Liew, J. Y. R., D. W. White, and W. F. Chen (1992). "Beam-Columns," in *Constructional Steel Design, An International Guide,* P. J. Dowling et al., eds., London: Elsevier, pp. 105–132.

Liew, J. Y. R., D. W. White, and W. F. Chen (1994). "Notional Load Plastic Hinge Method for Frame Design," *Journal of Structural Engineering,* ASCE, Vol. 120, No. 5, pp. 1434–1454.

Mondkar, D. P., and G. H. Powell (1975). *ANSR-I—General Purpose Computer Program for Analysis of Nonlinear Structural Response,* Report No. EERC 75-37, Berkeley: Earthquake Engineering Research Center, University of California.

Orbison, J. G. (1982). *Nonlinear Static Analysis of Three-Dimensional Steel Frames,* Report No. 82-6, Ithaca, NY: Department of Structural Engineering, Cornell University.

Standards Australia (1990). *Steel Structures,* AS4100, Sydney.

White, D. W. (1993). "Plastic Hinge Methods for Advanced Analysis of Steel Frames," *Journal of Constructional Steel Research,* Vol. 24, No. 2, pp. 121–152.

White, D. W., J. Y. R. Liew, and W. F. Chen (1991). *Second-Order Inelastic Analysis for Frame Design: A Report to SSRC Task Group 29 on Recent Research and the Perceived State-of-the-Art,* Structural Engineering Report No. CE-STR-91-12, West Lafayette, IN: Purdue University.

Zhou, S. P., L. Duan, and W. F. Chen (1990). "Comparison of Design Equations for Steel Beam-Columns," *Structural Engineering Review,* Vol. 2, No. 1, pp. 45–53.

Ziemian, R. D. (1990). "Advanced Methods of Inelastic Analysis in the Limit States Design of Steel Structures," Ph.D. dissertation, School of Civil and Environmental Engineering, Cornell University, Ithaca, NY.

8

SECOND-ORDER REFINED PLASTIC-HINGE ANALYSIS

8.1 INTRODUCTION

To qualify for advanced analysis, the analysis model must directly or indirectly account for the effects of material and geometrical nonlinearities, distributed plasticity as well as member and frame initial imperfections. As demonstrated in Chapter 7, a simple way to approximate these effects is through the use of notional loads. Although notional loads indirectly account for the effects of distributed plasticity and initial imperfections on the bending deformations, the effects on axial deformations are not captured in systems dominated by high axial-force actions. This aspect has been shown to be insignificant in the ultimate strength prediction, however, there are other aspects of the notional-load plastic-hinge approach that would lead the engineer to consider other more accurate methods for advanced analysis using the plastic-hinge-based approach.

First, the specification of notional loads (or the equivalent geometric imperfections) in the analysis model can be cumbersome. Although this procedure can be automated in a computer-aided procedure, the aspect on how the system notional forces and member notional forces for members subjected to large axial forces can be combined to produce the worst-case scenario remains to be discussed. Second, the "exact" load-deformation and force-distribution/redistribution behavior of the members and the system still cannot be modeled accurately by the notional-load-plastic hinge method. This is because in the elastic-plastic-hinge model, the inelastic behavior of a cross-section involves an abrupt transition from the elastic state to the fully yielded state. The element under consideration exhibits a sudden stiffness reduction upon the formation of a plastic hinge. Furthermore, the notional-load method is applicable only for regular frameworks whereby story-by-story types of frame configuration exist. For irregular frameworks in which the definition of story cannot be defined, then notional loads have to be applied sparingly. Last, the elastic-plastic-hinge model does not represent the degradation of flexural and axial stiffnesses due to distributed yielding and the associated second-order inelastic P–δ effects within the members. This approach, therefore, often overestimates the stiffness of a member loaded into the inelastic range (White, Liew, and Chen, 1991, 1993; Liew, White, and Chen, 1993a, 1993b; White, 1993; Chen and Toma, 1994). Although the limiting strength can be predicted conservatively to a large extent by using the notional-load approach, the behavioral response predicted by the elastic-plastic-hinge model is not smooth as compared to the plastic-zone solutions, which exhibit a smooth transition from the elastic state to the fully yielded state. Based on this aspect alone, it might be considered worth to investigate an advanced analysis procedure that can model the phenomenological behavior of the member. This procedure must provide the same efficiency as the plastic-hinge method in terms of computer usage and implementation.

This chapter presents an alternate method for advanced analysis called the *refined plastic-hinge approach*. This approach is based on the following two improvements to the elastic-plastic-hinge model:

1. A column tangent-modulus model E_t is used in place of the elastic modulus E to represent the distributed plasticity along the length of a member due to axial-force effects. The member inelastic stiffness, represented by the member axial and bending rigidities E_tA and E_tI, respectively, is assumed to be the function of axial load only. In other words, E_tA and E_tI can be thought of as the properties of an effective core of the section, considering column action only. The tangent-modulus concept was earlier explored by Beedle and Tall (1960) and Tall (1974) because of the effect of early yielding in the cross-section due to residual stresses, which were believed to be the cause for the low strength of inelastic column buckling. The tangent-modulus approach also has been utilized in previous work (Orbison et al., 1982; Liew, 1992; White et al., 1993) to improve the accuracy of the elastic-plastic-hinge approach for structures in which members are subjected to large axial forces.
2. Distributed plasticity effects associated with flexure is captured by gradually degrading the member stiffness at the plastic-hinge locations as yielding progresses under increasing load as the cross-section strength is approached. Several models of this type have been proposed in recent literature based on extensions to the elastic-plastic-hinge approach (Powell and Chen, 1986; Al-Mashary and Chen, 1991; Deierlein, Zhao, and McGuire, 1991; King, White, and Chen, 1992) as well as the tangent modulus inelastic-hinge approach (Liew et al., 1993a, 1993b;

White et al., 1993). The rationale of modeling stiffness degradation associated with both axial and flexural actions is that the tangent-modulus model represents the column-strength behavior in the limit of pure axial compression, and the plastic-hinge stiffness-degradation model represents the beam behavior in pure bending; thus, the combined effects of these two approaches should also satisfy the cases in which the member is subjected to combined axial compression and bending.

It is shown in what follows that with the previous two improvements, the refined plastic-hinge model can be used with sufficient accuracy to provide a quantitative assessment of a member's performance up to failure. The essential concepts behind these refinements to the elastic-plastic-hinge model are discussed in detail. However, before describing this method, it is useful to discuss some of the desirable attributes that need to be considered in the development of the refined plastic-hinge procedure.

8.2 ATTRIBUTES FOR THE REFINED PLASTIC-HINGE ANALYSIS

The following are attributes considered in the present work for development of the refined plastic-hinge model:

1. The analysis should be capable of capturing the characteristic behavior of a wide range of structural systems and member types. For instance, the analysis should represent, with sufficient accuracy, the strength behavior of (a) strong- and weak-axis bending action of rolled sections, (b) braced and unbraced beam-columns, (c) members whose designs are governed by elastic or inelastic instability, and (d) beam-columns subjected to axial and transverse forces.
2. The analysis should represent both the P–Δ and P–δ effects, including the instability effects from leaning columns. It should predict the inelastic response due to distributed plasticity, including the effects of welding residual stresses and geometric imperfections, for a wide range of load combinations.
3. The analysis should be fairly simple to implement. The model should predict, conservatively, the in-plane stability and strength of isolated beam-column members, and other benchmark frames. The analytical results should not be more than 5% unconservative when compared to the benchmark solutions obtained from second-order plastic-zone analysis.
4. The strengths, force deformations, and internal-force distributions predicted by the model for various types of member and systems should be reasonably close to those predicted by "exact" benchmark solutions.

5. The element formulation should reduce to a well-recognized behavior model in the limits of pure beam and pure column action. For the case of a member loaded by a pure axial load, the element inelastic stiffness should be close to that associated with the inelastic flexural rigidity $E_t I$ implied by the AISC LRFD column-strength equations.
6. The member forces calculated by the analysis model should not violate the cross-section plastic strength associated with the condition of full plasticity. The model should accommodate force-point movement on the plastic strength surface.
7. The beneficial effect of material strain hardening should not be relied upon.

The following section presents the refined beam-column model that satisfies the preceding attributes. The implementation of the method is based on the AISC LRFD specification. Provisions pertinent to other specifications may also be incorporated if required.

8.3 APPROXIMATE THE EFFECTS OF DISTRIBUTED YIELDING

As explained in the desired attribute for advanced analysis, the distributed plasticity effects associated with flexure must be captured to obtain an accurate physical model of beam-column behavior. This can be accomplished by representing the yielding effect at the plastic hinges starting from first-yield to full-yield states. Element stiffness will start to degrade once yielding is detected, and it progresses under increasing load until the full cross-section strength is reached. Because the distributed yielding is influenced largely by member initial imperfections (member out of straightness and residual stresses) as well as by axial force, it is proposed that these distributed plasticity effects be modeled by an effective stiffness approach for the calculation of column strength. The detailed formulation of the model is described in what follows.

8.3.1 Column Effective Stiffness

8.3.1.1 AISC LRFD Column-Strength Approach For columns subjected to pure compression, the AISC LRFD specification provides the following formula for the evaluation of axial compressive strength:

$$P_n = P_y(0.658^{\lambda_c^2}), \qquad \lambda_c \leq 1.5 \tag{8.3.1}$$

$$P_n = P_y\left(\frac{0.877}{\lambda_c^2}\right) \qquad \lambda_c > 1.5 \tag{8.3.2}$$

where

P_n = nominal column strength
P_y = axial load at full yield
λ_c = normalized column slenderness ratio defined by $\lambda_c = (KL/\pi r)\sqrt{F_y/E}$
F_y = material yield stress
E = modulus of elasticity
KL/r = effective slenderness ratio

The column tangent modulus E_t can be evaluated based on the inelastic stiffness reduction procedure given in the AISC LRFD manual for the calculation of inelastic column strength. The ratio of the tangent modulus to the elastic modulus, E_t/E, is defined as (Liew et al., 1991, 1992):

$$\frac{E_t}{E} \approx \frac{(P_n)_{\text{inelastic}}}{(P_n)_{\text{elastic}}} = \frac{0.658^{\lambda_c^2}}{0.877/\lambda_c^2} \leq 1.0 \tag{8.3.3}$$

For an inelastic column (i.e., $\lambda_c < 1.5$), the normalized column slenderness ratio λ_c corresponding to $P_n = P_u$ may be obtained from Eq. (8.3.1) as

$$\lambda_c = \sqrt{\frac{\ln (P_u P_y)}{\ln 0.658}} \tag{8.3.4}$$

where P_u is the required axial strength for the member under consideration. The value of λ_c is less than 1.5 for columns that fall in the inelastic range. By using Eq. (8.3.4), a corresponding λ can be obtained for a given value of P_u/P_y. The inelastic stiffness-reduction factor can then be calculated by substituting Eq. (8.3.4) into Eq. (8.3.3). The resulting equations are

$$\frac{E_t}{E} = \begin{cases} 1.0, & \text{for } P \leq 0.39P_y \\ -2.7243 \dfrac{P}{P_y} \ln\left(\dfrac{P}{P_y}\right), & \text{for } P > 0.39P_y \end{cases} \tag{8.3.5}$$

Because this E_t model is derived from the LRFD column-strength formula, it implicitly includes the effects of residual stresses as well as the member initial out of straightness in modeling the member effective stiffness (Liew, 1992). Equation (8.3.5) is plotted in Fig. 8.1. The tangent modulus is reduced from the elastic value for $P > 0.39P_y$.

The compressive axial load-displacement relationship for an element can be derived based on the following equations:

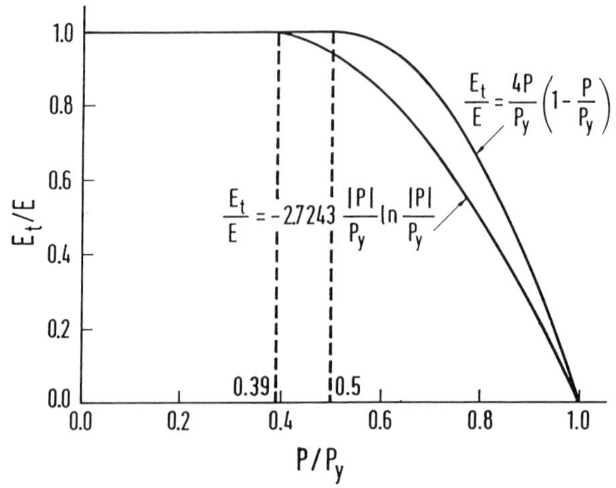

Fig. 8.1 Inelastic stiffness reductions for axial-force effect.

$$e = \begin{cases} \dfrac{PL}{EA}, & \text{for } P \leq 0.39P_y \\ \dfrac{0.39P_y L}{EA} + \displaystyle\int_{0.39P_y}^{P} \dfrac{L}{AE_t} dP, & \text{for } P > 0.39P_y \end{cases} \tag{8.3.6}$$

Upon substituting from Eq. (8.3.5) for E_t and carrying out the integration in Eq. (8.3.6), the normalized axial force-deformation relationship can be written as

$$\frac{P}{P_y} = \begin{cases} \dfrac{\varepsilon}{\varepsilon_y}, & \text{for } P \leq 0.39P_y \\ \exp\{-0.9416 \exp[2.7243(0.39 - \varepsilon/\varepsilon_y)]\}, & \text{for } P > 0.39P_y \end{cases} \tag{8.3.7}$$

where $\varepsilon/\varepsilon_y = Ee/(\sigma_y L)$ is the total normalized axial strain in the member.

8.3.1.2 CRC Column-Strength Approach The column tangent modulus also can be evaluated based on the Column Research Council (CRC) column equations (Galambos, 1988). The CRC column equations may be written in terms of nondimensional quantities P/P_y and λ_c given by

$$P_n = P_y(1 - \tfrac{1}{4}\lambda^2), \qquad \lambda_c \leq \sqrt{2} \tag{8.3.8}$$

$$P_n = P_y\lambda^{-2}, \qquad \lambda_c > \sqrt{2} \tag{8.3.9}$$

For a column in the inelastic range (i.e., $\lambda_c \leq \sqrt{2}$), the CRC column equations implicitly account for compressive residual stresses of $\sigma_{cr} = 0.5\sigma_y$. This value is chosen as a conservative estimate of maximum value of compressive residual stress present in hot-rolled, wide-flange sections, which is about 0.3 σ_y. Figure 8.2 compares the column strength predicted by the CRC and LRFD column equations. It can be observed that the LRFD column-strength curve is below the curve predicted by the CRC column equations. This is because the CRC column equations are derived based on columns with perfectly straight geometry, whereas the LRFD column curve is generated based on imperfect columns with initial out of straightness at midlength equal to $L/1500$ (Chen and Lui, 1987).

The column tangent modulus E_t also may be evaluated based on the CRC column equation as follows:

$$\frac{E_t}{E} \approx \frac{(P_n)_{\text{inelastic}}}{(P_n)_{\text{elastic}}} = \frac{1 - \lambda_c^2/4}{\lambda_c^{-2}} \leq 1.0 \tag{8.3.10}$$

λ_c may be expressed in terms of P/P_y from Eq. (8.3.8). When substituted into Eq. (8.3.10), the resulting E_t/E expressions become

$$\frac{E_t}{E} = \begin{cases} 1.0, & \text{for } P \leq 0.5P_y \\ \dfrac{4P}{P_y}\left(1 - \dfrac{P}{P_y}\right), & \text{for } P > 0.5P_y \end{cases} \tag{8.3.11}$$

Equation (8.3.11) is also plotted in Fig. 8.1. The tangent-modulus stiffness reduction becomes effective when the axial force reaches a value of $P =$

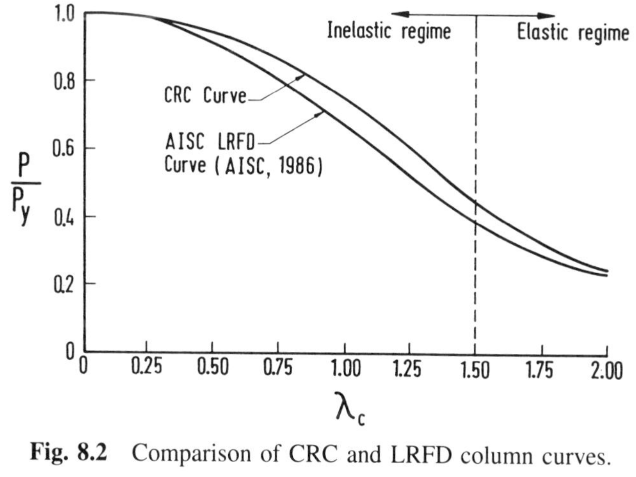

Fig. 8.2 Comparison of CRC and LRFD column curves.

$0.5P_y$. The total axial displacement based on the E_t model in Eq. (8.3.11) may be derived as

$$e = \begin{cases} \dfrac{PL}{EA}, & \text{for } P \leq 0.5P_y \\ \dfrac{0.5P_yL}{EA} + \displaystyle\int_{0.5P_y}^{P} \dfrac{L}{AE_t} dP, & \text{for } P > 0.5P_y \end{cases} \tag{8.3.12}$$

Upon substituting from Eq. (8.3.11) for E_t and carrying out the integration, the normalized axial force-deformation relationship can be written as

$$\frac{P}{P_y} = \begin{cases} \dfrac{\varepsilon}{\varepsilon_y}, & \text{for } P \leq 0.5P_y \\ \dfrac{1}{1 + \exp(2 - 4\varepsilon/\varepsilon_y)}, & \text{for } P > 0.5P_y \end{cases} \tag{8.3.13}$$

The normalized axial force–strain relationships described by Eqs. (8.3.7) and (8.3.13) are shown in Fig. 8.3. It should be noted that the normalized axial stress–strain relationships of an element described by the tangent moduli in Eqs. (8.3.7) and (8.3.13) are nonlinear in nature, whereas for the conventional elastic-plastic-hinge model, a linear elastic-perfectly plastic normalized axial stress–strain relationship is tacitly assumed. Also, for a member subjected to tensile axial force, only the residual-stress effects should be represented in the member stiffness degradation. In this case, the CRC tangent modulus would be more appropriate.

The main difference between the CRC E_t and the LRFD E_t is that the former considers only the residual-stress effects in modeling the column ef-

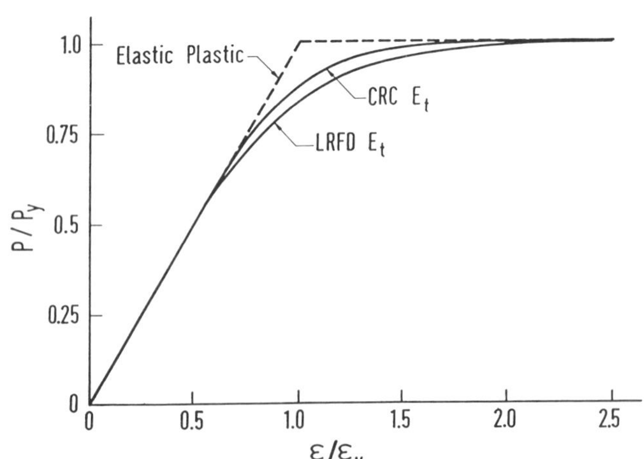

Fig. 8.3 Normalized axial-force-displacement relationship.

fective stiffness, whereas the latter is based on LRFD column-strength equations that account for both the effects of initial geometric imperfections and residual stresses. It is important to note from Fig. 8.3 that an axial strain greater than $2\varepsilon_y$ must be achieved to fully develop the squash load (P_y) of an element modeled by the tangent-modulus models.

The tangent-modulus approach is, of course, a behavioral model that cannot compute exactly the effects of initial imperfections on the distribution of forces in a structural system. If one wishes to compute these effects accurately, the initial geometrical imperfections need to be modeled explicitly in the analysis. In such cases, the CRC E_t is the suitable choice.

8.3.2 Beam-Column Stiffness Degradation

Recent research (Liew et al., 1993a, 1993b) has found that for members subjected to significant in-plane bending, the modeling of the member stiffness equation using only the tangent-modulus approach is not sufficient to represent the gradual stiffness degradation as yielding progresses through the volume of the member. Additional distributed plasticity effects in the beam-column member can be attributed to the bending action. These effects may be represented by modifying the basic elastic-plastic-hinge model such that the member stiffness degrades gradually from the stiffness associated with the onset of yielding to that associated with the formation of plastic hinge. This approach, which is called the refined plastic-hinge method, has been shown to be superior than the conventional elastic-plastic-hinge approach in representing the inelastic behavior of many types of frame structures.

In the refined plastic-hinge approach, the element stiffness is assumed to degrade according to a prescribed function after the element end forces exceed a predefined initial yield function. When the cross-section plastic strength is reached, the section at the plastic hinge is modeled as a real hinge with an applied moment M_{pc}, and the element tangent stiffness is adjusted to account for the presence of plastic hinge.

8.3.2.1 Effect of Plastification at End A Only To represent a gradual transition from the elastic stiffness at the onset of yielding (Eq. 6.2.16) to the stiffness associated with a full plastic hinge at end A (Eq. 6.4.5), the element incremental force-displacement relationships may be written as

$$\begin{Bmatrix} \dot{M}_A \\ \dot{M}_B \end{Bmatrix} = \frac{E_t I}{L} \begin{bmatrix} \phi_A S_1 & \phi_A S_2 \\ \phi_A S_2 & S_1 - \dfrac{S_2^2}{S_1}(1 - \phi_A) \end{bmatrix} \begin{Bmatrix} \dot{\theta}_A \\ \dot{\theta}_B \end{Bmatrix} \tag{8.3.14}$$

The terms S_1 and S_2 in this equation are the conventional beam-column stability functions, but with E_t used in place of the elastic modulus. The term

ϕ_A is a scalar parameter that allows for gradual inelastic stiffness reduction of the element associated with the effect of plastification at end A. This term is equal to one when the element is elastic, and it is zero when a plastic hinge is formed at end A. Parameter ϕ may be assumed to vary according to a prescribed function:

$$\phi = f(\alpha) = f(M, P) \tag{8.3.15}$$

where α is a force-state parameter that measures the magnitude of axial force P and bending moment M at the element end. By adopting the cross-section strength equations from the AISC LRFD (AISC, 1986), term α may be expressed as

$$\alpha = \frac{P}{P_y} + \frac{8}{9}\frac{M}{M_p}, \qquad \text{for } \frac{P}{P_y} \geq \frac{2M}{9M_p} \tag{8.3.16}$$

$$\alpha = \frac{P}{2P_y} + \frac{M}{M_p}, \qquad \text{for } \frac{P}{P_y} < \frac{2M}{9M_p} \tag{8.3.17}$$

Initial yielding is assumed to occur based on a yield surface that has the same shape as the plastic strength surface. The plastic strength surface ($\alpha = 1.0$) and initial yield surface denoted by $\alpha = 0.5$ are shown in Fig. 8.4. If the state of forces is changed in such a manner that the force point moves inside or along the initial yield surface, the element is assumed to remain fully elastic

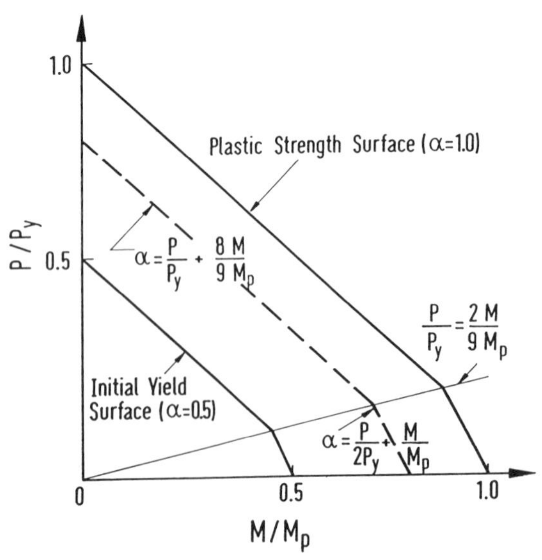

Fig. 8.4 Initial yield and plastic strength surfaces for refined plastic-hinge analysis.

with no stiffness reduction. If the force point moves beyond the initial yield surface, the element stiffness is reduced to account for the effect of plastification at the element end.

It is important to demonstrate that only a simple relationship for ϕ (Eq. 8.3.15) is required to "adequately" describe the degradation in stiffness associated with distributed plasticity effects. Although more complicated expressions for ϕ could be developed and proposed, a simple expression for ϕ is needed for keeping the analysis model simple and straightforward. Also, the element model based on this approach should satisfy many aspects of the desirable attributes for inelastic beam-column elements outlined in Section 8.2.

Figure 8.5 shows three possible functions tested in the present work. If a parabolic stiffness-reduction function is used, a gradual stiffness degradation is observed when the force state moves from the initial yield surface (α = 0.5) toward the plastic strength surface (α = 1.0). If a linear function for α is used, the change in the beam-column flexural stiffness is more abrupt.

8.3.2.2 Effect of Plastification at End* B *Only To represent a gradual transition from the elastic stiffness at the onset of yielding, Eq. (6.2.16), to the stiffness associated with a full plastic hinge at end B, Eq. (6.4.6), the element incremental force-displacement relationships may be written as

$$\begin{Bmatrix} \dot{M}_A \\ \dot{M}_B \end{Bmatrix} = \frac{E_t I}{L} \begin{bmatrix} S_1 - \dfrac{S_2^2}{S_1}(1 - \phi_B) & \phi_B S_2 \\ \phi_B S_2 & \phi_B S_1 \end{bmatrix} \begin{Bmatrix} \dot{\theta}_A \\ \dot{\theta}_B \end{Bmatrix} \tag{8.3.18}$$

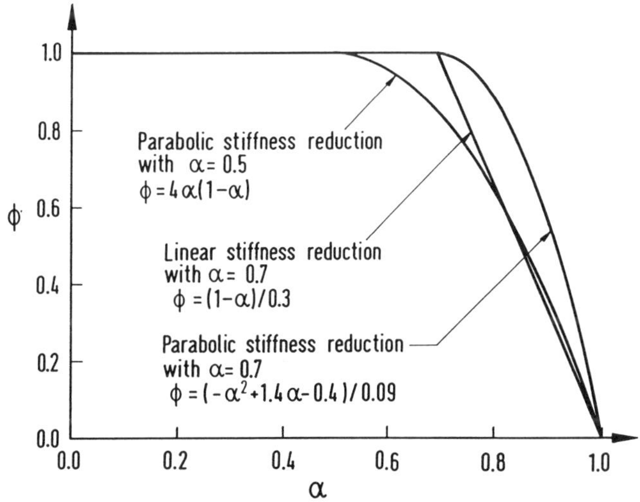

Fig. 8.5 Stiffness-reduction models for refined plastic-hinge analysis.

Equations (8.3.14) and (8.3.18) represent the gradual plastification effects at one end of the beam-column element only.

8.3.2.3 Effect of Plastification at Both Ends For the general case where plastic hinges may form at both ends of an element, the incremental force-displacement relationships may be written as

$$\begin{Bmatrix} \dot{M}_A \\ \dot{M}_B \end{Bmatrix} = \frac{E_t I}{L} \begin{bmatrix} \phi_A \left(S_1 - \dfrac{S_2^2}{S_1}[1 - \phi_B] \right) & \phi_A \phi_B S_2 \\ \phi_A \phi_B S_2 & \phi_B \left(S_1 - \dfrac{S_2^2}{S_1}[1 - \phi_A] \right) \end{bmatrix} \begin{Bmatrix} \dot{\theta}_A \\ \dot{\theta}_B \end{Bmatrix} \tag{8.3.19}$$

The following should be noted:

1. When $1 > \phi_A > 0$ and $1 > \phi_B > 0$, Eq. (8.3.19) accounts for the effects of partial plastification at both ends of the element.
2. When $\phi_A = \phi_B = 1$, both ends are fully elastic. Equation (8.3.19) reduces to the stiffness equations shown in Eq. (6.2.16).
3. When $\phi_A = 1$ and $1 > \phi_B > 0$, Eq. (8.3.19) represent the state at which end A is elastic. Equation (8.3.19) then reduces to Eq. (8.3.18).
4. When $\phi_B = 1$ and $1 > \phi_A > 0$, end B is elastic, and end A is partially yielded. Equation (8.3.19) then reduces to Eq. (8.3.14).

In a similar formulation proposed by King et al. (1992), it is assumed that once a plastic hinge is formed at the element end, the moment at the plastic hinge remains unchanged as the axial force is increased. However, as postulated in Section 8.2, neglecting force-point movement on the plastic strength surface may violate the cross-section plastic strength, which is not acceptable for maximum strength analysis of members. In the present refined plastic-hinge formulation, the force-point movement on the plastic strength surface is handled in the same way as in the elastic-plastic-hinge formulation in Section 6.4, and the detailed formulation is reported in Liew (1992). Figure 8.6 shows that when the axial force supported by a beam-column increases by a magnitude of ΔP after a plastic hinge has formed, the moment at the plastic hinge must decrease by ΔM_{pc}.

8.4 NUMERICAL IMPLEMENTATION

The refined plastic-hinge model is implemented using an automatic load-increment procedure. To prevent plastic hinges from forming within a constant-stiffness load increment, load-step sizes less than or equal to the specified increment magnitude are internally computed such that plastic

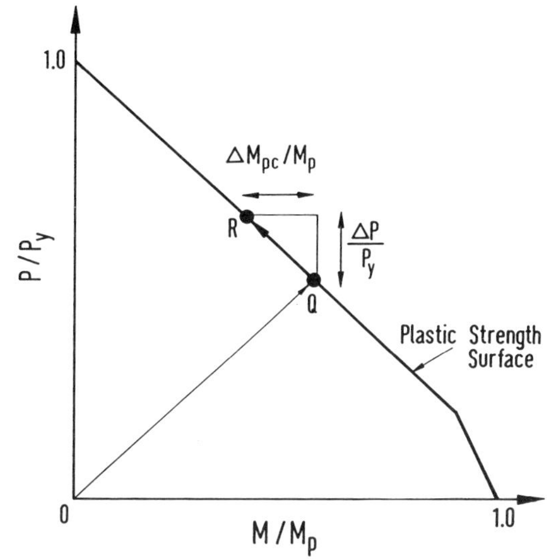

Fig. 8.6 Equilibrium correction for force-point movement on the plastic strength surface.

hinges form only after the load increment. Thus, the subsequent element stiffness formulation accounts for the stiffness reduction due to the presence of plastic hinges. For elements that are partially yielded at their ends, a limit is placed on the magnitude of the increment in the element end forces. Increments of these force components that are relatively large, compared to their respective plastic strengths, may indicate that the path taken from the initial yield surface to the plastic strength surface may have deviated excessively from the "exact" path, upon which the calculation of element effective stiffness may lead to some error. The analysis program automatically scales an attempted load increment when the change in the element stiffness parameter ($\Delta\phi$) exceeds a predefined tolerance. A default tolerance value of $\Delta\phi = 0.1$ has been specified in the solution scheme to suppress the error that may be incurred during the computation.

8.5 ILLUSTRATIVE EXAMPLES

8.5.1 Fixed-Base Portal Frame Bending about the Strong Axis

The fixed-base portal frame shown in Fig. 8.7 was analyzed by Vogel (1985) using a plastic-zone method. Plastic-zone solutions were generated for this frame based on the residual stress pattern specified in ECCS (1983) with a maximum magnitude of $0.3F_y$ in compression at the flange tips. The sway imperfection for this simple portal frame is assumed to be $\psi_0 = 1/400$, as

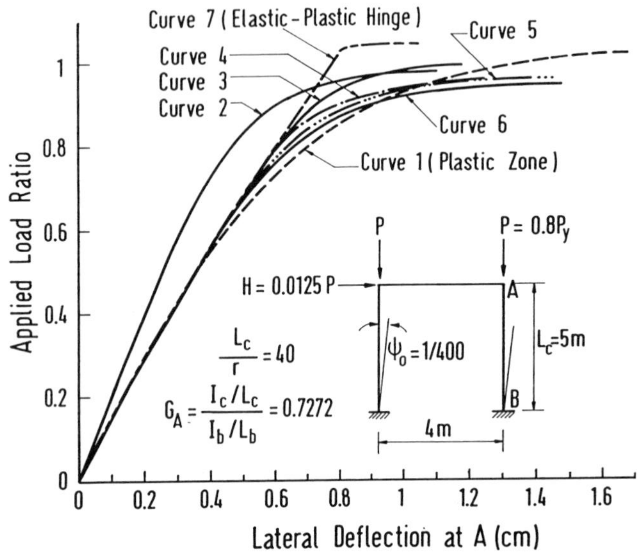

Fig. 8.7 Load-displacement traces of a fixed-base portal frame (see Table 8.1 for different stiffness-reduction schemes employed by the refined plastic-hinge analyses).

shown in Fig. 8.7. The structure is loaded by vertical forces P at the top of the columns and by a horizontal force H at the story level. The frame is analyzed with all the loads applied proportionally. The values of loads associated with an applied load ratio of 1.0 are shown in the figure. The columns in the frame are bending about the strong axis. The load-deflection curves generated by the elastic-plastic hinge and refined plastic-hinge analyses are compared with the "exact" plastic-zone curve by Vogel (1985) in Fig. 8.7.

Although the elastic-plastic-hinge analysis can predict the maximum strength of this frame quite accurately, the computed deflection at the top of the frame is somewhat in error. The load-displacement curve generated by the elastic-plastic-hinge analysis considering the effect of column out of plumbness is shown as curve 7 in Fig. 8.7. To improve the prediction of the force distribution and the magnitude of deformations, the stiffness degradation due to distributed yielding must be considered in the analysis.

To obtain a preliminary insight into ability of the refined plastic-hinge analysis to predict the inelastic response, the frame shown in Fig. 8.7 is analyzed by using several stiffness-reduction schemes as outlined in what follows:

1. Parabolic stiffness reduction with initial yield surface at $\alpha = 0.5$ in which

$$\phi = 4\alpha(1 - \alpha), \qquad \text{for } \alpha > 0.5 \tag{8.5.1}$$

2. Parabolic stiffness reduction with initial yield surface at $\alpha = 0.7$ in which

$$\phi = \frac{-\alpha^2 + 1.4\alpha - 0.4}{0.09}, \qquad \text{for } \alpha > 0.7 \tag{8.5.2}$$

3. Linear stiffness reduction with initial yield surface at $\alpha = 0.7$ in which

$$\phi = \frac{1 - \alpha}{0.3}, \qquad \text{for } \alpha > 0.7 \tag{8.5.3}$$

These functional expressions are plotted in Fig. 8.5. Also, both the CRC and LRFD tangent-modulus equations, Eqs. (8.3.5) and (8.3.11), have been used to approximate the distribution of yielding along the member length due to the axial-force effect. The different schemes employed in this study are summarized in Table 8.1. The resulting load-displacement curves from the different refined plastic-hinge analyses are shown in Fig. 8.7.

The first attempt was to use the LRFD tangent modulus (LRFD E_t) without modeling the sway imperfections explicitly. The element end stiffness in this analysis was assumed to degrade parabolically starting at initial yield surface $\alpha = 0.5$. The resulting load-displacement curve is shown as curve 2 in Fig. 8.7. Comparison of this curve with the plastic-zone results shown as curve 1 indicates that if one wishes to compute accurately the effects of sway imperfections on the overall frame responses, the sway imperfections must be modeled explicitly in the analysis. Although the LRFD E_t model implicitly accounts for the effect of member imperfections, explicit modeling of sway imperfections is still required for the analysis to predict accurately the load-displacement response of the structure. This is particularly true if accurate calculation of force distributions in the imperfect frame is required. It can be

TABLE 8.1 Different Stiffness-Reduction Schemes Employed by the Refined Plastic-Hinge Analyses in Fig. 8.7

Curve No.	Analysis Attributes	Sway Imperfection, ψ_0	Tangent Modulus, E_t	Initial Yield Surface	Degradation Function, ϕ
1	PZ	1/400	—	—	—
2	RPH	0	LRFD	$\alpha = 0.5$	Parabolic, Eq. (8.5.1)
3	RPH	1/400	CRC	$\alpha = 0.7$	Parabolic, Eq. (8.5.2)
4	RPH	1/400	CRC	$\alpha = 0.7$	Linear, Eq. (8.5.3)
5	RPH	1/400	CRC	$\alpha = 0.5$	Parabolic, Eq. (8.5.1)
6	RPH	1/400	LRFD	$\alpha = 0.5$	Parabolic, Eq. (8.5.1)
7	EPH	1/400	—	—	—

PZ = Plastic-zone analysis. RPH = Refined plastic-hinge analysis. EPH = Elastic-plastic-hinge analysis.

observed from Fig. 8.7 that if the sway imperfections (or column out of plumbness) are modeled explicitly in the analysis, the consequent load-displacement curve is curve 6, which compares well with the plastic-zone curve (curve 1). The maximum strength indicated by curve 6 is about 7% below the maximum strength predicted by the plastic-zone analysis (curve 1). This is because the refined hinge approach using the LRFD E_t implicitly accounts for member initial out-of-straightness effects in the computation of the frame's maximum strength, whereas these effects have not been considered in the plastic-zone analysis through explicit modeling of member initial out of straightness.

Figure 8.7 shows that the maximum loads predicted by the refined plastic-hinge analyses with different end stiffness-degradation functions are always lower than the plastic-zone result. The greater the stiffness reduction employed by the refined hinge model (e.g., the use of the LRFD E_t and parabolic stiffness function with initial yield surface $\alpha = 0.5$), the smaller the predicted maximum load and the softer the predicted response of the frame. The results obtained using the refined hinge analysis based on the use of the CRC E_t model and the initial yield surface with $\alpha = 0.5$ (curve 5) give the best correlation with the plastic-zone solutions.

8.5.2 Pinned-Base Portal Frame Bending about the Weak Axis

The portal frame shown in Fig. 8.8 was analyzed by Kanchanalai (1977) using plastic-zone methods. The columns are subjected to weak-axis bending. Kanchanalai's results were generated based on an assumed maximum compressive residual stress of $0.3F_y$. No initial geometric imperfections were specified for this frame. The gravity loads were applied first and held constant at $P = 0.5P_y$ and horizontal load H was applied to collapse and through the postcollapse range.

In the present work, the same frame is analyzed using the elastic-plastic-hinge and refined plastic-hinge methods. The refined plastic-hinge analyses are carried out based on the following stiffness-reduction schemes:

1. CRC E_t, Eq. (8.3.11), and parabolic end stiffness function with initial yield surface $\alpha = 0.5$, Eq. (8.5.1).
2. LRFD E_t, Eq. (8.3.5), and parabolic end stiffness function with initial yield surface $\alpha = 0.5$, Eq. (8.5.1).

The weak-axis load-displacement curves have been generated using the elastic-plastic-hinge and refined plastic-hinge analyses. The lateral load-displacement curves from these hinge-based analyses are compared with the "exact" plastic-zone curves in Fig. 8.8.

It is observed that the difference between the maximum loads obtained from the elastic-plastic-hinge and the plastic-zone analyses is quite significant. The elastic-plastic-hinge result for the maximum value of H is 45% greater

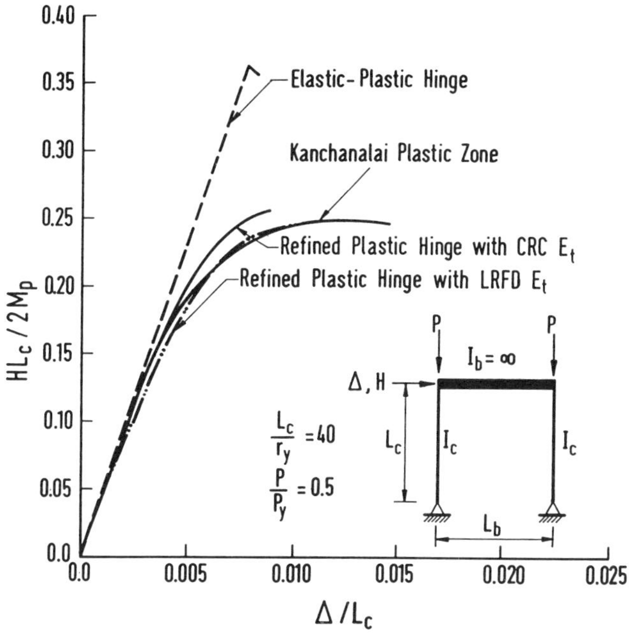

Fig. 8.8 Load-displacement characteristics of a pinned-base portal frame: Refined plastic-hinge analysis with initial yield surface $\alpha = 0.5$ and parabolic stiffness degradation.

than the plastic-zone results for the weak-axis bending. The comparisons also show that the refined plastic-hinge result based on the CRC E_t approach is conservative and sufficiently accurate in predicting the strong-axis inelastic response of this simple portal frame. The load-displacement response for the weak-axis bending case is slightly "stiffer" than the "exact" curve as the maximum load is approached.

The load-displacement curves obtained from the refined plastic-hinge analyses based on the LRFD E_t model exhibit larger deflection for the same applied load. The predicted lateral-load moment in the column at the frame's limit of resistance is about few percent lower than the plastic-zone result. The reason for this conservativeness is that the LRFD tangent modulus partially accounts for geometric imperfection effects because LRFD E_t is less than the elastic modulus when P/P_y is greater than 0.39. In other words, although the residual stress implicit in the LRFD column strength is approximately equal to $0.3F_y$, the onset of yielding assumed by the LRFD E_t model occurs when the axial force in the column reaches $P = 0.39P_y$. For the purpose of comparison with the plastic-zone solutions generated for the frame without geometric imperfections, the use of the CRC tangent-modulus model is more appropriate because this expression does not account directly for member out of straightness in its derivation. However, the disadvantage of using the CRC tangent modulus for design is that explicit modeling of member geometric imperfections is needed for the analysis to satisfy the design specification requirements regarding the fabrication and erection tolerances for compression members. Of course, the procedure for specifying geometric imperfections can be automated in the computer-aided analysis procedure.

One important conclusion can be reached by again studying the curves shown in Fig. 8.8. Because the axial forces in the columns are $P = 0.5P_y$, the tangent modulus evaluated based on the CRC E_t does not have any effect on the prediction of the inelastic behavior of this frame. The inelastic stiffness reduction in the model of the columns is associated with bending action only. Based on the assumed initial yield surface ($\alpha = 0.5$) and the parabolic stiffness-reduction function (Eq. (8.5.1)), the refined plastic-hinge approach is significantly more accurate in predicting the inelastic load-displacement behavior of this frame when compared to the elastic-plastic hinge method. It gives conservative predictions of the maximum strengths of the frames compared to the "exact" plastic-zone solutions.

8.5.3 A Fixed-End Beam Loaded at the One-Third Point

The fixed-end beam shown in Fig. 8.9 is loaded with a concentrated lateral load at the one-third point along the span of the beam. The normalized load-

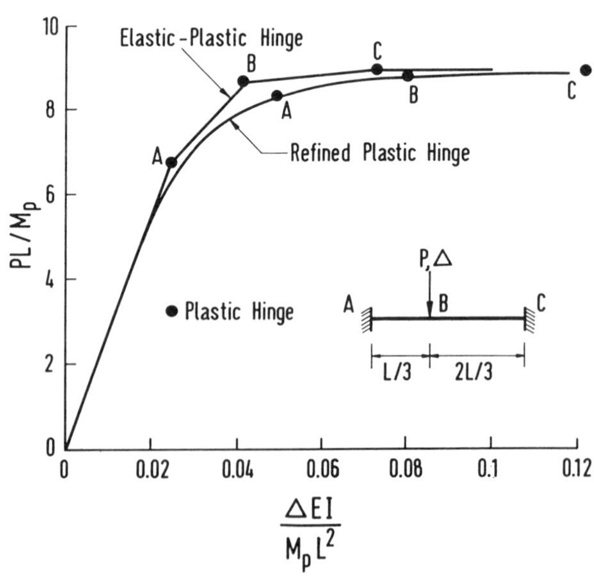

Fig. 8.9 Load-displacement characteristics of a fix-ended beam: Plastic-hinge versus refined plastic-hinge analysis.

displacement curve of this beam is determined by using the elastic-plastic-hinge and refined plastic-hinge approaches. The purpose of this study is to contrast the inelastic behavior predicted by the conventional elastic-plastic-hinge and refined plastic-hinge analyses for member loaded by in-plane bending only. The refined plastic-hinge analyses are performed based on an assumed initial yield surface with $\alpha_0 = 0.5$ and the parabolic stiffness-degradation function given by Eq. (8.5.1). The column tangent modulus in the refined plastic-hinge analysis is not effective because the axial load in the beam is negligibly small. The normalized load-displacement curves obtained from the analyses are shown in Fig. 8.9. The beam is subdivided into two elements at the load point. The collapse load predicted by the elastic-plastic-hinge analysis is $P = 8.91M_p/L$, which corresponds to the formation of a beam mechanism.

The plastic-hinge methods should be conservative in predicting the collapse load of this fixed-end beam of compact cross-section, because the effect of strain hardening is neglected. Although the elastic-plastic-hinge analysis is successful in predicting the collapse load of the beam, the load-deflection curve is somewhat different from that predicted by the refined plastic-hinge method. The load-deflection curve generated by the elastic-plastic-hinge analysis is piecewise linear because the plastic hinges form abruptly. The load-defection curve from the refined plastic-hinge analysis exhibits a smooth transition of stiffness during the entire loading process. The actual load-deflection behavior of the beam is expected to be smooth because the presence of residual stresses in the beam as well as possible welding stresses at the fixed beam ends result in early yielding of the section and result in smoother degradation of member stiffness. It is noted that the deformations of the beam in the inelastic range predicted by the refined plastic-hinge analysis are larger than those predicted by the plastic-hinge method.

In the elastic-plastic hinge analysis, the first plastic hinge is formed at the beam's fixed end, denoted by point A in the load-displacement curves in Fig. 8.9. The applied load corresponding to the formation of first plastic hinge is $P = 6.75M_p/L$. The elastic-plastic-hinge analysis neglects the effects of residual stresses and gradual plastification in the beam, so the distribution of internal forces during the loading process is somewhat different from that predicted by the refined plastic-hinge and plastic-zone methods. Because of the assumptions that the cross-section behavior is elastic-perfectly plastic, the analysis tends to overpredict the stiffness at the critical location of the beam. The elastic stiffness at the critical location of the beam attracts greater forces until the forces reach the cross-section plastic strength at which a plastic hinge is formed. Inelastic force redistribution does not occur until the first plastic hinge in the beam is formed. However, this inaccuracy in the prediction of the inelastic forces does not have any influence on the prediction of the ultimate strength of the beam. This is because the second-order effect in the beam is virtually absent, and beam failure is due to the formation of a beam mechanism.

In the refined plastic-hinge analysis, the first plastic hinge is also formed at point A in the beam. The load corresponding to the formation of the first plastic hinge is $P = 8.31M_p/L$, which is slightly higher than that predicted by the plastic-hinge analysis. The formation of the first plastic hinge is delayed because the inelastic force redistribution occurs far before the attainment of full-section plastic capacity at A. The plastification effects at A tend to shed load to other less critical locations of the member when the cross-section forces at A exceed the forces implied by the initial yield surface.

It is important to mention that strain-hardening effects also have important influence on the inelastic behavior of a beam subjected to a moment gradient along the length. However, further research is still required to study the strain-hardening behavior of general beam-column problems under different loading conditions, and to categorize the beneficial effect of strain hardening versus the negative effects of local and lateral-torsional buckling for practical design uses.

8.6 VERIFICATION STUDIES

8.6.1 Sway Frames

Figures 8.10 through 8.12 compare the in-plane strength curves obtained by the second-order elastic-plastic-hinge and refined plastic-hinge analyses with the plastic-zone results from Kanchanalai (1977). Because Kanchanalai's plastic-zone solutions were used as benchmarks for the development of the AISC LRFD beam-column equations (Yura, 1988; Liew et al., 1991), any analysis methods that can match these benchmarks are expected to satisfy the requirements for two-dimensional advanced inelastic analysis, for which separate specification member-capacity checks are not required.

For comparison with Kanchanalai's plastic-zone solutions, the refined plastic-hinge analysis is performed using the CRC tangent modulus, which represents the effects of residual stress only. Also, for all the verification studies that follow, the refined plastic-hinge analyses are performed using the parabolic stiffness-degradation function, defined by Eq. (8.5.1). In addition, a curve that denotes the initial onset of yielding in the beam-column member is plotted in Figs. 8.10 through 8.12. This curve is determined by the second-order analysis considering the strong-axis bending and the effect of residual stress with a maximum compressive magnitude of $0.3F_y$ at the flange tip.

For both the strong- and weak-axis cases in Figs. 8.10 through 8.12, the refined plastic-hinge approach is not more than 4% unconservative compared with the plastic-zone results for all ranges of axial force versus moment. The refined plastic-hinge analysis is particularly accurate for the frame shown in Fig. 8.12 in which the elastic-plastic-hinge analyses are most in error. In this case, the elastic-plastic-hinge model overpredicts the "exact" weak-axis

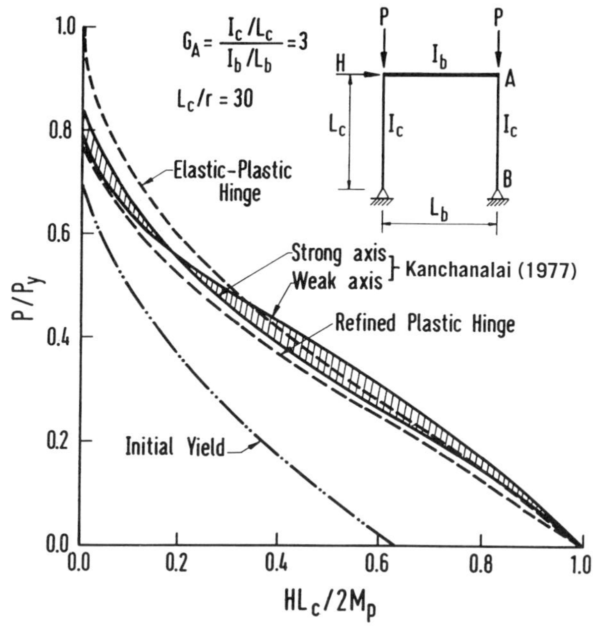

Fig. 8.10 Comparison of strength curves for a portal frame with $G_A = 3$ and $L_c/r = 30$.

strengths by a maximum of 30%, whereas the refined plastic-hinge model gives excellent strength predictions with errors not more than 1% unconservative.

When the axial force in the frame member is large, the use of the tangent modulus E_t alone in the elastic-plastic-hinge analyses gives a good estimate of the ultimate strengths. However, at the intermediate axial-load range, this model somewhat overestimates the "exact" strength. In this load range, additional distributed plasticity effects in the member are associated with bending actions. Therefore, the use of E_t alone in the elastic-plastic-hinge analyses is not sufficient to represent the member stiffness degradation, particularly when the bending moment in the member is not insignificant. This is evidenced in Fig. 8.11 in which the elastic-plastic-hinge model overestimates the "exact" strengths for both the strong- and weak-axis cases when the axial load is less than $0.5P_y$. For these cases, the use of the CRC tangent-modulus model alone does not improve the strength prediction because the CRC tangent-modulus formula is not effective when $P < 0.5P_y$. For general loading conditions, the use of the plastic-hinge stiffness-degradation model as well as the CRC tangent-modulus model provides an overall improvement in the elastic strength predictions.

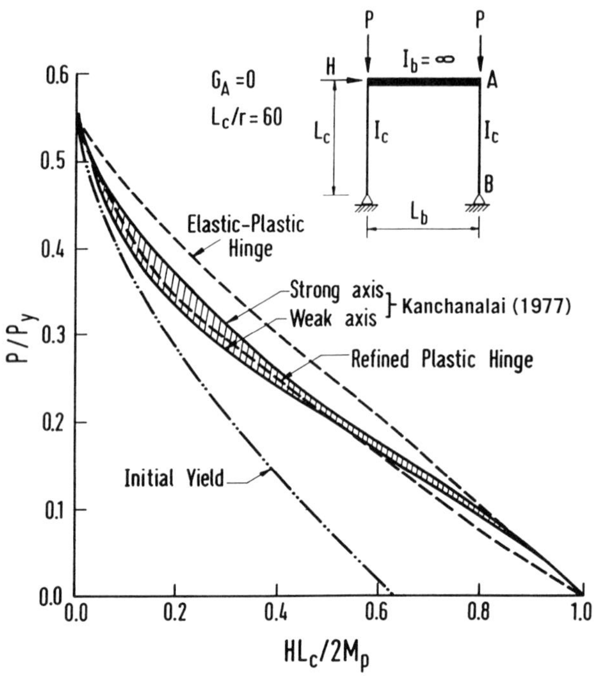

Fig. 8.11 Comparison of strength curves for a portal frame with $G_A = 0$ and $L_c/r = 60$.

8.6.2 Axially Loaded Columns

One may argue that the refined plastic-hinge model may underestimate the inelastic flexural stiffness of the member compared to the stiffness implied by the tangent-modulus concept for pure axial-load cases. This is because the element inelastic flexural stiffness is reduced by the column tangent modulus and by the plastic-hinge stiffness-degradation model. However, this aspect appears to have little consequence on strength prediction for columns subjected to an axial load only.

Figure 8.13 shows the results obtained from the refined plastic-hinge analysis using two elements per column. A W8 $\times$ 31 column section bending about the strong axis is assumed, and the column initial out of straightness with a maximum midlength deflection equal to $L/1500$ is explicitly modeled. This out-of-straightness magnitude is equivalent to that assumed by the LRFD column-strength curve. The inelastic column-strength curve generated by the refined plastic-hinge analyses is compared with the AISC LRFD column-strength curve. The comparison shows that the column-strength curve generated by the refined plastic-hinge analysis is not more than 5% unconservative for columns with slenderness parameters up to $\lambda_c = 1.5$. This

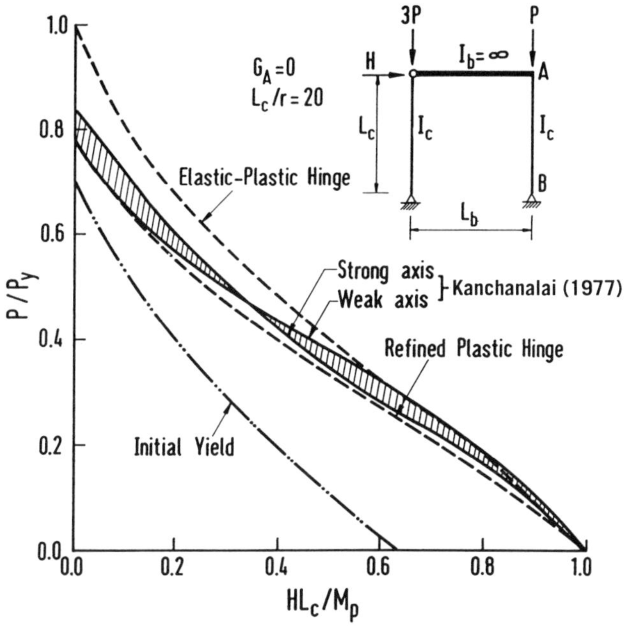

Fig. 8.12 Comparison of strength curves for a leaning column frame with $G_A = 0$ and $L_c/r = 20$.

concludes that the double stiffness-reduction schemes employed by the refined plastic-hinge model have little consequence on strength prediction for columns subjected to a pure axial load.

Figure 8.13 also compares the AISC LRFD column-strength curve with the strength curves generated by inelastic buckling analyses using the CRC and LRFD tangent moduli, and by the refined plastic-hinge approaches. In the same figure, the column capacity from the refined plastic-hinge analysis using one element is obtained by calculating the load level at which the determinant of the element tangent-stiffness matrix is zero. The column strength obtained by this approach is essentially the same as that predicted by the inelastic buckling analysis with the use of the column tangent modulus.

It can be observed that both tangent-modulus inelastic buckling analyses overpredict the column capacity implied by the AISC LRFD column equations (AISC, 1986). The maximum error associated with the CRC E_t inelastic buckling analysis is about 14%. This maximum error corresponds to the column with a slenderness parameter of unity, at which value residual stresses and member initial out-of-straightness effects interact to produce the greatest reduction in strength from the theoretical value for a perfect column. However, for columns with slenderness parameters λ_c less than or equal to 0.4, the maximum error from the CRC E_t inelastic buckling analysis is not more

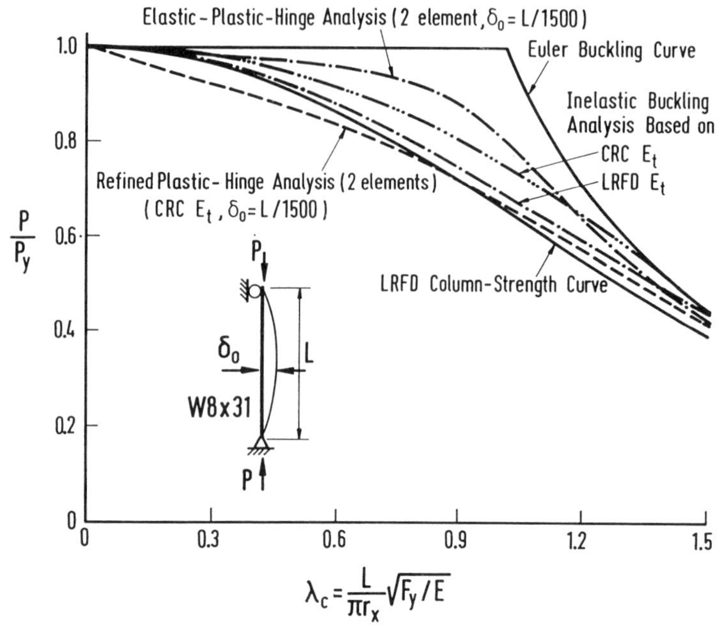

Fig. 8.13 Comparison of strength curves for an axially loaded pinned-end column.

than 5% unconservative compared to the AISC LRFD column-strength curve. These results suggest that for columns with $\lambda_c \leq 0.4$, the solutions are not sensitive to geometric imperfection effects.

If the inelastic buckling analysis is performed using the LRFD tangent modulus (LRFD E_t), the resulting strength curve compares well with the LRFD column-strength curve, and the error is not more than 5% unconservative if λ_c is less than 1.0. The error in strength prediction, however, increases from 5% at $\lambda_c = 1.0$ to 14% at $\lambda_c = 1.5$.

When inelastic enalysis is used for frame design, the AISC LRFD places a limit on the axial force by requiring that $\lambda_c \leq 1.5$ and $P \leq 0.75P_y$ for sway columns and $P \leq 0.8P_y$ for nonsway columns. Also in the LRFD provisions, explicit modeling of geometric imperfections is not required in the analysis of steel frame structures. By adopting the same design principles, it appears that the inelastic analyses are sufficiently accurate for columns with $\lambda_c < 0.4$, for which explicit modeling of member initial imperfections is not necessary. The refined plastic-hinge analysis based on the use of one element for the column with $\lambda_c < 0.4$ will never be more than 5% unconservative. For columns with λ_c values between 0.4 and 1.0, the LRFD E_t should be used in the inelastic analysis so that the member imperfection effects can be accounted for implicitly by the tangent modulus. However, for columns with $\lambda_c > 1.0$, explicit modeling of member initial out of straightness is still required for a proper evaluation of column strength. In this case, the use of the CRC E_t model would be more appropriate.

It is important to emphasize that the conventional elastic-plastic-hinge method overpredicts the column strength up to a maximum error of 22.5%. This error, however, reduces to less than 5% in the range where $\lambda_c \leq 1.0$, when the "one-element" refined plastic-hinge method with LRFD E_t is employed.

8.6.3 Isolated Beam-Columns

8.6.3.1 Strong-Axis Bending The plastic-zone solutions generated by Zhou, Duan, and Chen (1990) (see Section 7.2.2) are included in the present investigation to assess the performance of the elastic-plastic-hinge and refined plastic-hinge approaches for beam-columns subjected to equal end moments producing single-curvature bending. The plastic-zone strength curves are shown in Fig. 8.14, where they are compared with the elastic-plastic-hinge and refined plastic-hinge solutions. Both the plastic-hinge-based results in Fig. 8.14 were generated based on the use of two elements per member, and the analyses explicitly model the member initial out of straightness, which has a maximum magnitude of $\delta_0 = L/1000$ at midlength. The refined plastic-hinge model is based on the use of the CRC column tangent modulus because member initial out of straightness is modeled explicitly in the analysis.

It can be observed from Fig. 8.14 that for beam-columns with L/r equal to 20 and P less than $0.2P_y$, the elastic-plastic-hinge analysis is sufficiently accurate for the prediction of the ultimate strengths of these beam-columns. Otherwise, the elastic-plastic-hinge analyses always overpredict the capacities of the beam-columns by more than 5%.

The refined plastic-hinge approach using the CRC E_t approach is successful in predicting the limiting strength of the beam-columns bent in single curvature, as shown in Fig. 8.14. For most cases, the refined plastic-hinge analyses overpredict the maximum strengths of the beam-columns, but the maximum error is not more than 7%. This maximum error corresponds to beam-columns with $L/r = 80$ and 100.

It should be noted that when the beam-column is in single-curvature bending, the maximum second-order moment is always at the midlength of the column. Therefore, the use of two beam-column elements is sufficient for the refined plastic-hinge analysis to capture the beam-column in-plane strength behavior. Although the use of more elements per member would improve the strength predictions, the improvements (for the case of single-curvature bending) are less than 1%.

Additional results are in Fig. 8.15 in which the refined plastic-hinge analyses were performed using the LRFD tangent modulus instead of the CRC tangent modulus. Because the LRFD E_t model accounts implicitly for member initial out of straightness with maximum magnitude $\delta_0 = L/1500$, it could be stated that the member imperfection value of $\delta_0 = L/1000$ for the actual

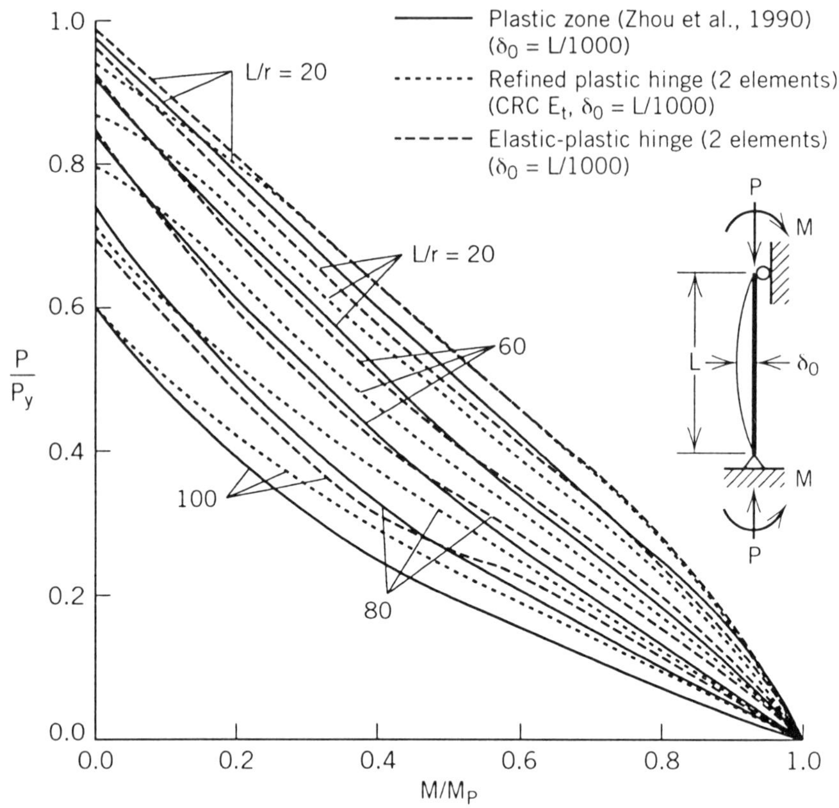

Fig. 8.14 Comparison of strength curves from plastic-hinge-based analyses with "exact" strength curves (Zhou et al., 1990): single-curvature bending.

beam-column used in the plastic-zone analysis is not fully accounted for by the LRFD tangent-modulus model. To compensate for this "extra" geometrical out-of-straightness imperfection, the analysis is carried out assuming the member is initially crooked with $\delta_0 = L/3000$. This δ_0 value is obtained by subtracting the actual value of $\delta_0 = L/1000$ from the assumed value of $\delta_0 = L/1500$. The results obtained from this approach are compared with the plastic-zone results and the LRFD beam-column strength curves in Fig. 8.15. The LRFD beam-column strength curves are obtained based on the LRFD beam-column interaction equations in which the required second-order moment in these equations is computed based on second-order elastic analysis. Also, an effective length factor of 1 is used in these equations for the evaluation of the beam-column strength because the member is pinned at both ends.

The results in Fig.8.15 show that the strength predicted by the refined plastic-hinge analysis are generally greater than the plastic-zone results. How-

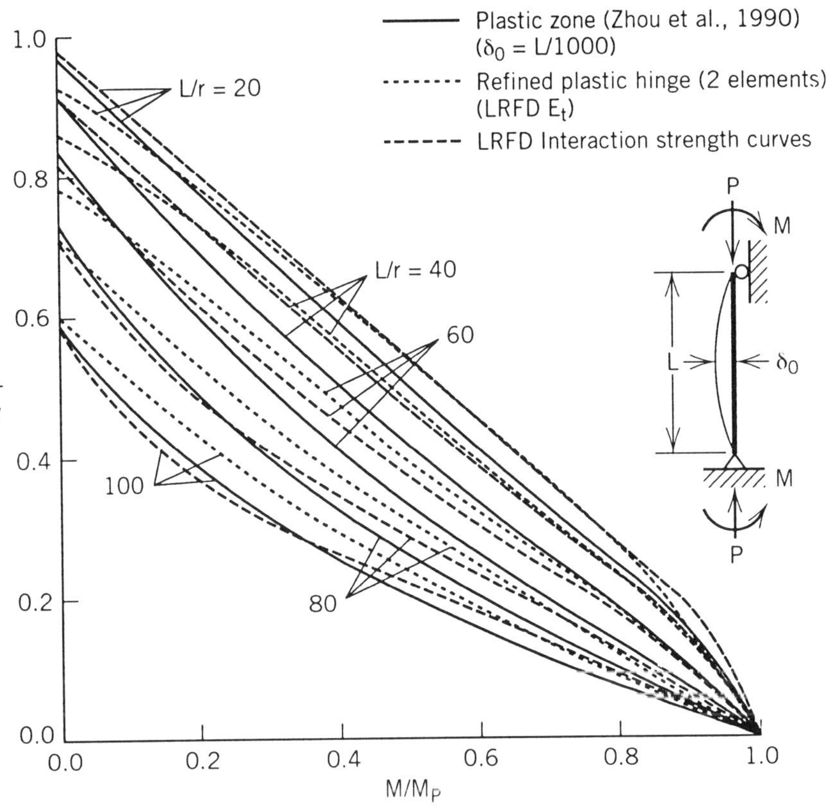

Fig. 8.15 Comparison of beam-column strength curves from refined plastic-hinge analysis (LRFD E_t) with "exact" strength curves (Zhou et al., 1990) and AISC LRFD equations: single-curvature bending.

ever, these results compare well with the AISC LRFD beam-column strength curves. In summary, the maximum error from the refined plastic-hinge analysis is about 7% unconservative compared to the plastic-zone solutions. The maximum error is, however, less than 5% in comparison with the code provision for beam-column design in AISC LRFD. Therefore, it can be concluded that the modeling of member initial out of straightness is not necessary for the refined plastic-hinge analysis if the LRFD tangent-modulus model is employed in the analysis.

8.6.3.2 Weak-Axis Bending Figure 8.16 shows the strength curves for beam-columns subjected to unequal end moments producing reverse-curvature bending. Because the maximum moment does not occur at midlength, it was decided to use four elements per member in the plastic-hinge analyses.

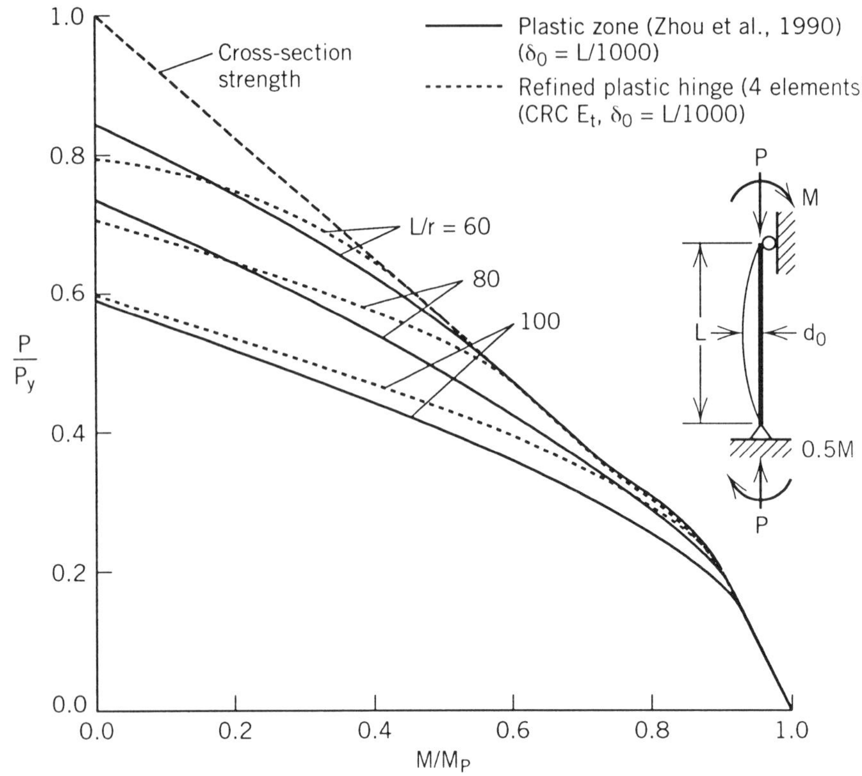

Fig. 8.16 Comparison of beam-column strength curves from refined plastic-hinge analysis (CRC E_t) with "exact" strength curves (Zhou et al., 1990): reversed-curvature bending.

The refined plastic-hinge analysis is carried out using the CRC tangent modulus, and member initial out of straightness of $\delta_0 = L/1000$ is included explicitly in the analysis.

The refined plastic-hinge analysis again is shown to be accurate enough in predicting the ultimate strengths of beam-columns subjected to reversed-curvature bending. The maximum unconservative error is not more than 5%. The overall errors are smaller for members subjected to double-curvature bending because the stability of these members is less critical, and the stability is less affected by member initial imperfections and distributed plasticity. For reversed-curvature bending, the moment gradient effect tends to improve the stability and strength of the beam-column compared to the same member that is subjected to uniform bending.

When the beam-column is subjected to unequal moments subjecting it to bend in reversed curvature, the location of the maximum second-order inelastic moment varies with the magnitude of applied axial force and the ratio of end moments. The location of the peak moment in the member is difficult to obtain analytically because it is load-path-dependent. Moreover, the member is initially crooked, and the spread of plasticity within the member would ultimately influence the deflected shape of the member. Therefore, at least four elements are required for the refined plastic-hinge analysis to capture the inelastic behavior of this type of beam-column members.

8.7 ANALYSIS OF NONSWAY FRAMES

The plastic-zone solutions for braced frames reported by Chen, Duan, and Zhou (1990) are used as the benchmarks to assess the capability of the refined plastic-hinge and elastic-plastic-hinge approaches in predicting the limit state behavior of nonsway frames. These frames are loaded primarily on the beam and on the columns, subjecting the columns to single- and double-curvature bending. The maximum strengths of the frames obtained from the inelastic analyses are contrasted with the results obtained by the LRFD elastic-analysis approach. These studies provide evidence that supports the use of inelastic analysis in the design of two-dimensional braced frames.

More specifically, the investigation focuses on the following:

1. Providing further insight into the behavior of redundant structural systems at their strength limit states
2. Establishing the reliability and accuracy of the proposed inelastic analyses intended for use in design by comparing results with those obtained by the more accurate, but more complex, plastic-zone analysis
3. Comparing the strength limit state behavior of nonsway frames obtained from inelastic analyses with the results obtained using the conventional LRFD elastic-analysis/design method
4. Comparing the force distribution/redistribution and yielding characteristics in a system modeled using the various inelastic-analysis methods
5. Providing evidence that would support the use of inelastic analysis in the design of planar frames without the need of member-capacity checks

Detailed studies of sway frames are given in Section 8.8.

8.7.1 Description of Frames and Loading

The dimensions and member sizes for the two portal frames in which sidesway is prevented are shown in Fig. 8.17. The beams and columns in these frames are rigidly connected about their strong-axis bending direction. All the member cross-sections are assumed to be fully compact, and the members are fully braced to prevent out-of-plane deformations. The frames are sub-

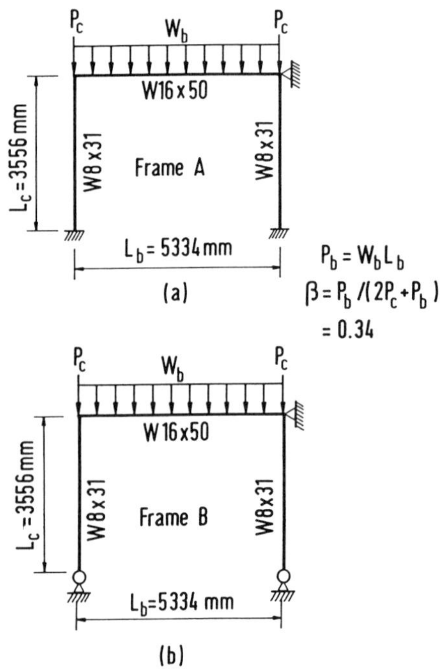

Fig. 8.17 Geometry and loading pattern. (a) A braced frame with fixed supports. (b) A braced frame with pinned supports.

jected to different loading patterns with β values ranging from 0.0 to 1.0, where β is the ratio of the load on the beam to the total load on the frame. The column slenderness ratio for Type A and Type B frames shown in Fig. 8.17 is $L_c/r = 40.3$, and the relative stiffness of the column to the beam is $G = I_c L_c / I_b L_b = 0.25$. The only difference between the Type A and Type B frames is the support conditions. A Type A frame has fixed supports, whereas a Type B frame is simply supported at the columns' bases.

In the plastic-zone analysis (Chen et al., 1990), all the cross-sections in the beam and columns contained initial residual stresses with a maximum compressive magnitude of $0.3F_y$ at the flange tips. The column members were assumed to have an initial out of straightness that varied sinusoidally with a maximum in-plane deflection, $\delta_0 = L/1000$, at midlength. The beam was assumed to be perfectly straight. Material strain hardening was not considered in the plastic-zone analyses.

In using the plastic-hinge-based analyses, all loads are applied proportionally. The beam is modeled by eight discrete elements to approximate the distributed loads as discrete point loads acting on the beam. The column is discretized into four elements in order to detect possible formation of plastic hinges between the column ends, and also to approximate the sinusoidal initial out of straightness of the member. In using the LRFD elastic-analysis/design approach, the effective-length factors for the braced frames are evaluated based on the alignment chart formula (AISC, 1986).

8.7.2 Summary and Comparison of Results

8.7.2.1 Maximum Strengths The plastic hinge-based analyses and the elastic LRFD approach are compared with the plastic-zone analysis with respect to the maximum strength of the frames. The column forces (at the limiting strength) calculated by these four analysis approaches are shown in Tables 8.2 and 8.3 for the pinned-base frames and fixed-base frames, respectively. The results are discussed in what follows.

1. For the frames with no load acting on the beam (i.e., $\beta = 0.0$), the maximum strengths predicted by the LRFD equations are very close to the results obtained by the inelastic analyses. The limit loads of the braced frames subjected to gravity load acting on the columns can only be calculated directly from the LRFD column equations based on the effective-length factors evaluated from the alignment chart. The beam-column equations predict the strength of the frames rather accurately because the maximum strengths of these frames are controlled by column instability; the beam remains elastic when the columns reach their limit loads.
2. The maximum strengths predicted by the LRFD equations become increasingly conservative as the proportion of the load on the beam is increased from $\beta = 0.0$ to 0.2. As β is increased from 0.2 to 1.0, the difference between the LRFD and the "exact" solutions decreases, but the LRFD equations remain conservative for all cases. The conservative nature of the LRFD equations is because the inelastic-force redistribution is not considered in the elastic analysis. For β values ranging from 0.0 to 0.2, the inelastic analyses show that forces are redistributed from the columns to the beam due to inelastic action in the columns. In these cases, the beam is restraining the columns from buckling. However, for frames loaded with higher values of β (>0.20), the beam becomes inelastic before the columns reach their maximum strengths. Therefore, the amount of inelastic-force redistribution from the columns to the beam is reduced. In other words, the beam provides less and less restraints to the columns as the loads increase.
3. If load parameter β is greater than 0.5, the maximum strengths of frames A and B predicted by the refined plastic-hinge analyses are same as those obtained from the elastic-plastic-hinge analyses. This is because the failure loads of these frames are associated with the formation of a plastic collapse mechanism in the beam. For all cases, the results obtained by the two plastic hinge-based analyses compare well with the

TABLE 8.2 Comparison of Maximum Strengths of Fixed-Base Frames (Frame A)

β (1)	LRFD, P_L/P_y (2)	Elastic-Plastic Hinge, P_{ep}/P_y (3)	Refined Plastic Hinge, P_r/P_y (4)	Plastic Zone, P_{pz}/P_y (5)	Error P_L/P_{pz} (2)/(5)	Error P_{ep}/P_{pz} (3)/(5)	Error P_r/P_{pz} (4)/(5)
0.00	0.974	0.982	0.975	0.970	1.004	1.007	1.005
0.05	0.850	0.941	0.934	0.954	0.891	0.986	0.979
0.10	0.755	0.935	0.928	0.933	0.809	1.002	0.995
0.20	0.616	0.918	0.924	0.893	0.690	1.028	1.035
0.34	0.486	0.637	0.637	0.660	0.736	0.965	0.965
0.40	0.446	0.554	0.555	0.568	0.785	0.977	0.977
0.50	0.393	0.457	0.456	0.468	0.840	0.977	0.977
0.60	0.350	0.389	0.389	0.399	0.877	0.975	0.975
0.80	0.288	0.299	0.299	0.307	0.938	0.974	0.974
1.00	0.235	0.244	0.244	0.250	0.940	0.976	0.976

P_y is the squash load.

TABLE 8.3 Comparison of Maximum Strengths of Pinned-Base Frames (Frame B)

β (1)	LRFD, P_L/P_y (2)	Elastic-Plastic Hinge, P_{ep}/P_y (3)	Refined Plastic Hinge, P_r/P_y (4)	Plastic Zone, P_{pz}/P_y (5)	Error		
					P_L/P_{pz} (2)/(5)	P_{ep}/P_{pz} (3)/(5)	P_r/P_{pz} (4)/(5)
0.00	0.944	0.974	0.960	0.951	0.992	1.024	1.009
0.34	0.536	0.638	0.637	0.645	0.831	0.989	0.988
0.50	0.439	0.459	0.459	0.466	0.942	0.985	0.985
1.00	0.220	0.244	0.244	0.250	0.880	0.976	0.976

plastic-zone solutions. The maximum error in the maximum load predictions is less than 4%. The LRFD elastic-analysis approach is rather conservative for the fixed-base frames for β ranges from 0.05 to 0.60. The maximum conservative error associated with the elastic-analysis procedures is about 30% (see Table 8.3).

4. For the case $\beta = 0.0$, the results in Tables 8.2 and 8.3 suggest that the limit load for frame A predicted by the inelastic-hinge-based analyses and plastic-zone analysis is larger than that for frame B. This is obvious because the fixed supports should enhance considerably the overall stability of the frame. The fixed-base frame (frame A) is capable of distributing forces to the fixed supports; therefore, the maximum strength of this frame is expected to be higher. This behavior is correctly reflected by the LRFD interaction equations through the use of the effective-length factor. For the load cases $\beta = 0.34$ and 0.5, the limit loads predicted by the inelastic-hinge-based analyses and plastic-zone analysis for the fixed-base and pinned-base frames are almost identical. This is because the failure modes of these frames are associated with the formation of a beam mechanism. However, the LRFD elastic-analysis predicts greater strength for the pinned-base frame than for the frame with fixed bases. This is because the maximum second-order elastic moments in the columns of the pinned-base frame are smaller than those of the fixed-base frame. These studies suggest that the specification member-capacity checks based on elastic analysis are limited in their ability to provide a "true" assessment of the maximum strengths of these frames.

8.7.2.2 Force Distribution and Predicted Modes of Failure Figures 8.18a and 8.18b show the moment and axial force diagrams at the limit point for frames A and B with load parameter $\beta = 0.34$. In general, the axial forces in the columns are approximately proportional to the uniform load tributary to each column. The axial forces in the beams are negligible, therefore, they are not shown in the figures. The moment distributions at the frames' limits of resistance predicted by the refined plastic-hinge-analyses are found to be almost identical with the results from the elastic-plastic-hinge analyses. These results are in general agreement with those predicted by the plastic-zone analysis.

The second-order plastic-zone analysis indicates that the failure of the frames subjected to the load case $\beta = 0.34$ is due to column instability with significant yielding occurring at the midspan of the beam (Chen et al., 1990). The limit loads obtained from the plastic-hinge-based analyses are associated with the formation of plastic hinges at the top of the columns and at the midspan of the beam (see Figs. 8.18a and 8.18b). On the contrary, the limit loads predicted by the LRFD elastic-analysis/design approach are associated with column instability only—the beam remains elastic when the columns reach their limiting strengths. This is an obvious shortcoming of the elastic-analysis approach because the beneficial effect of force redistribution in the

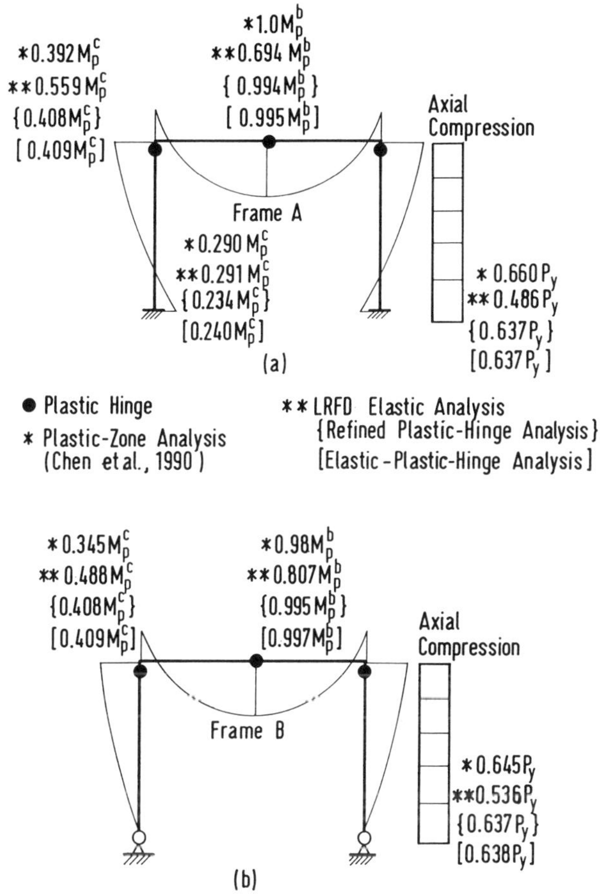

Fig. 8.18 Moment and axial-force diagrams. (a) Fixed-support frame. (b) Pinned-support frame.

redundant frame is not utilized for the assessment of system strength. In addition, the LRFD elastic-analysis/design approach predicts greater load capacity for the simply supported frame than for the fixed-base frame. This is because, in the fixed-base frame, the elastic-analysis/member-capacity check approach penalizes the columns that attract greater elastic forces, although the columns can carry more loads by distributing the forces to the beam through inelastic force redistribution.

8.7.2.3 ***Force-Point Traces*** The force-point traces shown in Figs. 8.19a and 8.19b represent the change in magnitude and direction of forces at selected cross-sections during the loading history. The force-distribution re-

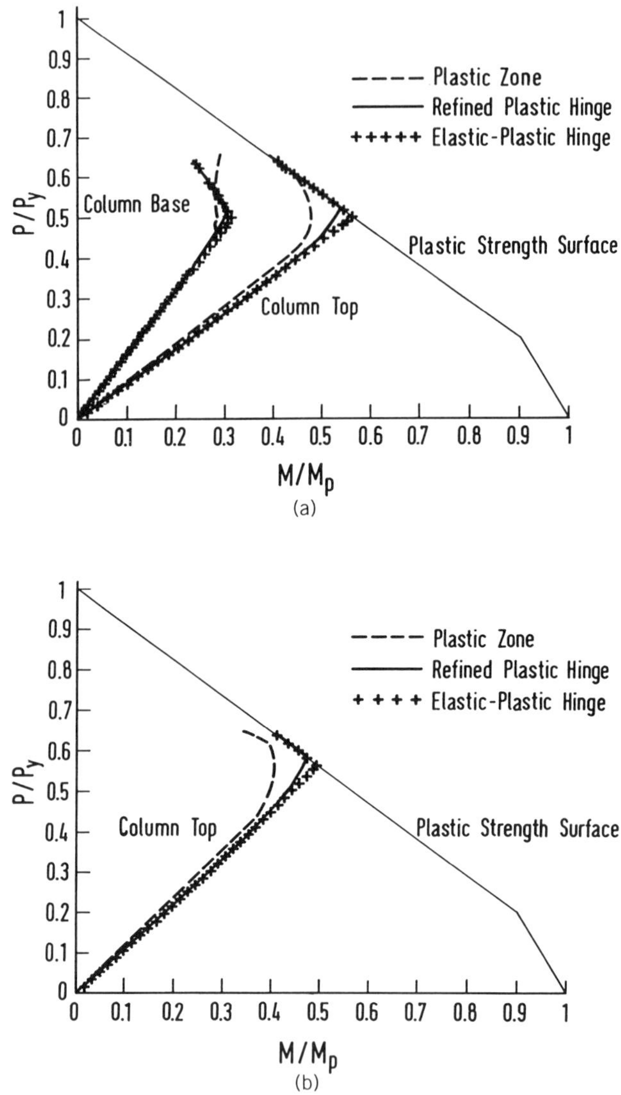

Fig. 8.19 Comparison of force-point traces. (a) Fixed-support frame. (b) Pinned-support frame.

sponse curves from the elastic-plastic-hinge analyses show an abrupt change in the force-point trace after the formation of plastic hinges at the top of the columns. As additional load is applied to the structure, the hinged columns carry more axial load by redistributing their bending moments to the beam. Finally, a plastic hinge occurs at the midspan of the beam, leading to the formation of a collapse mechanism. The plastic-zone response curves in Figs. 8.19a and 8.19b show a gradual force redistribution before the plastic strength is reached. The response curve obtained by the refined plastic-hinge analysis is between the two response curves predicted by the previous inelastic analyses. Although the predicted force-redistribution response curves are different from these analyses, the limit loads obtained from the analyses are in good agreement.

8.8 DETAILED STUDY OF SWAY FRAMES

Vogel (1985) presented a set of frames intended for use in judging the accuracy and reliability of computer programs. The results of plastic-zone analysis are included as standards for comparison. Ziemian (1989) studied the same frames using elastic-plastic-hinge and plastic-zone analyses. The plastic-zone analysis program used in Ziemian's work is developed by White (1985). The results from Ziemian's work supplement the results presented by Vogel by providing the detailed force distribution in the system not included in Vogel's work.

The set of sway frames shown in Figs. 8.20–8.22 is used to demonstrate the merits of applying the inelastic analyses proposed in Chapters 7 and 8 in relation to their acceptability for use in frame design. The notional-load plastic-hinge and the refined plastic-hinge analyses are compared to the plastic-zone analyses in terms of force distribution and system behavior. These

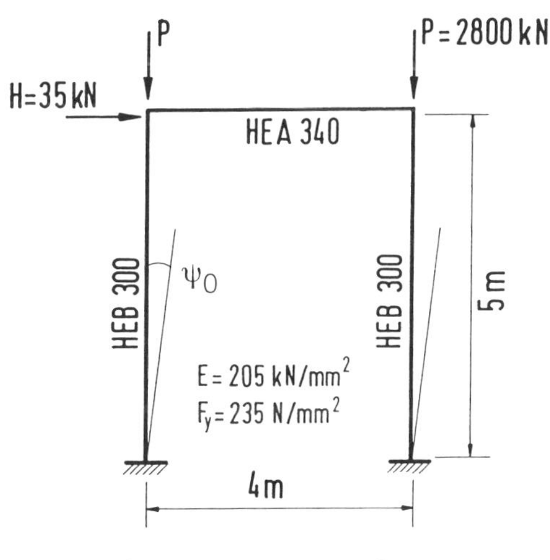

Fig. 8.20 Calibration frame: portal frame.

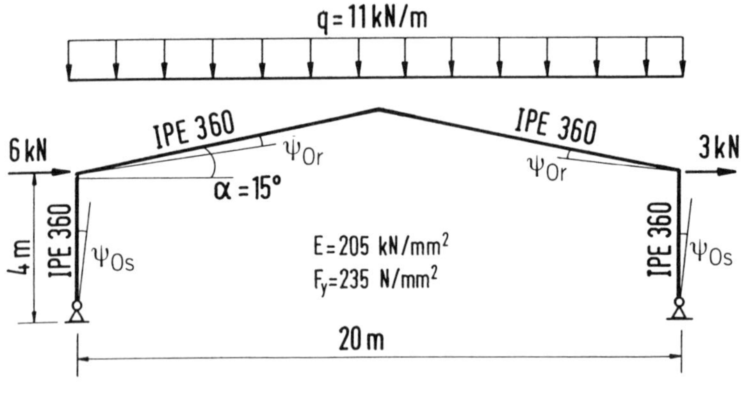

Fig. 8.21 Calibration frame: gable frame.

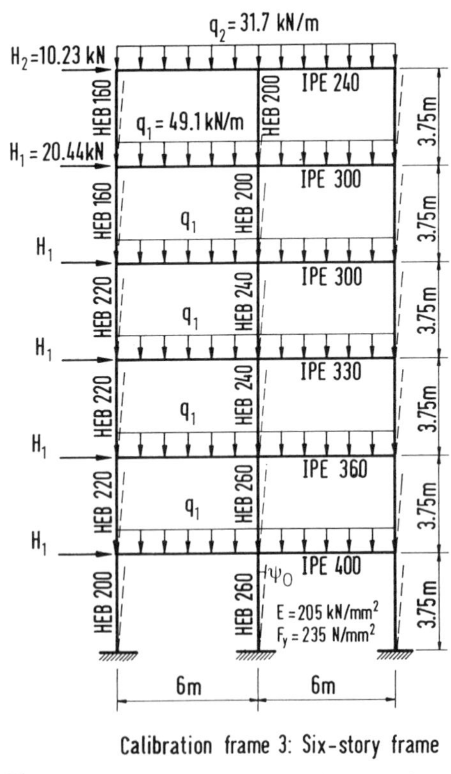

Fig. 8.22 Calibration frame: six-story frame.

analysis methods are investigated for reliability and accuracy by comparing the results with those obtained by the plastic-zone analyses (Vogel, 1985; Ziemian, 1989). The purpose is to establish the validity of the inelastic-hinge-based methods so that they can be used in routine design for member proportioning without the need for separate member-capacity checks.

8.8.1 Frames, Loading, and Analysis Details

Vogel (1985) selected and recommended three types of frames for the calibration of second-order inelastic analysis:

1. Portal frame
2. Gable frame
3. Six-story frame

The configurations of the calibration frames are shown in Figs. 8.20–8.22. All members of the frames are made of hot-rolled sections. The cross-section dimensions and properties of the frame members are summarized in Table 8.4. The detailed section dimensions referred to in the table are shown in Fig. 8.23. Also included in Fig. 8.23 are the residual stress patterns assumed in Vogel's plastic-zone analysis. According to the ECCS recommendations, when the section depth-to-width ratio d/b is greater than 1.2, the maximum compressive residual-stress magnitude to be used in the plastic-zone analysis is $F_{rc} = 0.3F_y$. When the ratio of d/b is less than 1.2, a larger compressive residual-stress magnitude of $0.5F_y$ should be used instead. These residual stresses are assumed to be distributed uniformly along the length of the frame members. In addition, initial sway imperfections are also specified explicitly in the plastic-zone analyses. The magnitudes of column-sway imperfections for the calibration frames are also summarized in Table 8.4.

The following simplified analysis methods are used to assess the strength limit states of the calibration frames considered in this section.

- *LRFD Elastic-Analysis/Design Approach:* A direct second-order elastic analysis is used for determining the member maximum forces. The maximum forces in every member are then checked against the LRFD interaction equations for increasing levels of applied load until the limit load is reached, at which point the maximum forces at the most critically load member satisfy the LRFD limiting strength equations. The resistance factors in the LRFD member strength equations are taken as 1.0 for comparison with the inelastic solutions.
- *Elastic-Plastic-Hinge Approach:* The conventional plastic-hinge method is used to model material nonlinear behavior. Initial geometric imperfections, similar to those assumed in the "exact" plastic-zone analysis, are explicitly modeled in this analysis.

TABLE 8.4 Cross-Section Properties of Members and Initial Sway Imperfections

Members	A (cm^2)	I_x (cm^4)	I_y (cm^4)	Calibration Frames	ψ_0 or ψ_{0s}*	Initial Sway Deflection** (mm)
IPE240	39.1	3,892	284	Portal frame	1/400	12.5
IPE300	53.8	8,356	604			
IPE330	62.6	11,770	788			
IPE360	72.7	16,270	1,043			
IPE400	84.5	23,130	1,318	Gable frame	Columns 1/300	13.3
HEB160	54.3	2,492	889		Roof beams 1/432	24.0
HEB200	78.1	5,696	2,003			
HEB220	91.0	8,091	2,843			
HEB240	106.0	11,260	3,923	6-story frame	1/450	50.0
HEB260	118.0	14,920	5,135			
HEB300	149.0	25,170	8,563			
HEA340	133.0	27,690	7,346			

I_x, I_y is the moment of inertia about strong and weak axes. A is the cross-section area.
*ψ_0 is the angle of initial sway imperfection. **The initial sway deflection is measured at the top of the frame.

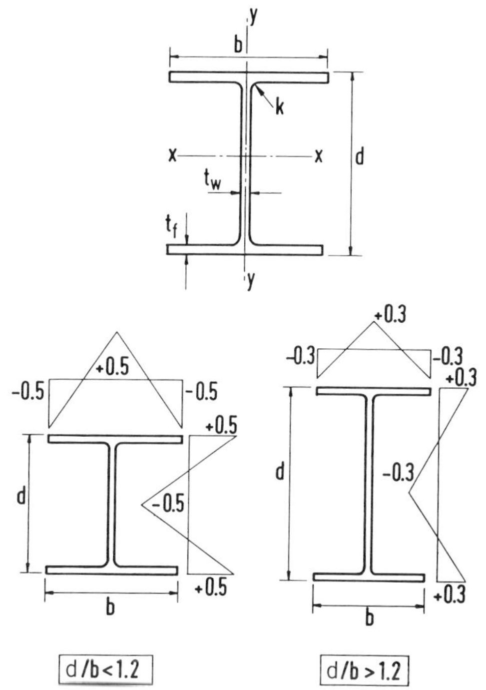

Fig. 8.23 Residual-stress patterns and dimensions for hot-rolled wide-flange sections used in Vogel's benchmark frames.

- *Refined Plastic-Hinge Approach:* This method employs the CRC tangent modulus and parabolic stiffness degradation function to model the effective stiffness of the member. Initial geometric imperfections, as specified in the plastic-zone analysis, are explicitly modeled in the refined plastic-hinge analysis.

The discretization of frame members used in modeling each calibration frame is shown in Table 8.5. The maximum strength results from the preceding analyses are summarized in Table 8.6 and discussed in the following sections.

TABLE 8.5 Element Discretization

Model Attribution	Portal Frame	Gable Frame	Six-Story Frame
No. of elements per beam	1	20	4
No. of elements per column	1	1	1

TABLE 8.6 Comparison of Maximum Strengths

		Portal Frame	Gable Frame	Six-Story Frame
Plastic-zone analysis	Vogel	1.02	1.07	1.12
	Ziemian	1.00	1.07	1.18
Refined plastic hinge		0.960	0.950	1.118
Notional-Load Plastic Hinge		0.999	1.037	1.097
LRFD elastic analysis		0.999	—	0.765

8.8.2 Portal Frame

Lateral load-displacement curves from the three simplified analyses and Vogel's plastic-zone analysis are plotted and compared in Fig. 8.24. The load-deflection curves predicted by the LRFD elastic analysis show a near-linear limit behavior. The load-deflection curves predicted by the inelastic analysis do not exhibit such behavior. Instead, the stiffness of the frame degrades gradually as the inelastic limit load is approached. This is because this portal frame is very sensitive to sway imperfections, and the large axial loads on the columns produce significant P–Δ effects, which severely degrade the stiff-

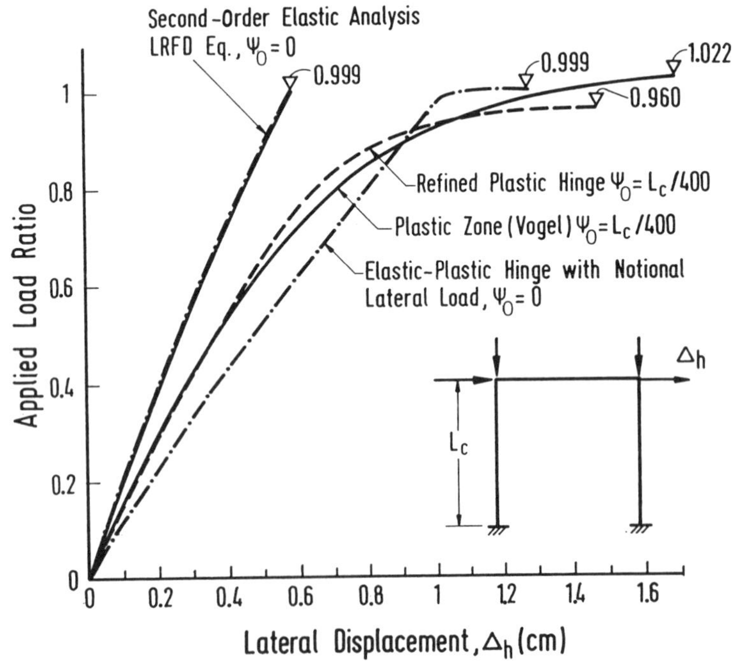

Fig. 8.24 Comparison of load-displacement curves for the portal frame.

ness of the columns. Although the maximum load predicted by the LRFD elastic analysis does not differ from the inelastic analyses, the stiffness is quite different, especially in the transition range from elastic to plastic. The use of the LRFD elastic approach underestimates the lateral drift of this simple portal frame.

The notional-load plastic-hinge method predicts larger frame drift than would be expected in the initial portion of the load-deflection response, and it predicts a smaller drift at failure of the frame. The load-deflection curve predicted by the refined plastic-hinge analysis is essentially the same as the plastic-zone curve, but the maximum load is about 6% lower than the plastic-zone solution.

Figure 8.25 shows a force-point trace of the normalized axial force P/P_y and bending moment M/M_p at the base of the right-hand column. In notional-load plastic-hinge analysis, the failure of the frame is driven by excessive sidesway due to the loss of stiffness after the formation of plastic hinges at the fixed supports. The failure mode predicted by the refined plastic-hinge analysis is associated with the inelastic instability of the columns.

Figures 8.26a–8.26c show the yielded zones predicted by the various inelastic models. In Vogel's plastic-zone analysis, about 60% of the cross-section area in the columns is yielded under compression (Fig. 8.26a), and the distributed plasticity in the columns is dominated mainly by the axial-

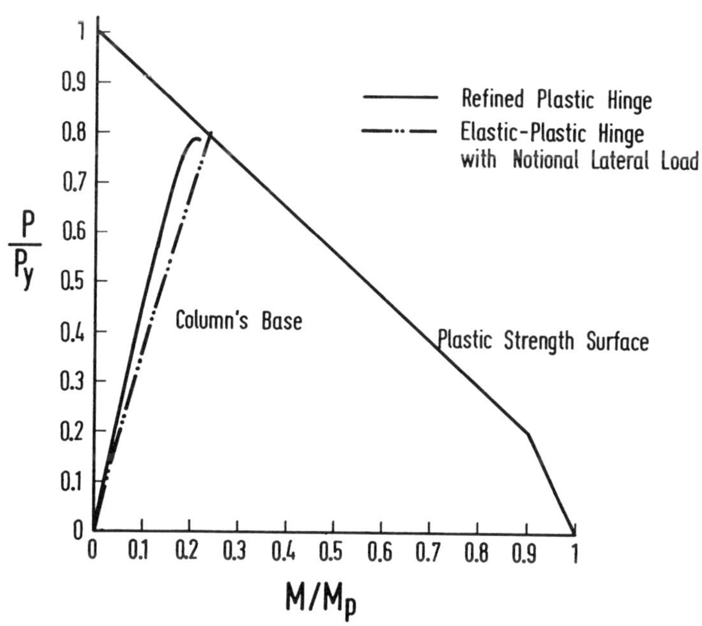

Fig. 8.25 Comparison of the force-point trace for the portal frame.

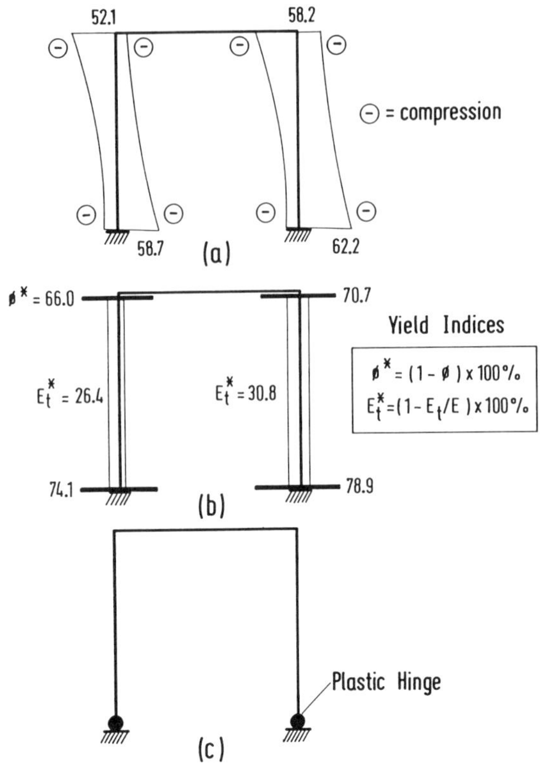

Fig. 8.26 Comparison of yield zones for the portal frame.

force action. The yielded zones predicted by the elastic-plastic-hinge analysis are concentrated as plastic hinges at the fixed supports (Fig. 8.26c). The regions in the frame other than the plastic hinges remain elastic. In the refined plastic-hinge analysis, the following yield indices are defined to estimate the degree of yielding in the frame members:

$$\phi^* = (1 - \phi) \times 100\% \tag{8.8.1}$$

$$E_t^* = (1 - E_t/E) \times 100\% \tag{8.8.2}$$

where ϕ, which is defined by Eq. (8.5.1), is the stiffness parameter associated with the effect of yielding at the element's end. E_t is the CRC tangent modulus. Parameter ϕ^* in Eq. (8.8.1) is a yield index that shows the degree of yielding at the element's end. $\phi^* = 1$ indicates that the member end is fully yielded, whereas $\phi^* = 0$ signifies an elastic end. E_t^* in Eq. (8.8.2) indicates the degree of distributed yielding associated with the axial force in the member. Although the definitions of these yield indices do not give an "exact" representation of the degree of yielding predicted by the plastic-zone analysis, they provide useful information for locating the critical sections or highly yielded regions in the structures. Figure 8.26b shows that the columns are significantly softened at their ends with the ϕ^* index ranging from 66.0 to 79.0%. The column flexural stiffness is approximately 70% effective, which is represented by a distributed yield index $E_t \approx 30.0\%$.

Figures 8.27a–8.27b show the moment diagram and axial forces at the limit point obtained by the inelastic-hinge-based models and the plastic-zone analysis (Ziemian, 1989). Although the axial-force distribution predicted by the notional-load plastic-hinge analysis agrees with the plastic-zone results, the bending moments at the element ends are slightly higher. The maximum error associated with this analysis is about 20%. The relative distribution of forces predicted by the refined plastic-hinge analysis is in general agreement with that obtained by the plastic-zone approach. The error involved in computing the internal force distribution is less than 8%.

8.8.3 Gable Frame

In the hinge-based analyses, uniformly distributed loads are applied as concentrated loads at the nodal points of each roof girder span. In Fig. 8.28, the

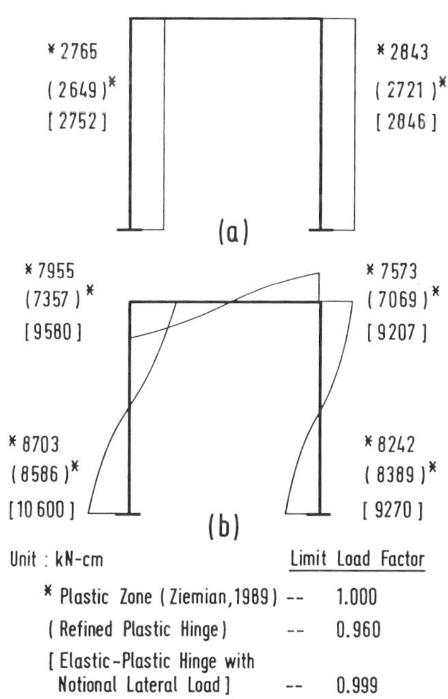

Fig. 8.27 Comparison of force distributions for the portal frame.

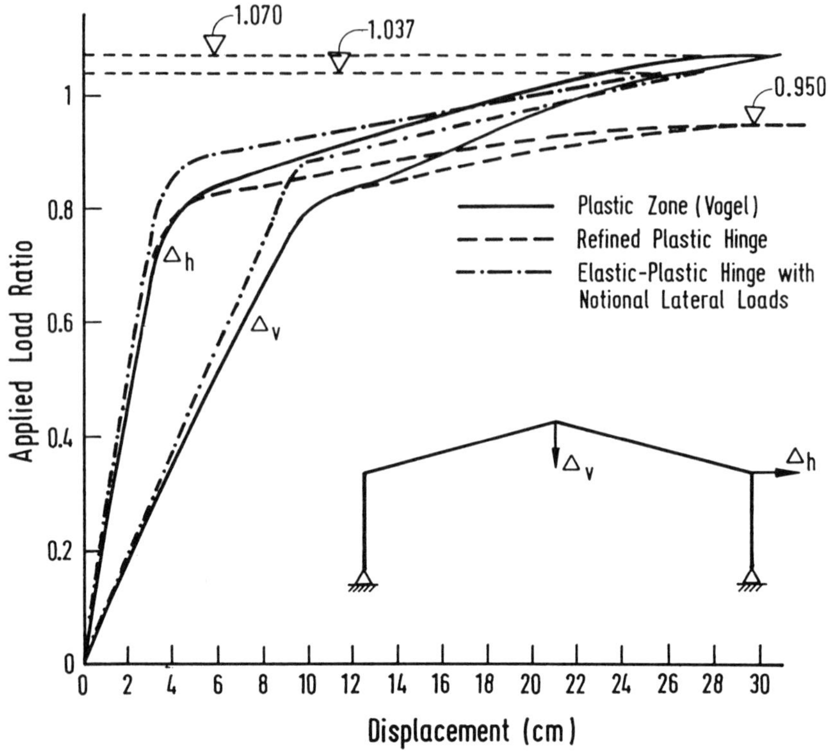

Fig. 8.28 Comparison of load-displacement curves for the gable frame.

load-displacement curves obtained by the two inelastic-hinge-based analyses are compared with the results of Vogel's plastic-zone analysis. The plastic-zone analysis response curves start to deviate from the inelastic-hinge-based curves after reaching an applied load ratio of about 0.8. After reaching this load ratio, the lateral deflection of the right column begins to increase dramatically with small increases in the applied loads. Both the refined-plastic hinge and plastic-zone analyses predict smooth response curves because the effects of gradual yielding of cross-sections have been modeled in these analyses. But the maximum strength obtained by the plastic-zone analysis is higher than that computed by the refined hinge analysis. This is because strain-hardening effects have been included in Vogel's plastic-zone analysis, because when the frame is loaded to its limit of resistance and the deflections become large, the effect of strain hardening at the right knee of the gable frame becomes significant. This strain-hardening effect increases the lateral-load resistance of this frame. However, this beneficial effect has been neglected in the hinge-based analyses.

In the notional-load plastic-hinge approach, an equal proportion of the 5% total gravity loads is applied horizontally to the top of the left- and right-hand columns. The frame's response predicted by the notional-load plastic-hinge analysis is slightly stiffer than the more "exact" response curve before reaching the limit load. Although this analysis approach approximates the column-sway imperfections partly through the use of the notional lateral loads, the initial geometric imperfections in the roof girders (see Table 8.4) are not modeled by the notional loads. Nevertheless, the limit load predicted by this model compares well with the plastic-zone solution.

The percentage of section area yielded at the limit load predicted by Vogel's plastic-zone analysis is shown in Fig. 8.29a. The locations of high-yield zones are consistent with the ϕ^* index from the refined plastic-hinge analysis, as shown in Fig. 8.29b. Both the hinge-based analyses predict a plastic hinge at the top of the right-hand column.

Axial-force and bending moment diagrams at the limit points are shown in Figs. 8.30a and 8.30b. The plastic-zone solutions shown in these figures are based on the analyses by Ziemian (1989). The distribution of forces obtained by these three analyses are in good agreement—except at the right-

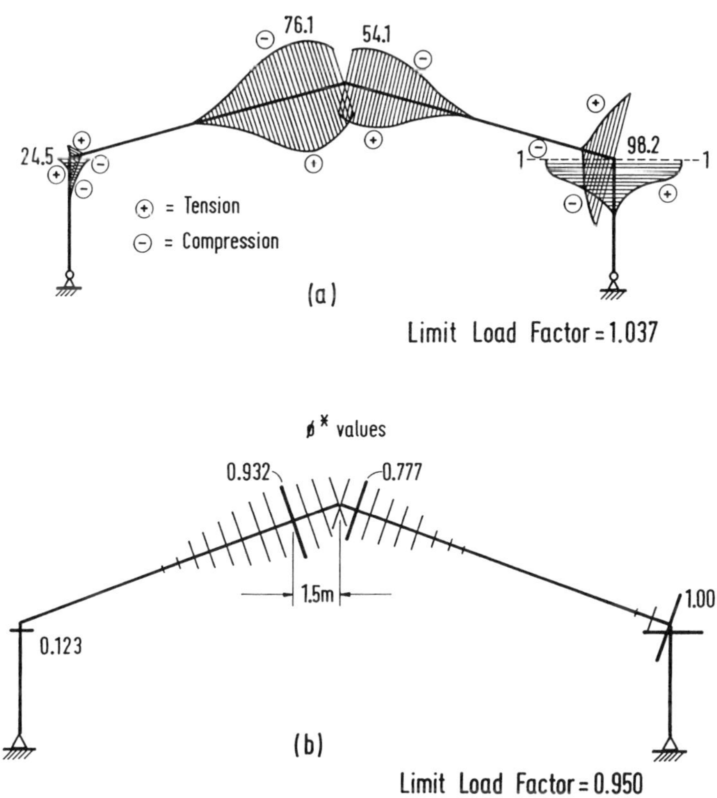

Fig. 8.29 Comparison of yield zones for the gable frame.

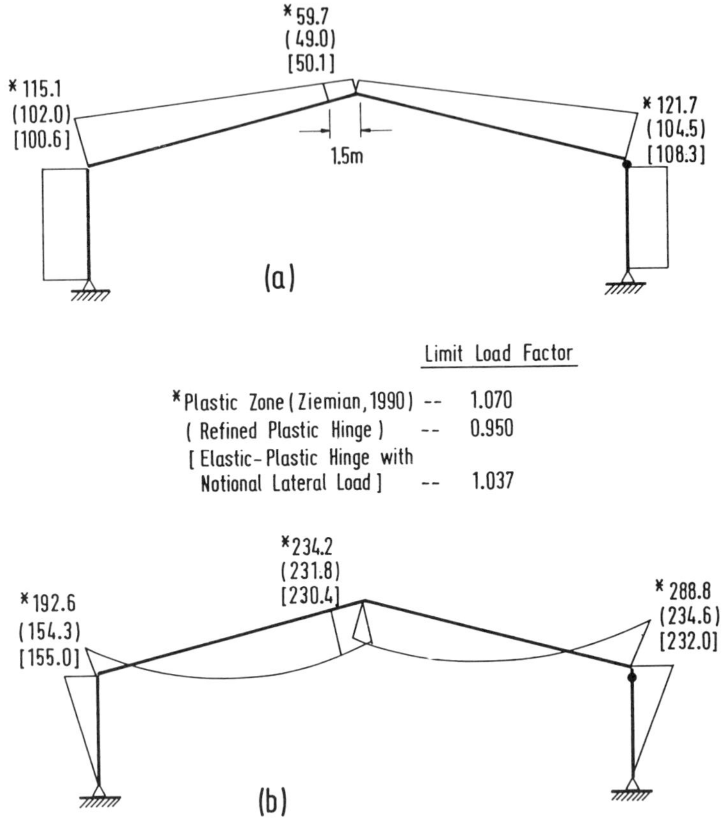

Fig. 8.30 Internal force distribution for the Gable frame. (a) Axial force (kN). (b) Bending moment (kN-m).

hand column where the plastic-zone analysis predicts higher actions because of the strain-hardening effect.

8.8.4 Six-Story Frame

Lateral load-displacement curves for the fourth and sixth stories are shown in Fig. 8.31, in which the results from the notional-load plastic hinge-based analyses are compared with Vogel's plastic-zone results. Figure 8.31 shows that all the inelastic analysis methods predict essentially the same limit load for this frame. The maximum frame resistance is reached at a load parameter of 1.111 in Vogel's plastic-zone study, at 1.118 for the refined plastic-hinge analysis, and 1.097 for the notional plastic-hinge analysis. The maximum difference between these limit loads is less than 2%. These results support earlier studies (Ziemian, 1989; King et al., 1992; White et al., 1991) that

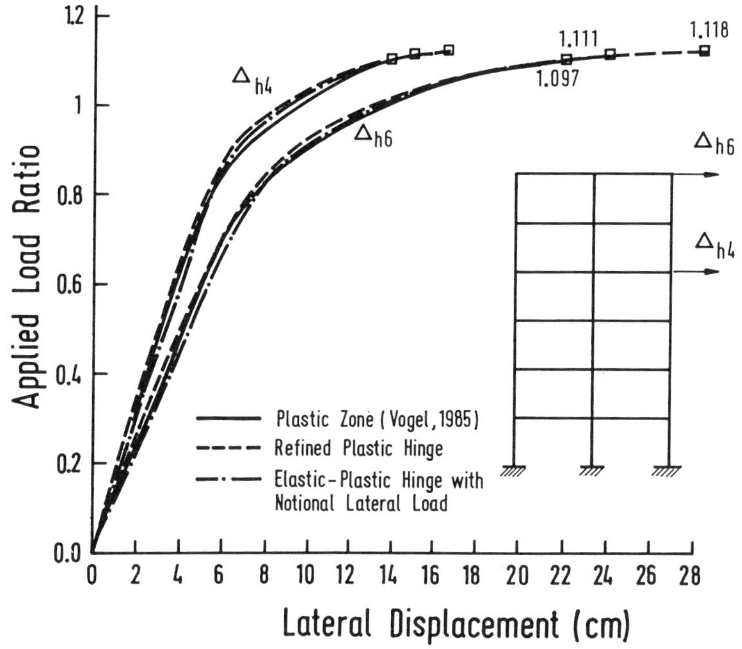

Fig. 8.31 Comparison of load-displacement curves for the six-story frame.

when the overall nonlinear behavior of the frame is dominated by inelastic action in the beams, the plastic-hinge based models generally gives sufficient representation of the overall frame behavior.

Axial force and bending moments at selected locations in the frame, computed based on the three inelastic analyses at the frame's limit of resistance, are compared in Figs. 8.32 and 8.33. The "exact" solutions shown in these figures are based on Ziemian's plastic-zone analysis in which the frame's limit of resistance is reached at a load factor of 1.180, which is slightly higher than the limit load predicted by Vogel's plastic-zone analysis. The distribution of forces predicted by all the plastic-hinge-based analyses are remarkably close. Included in Fig. 8.33 are the locations of plastic hinges detected by the refined plastic-hinge and notional-load plastic-hinge analyses. The notional-load plastic-hinge analysis detects a total of 21 plastic hinges in comparison with the refined plastic-hinge approach, which detects only 15 plastic hinges.

8.8.5 Comparison with the AISC LRFD Elastic-Analysis Approach

An attempt is also made to evaluate the limit of resistance of the Vogel frame by the LRFD elastic-analysis/design approach. For this six-story frame, the

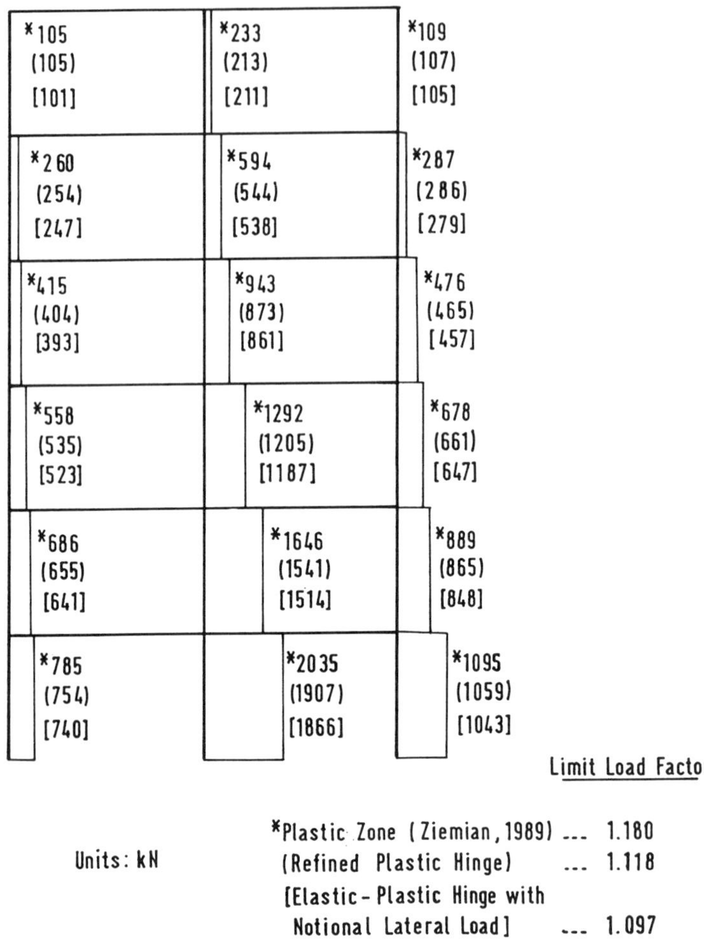

Fig. 8.32 Comparison of axial-force distribution in the six-story frame.

effective lengths of the columns are evaluated based on the modified effective-length factor K_i, which is given as (Liew et al., 1991):

$$K_i = \sqrt{\frac{I_i}{P_{ui}} \frac{\Sigma P_u}{\Sigma(I/K_n^2)}}, \qquad P_{ui} > 0 \tag{8.8.1}$$

where

I_i = moment of inertia in the plane of bending
K_i = effective-length factor with story stability effect for ith rigid column
K_n = member effective-length factor calculated from alignment chart for unbraced frame (AISC, 1986)

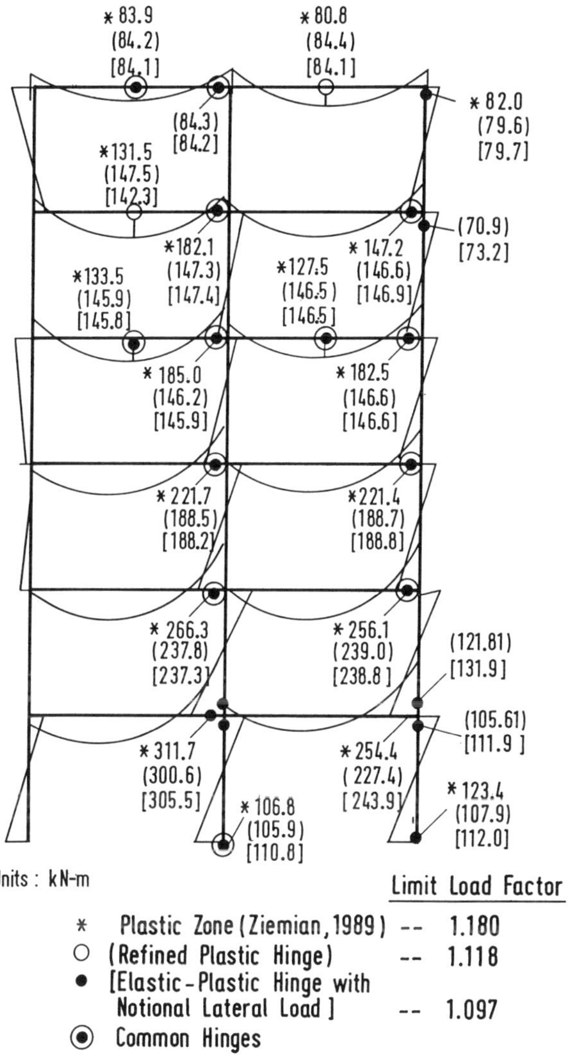

Fig. 8.33 Comparison of bending moments in the six-story frame.

P_{ui} = required axial compression strength for ith rigid column
ΣP_u = required axial compression strength for all columns in a story

The K-factor results for the calibration frame are shown in Fig. 8.34. Also included in this figure are the beam-column strengths evaluated based on the LRFD beam-column interaction equations. The limit load of this frame is reached at a load parameter of 0.765 at which the most critically loaded

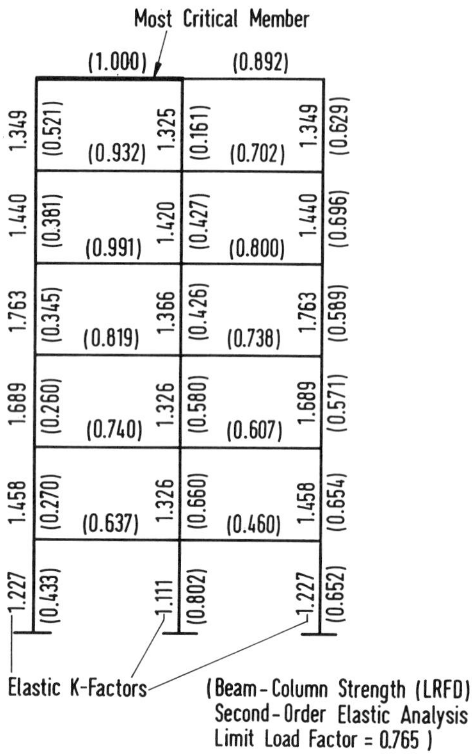

Fig. 8.34 *K*-factors and beam-column strengths from LRFD second-order elastic/member-capacity checks approach.

member (the roof beam) reaches its limiting strength of 1.0. The results in Fig. 8.34 suggest that the LRFD elastic-analysis/design procedure is conservative. It can be observed that several lower-floor beams and almost all the columns in the frame have reserve capacities that have not been utilized in the evaluation of the overall system strength. These reserve capacities, which can only be realized through inelastic force redistribution after the strengths of the most critically loaded members are reached, are quite substantial for a highly redundant system.

The main limitation of the member design approach is that it cannot predict the actual failure mode of a redundant structure. On the other hand, advanced techniques based on plastic-hinge-based analysis can predict the mode of failure associated with the effects of inelasticity and instability.

8.9 DESIGN IMPLICATIONS AND RECOMMENDATIONS

For typical building frames analyzed by the conventional elastic-plastic-hinge method, the forces at some plastic-hinge locations would be overpredicted and the forces at some less critical locations of the frame would be underestimated. The stiffnesses of the more critical members in the frame are overpredicted because gradual yielding is not accommodated in the analysis. Thus, the more critical members tend to attract greater forces, and these forces lead to the formation of plastic hinges at the peak moment points in these members. However, if the gradual yielding in the member is simulated in the analysis, the resistance of loads is distributed more uniformly among all members. Also, inelastic-force redistribution in the system is accommodated more gradually without having to experience abrupt changes in stiffness as modeled by the conventional elastic-plastic-hinge analysis. It is expected that smaller or equal numbers of plastic hinges will be predicted in a redundant structure by the refined plastic-hinge analysis than by the elastic-plastic-hinge analysis, because the predicted inelastic-force redistribution is significant prior to reaching the cross-section plastic strength of the component members.

In a local subassemblage where columns are slender ($\lambda_c > 0.4$) and subjected to high axial loads ($P > 0.5P_y$), the prediction of the distribution of loads and the magnitude of deformations by the elastic-plastic-hinge method is expected to be less accurate than the refined plastic-hinge method. If the second-order effects in the members are significant, errors in the estimation of the local distribution of forces and deformations are expected to have some influence on the prediction of the overall system strength. However, not all members in a large structural system are loaded to their individual maximum strengths prior to reaching the system's maximum strength. Therefore, the errors in the estimation of the local effects have very little influence on overall system strength and stability, particularly for a highly redundant structure.

Yet, there are cases where possible local failures of members may have significant effects on overall system stability. For these cases, the direct assessment of system and member strengths using second-order elastic-plastic-hinge analysis may yield results on the unconservative side. On the other hand, the refined plastic-hinge analysis has general validity, and, therefore, it is suitable for use in the analysis of a wide range of structural systems. Numerical studies have demonstrated that the refined plastic-hinge analysis satisfies the requirements for an advanced analysis.

Design recommendations for use of the refined plastic-hinge analysis are outlined in what follows:

1. Initial column out of plumbness should be specified in the analysis. The AISC erection tolerance of $\Delta = L/500$ may be specified in the analysis to account for the effects of member out of plumbness on the force distribution in the structural system. However, one would not expect all the columns in a building to be at their maximum permitted out of plumbness leaning in the same direction. ECCS (1991) recommends that column out of plumbness may be reduced by the following expression in the analysis of sway frames using plastic-zone analysis:

$$\Delta_i = \left(1 + \frac{1}{n_s}\right)\frac{L}{600} \tag{8.9.1}$$

where L is the story height, and n_s is the number of columns in the story. In using the refined plastic-hinge analysis, the frame might be analyzed for an out of plumbness specified as given in Eq. (8.9.1) in the plane of each story of the frame. The column out of plumbness should be applied in conjunction with the design lateral load to cause additional frame drift. Schematic diagrams shown in Fig. 8.35 demonstrate how initial sidesway deflections may be applied to irregular frameworks.

2. Explicit modeling of residual stress effects is not required in the refined plastic-hinge analysis because these effects have been included implicitly in the CRC tangent modulus. However, explicit modeling of member initial out of straightness is still required if the axial force in the member is significant. The Eurocode 3 (1992) procedure for specifying member initial out of straightness may be adopted in the present refined plastic-hinge analysis. In the AISC LRFD format, for a member with axial force $P > 0.25\,\pi^2 EI/L^2$, initial out of straightness with a maximum midspan deflection $\delta_0 = L/1500$ should be specified.
3. A parabolic stiffness-degradation function based on an initial yield surface that is the same shape and one-half the size of the plastic strength surface is recommended (see Fig. 8.4). Although the stiffness-degradation function used in the numerical analysis is a phenomenological model, it produces sufficiently accurate results in comparison with the plastic-zone solutions.

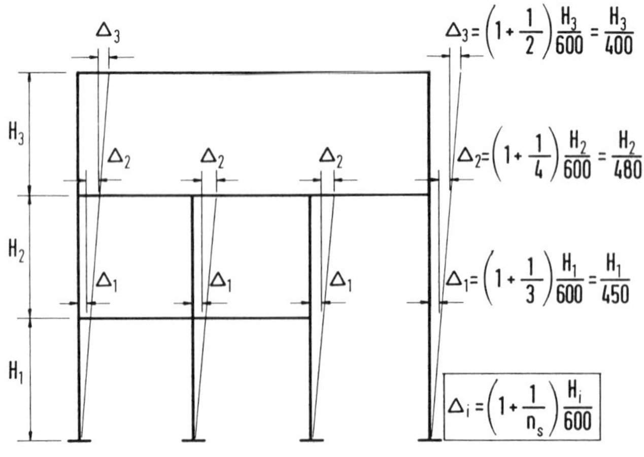

Fig. 8.35 Modeling of sidesway imperfections.

4. In current limit states design practice, the structure must be designed in such a way that the factored structural resistance ϕR is greater than the factored loads γQ_i, written as

$$\phi R \geq \Sigma \gamma_i Q_i \tag{8.9.2}$$

where ϕ and γ are the resistance and load factors, respectively, and subscript i is the number of load combinations under consideration. The present AISC LRFD (AISC, 1986) adopts resistance factors of 0.85 and 0.9 for column and beam designs, respectively, whereas the British Standard Institution (BSI, 1990) adopts a resistance factor of 1.0 for beam-column design in exchange for larger load factors. The authors propose that for limit states design, a resistance factor of $\phi = 0.9$ may be used for direct inclusion in the advanced analysis using a reduced yield strength ϕF_y. For second-order plastic-hinge analysis, design yield surfaces consisting of design resistances ϕP_y and ϕM_p, as shown in Fig. 8.36, can be used instead. Also, the reduced yield strength of $0.9F_y$ is currently adopted by Standards Australia (1990) for plastic-zone analysis of steel structures.

5. The use of the LRFD cross-section strength equations, Eqs. (6.4.1) and (6.4.2) in Chapter 6, is adequate and conservative for predicting both strong- and weak-axis in-plane strength behavior. Although separate in plane member-capacity checks are not required, all members should be

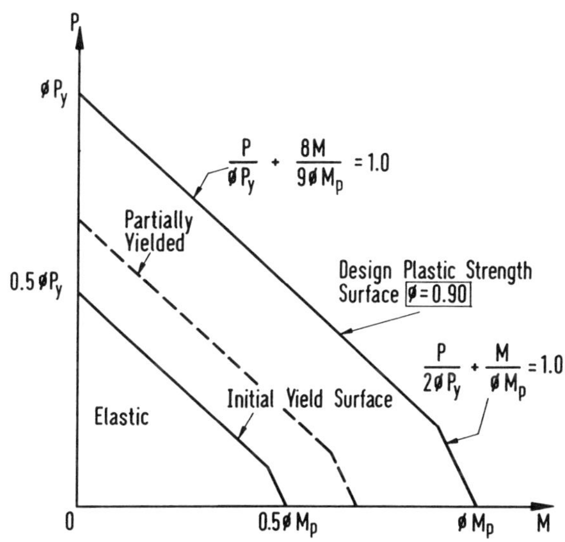

Fig. 8.36 Design yield surfaces with reduction for material yield strength.

fully compact and adequately braced to prevent lateral-torsional buckling. Members contain plastic hinges must satisfy the cross-section compactness checks for stocky plate elements. This requirement ensures that the plastic moment capacity, reduced for axial force, can be maintained with adequate plastic-hinge rotation without premature failure due to local buckling.

8.10 CONCLUSIONS

It is appropriate to conclude that the refined plastic-hinge method provides an improved representation of inelastic beam-column performance, neglecting the possible beneficial effects of strain hardening. It should be noted that the state of stresses in actual structures is rather complicated. The individual structural elements may have localized residual stresses that are greater than one-third the yield stress. In connecting the components, additional local stresses are produced by welding, and by other fabrication and erection methods, and by lack of fit. Because of these factors, it is inevitable that local yielding occurs in most steel frames, particularly at the member's connecting ends, by the time full service loads are reached. The refined plastic-hinge model can be said to represent these effects better than the elastic-plastic-hinge approach. Furthermore, suitable tangent-modulus formulas can be derived for use in the refined hinge analysis to predict the axial strength of columns of different classifications (e.g., the SSRC multiple-column curves; Galambos, 1988). These code equations are obtained by extensive calibration with both experimental and theoretical studies of columns with initial imperfections. Through the development of equivalent E_t values based on the appropriate column-strength equations, the refined plastic-hinge method correlates well with the appropriate specification column strengths.

Several important conclusions can be drawn from the numerical examples with regard to the accuracy of the conventional elastic-plastic-hinge and refined plastic-hinge methods for advanced inelastic frame analyses.

1. For lightly loaded frame members, particularly beam members, the second-order elastic-plastic-hinge analyses should provide reliable results. In addition, this method is sufficiently accurate in predicting the failure load of a frame either at the elastic buckling load or at the collapse load associated with the formation of a plastic mechanism. However, conventional plastic-hinge analysis does not perform adequately for structural systems subjected to a high gravity load, and for structural systems that have little reserve capacity for inelastic redistribution of forces. Hence, it cannot be classified as an advanced analysis technique in general.
2. The second-order refined plastic-hinge analysis enables the prediction of a column strength that is close to that implied by the code require-

ments for column design. Also, it is accurate in predicting the collapse load of a beam member whose failure is governed by the formation of a plastic collapse mechanism. For the sway frame examples studied by Liew (1992), the refined plastic-hinge approach gives conservative results and provides a reasonable degree of accuracy. For the isolated beam-column examples, the maximum unconservative error is not more than 5% compared to the LRFD beam-column strength results. The refined plastic-hinge method can capture strength and stability phenomena that range from isolated members and simple portal frame systems to entire multistory systems. Therefore, it can be classified as an advanced analysis method. The analysis method has been calibrated against the LRFD member strength equations; therefore, its development is consistent with the LRFD format.

3. The present studies show that the elastic analysis procedures based on specification member-capacity checks are limited in their ability to provide true assessment of the maximum strength behavior of redundant structural systems. The effective-length-factor concept employed by the LRFD beam-column equations does not represent the behavior of a frame after redistribution of forces occur. Therefore, its application is valid only when the frame is essentially elastic. In general, the LRFD elastic-analysis/design approach is conservative. However, several benchmark studies in Liew (1992) demonstrate that the strength limit state predicted by this approach may be too conservative compared to the "true" strength of a redundant frame. The conservativeness of the LRFD elastic-analysis/design approach, in general, increases with the degree of frame redundancy.
4. Several particular advantages could result from the use of the refined plastic-hinge analysis:
   - The analysis is formulated based on the concepts of structural stability and plasticity. It rigorously accounts for strength and stability. It provides a more accurate prediction of possible modes of failure and serviceability displacements. The overall structure will have a more uniform level of safety.
   - The method strikes a balance between simplicity and accuracy, hence, it is a feasible method for the design office. By incorporation of the advanced analysis into modern computer-aided engineering software, the design effort involved in individual member strength and stability checks can be reduced and replaced by direct and vigorous checks of the interdependent member and system capacities.
   - The analysis method can couple with a computer-aided software environment; thus, it enables the engineer to investigate more directly the physical performance and actual failure modes of a design. Computer-aided software can provide instant visual understanding of the structural performance, and enhancement of design can be prescribed

in a more direct manner. The computer takes on the tedious calculations and checks, data manipulation, and design documentation, while the engineer can focus on detail evaluation of overall design requirements for strength, stability, and serviceability. This will be the state-of-the-art design approach to serve the structural engineering profession into the next century.

REFERENCES

AISC (1986). *Load and Resistance Factor Design Specifications for Structural Steel Buildings,* Chicago: American Institute of Steel Construction.

Al-Mashary, F., and W. F. Chen (1991). *Simplified Second-Order Inelastic Analysis for Steel Frames, The Structural Engineer, Journal of the Institution of Structural Engineers,* Vol. 69, No. 23, pp. 395–399.

Beedle, L. S., and L. Tall (1960). "Basic Column Strength," *Proceedings of the ASCE,* Vol. 86, No. ST7, pp. 139–173.

BSI (1990). *BS5950: Structural Use of Steelwork in Building, Part 1: Code of Practice for Design in Simple and Continuous Construction: Hot Rolled Section,* London: British Standards Institution.

Chen, W. F., Duan, L., and Zhou, S. P. (1990). "Second-Order Inelastic Analysis of Braced Portal Frames—Evaluation of Design Formulae in LRFD and GBJ Specification," *Steel Structures,* Vol. 1, No. 1, pp. 5–15.

Chen, W. F., and E. M. Lui (1987). *Structural Stability: Theory and Implementation,* New York: Elsevier.

Chen, W. F., and S. Toma (1994). *Advanced Analysis of Steel Frames: Theory, Software, and Applications,* Boca Raton, FL: CRC Press.

Deierlein, G. G., Y. Zhao, and W. McGuire (1991). "A Discrete Model for Gradual Plastification and Nonlinear Connection Response in Three-Dimensional Framed Structures," in *Annual Technical Session Proceedings,* Bethlehem, PA: Structural Stability Research Council, Lehigh University, pp. 423–432.

Eurocode 3 (1992). *Design of Steel Structures: Part 1.1—General Rules and Rules for Buildings,* National Application Document for use in the UK with ENV1993-1-1: 1991, Draft for Development.

ECCS (1983). *Ultimate Limit State Calculation of Sway Frames with Rigid Joints,* Technical Working Group 8.2, Systems, Publication No. 33, Brussels: European Convention for Constructional Steelwork.

ECCS (1991). *Essentials of Eurocode 3 Design Manual for Steel Structures in Building,* ECCS Advisory Committee 5, No. 65, Brussels: European Convention for Constructional Steelwork.

Galambos, T. V., ed. (1988). *Guide to Stability Design Criteria for Metal Structures,* 4th Ed. New York: Wiley.

Kanchanalai, T. (1977). *The Design and Behavior of Beam-Columns in Unbraced Steel Frames,* AISI Project No. 189, Report No. 2, Austin: Civil Engineering/Structures Research Laboratory, University of Texas.

King, W. S., D. W. White, and W. F. Chen (1992). "On Second-Order Inelastic Methods for Steel Frame Design," *Journal of Structural Engineering,* ASCE, Vol. 118, No. 2, pp. 408–428.

Liew, J. Y. R. (1992). "Advanced Analysis for Frame Design," Ph.D. dissertation, School of Civil Engineering, Purdue University, West Lafayette, IN.

Liew, J. Y. R., D. W. White, and W. F. Chen (1991). "Beam-Column Design in Steel Frameworks—Insight on Current Methods and Trends," *Journal of Constructional Steel Research,* Vol. 18, No. 4, pp. 269–308.

Liew, J. Y. R., D. W. White, and W. F. Chen (1992). "Beam-Columns," in *Constructional Steel Design, An International Guide,* P. J. Dowling et al., eds., London: Elsevier, pp. 105–132.

Liew, J. Y. R., D. W. White, and W. F. Chen (1993a). "Second-Order Refined Plastic Hinge Analysis for Frame Design: Part 1," *Journal of Structural Engineering,* ASCE, Vol. 119, No. 11, pp. 3196–3216.

Liew, J. Y. R., D. W. White, and W. F. Chen (1993b). "Second-Order Refined Plastic Hinge Analysis for Frame Design: Part 2," *Journal of Structural Engineering,* ASCE, Vol. 119, No. 11, pp. 3217–3237.

Orbison, J. G. (1982). *Nonlinear Static Analysis of Three-Dimensional Steel Frames,* Report No. 82-6, Ithaca, NY: Department of Structural Engineering, Cornell University.

Powell, G. H., and P. F. Chen (1986). "3D Beam-Column Element with Generalized Plastic Hinges," *Journal of Engineering Mechanics,* ASCE, Vol. 112, No. 7, pp. 627–641.

Standards Australia (1990). *Steel Structures,* AS4100, Sydney.

Tall, L., ed. (1974). *Structural Steel Design,* 2nd ed., New York: Ronald Press.

Vogel, U. (1985). "Calibrating Frames," *Stahlbau,* Vol. 10, pp. 295–301.

White, D. W. (1985). "Material and Geometric Nonlinear Analysis of Local Planar Behavior in Steel Frames Using Iterative Computer Graphics," M.Sc. thesis, Cornell University, Ithaca, NY.

White, D. W. (1993). "Plastic Hinge Methods for Advanced Analysis of Steel Frames," *Journal of Constructional Steel Research,* Vol. 24, No. 2, pp. 121–152.

White, D. W., J. Y. R. Liew, and W. F. Chen (1991). *Second-Order Inelastic Analysis for Frame Design: A Report to SSRC Task Group 29 on Recent Research and the Perceived State-of-the-Art,* Structural Engineering Report No. CE-STR-91-12, West Lafayette, IN: Purdue University.

White, D. W., J. Y. R. Liew, and W. F. Chen (1993). "Toward Advanced Analysis in LRFD," in *Plastic Hinge Based Methods for Advanced Analysis and Design of Steel Frames—An Assessment of the State-of-the-Art,* D. W. White and W. F. Chen, eds., Bethlehem, PA: Structural Stability Research Council, Lehigh University, pp. 95–173.

Yura, J. A. (1988). *Elements for Teaching Load and Resistance Factor Design: Combined Bending and Axial Load,* Austin: University of Texas.

Zhou, S. P., L. Duan, and W. F. Chen (1990). "Comparison of Design Equations for Steel Beam-Columns," *Structural Engineering Review,* Vol. 2, No. 1, pp. 45–53.

Ziemian, R. D. (1989). "Verification Study," draft version, Ithaca, NY: Cornell University.

9

DESIGN OF SEMI-RIGID FRAMES USING ADVANCED ANALYSIS

9.1 INTRODUCTION

In the design of steel frame structures, it is commonly assumed that the joints either provide full continuity of moments and rotations or that they are incapable of transmitting moments, thereby permitting free rotation. These connections are regarded as either "pinned" or "rigid." Modern codes, such as the AISC LRFD and Eurocode 3, now recognize the method of semi-rigid design that allows for connection flexibility in the analysis and design of steel frames. The need for such change arises because real behavior of steelwork connections, as observed in tests, falls somewhat between the two extremes of pinned and rigid. Thus it is more appropriate to describe these connections as "semi-rigid" or "partially restrained."

In the design of semi-rigid frames, it is important to appreciate the behavior of the joints themselves. This requires an understanding of the load applied to the joints (moment, shear, axial force, torque, and bimoment) and the corresponding deformations. However, because there is very little research on three-dimensional joint behavior, and there is virtually no study on the moment–shear–axial-force interaction for the in-plane behavior, the problem reduces to the ability to represent joint moment–rotation curves either through analytical means or by curve fitting of test results.

One common problem for semi-rigid frame design is that the connection moment–rotation behavior is mostly nonlinear, and the structures with semi-rigid connections must be analyzed under different load combinations to ensure that they satisfy the limit states of strength and serviceability. The moment-amplification factors B_1 and B_2 contained in Chapter H of the AISC

LRFD (AISC, 1993) specification cannot model the reduction in stiffness in semi-rigid connections due to a static increasing load. The expressions for the moment-amplifications factors (B_1 and B_2 factors) can be developed for frames with pinned and/or rigid connections; they are not directly applicable for frames with nonlinear connection behavior. However, one may argue that the current AISC LRFD procedure still may be used in the design of semi-rigid frames by assuming linear moment-rotation relationship of the connections. Global analysis can be carried out using elastic first-order analysis. The design forces are then evaluated by amplifying the member forces based on the B_1 and B_2 amplification factors. The column effective-length factor (K-factor) is then evaluated for the beam-column member considering the restraining effects of the end connections and the adjoining members. However, the detailed guidelines for calculating the values for B_1, B_2, and K-factors for beam-column design, considering the effects of semi-rigid frame actions, are not provided in LRFD. The accuracy of these factors for representing the interaction effects of the system frame on its component members is also questionable.

Because connection flexibility and member instability are closely related to each other, their interaction effects can have significant effects on the overall frame's performance. Therefore, it is more realistic and straightforward to develop advanced analysis method considering all these effects directly in the analysis. Also, to make the advanced analysis method more attractive for design use, separate specification member-capacity checks should be eliminated in the advanced analysis/design process.

In order to design steel frames with semi-rigid connections, engineers need to be assured that they understand the effects of connections on the structure's performance as a whole. To achieve this, it is essential to develop a readily understood analysis/design approach that can provide reliable, economic, and safe design. Also, this design approach should provide an economic trade-off when compared with other more conventional design methods. The main obstacles encountered in the implementation of semi-rigid frame design are (1) connection uncertainties; (2) the need for a reliable and general connection moment–rotation model; (3) the development of efficient analysis methods; and (4) concerns on strength, stability, and serviceability limit state conditions. Much has been written about these obstacles (Stelmack, Marley, and Gerstle, 1986; Heish, 1990; ECCS, 1992; Liew, White, and Chen, 1993a, 1993b; Ammerman and Leon, 1987).

The draft version of Eurocode 3 (1992) provides some guidelines to address connection behavior in the design of sway and nonsway steel building frames. Included in Eurocode 3 are some basic criteria for the classification and modeling of moment–rotation of beam-to-column connections (Bijliaard and Steenhuis, 1991; Eurocode 3, 1992). Although recent specifications (AISC, 1993; BSI, 1990; Standards Australia, 1990) allow the consideration of semi-rigid behavior of connections in the design of steel frames, specific guidelines for design are still required.

This chapter addresses many of the previous issues related to semi-rigid frame design using an advanced analysis technique. The analysis and design approach is based on second-order plastic-hinge-based approach in line with the AISC LRFD format (Liew et al., 1993c, 1993d). The connections are modeled using a nonlinear spring model based on a three-parameter power model. The design of connections and frames is integrated and iterative in nature, and the analysis accounts for the geometric nonlinearities and distributed plasticity effects in frame elements, and the nonlinear moment–rotation (M–θ_r) behavior of connections.

Sections 9.2 and 9.3 present procedures for calculating the key parameters used to describe the moment–rotation (M–θ_r) curves of various types of angle connections. The size parameters that affect the M–θ_r behavior of these connections are expressed in terms of a set of standardized variables. For convenience, this set of variables is expressed in nondimensional form. Analysis procedures to calculate the connection parameters are outlined for practical design use. Design aids are then generated in which salient size parameters that affect the M–θ_r behavior of the connection can be identified. A user-friendly computer program has been developed to compute the ultimate moment and initial stiffness of connections composed of angles; a detailed description of the computer program is given in Chapter 11.

Two schemes by which connections can be classified in terms of strength, stiffness, and rotation capacity are reviewed in Section 9.4. The design implications of these schemes are discussed. Section 9.5 describes an advanced analysis technique by which connection flexibility can be accounted for in the second order refined plastic hinge analysis. In Section 9.6, a design procedure is proposed for the analysis and design of steel frames and connections.

A case study is presented in Section 9.7. A two-dimensional semi-rigid frame is analyzed and designed using the refined plastic-hinge analysis. The frame is evaluated with respect to (1) strength limit states, (2) serviceability limit states, and (3) connection rotation capacity. The pertinent use of the analysis and design concepts outlined in Section 9.6 is illustrated. This simple frame example represents only a first step toward the use of advanced analysis for designing semi-rigid frames. The designs of more substantial structures should follow the same principles, but the design variables and constraints are expected to increase, thus increasing the complexity of the design process.

Section 9.8 provides some insight on use of bracing to reduce frame drift so that the serviceability limit state can be satisfied in designing semi-rigid frames with flexible joints. The effects of support fixity on the behavior of semi-rigid frames are examined in Section 9.9. The advantages of using semi-rigid construction are discussed in Section 9.10. Conclusions and directions for further research are provided at the end of the chapter.

9.2 MODELING CONNECTION MOMENT–ROTATION BEHAVIOR

A large amount of research has been reported on methods of accounting for connection flexibility in the analysis and design of planar frame structures (some recent works include Ackroyd, 1985; Lui, 1985; Stelmack et al., 1986; Yu and Shanmugam, 1986; Chen, 1987, 1988; Goto and Chen, 1987; Wu, 1988; Narayanan, 1989; Nethercot, 1989; Nethercot and Zandonini, 1989; Barakat and Chen, 1990, 1991; Dhillon and Abdel-Majid, 1990; Heish, 1990; Chen and Lui, 1991; Gibbons, Kirby, and Nethercot, 1991; Goto, Satsuki, and Chen, 1991; Jaspart, Gerardy, and Maquoi, 1991; Al-Bermani and Kitipornchai, 1992; Chen, 1993; Liew et al., 1993a, 1993b; Chen and Toma, 1994). Comprehensive state-of-the-art summaries on semi-rigid frame behavior and design are given in three monographs edited by Chen (1987, 1988, 1993) and in a monograph on structural connections edited by Narayanan (1989).

Most of the analytical and experimental work performed so far investigate only the in-plane flexural deformation behavior of connections. Very few investigations have been made on the analysis and behavior of three-dimensional semi-rigid frames (Heish, 1990; Gibbons et al., 1991; Al-Bermani and Kitipornchai, 1992). A connection in a building framework can be subjected to various force combinations such as moments, shears, axial forces, torques, bimoments, and their corresponding deformations. However, because there is very little research on three-dimensional joint behavior and there is virtually no study on the moment–shear–axial-force interaction, the current state of the art on connection analysis focuses primarily on the in-plane behavior represented by the connection's moment–rotation curve.

Many connection models have been proposed to describe the moment–rotation relationships of steelwork connections used in building construction. Literature reviews of these models are provided in Nethercot and Zandonini (1989), Chen and Lui (1991), and Chen (1993). The present work adopts a three-parameter power model proposed by Kishi and Chen (1990) to estimate the moment–rotation behavior of semi-rigid frames. The selection of this connection model is guided by its simplicity and robustness for representing the basic behavior of typical beam-to-column connections, and for ease of implementation in the second-order analysis program. The generalized equation of the power model has the form

$$m = \frac{\theta}{(1 + \theta^n)^{1/n}}, \qquad \text{for } \theta > 0 \text{ and } m > 0 \tag{9.2.1}$$

or, equivalently,

$$\theta = \frac{m}{(1 - m^n)^{1/n}}, \qquad \text{for } \theta > 0 \text{ and } m > 0 \tag{9.2.2}$$

where

$m = M/M_u$
M = connection moment
M_u = ultimate moment capacity of the connection
n = shape parameter
R_{ki} = initial connection stiffness
$\theta = \theta_r/\theta_0$
θ_r = relative rotation between beam and column
θ_0 = reference plastic rotation, M_u/R_{ki}

The connection tangent stiffness R_{kt} at any arbitrary rotation $|\theta_r|$ can be evaluated by differentiating M with respect to $|\theta_r|$. The resulting connection tangent stiffness is

$$R_{kt} = \frac{dM}{d|\theta_r|_{|\theta_r|}} = \frac{M_u}{\theta_0(1 + \theta^n)^{1+1/n}} \tag{9.2.3}$$

when the connection is loaded, and it is equal to

$$R_{ki} = \frac{dM}{d|\theta_r|_{|\theta_r=0|}} = \frac{M_u}{\theta_0} = R_{ki} \tag{9.2.4}$$

when the connection is unloaded.

Equation (9.2.1) or (9.2.2) has the shape shown in Fig. 9.1. The idealized rigid and pinned connections correspond to the vertical and horizontal axes, respectively. The connection model exhibits several distinctive features; particularly significant are the following:

1. The curve is nonlinear with decreasing stiffness (as indicated by the slope of the curve) as rotation increases. It can be observed that for larger values of power index n, the transition from the initial stiffness, R_{ki}, to the final curve of maximum moment, M_u, is more abrupt. If n is infinity, the model becomes a bilinear curve that has initial connection stiffness, R_{ki}, and ultimate moment capacity, M_u.
2. The model may plot curves anywhere on the M–θ space. Thus, it can represent connections of varying stiffness and strength.
3. The connection stiffness decreases as rotation increases, but in the reverse loading direction, it approximates the initial stiffness.
4. A clear plateau is observed as the ultimate strength of the connection is reached. Thus, the model is suitable for connections without signifi-

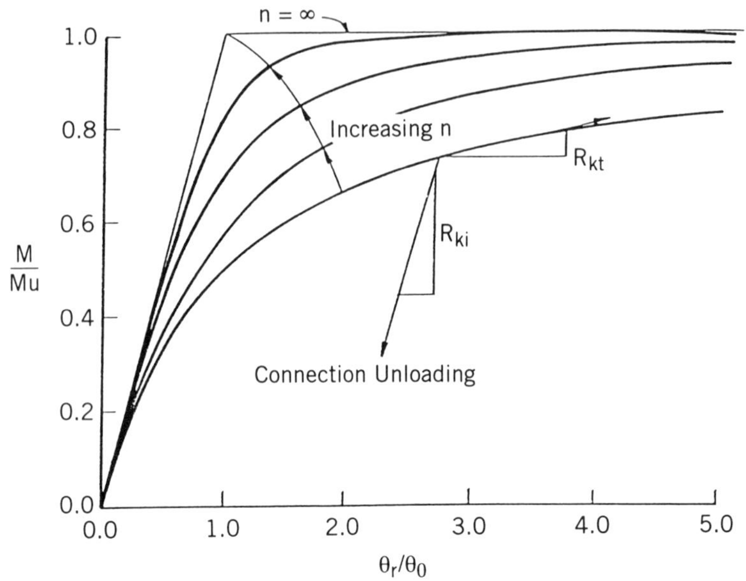

Fig. 9.1 Moment–rotation behavior of connections.

cant strain-hardening behavior. It may also be used conservatively for connections that exhibit strain hardening in the large deformation range.

The principle merit of the proposed model is that it allows the designer to execute the nonlinear structural analysis quickly and accurately. This is because the connection tangent stiffness and rotation can be determined directly from Eqs. (9.2.3) and (9.2.4) without iteration. Also, the connection parameters in Eq. (9.2.1) or (9.2.2) can be determined analytically based on the connection configuration, thus making it more appropriate for practical use.

It should be noted that the strain-hardening region in the M–θ_r curve at a relatively large deformation has not been considered in the proposed connection model. In fact, the ability of a beam-to-column connection to develop significant strain hardening is dependent on several factors, such as gauge distances of various component plate elements and the effects of local yielding and stability of the connecting elements in a large deformation. At the present time (1994), an analytical procedure to quantify the precise effect of strain hardening on the full moment–rotation characteristics of beam-to-column connections is not well enough developed for direct implementation in design. However, connection moment–rotation models considering strain-hardening effects at relatively large rotations are available (Wu, 1988; Heish, 1990). The parameters for these connection models are normally obtained from calibration with the experimental data.

9.3 ANGLE CONNECTIONS

This section illustrates the design of four of the most commonly used semi-rigid connections: (1) single web-angle connections, (2) double web-angle connections, (3) top- and seat-angle connections, and (4) top- and seat-angle connections with double web cleats. The LRFD manual includes many tables that make the design of these connections simple. The aim of this section is to provide a step-by-step procedure for calculating the connection initial stiffness and ultimate moment capacity. These connection parameters are required in a second-order refined plastic-hinge analysis to provide strength and serviceability assessments of the semi-rigid frame under consideration.

9.3.1 Standardized Parameters for Angle Connections

For convenience, the following nondimensional parameters are used to describe the dimensions of angle connections (Fig. 9.2):

$$\beta = g_c/l, \qquad \gamma = l/t, \qquad \delta = d/t$$
$$\kappa = k/t, \qquad \omega = W/t, \qquad \rho = t_w/t_t \tag{9.3.1}$$

where

- d = beam depth
- g_c = distance from edge of the angle to center of the bolt line
- k = distance from angle heel to toe of the fillet
- l = length of angle
- t = thickness of the angle; subscripts w and t may be used to refer to web angle and top angle, respectively
- W = nut diameter

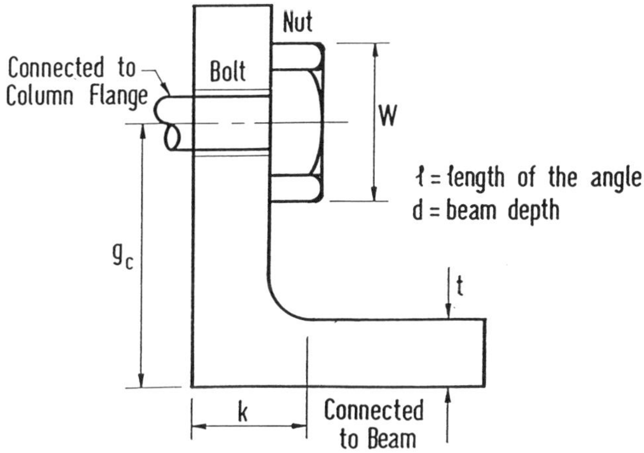

Fig. 9.2 Standard dimensions of an angle connection.

The following parameters define the stiffness and strength of an angle leg:

$$
\begin{aligned}
I_0 &= t^3/12, \text{ moment of inertia per unit length} \\
M_0 &= F_y t^2/4, \text{ plastic moment per unit length} \\
V &= \text{shear force} \\
V_0 &= F_y t/2, \text{ plastic shear capacity per unit length} \\
\xi &= V/V_0, \text{ nondimensional shear-force parameter}
\end{aligned}
$$

Subscripts t, s, and w, which refer to the top, seat, and web angle, respectively, may be applied to the parameters in Eq. (9.3.1) to describe the geometry and stiffness of the connection subassemblage.

The following sections provide a step-by-step procedure to calculate the initial stiffness, R_{ki}, and the ultimate moment capacity, M_u, of the four types of angle connections. Design aids are generated for design office use. Based on the characteristic results obtained for R_{ki} and M_u, salient size parameters that affect the M–θ_r behavior of the connections can be identified. By expressing the moment–rotation behavior of a connection as a function of its size parameters, the expression for the M–θ_r is more general and can be applied to similar connections of different sizes.

9.3.2 Single and Double Web-Angle Connections

9.3.2.1 Design Aspects Single web-angle connections consist of an angle either bolted or welded to both the column flange and beam web, as shown in Fig. 9.3a. Design of this type of connection is accomplished by assuming a model and the associated components (bolts, welds, angles, plates) that are proportioned to satisfy the factored reaction force of the beam it is connecting. The connection must have enough rotation capacity at the factored-load level to accommodate the simple beam end rotation that it is connecting.

Double web-angle connections consist of two angles either bolted or welded to both the column flanges and beam web, as shown in Fig. 9.3b. Generally, the moment capacity of the double web-angle connection is about double the value of the single web-angle connection. The use of double web-angle connections connected by high-strength bolts is gaining popularity because of the ease of fabrication and installation. The procedure for designing a double web-angle connection is given follows:

1. From the beam reaction, determine the number of bolts required (LRFD Specification, Section J3.3). The design must ensure that the total shear reaction divided by the number of bolts must not exceed the specific design strength of the particular bolt used. For double web-angle connections, the bolts in the beam web are in double shear.

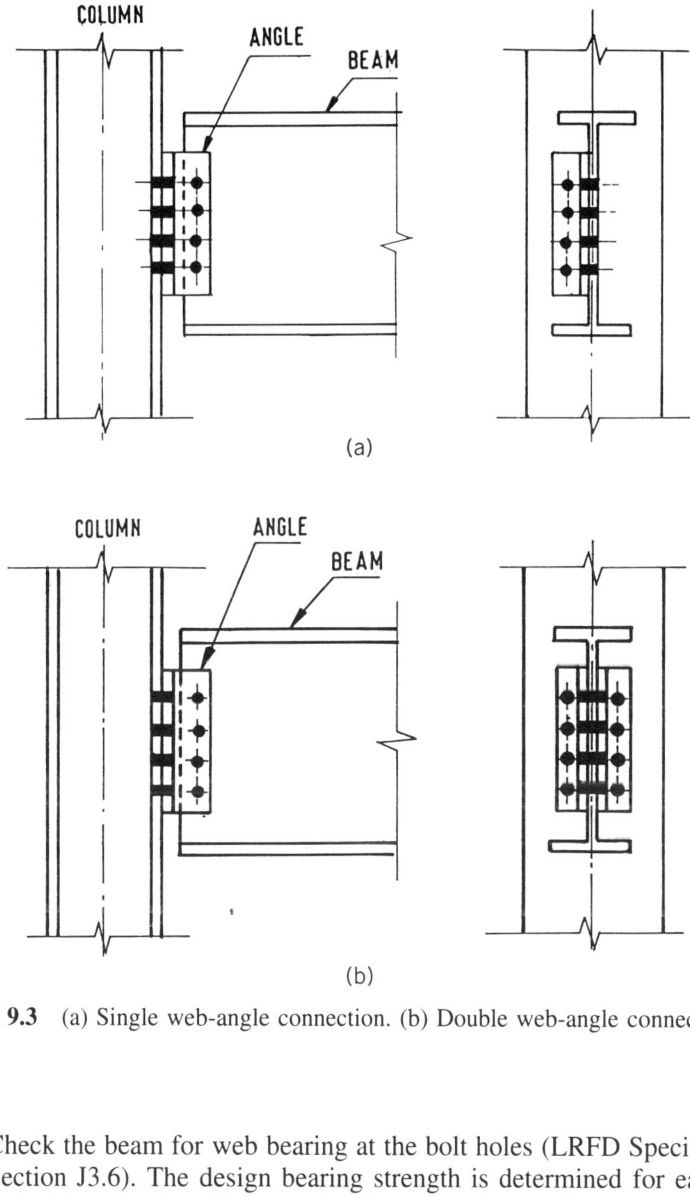

Fig. 9.3 (a) Single web-angle connection. (b) Double web-angle connection.

2. Check the beam for web bearing at the bolt holes (LRFD Specification, Section J3.6). The design bearing strength is determined for each bolt and added together. The total bearing strength must be greater than the factored reaction force from the beam. Bolts at the edge may have a smaller capacity due to failure by tear rather than by direct bearing of the web material.
3. Check the angle for bolt bearing. This check is similar to the check for the beam web.

4. Check the beam web for block shear. This is required only when the beam is coped. Both the shear yield and tension fracture and the tension yield and shear fracture conditions must be checked.
5. Check the angles for shear fracture through the net section.
6. An additional check is also required when the beam is subjected to axial force in addition to gravity load. In this case, the connection must be designed to resist the interaction effects of shear and axial force.

9.3.2.2 Initial Stiffness The initial elastic stiffness of a web angle may be derived based on the following assumptions (Kishi et al., 1988a, 1988b):

1. The effects of shear force on the connection deformation are ignored.
2. The part of the angle fastened to the column flange behaves in a linearly elastic manner, and the leg of the angle connected to the beam behaves as a rigid body.
3. The connection deformation is small.
4. The part of the angle fastened to the column flange acts as a moderately thick plate in which a fixed support is assumed to be located at the fastener-nut edge attached to the column. The concentrated torsional moment is in equilibrium with the connection moment acting at the free-edge line, which is the edge of the angle plate connected to the beam web.

Based on the foregoing assumptions and applying elastic bending-torsion theory to the part of the angle connected to the column, the normalized initial stiffness of a *single* web-angle connection may be derived as (Kishi et al., 1993; Liew et al., 1993a):

$$\frac{R_{kiw}}{EI_{0w}} = \frac{12\alpha\cosh(\alpha\beta'_w)}{7.8[\alpha\beta'_w\cosh(\alpha\beta'_w) - \sinh(\alpha\beta'_w)]} \tag{9.3.2}$$

where EI_{0w} is the bending stiffness per unit length of the angle leg adjacent to the column face, $\alpha = 4.2967$ when the Poisson's ratio is 0.3, and β'_w is defined as

$$\beta'_w = \beta_w - \frac{1}{\gamma_w}\left(\kappa_w + \frac{\omega_w}{2}\right) \tag{9.3.3}$$

Equation (9.3.2) is plotted in Fig. 9.4 for β'_w ranging from 0.0 to 0.5. The initial connection stiffness does not change significantly when β'_w is greater than 0.5.

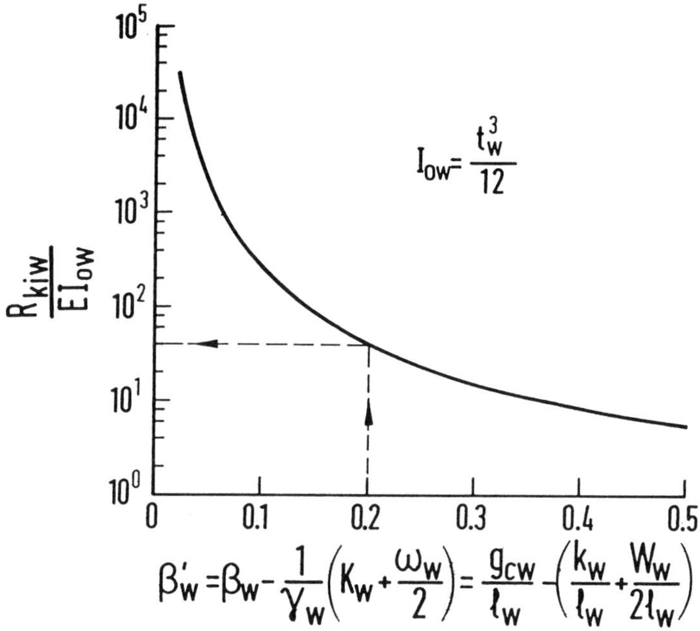

Fig. 9.4 Characteristic curve for calculating initial stiffness of single web-angle connections.

9.3.2.3 Ultimate Moment Capacity The ultimate moment capacity of a web-angle connection is derived by assuming that an idealized elastic-plastic collapse mechanism is developed in the assembly angles. The connection moment capacity is computed based on a plastic yield-line theory considering the interaction of bending moment and shear. The relationship between the plastic moment and shear force may be obtained based on the Tresca yield criterion. The result after some simplification is a fourth-order equation, which may be written as

$$\xi_w^4 + \beta_w^* \xi_w - 1 = 0 \tag{9.3.4}$$

where

$$\beta_w^* = \beta'_w \gamma_w + \frac{\omega_w}{2} \tag{9.3.5}$$

Substituting β'_w from Eq. (9.3.3) into Eq. (9.3.5), β_w^* may be simplified as

$$\beta_w^* = \beta_w \gamma_w - \kappa_w \tag{9.3.6}$$

If the assumption is made that the shear force is distributed linearly along the height of the angle, the normalized ultimate moment capacity of the connection can be derived as (Kishi et al, 1993):

$$\frac{M_{uw}}{M_{0w}t_w} = \frac{(2\xi_w + 1)}{3}\gamma_w^2 \tag{9.3.7}$$

where the term ξ_w can be solved from Eq. (9.3.4) for a known value of β_w^*.

The connection initial stiffness and ultimate moment from Eqs. (9.3.2) and (9.3.7) are for a *single* web angle only. The corresponding values for a double web-angle connection are simply *twice* the value evaluated from these equations.

The procedures for calculating the ultimate moment of a web-angle connection are shown in what follows (Liew et al., 1993a):

1. For a given set of connection data, g_{cw}, l_w, and k_w, determine the dimensionless parameters β_w, and κ_w from Eq. (9.3.1).
2. Determine β_w^* from Eq. (9.3.6).
3. Determine ξ_w from Eq. (9.3.4) using a simple iteration procedure.
4. Finally, evaluate $M_{ow} = F_y t_w^2/4$, and solve for M_{uw} from Eq. (9.3.7).

Based on the foregoing procedures, characteristic curves for web-angle connections, which show the relationships between the normalized moment capacity $M_{uw}/(M_{0w}t_w)$ and β_w, can be generated, as shown in Fig. 9.5. These curves are shown for $\kappa_w = 1.50$ and 2.00, and γ_w from 3.0 to 21.0. It is observed that the moment capacity is not sensitive to γ_w if γ_w is less than 6.0. Also the κ_w value has negligible influence on the ultimate moment capacity of the connection for all values of γ_w except when γ_w is greater than 9.0 and γ_w is less than 0.4. These curves can be used to determine the ultimate moment capacity of a web-angle connection once the parameter γ_w is determined in step 1 of the preceding procedures.

9.3.2.4 Example 1: Double Web-Angle Connection Figure 9.6 shows an example of a double web-angle connection with $g_{cw} = 3$ in., $l_w = 7$ in., $t_w = 7/16$ in., $k_w = 7.8$ in., and $W_w = 1\text{-}7/16$ in. Assuming the material yield strength $F_y = 36$ ksi and the modulus of elasticity $E = 29{,}000$ ksi, determine the connection initial stiffness and the ultimate moment capacity.

Initial Connection Stiffness, R_{kiw}. The nondimensional parameters may be computed for the web-angle connection as

$$\beta_w = \frac{g_{cw}}{l_w} = 0.428571, \qquad \gamma_w = \frac{l_w}{t_w} = 16.0, \qquad \kappa_w = \frac{k_w}{t_w} = 2.0$$

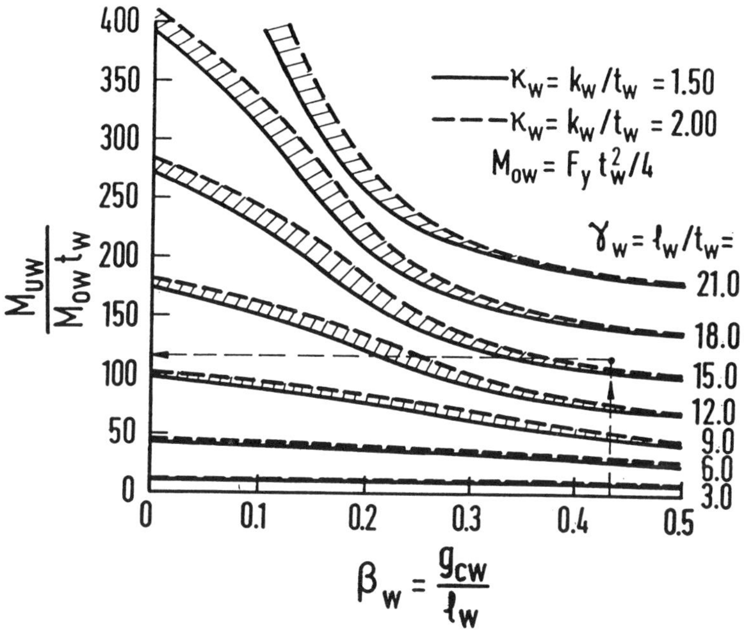

Fig. 9.5 Characteristic curves for calculating ultimate moment capacity of single web-angle connections.

$$\omega_w = \frac{W_w}{t_w} = 3.2857, \qquad I_{0w} = 0.006978$$

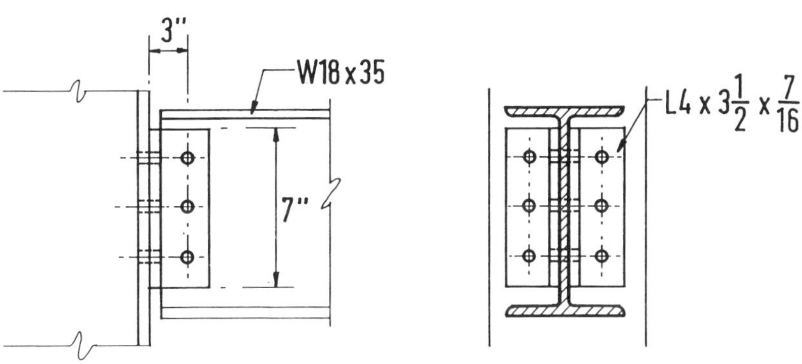

Fig. 9.6 Example 1: double web-angle connections.

From Eq. (9.3.3), determine

$$\beta'_w = \beta_w - \frac{1}{\gamma_w}\left(\kappa_w + \frac{\omega_w}{2}\right) = -0.200893$$

From Eq. (9.3.2),

$$\frac{R_{kiw}}{EI_{0w}} = \frac{12\alpha \cosh(\alpha\beta'_w)}{7.8[\alpha\beta'_w \cosh(\alpha\beta'_w) - \sinh(\alpha\beta'_w)]} = 39.939$$

$$\Rightarrow R_{kiw} = EI_{0w} \times 39.939 = 8083 \text{ kips-in. (single web-angle connections)}$$

This solution is for a single web angle. For a double web-angle connection, R_{kiw} is twice the value, which is 16,166 kips-in.

Ultimate Moment Capacity, M_u. For $\beta_w = 0.428571$, $\gamma_w = 16.0$, and $\kappa_w = 2.0$, determine

$$\beta^*_w = \beta_w \gamma_w - \kappa_w = 4.857136$$

Evaluate ξ_w iteratively from Eq. (9.3.4) of which $\xi_w = 0.20548$. From Eq. (9.3.7),

$$\frac{M_{uw}}{M_{0w}t_w} = \frac{2\xi_w + 1}{3}\gamma_w^2 = 120.4$$

Alternatively, the preceding value can also be obtained from Fig. 9.5 based on the computed values of $\beta_w = 0.428571$, $\gamma_w = 16.0$, and $\kappa_w = 2.0$.

Finally, evaluate $M_{ow}t_w = 0.7537$, and then solve for M_{uw}:

$$M_{uw} = 90.74 \text{ kips-in. (single web connection)}$$

For a double web-angle connection, $M_{uw} = 2 \times 90.74 = 181.5$ kips-in.

9.3.3 Top- and Seat-Angle Connections

9.3.3.1 Design Aspects A typical top- and seat-angle connection is shown in Fig. 9.7. The AISC ASD specification describes the top- and seat-angle connections as follows: (1) The seat angle transfers only vertical reactions and should not provide significant restraining moments to the beam. (2) The top angle is provided to enhance the lateral stability of the beam. It is not supposed to carry any gravity loads. Design loads for these connections are tabulated in the LRFD Manual.

For unstiffened seated connections, two limit states must be checked in calculating the design strength of the connections: (1) bending of the seat angle and (2) yielding of the beam web. For beams with slender webs, an

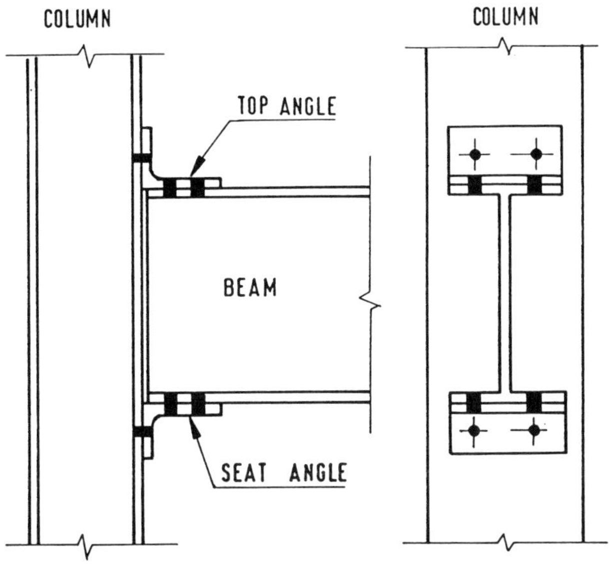

Fig. 9.7 Top- and seat-angle connections.

additional check for web crippling (LRFD Specification, Eq. K1-5) is also required.

Top- and seat-angle connections are used commonly to frame a beam into the weak axis of a column. It can rotate sufficiently at ultimate load, without fracturing or imposing a significant moment on the column. However, experimental investigations by Hechtman and Johnson (1947) have shown that these connections can resist some end moment. Therefore, this type of connection can be used for Partial Rigid (PR) construction in the AISC specification.

9.3.3.2 Initial Stiffness Based on the experimental observations resported in Hechtman and Johnson (1947), Altman et al. (1982), and Azizinamini, Bradburn, and Radziminski (1985), the initial elastic stiffness R_{kits} of this connection type may be assumed to behave in the following manner:

1. The center of rotation for the connection is at the edge of the angle leg attached to the compression flange of the beam, and located at the end of the beam.
2. The top angle acts as a cantilever beam in which a fixed support is assumed to be at the fastener-hole edge closest to beam flange, and in the leg attached to the column face.
3. The bending rigidity of the angle leg at the center of rotation is so small that it can be neglected.

If these assumptions are employed, it can be shown that the normalized initial connection stiffness is given as

$$D_{ts} = \frac{R_{kits}}{EI_{0t}} \frac{1}{(1 + \delta_t)^2} = \frac{3}{\beta_t'[\gamma_t^2(\beta_t')^2 + 0.78]} \tag{9.3.8}$$

where

$$\beta_t' = \beta_t - \frac{1}{2\gamma_t}(1 + \omega_t) \tag{9.3.9}$$

Equation (9.3.8) is plotted in Fig. 9.8 for β_t' ranging from 0.0 to 0.5 and for γ_t from 3.0 to 21.0. It is observed that the normalized initial connection stiffness drops drastically as β_t' and γ_t increase.

The step-by-step procedure for determining the initial connection stiffness is given as follows (Liew et al., 1993a):

1. For given values of g_{ct}, l_t, t_t, and W_t, determine β_t, γ_t, and ω_t from Eq. (9.3.1).
2. Evaluate β_t' from Eq. (9.3.9).
3. Determine D_{ts} from Eq. (9.3.8) for the known values of γ_t and β_t'.
4. For a given beam depth d, evaluate $\delta_t = d/t_t$.
5. Finally, evaluate $I_{0t} = t_t^3/12$, and solve for R_{kits} from Eq. (9.3.8).

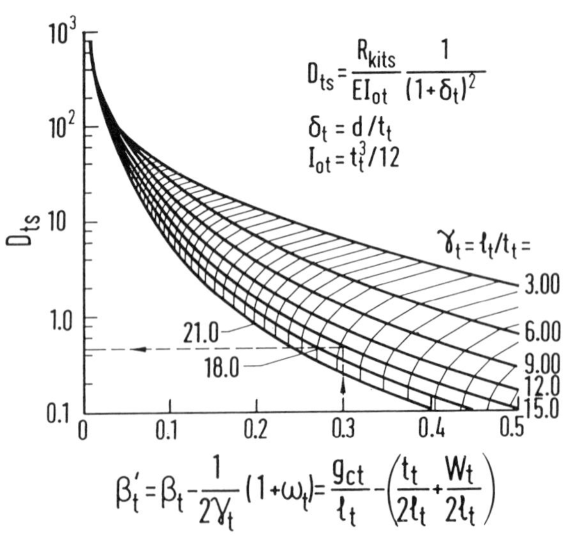

Fig. 9.8 Characteristic curves for calculating the initial stiffness of the top- and seat-angle connections.

9.3.3.3 Ultimate Moment Capacity The ultimate moment capacity of the connection is derived using the work equation for an assumed mechanism in conjunction with the yield criterion for flexural-shear interaction. The corresponding fourth-order equation with respect to $\xi_t = V/V_0$ may be written as

$$\xi_t^4 + \beta_t^* \xi_t - 1 = 0 \tag{9.3.10}$$

where

$$\beta_t^* = \beta_t' \gamma_t - \kappa_t \tag{9.3.11}$$

and β_t' is defined in Eq. (9.3.9).

The ultimate moment capacity is obtained by summing the moments about the center of rotation of the connection. It may be written in a dimensionless form as

$$\frac{M_{uts}}{M_{0t}t_t} = \gamma_t\{1 + \xi_t[1 + \beta_t^* + 2(\kappa_t + \delta_t)]\} \tag{9.3.12}$$

The ξ_t term in Eq. (9.3.12) can be solved iteratively from Eq. (9.3.10) for a predetermined value of β_t^*.

Equation (9.3.12) is plotted in Fig. 9.9 for assumed values of $\kappa_t = 1.50$ and 2.00, and for γ_t varying from 3.0 to 21.0. Also, Fig. 9.9 is generated based on $\delta_t = 6.00$. It is observed that when $\beta_t^* > 4.0$, the ultimate moment

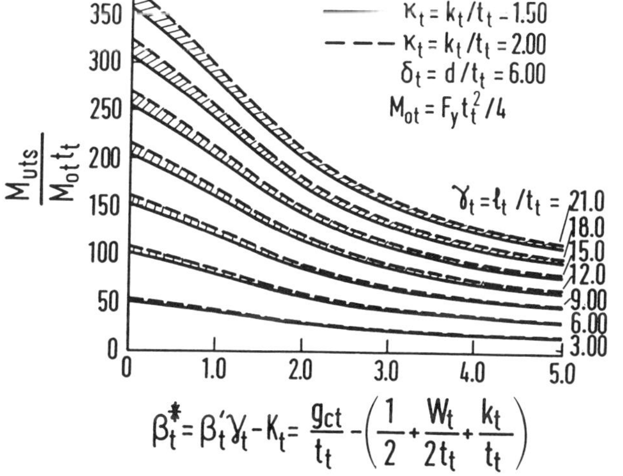

Fig. 9.9 Characteristic curves for calculating the ultimate moment capacity of the top- and seat-angle connections.

capacity is not very sensitive to the change in β_t^* and κ_t values. Also, to provide substantial bending resistance, the connection should be detailed with $\gamma_t \geq 9.0$ and $\beta_t^* \leq 3.0$. Because the curves shown in Fig. 9.9 are not very sensitive to κ_t values, it is suggested that the ultimate moment strength curve based on $\kappa_t = 1.50$ may be used conservatively for the determination of general connection moment capacities.

The procedure for determining the ultimate moment capacity is summarized as (Liew et al., 1993a):

1. For known values of g_{ct}, l_t, t_t, k_t, and W_t, determine β_t, γ_t, κ_t, and ω_t from Eq. (9.3.1).
2. Evaluate β_t' from Eq. (9.3.9).
3. Solve for β_t^* from Eq. (9.3.11).
4. Evaluate δ_t from Eq. (9.3.1) for a known value of beam depth d.
5. Finally, determine M_{uts} from the appropriate curve shown in Fig. 9.9. Alternatively, the exact M_{uts} value may be obtained from Eq. (9.3.12) by solving ξ_t iteratively first from Eq. (9.3.10).

9.3.3.4 Example 2: A Top- and Seat-Angle Connection Figure 9.10 shows an example of a top- and seat-angle connection with $g_{ct} = g_{cs} = 3$ in. $l_t = l_s = 7$ in., $t_t = t_s = 7/16$ in., $k_t = k_s = 7/8$ in., and $W_t = W_s = 1\text{-}7/16$ in. Assuming a material yield strength $F_y = 36$ ksi, a modulus of elasticity $E = 29{,}000$ ksi, and a beam depth $d = 17.7$ in. (for W18 $\times$ 35), determine the connection initial stiffness and the ultimate moment capacity.

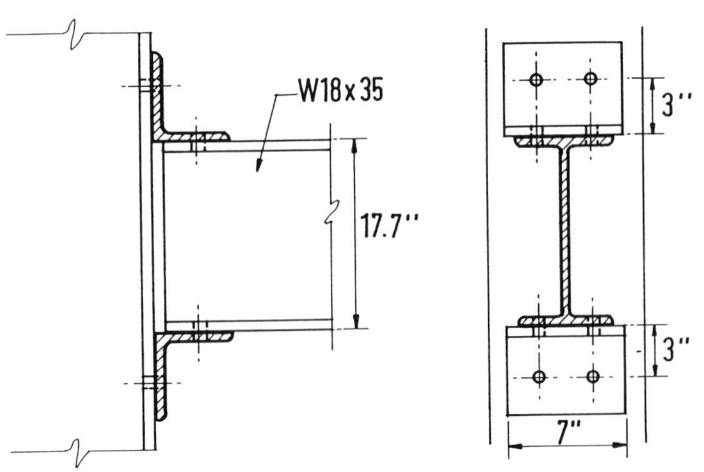

Fig. 9.10 Example 2: top- and seat-angle connections.

Initial Connection Stiffness, R_{kiw} The nondimensional parameters may be computed for the web-angle connection as

$$\beta_t = \frac{g_{ct}}{l_t} = 0.428571, \qquad \gamma_t = \frac{l_t}{t_t} = 16.0, \qquad \kappa_t = \frac{k_t}{t_t} = 2.0$$

$$\omega_t = \frac{W_t}{t_t} = 3.285714, \qquad I_{0t} = 0.006978, \qquad \delta_t = \frac{d}{t_t} = 40.457146$$

From Eq. (9.3.9),

$$\beta_t' = \beta_t - \frac{1}{2\gamma_t}(1 + \omega_t) = 0.294643$$

Evaluate D_{ts} either from Eq. (9.3.8) or from Fig. 9.8 as

$$D_{ts} = \frac{R_{kits}}{EI_{0t}} \frac{1}{(1 + \delta_t)^2} = \frac{3}{\beta_t'[\gamma_t^2(\beta_t')^2 + 0.78]} = 0.4426$$

Finally, determine the connection initial stiffness:

$$R_{kits} = EI_{0t}(1 + \delta_t)^2 \times 0.4426 = 153{,}943 \text{ kips-in.}$$

Ultimate Moment Capacity. For $\beta_t' = 0.428571$, $\gamma_t = 16.0$, and $\kappa_t = 2.0$, determine

$$\beta_t^* = \beta_t' \gamma_t - \kappa_t = 2.714286$$

Evaluate ξ_t from Eq. (9.3.10) of which $\xi_t = 0.361945$.
For $\delta_t = 40.457146$, solve for

$$\frac{M_{uts}}{M_{0t}t_t} = \gamma_t(1 + \xi_t[1 + \beta_t^* + 2(\kappa_t + \delta_t)]) = 529.2587$$

Finally, for $M_{0t}t_t = 1.722656 \times 0.4375 = 0.753662$, the ultimate moment can be computed as

$$M_{uts} = 529.2587 \times M_{0t}t_t = 398.88 \text{ kips-in.}$$

9.3.4 Top- and Seat-Angle Connections with Double Web Cleats

9.3.4.1 General A typical top- and seat-angle connection with web angles is shown in Fig. 9.11. Double web cleats are used to enhance the restraint characteristics of the top- and seat-angle connections. Conventional design

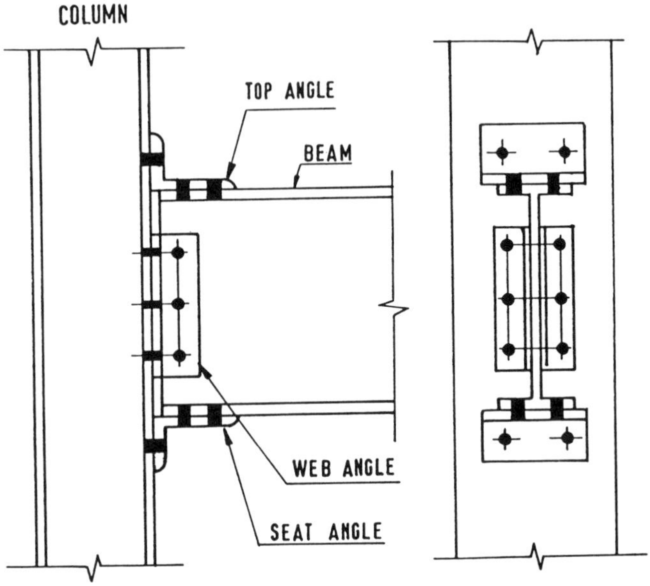

Fig. 9.11 Top- and seat-angle connections with web angles.

normally assumes that the top and seat angles are designed to resist moment, whereas the web cleats are designed to resist shear reaction only. The rotational performance of the joint depends mainly on the pullout behavior of the top angle, although an important contribution at large rotations may be provided by the web angles as well. The top angles are subjected to tension forces, which they resist through shear and bending of the leg fastened to the column flange. The available stiffness may be limited due to flexibility of the angle and the column flange if the latter is not stiffened.

The procedure for calculating the initial stiffness and moment resistance of this type of connections is given in what follows.

9.3.4.2 Initial Stiffness Based on the experimental observations from Altman et al. (1982) and Azizinamini et al. (1985), the initial connection stiffness of this type of connection may be modeled with the following assumed behavior:

1. The beam and the angle legs attached to the beam act as rigid members.
2. The center of rotation for the connection is at the face of the angle leg attached to the compressive flange of the beam, and at the end of the beam.
3. The top angle acts as a cantilever beam similar to that of the top- and seat-angle connection.

4. The web angle acts as a cantilever beam similar to the behavior of the top angle.
5. The resisting moment of the angle at the center of rotation is neglected.

Based on these assumptions, it can be shown that the initial connection stiffness R_{ki} can be written as

$$\frac{R_{ki}}{EI_{0t}} = \frac{R_{kits}}{EI_{0t}} + \frac{R_{kiw}}{EI_{0t}} \tag{9.3.13}$$

The first term at the right-hand side of Eq. (9.3.13) is the normalized initial connection stiffness for the top- and seat-angle connection in which R_{kits} can be obtained from Eq. (9.3.8) following the procedures discussed in Section 9.3.3. The second right-hand term of Eq. (9.3.13) is the initial connection stiffness contributed by the double web angles. It can be shown that R_{kiw} in Eq. (9.3.13) can be written as

$$D_w = \frac{R_{kiw}}{EI_{0t}} \frac{1}{\rho(1 + \delta_t)^2} = \frac{3}{2\beta'_w[\gamma_w^2(\beta'_w)^2 + 0.78]} \tag{9.3.14}$$

It should be noted that β'_w in Eq. (9.3.14) should be computed using the same form of equation as for the β'_t in Eq. (9.3.9). One *must not* use Eq. (9.3.3) to compute β'_w for the double web angles when they are used with a top- and seat-angle connection.

In view of Eqs. (9.3.8) and (9.3.14), Eq. (9.3.13) may be rewritten as

$$\frac{R_{ki}}{EI_{0t}} = (D_{ts} + \rho D_w)(1 + \delta_t)^2 \tag{9.3.15}$$

The procedure for determining the initial connection stiffness is given as follows (Liew et al., 1993a):

1. For known values of g_{ct}, l_t, t_t, and W_t for the top- and seat-angle connections, determine the nondimensional parameters β_t, γ_t, and ω_t from Eq. (9.3.1).
2. Similarly, for known values of g_{cw}, l_w, t_w, k_w, and W_w for the double web angles, determine β_w, γ_w, κ_w, and ω_w from Eq. (9.3.1).
3. Evaluate β'_t and β'_w from Eqs. (9.3.9). It should be noted that subscripts t and s are interchangeable in Eq. (9.3.9).
4. Evaluate δ_t from Eq. (9.3.1) based on known values of d and t_t.
5. Determine D_{ts} and D_w from Eqs. (9.3.8) and (9.3.14), respectively.
6. Finally, determine R_{ki} from Eq. (9.3.15) for the given values of ρ, δ_t, and EI_{0t}.

Characteristic curves, such as those shown in Figs. 9.12 and 9.13, may be generated based on this procedure. These figures may be used to determine the initial connection stiffness once the detailed geometry of the connection subassemblage is known.

9.3.4.3 Ultimate Moment Capacity The collapse mechanism of a top- and seat-angle connection with double web angles is assumed to be a combination of the collapse mechanisms of the double web-angle connection and the top- and seat-angle connection described in Section 9.3.2. The ultimate moment capacity M_u may be written as

$$\frac{M_u}{M_{0t}t_t} = \frac{M_{uts}}{M_{0t}t_t} + \frac{M_{uw}}{M_{0t}t_t} \tag{9.3.16}$$

The first term in the right-hand side of Eq. (9.3.16) represents the contribution of the ultimate moment capacity from the top and seat angles, and it can be determined from Eq. (9.3.12). The second right-hand term in Eq. (9.3.16) is the contribution of the ultimate moment capacity from the double web-angle connection. The ultimate connection for the *double* web angles, which are part of the combined angle assembly, is written as

$$\frac{M_{uw}}{M_{0t}t_t} = \gamma_w(1 + \xi_w)\rho^3\left(\frac{\gamma_w(\xi_w - 1)}{3(\xi_w + 1)} + \delta_w + \frac{1}{\rho}\right) \tag{9.3.17}$$

in which ξ_w can be determined from Eq. (9.3.4).

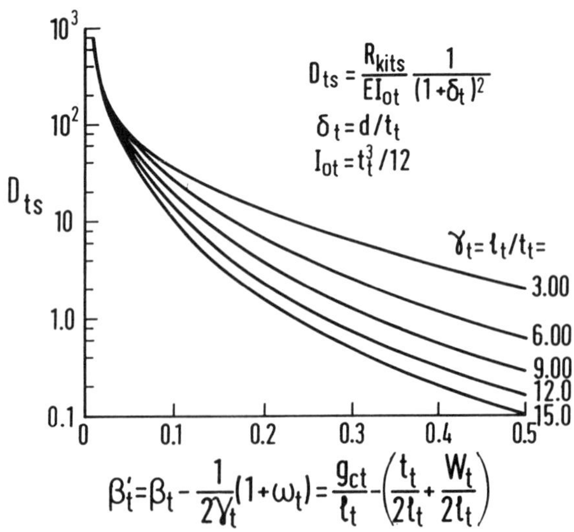

Fig. 9.12 Characteristic curves for calculating the initial stiffness of the top and seat angles when used with the double web-angle connections.

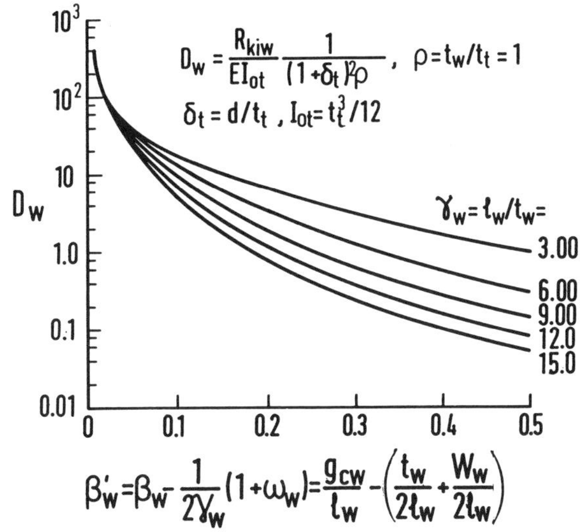

Fig. 9.13 Characteristic curves for calculating the initial stiffness of the double web angles when used with the top- and seat-angle connections.

The moment capacities contributed by the top and seat angles and the double web angles can be determined separately from the charts shown in Figs. 9.14 to 9.17, and they can be added to obtain the moment capacity of the connection assemblage. The procedures for generating design aids such as those shown in Figs. 9.14 through 9.17 for connections of this type are summarized in what follows (Liew et al., 1993a):

1. For known values of β_w, γ_w, and κ_w, determine β_w^* from Eq. (9.3.6).
2. Evaluate ξ_w from Eq. (9.3.4).
3. Solve for $M_{uw}/M_{0t}t_t$ from Eq. (9.3.17) for the known values of ρ and δ_w.
4. Similarly, for known values of β_t, γ_t, and ω_t, determine β_t' from Eq. (9.3.9).
5. Solve for β_t^* from Eq. (9.3.11).
6. Evaluate ξ_t from Eq. (9.3.10).
7. Solve for $M_{uts}/M_{0t}t_t$ from Eq. (9.3.12) for the known values of κ_t and δ_t.
8. Finally, evaluate EI_{0t}, and solve for M_u from Eq. (9.3.16).

9.3.4.4 Example 3: Combined Top- and Seat-Angle Connection with Double Web Cleats Figure 9.18 shows a connection that combines the double web-angle cleats in Example 1 with the top- and seat-angle connection in Example 2 to form a connection with dimensions given as follows:

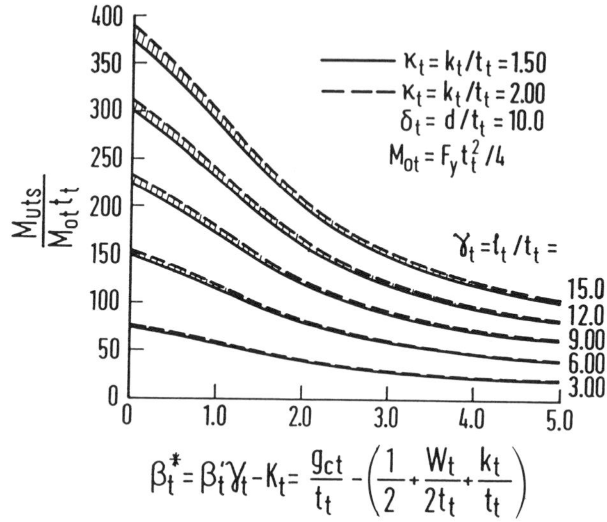

Fig. 9.14 Characteristic curves for calculating the ultimate moment capacity of the top- and seat-angle connections when used with the double web angles, $\delta_t = 10.0$.

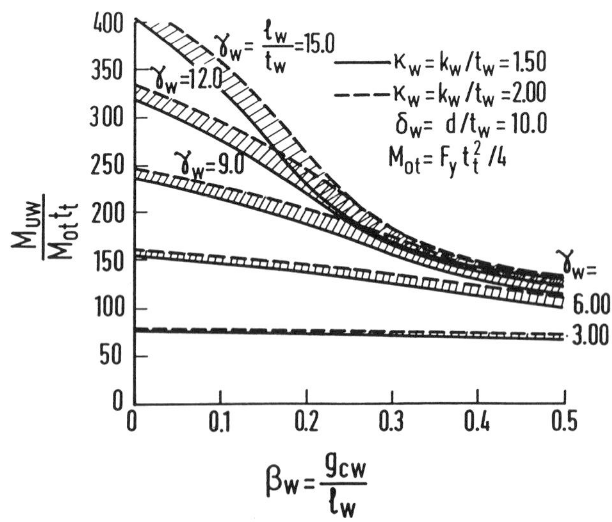

Fig. 9.15 Characteristic curves for calculating the ultimate moment capacity of the double web angles when used with the top- and seat-angle connections, $\delta_w = 10.0$.

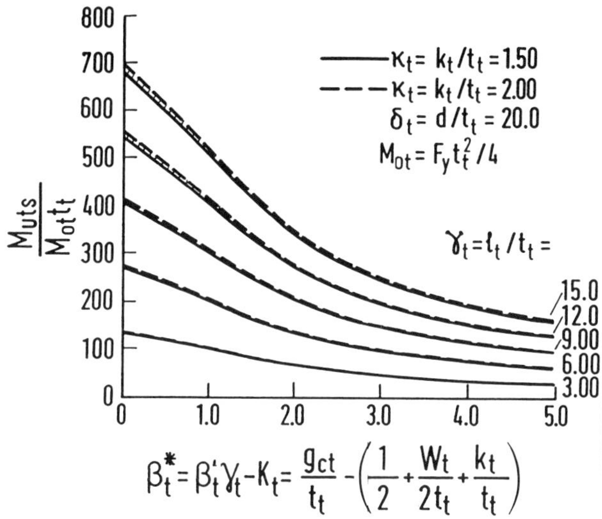

Fig. 9.16 Characteristic curves for calculating the ultimate moment capacity of the top and seat angles when used with the double web-angle connections, δ_t = 20.0.

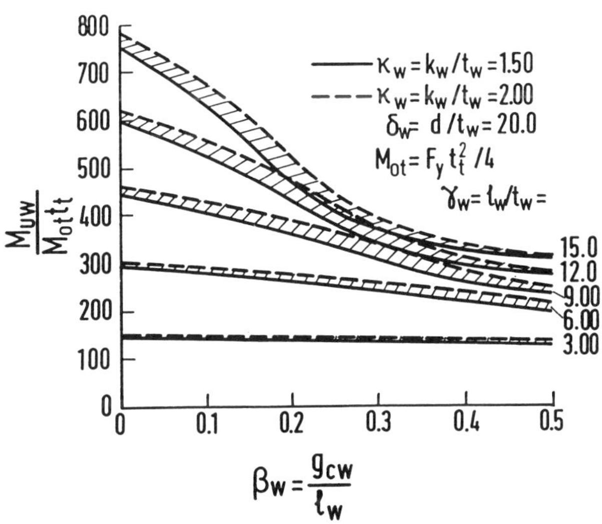

Fig. 9.17 Characteristic curves for calculating the ultimate moment capacity of the double web angles when used with the top- and seat-angle connections, δ_w = 20.0.

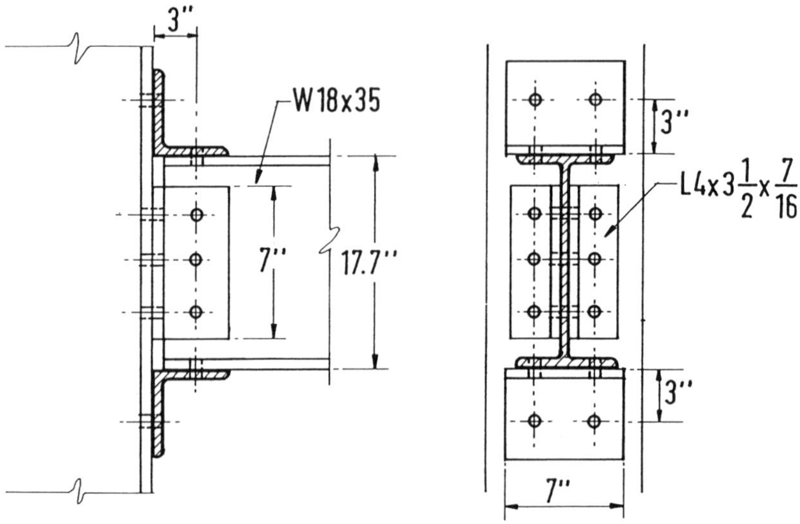

Fig. 9.18 Example 3: top- and seat-angle connection with double web angles.

1. For the double web angles, g_{cw} = 3 in., l_w = 7 in., t_w = 7/16 in., k_w = 7/8 in., and W_w = 1-7/16 in.
2. For the top- and seat-angle cleats, $g_{ct} = g_{cs}$ = 3 in., $l_t = l_s$ = 7 in., $t_t = t_s$ = 7/16 in., $k_t = k_s$ = 7/8 in., and $W_t = W_s$ = 1-7/16 in.

Assuming a material yield strength F_y = 36 ksi, a modulus of elasticity E = 29,000 ksi, and a beam depth d = 17.7 in. (for W18 × 35), determine the combined connection initial stiffness and the ultimate moment capacity.

Initial Connection Stiffness, R_{ki} The nondimensional parameters may be computed for the web-angle connection as

$$\beta_w = \frac{g_{cw}}{l_w} = 0.428571, \qquad \gamma_w = \frac{l_w}{t_w} = 16.0, \qquad \kappa_w = \frac{k_w}{t_w} = 2.0$$

$$\omega_w = \frac{W_w}{t_w} = 3.2857, \qquad I_{0w} = 0.006978$$

and for the top and seat angles as

$$\beta_t = \frac{g_{ct}}{l_t} = 0.428571, \qquad \gamma_t = \frac{l_t}{t_t} = 16.0, \qquad \kappa_t = \frac{k_t}{t_t} = 2.0$$

$$\omega_t = \frac{W_t}{t_t} = 3.285714, \qquad I_{0t} = 0.006978 \qquad \delta_t = \frac{d}{t_t} = 40.457146$$

From Eq. (9.3.9), determine β'_t and β'_w

$$\beta'_t = \beta_t - \frac{1}{2\gamma_t}(1 + \omega_t) = 0.294643$$

$$\beta'_w = \beta_w - \frac{1}{2\gamma_w}(1 + \omega_w) = 0.294643$$

Determine

$$D_{ts} = \frac{3}{\beta'_t[\gamma_t^2(\beta'_t)^2 + 0.78]} = 0.439780$$

and

$$D_w = \frac{3}{2\beta'_w[\gamma_w^2(\beta'_w)^2 + 0.78]} = 0.219890$$

Evaluate $\rho = t_w/t_t = 1.0$, $I_{0t} = t_t^3/12 - 0.006978$ in.^{3}, and $\delta_t - d/t_t -$ 40.457146, and, finally, compute the connection initial stiffness as

$$R_{ki} = EI_{0t}(D_{ts} + \rho D_w)(1 + \delta_t)^2 = 230{,}916 \text{ kips-in.}$$

Ultimate Moment, M_u For $\beta_w = 0.428571$, $\gamma_w = 16.0$, and $\kappa_w = 2.0$, determine

$$\beta_w^* = \beta_w \gamma_w - \kappa_w = 2.714286$$

Evaluate ξ_w iteratively from Eq. (9.3.4) of which $\xi_w = 0.361945$.
For $\rho = 1.0$, solve for

$$\frac{M_{uw}}{M_{0t}t_t} = \gamma_w(1 + \xi_w)\rho^3\left(\frac{\gamma_w(\xi_w - 1)}{3(\xi_w + 1)} + \delta_w + \frac{1}{\rho}\right) = 529.26$$

Similarly, for $\beta'_t = 0.428571$, $\gamma_t = 16.0$, and $\kappa_w = 2.0$, determine

$$\beta_t^* = \beta'_t \gamma_t - \kappa_t = 4.857143$$

Evaluate ξ_t from Eq. (9.3.10) of which $\xi_t = 0.20548$.
For $\delta_t = 40.457146$, solve for

$$\frac{M_{uts}}{M_{0t}t_t} = \gamma_t(1 + \xi_t[1 + \beta_t^* + 2(\kappa_t + \delta_t)]) = 1261.07$$

Evaluate $M_{0t}t_t = 0.753662$, and, finally, solve for

$$M_u = M_{0t}t_t\left[\frac{M_{uts}}{M_{0t}t_t} + \frac{M_{uw}}{M_{0t}t_t}\right] = 950.4 \text{ kips-in.}$$

9.3.5 Shape Parameter

Once the connection initial stiffness R_{ki} and the ultimate moment capacity M_u are evaluated for a given connection type (using the appropriate procedures presented in Sections 9.3.2–9.3.4, shape parameter n is determined such that the connection model given by Eq. (9.3.1) or (9.3.2) can best fit the experimental data.

Kishi et al. (1988a, 1988b) have performed statistical analyses of test data to obtain formulas for the n values of several angle connections. These formulas are summarized as follows:

1. For single web-angle connections, the formula for n, determined from a least-mean-square fit of 16 test curves, is

$$n = 0.520 \log_{10} \theta_0 + 2.291 \geq 0.70 \tag{9.3.18}$$

   where n is expressed only in terms of $\theta_0 = M_u/R_{ki}$ for each connection type.
2. For double web-angle connections, 14 test curves were used to determine n. The expression for n is

$$n = 1.322 \log_{10} \theta_0 + 3.952 \geq 0.57 \tag{9.3.19}$$

3. For top- and seat-angle connections, 15 test curves were used for curve fitting n. The expression obtained is

$$n = 2.003 \log_{10} \theta_0 + 6.070 \geq 0.30 \tag{9.3.20}$$

4. For top- and seat-angle connections with double web angles, 22 test curves were used for curve fitting n. The expression for n is

$$n = 5.483 \log_{10} \theta_0 + 14.745 \geq 0.80 \tag{9.3.21}$$

Figure 9.19 shows the results of n parameters in comparison with the experimental test results obtained from the connection data bank in Kishi and Chen (1990). It can be observed that the calibration of the n parameter is by

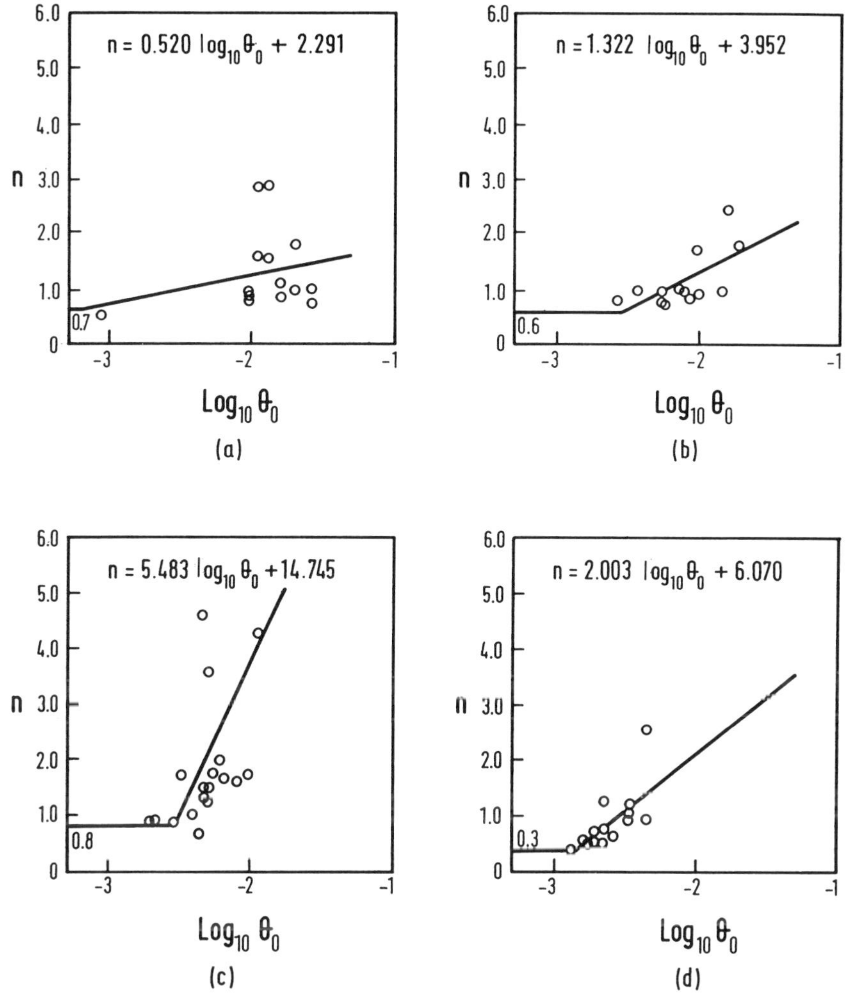

Fig. 9.19 Comparisons of shape parameter n with experimental test data (Kishi et al., 1988a, 1988b). (a) Single web-angle connections. (b) Double web-angle connections. (c) Top- and seat-angle connections with double web angles. (d) Top- and seat-angle connections.

no means conclusive because test data are not available for n smaller than the ranges indicated in Eqs. (9.3.18)–(9.3.21). It is, therefore, advisable to restrict connection designs to the range for which the connection behavior has been well established, unless this behavior can be demonstrated by tests or by other rational means.

In summary, for a given angle connection, the moment–rotation relationship can be obtained by the following procedure:

1. Determine the initial stiffness K_i.
2. Compute the ultimate moment capacity M_u.
3. Compute the reference rotation θ_0.
4. Obtain the design shape parameter n.
5. Determine the moment–rotation model using Eq. (9.2.1).

Once the connection behavior is well defined, then proceed to analyze the steel frame using the proposed second-order analysis.

Examples

1. For the double web cleats given in Example 1, R_{ki} = 8083 kips-in. and M_{uw} = 181.5 kips-in., for which θ_0 may be computed as

   $$\theta_0 = M_u/R_{ki} = 0.022180$$

   From Eq. (9.3.19),

   $$n = 1.322 \log_{10} \theta_0 + 3.952 = 1.374$$

2. For the top- and seat-angle connection given in Example 2, R_{ki} = 153,943 kips-in. and M_{uw} = 398.88 kips-in., for which θ_0 is computed as

   $$\theta_0 = M_u/R_{ki} = 0.0025910$$

   From Eq. 9.3.20,

   $$n = 2.003 \log_{10} \theta_0 + 6.070 = 0.889$$

3. For the combined top- and seat-angle connection with double web cleats given in Example 3, R_{ki} = 230,916 kips-in. and M_{uw} = 950.4 kips-in., for which θ_0 is computed as

   $$\theta_0 = M_u/R_{ki} = 0.004115782$$

   From Eq. 9.3.21,

   $$n = 5.483 \log_{10} \theta_0 + 14.745 = 1.665$$

9.4 CONNECTION CLASSIFICATIONS

9.4.1 General

In preliminary design, it is useful to estimate what influence the connection will have on the behavior of the structure without having to perform a nonlinear analysis. This can be achieved by categorizing the connection M–θ_r response curves into three basic categories: (1) simple connections (pinned), (2) rigid connections, and (3) semi-rigid connections. The first two categories are currently used in numerous analysis and design provisions, and the non-linear connection response is not considered in the overall analysis of the structure. However, the idealization of the connection moment–rotation behavior as perfectly pinned or fully rigid is strictly applicable for a small range of typical connection types. It is necessary to consider the connection flexibility for the third category to which many types of practical connections belong.

The background of the demarcation lines among the three general categories of connection behavior is somewhat quantitative. A useful guide has been proposed by Bijlaard and Zoetemeijer (1986) in which beam-to-column connections can be classified with respect to (1) rotational stiffness and (2) moment resistance. A nominally pinned connection must be capable of transmitting the forces calculated in design and accepting the resulting rotations. While transmitting the forces, it must not develop significant moment that might adversely affect members of the structure. The rotation capacity of a nominally pinned connection must be sufficient to enable all the necessary plastic hinges to develop under the design loads.

A rigid connection must be designed so that its deformation has no significant influence on the distribution of internal forces in the structure nor on the overall performance of structures. The design resistance of a rigid connection normally should at least equal the plastic strength of the members that it is connecting. The stiffness of a rigid connection must be such that, under the design loads, the rotations at the necessary plastic hinges do not exceed their rotation capacities.

Two connection classification systems, Eurocode 3 (1992) and Bjorhovde, Brozzetti, and Colson (1990), are reviewed in the following sections. These classification systems are intended for use in the analysis of steel structures consisting of semi-rigid beam-to-column connections. Their design implications and the design concepts are demonstrated through a complete design example given in Section 9.5.

9.4.2 Classification System in Eurocode 3

In the Eurocode 3 approach, connections are classified by (1) rigidity (i.e., stiffness, R_{ki}) and (2) strength (moment resistance, M_u).

9.4.2.1 Rotational Stiffness The rotational stiffness of a beam-to-column connection may be classified as (1) nominally pinned, (2) rigid, or (3) semi-rigid. From the stiffness viewpoint, the equations for the boundaries between rigid and semi-rigid were chosen so that the deformations of the connection must not reduce the resistance of the structure by more than 5%. In addition, Eurocode 3 differentiates between performance requirements for connections in braced and unbraced frames.

The following nondimensional parameters are used in Eurocode 3 for the classification of connections:

$$\overline{m} = \frac{M}{M_p}, \qquad \overline{\theta} = \frac{\theta_r}{\theta_p} \tag{9.4.1}$$

where $\theta_p = M_p/(ELI_b/L_b)$, I_b and L_b are the moment of inertia and length of the beam, respectively, and M_p is the full plastic moment capacity of the beam. The locations and boundaries for the classification of beam-to-column connection stiffness as semi-rigid are shown in Fig. 9.20. They are written as follows:

For unbraced frames:

$$\text{When } \overline{m} < \frac{2}{3} \quad \Rightarrow \quad \overline{m} < 25\overline{\theta} \tag{9.4.2}$$

$$\text{When } \frac{2}{3} < \overline{m} < 1.0 \quad \Rightarrow \quad \overline{m} < \frac{25\overline{\theta} + 4}{7} \tag{9.4.3}$$

The preceding equations are given for an unbraced frame and may be used only for frames in which every story satisfies

$$\frac{I_b/L_b}{I_c/L_c} \leq 1.0 \tag{9.4.4}$$

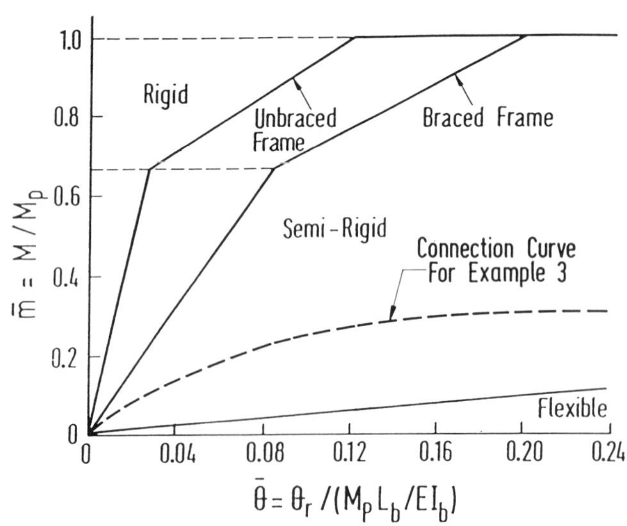

Fig. 9.20 Classification of connections according to Eurocode 3 (1992).

where I_b and I_c are the second moment of area of a beam and column, respectively, and L_b and L_c are the span of a beam from center to center of the columns and the story height for a column, respectively.

For braced frames, the boundary for a semi-rigid connections is defined as

$$\text{When } \overline{m} < \frac{2}{3} \quad \Rightarrow \quad \overline{m} < 8\overline{\theta} \tag{9.4.5}$$

$$\text{When } \frac{2}{3} < \overline{m} < 1.0 \quad \Rightarrow \quad \overline{m} < \frac{20\overline{\theta} + 3}{7} \tag{9.4.6}$$

A beam-to-column connection may be classified as nominally pinned if its rotational stiffness satisfies the condition

$$R_{ki} \leq \frac{EI_b}{2L_b} \tag{9.4.7}$$

9.4.2.2 Moment Resistance Beam-to-column connections may be classified as (1) nominally pinned, (2) full strength, and (3) partial-strength based on its moment resistance. A connection is classified as nominally pinned if its design moment resistance M is not greater than 0.25 times the plastic moment resistance of the connected beam M_p:

$$M \leq \frac{M_p}{4} \tag{9.4.8}$$

A connection may be classified as full strength if its design moment resistance is at least equal to the design plastic moment resistance of the connected beam:

$$M > M_p \tag{9.4.9}$$

A connection may be classified as partial strength if its design moment resistance is within the range of

$$M_p/4 < M \leq M_p \tag{9.4.10}$$

In addition to strength and stiffness classification, the connection must also be checked for sufficient rotation capacity. However, if the moment capacity of the beam-to-column connection is larger than 1.2 times the design plastic moment resistance of the connected beam, the connection rotation capacity need not be checked. This is because the plastic hinge will always form in the connected beam. In cases where plastic hinges may be developed in the connections, the connection rotation capacity needs to be checked if redistribution of moments is considered in the global analysis of the structure.

The connection example given in Section 9.3.4.4 is classified using the Eurocode procedure, and the resulting curve is shown in Fig. 9.20. The connection can be classified as semi-rigid and as partial strength.

9.4.3 Classification System in Bjorhovde et al.

Bjorhovde et al. (1990) proposed a classification system that uses a reference beam length of $5d$ to define the beam stiffness. The nondimensional parameters used in the classification of the connections are the same as those in Eq. (9.4.1), but θ_p is now defined as $M_p(ELI_b/5d)$, where d is the beam depth. The classification is based on the strength and stiffness of the connections with the boundary regions shown in Fig. 9.21. The three different regions in Fig. 9.21 are defined as follows:

1. Rigid connection:

$$\text{In terms of strength:} \quad \overline{m} \geq 0.7 \tag{9.4.11}$$

$$\text{In terms of stiffness:} \quad \overline{m} \geq 2.5\overline{\theta} \tag{9.4.12}$$

2. Semi-rigid connection:

$$\text{In terms of strength:} \quad 0.7 > \overline{m} > 0.2 \tag{9.4.13}$$

$$\text{In terms of stiffness:} \quad 2.5\overline{\theta} > \overline{m} > 0.5\overline{\theta} \tag{9.4.14}$$

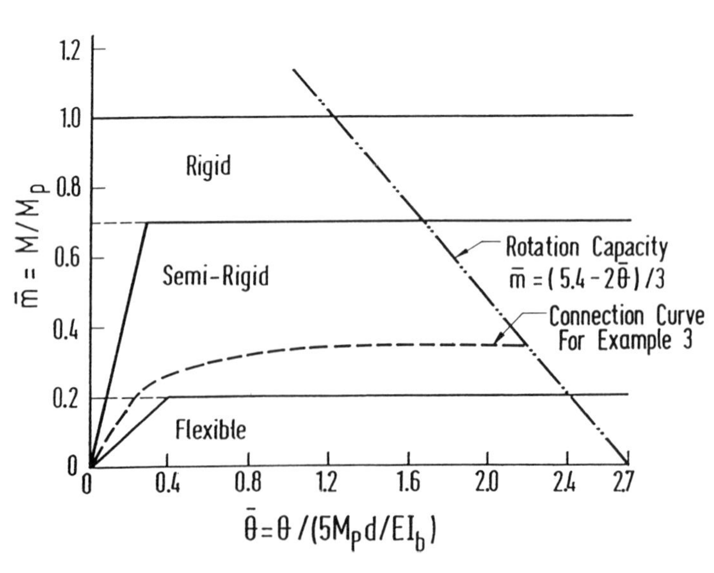

Fig. 9.21 Classification of connections according to Bjorhovde et al. (1990).

3. Simple connection:

$$\text{In terms of strength:} \qquad \overline{m} < 0.2 \tag{9.4.15}$$

$$\text{In terms of stiffness:} \qquad \overline{m} \leq 0.5\overline{\theta} \tag{9.4.16}$$

In addition, Bjorhovde et al. (1990) proposed an expression for calculating the rotation capacity of the connection based on calibration with test data. This simplified expression is written as

$$\overline{m} = \frac{5.4 - 2\overline{\theta}}{3} \tag{9.4.17}$$

According to this formula, the required rotation capacity of a beam-to-column connection depends on the ratio of the ultimate moment capacity of the connection to the fully plastic moment of the beam. It is also inversely proportional to the initial connection stiffness, R_{ki}. In other words, the smaller the initial connection stiffness, the larger the necessary rotation capacity. Equation (9.4.17) is plotted as shown in Fig. 9.21.

The connection example given in Section 9.3.4.3 is classified using the Bjorhovde et al. procedure, and the resulting curve is shown in Fig. 9.21. The connection may be classified as semi-rigid in terms of both the stiffness and strength.

9.4.4 Comparison of the Two Classification Systems

In summary, both Eurocode 3 (1992) and Bjorhovde et al. (1990) approaches introduce a nondimensional system of classification that compares the connection stiffness to the beam stiffness. The rotation axis is nondimensionalized by the beam stiffness. In defining the beam stiffness, two different concepts are adopted. The Eurocode 3 (1992) approach is based on the actual stiffness of the beam, and the Bjorhovde et al. (1990) approach is based on a standard reference length of the beam. In these two approaches, the moment axis of the connection response is nondimensionalized with reference to the plastic moment of the connected beam. The Eurocode 3 approach is more accurate if the layout and member details of the structural system are known in advance. The classification approach by Bjorhovde et al. is, however, necessary for cases where prior knowledge of the member details is not available. For both cases, it is necessary to know the entire range of the connection M–θ_r curve, either from tests for from analytical models, as described in Section 9.3.

Both classification systems are fairly rough, because there is a wide range of parameters that will affect the performance of connections. They are, however, convenient when appraising current knowledge, and they were adopted in this chapter for designing some semi-rigid frames.

9.5 ANALYSIS METHOD

9.5.1 Second-Order Refined Plastic-Hinge Analysis

The analysis method presented herein is based on second-order refined plastic-hinge analysis with modifications to include the effects of end connections in the beam-column member (Liew et al., 1993b). The refined plastic-hinge analysis represents the distribution yielding in the beam-column element loaded by an arbitrary set of end forces using a two-step procedure:

1. The LRFD column tangent-modulus model E_t is used in place of the elastic modulus E to represent the distributed plasticity along the length of a member due to axial-force effects. The formula for E_t is written as

$$E_t = -2.7243E\frac{P}{P_y}\ln\left(\frac{P}{P_y}\right), \qquad \text{for } P > 0.39P_y \qquad (9.5.1)$$

   Because this E_t model is derived from the LRFD column-strength formula, it implicitly includes the effects of residual stresses and initial out of straightness in modeling the member effective stiffness.
2. Distributed plasticity effects associated with flexure are captured by gradually degrading the member stiffness at the plastic-hinge locations as yielding progresses under increasing load as the cross-section strength is approached. For a general beam-column member in which distributed yielding may occur at both ends, the incremental force-displacement relationship is written as

$$\begin{Bmatrix} \dot{M}_A \\ \dot{M}_B \end{Bmatrix} = \frac{E_t I}{L}\begin{bmatrix} \phi_A\left[S_1 - \dfrac{S_2^2}{S_1}(1-\phi_B)\right] & \phi_A\phi_B S_2 \\ \phi_A\phi_B S_2 & \phi_b\left[S_1 - \dfrac{S_2^2}{S_1}(1-\phi_A)\right] \end{bmatrix}\begin{Bmatrix} \dot{\theta}_A \\ \dot{\theta}_B \end{Bmatrix} \qquad (9.5.2)$$

The terms S_1 and S_2 are the conventional beam-column stability functions, but with E_t used in place of the elastic modulus. The term ϕ (subscripts A and B denote the respective ends of the element) is a scalar parameter that allows for gradual inelastic stiffness reduction of the element associated with the effect of plastification at the element end. This term is equal to one when the element is elastic, and it is zero when a plastic hinge is formed at the end. Parameter ϕ is taken as

$$\phi = 4\alpha(1-\alpha), \qquad \text{for } \alpha > 0.5 \qquad (9.5.3)$$

where α is a force-state parameter that measures the magnitude of axial force P and bending moment M at the element end. The term α is expressed as

$$\alpha = \frac{P}{P_y} + \frac{8}{9}\frac{M}{M_p}, \qquad \text{for } \frac{P}{P_y} \geq \frac{2}{9}\frac{M}{M_p} \tag{9.5.4}$$

$$\alpha = \frac{P}{2P_y} + \frac{M}{M_p} \qquad \text{for } \frac{P}{P_y} < \frac{2}{9}\frac{M}{M_p} \tag{9.5.5}$$

Initial yielding is assumed to occur based on a yield surface that has the same shape as the plastic strength surface, as shown in Fig. 9.22. In this figure, the yield-surface function corresponding to $\alpha = 1.0$ represents the state of forces at which the cross-section is fully yielded. The yield-surface function corresponding to $\alpha = 0.5$ represents the state of forces at which initial yielding starts to occur. If the state of forces is changed in such a manner that the force point moves inside or along the initial yield surface, the element is fully elastic with no stiffness reduction. If the force point moves beyond the initial yield surface, the element stiffness is reduced to account for the effect of plastification at the element end. A detailed description of the aspect has been reported in Chapter 8. Recent books on plastic-hinge-based analysis and design (White and Chen, 1993; Chen and Toma, 1994; Chen and Sohal, 1995) also provide comprehensive guidelines, including computer programs for the design of multistory building framework.

In the refined plastic-hinge formulation, the AISC LRFD beam-column interaction equations for members of zero length are adopted as the cross-

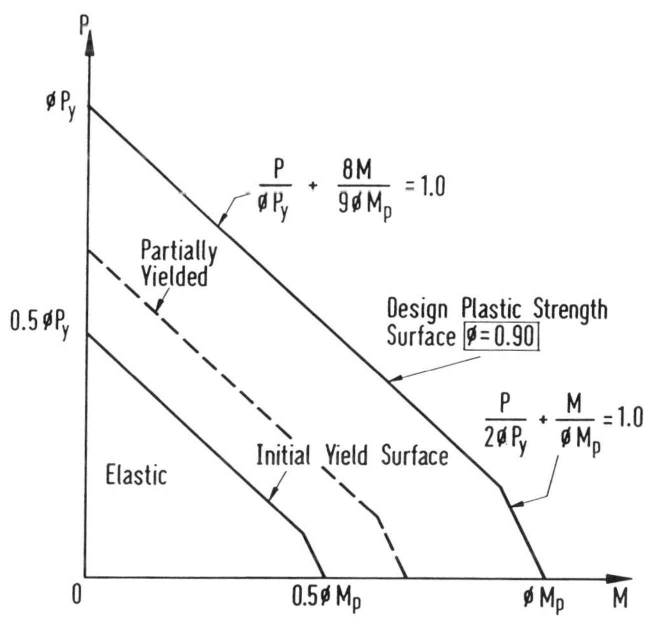

Fig. 9.22 Two-surface stiffness-degradation model for a second-order refined plastic-hinge analysis.

section plastic strength equations for representing both the strong- and weak-axis in-plane strength behavior. This cross-section plastic strength may be reduced by the appropriate resistance factors for use in limit states design. In the present study, the nominal plastic strength surface is reduced to a design yield surface based on a resistance factor of ϕ for both flexural bending and axial compression, as shown in Fig. 9.22. Different resistance factors for flexural bending and axial compression, such as those currently used in AISC LRFD for beam-column design, may also be used in the refined plastic-hinge analysis, if required.

9.5.2 Modification of Element Stiffness for the Presence of End Connections

The stiffness matrix of a beam-column element with semi-rigid connections attached at its ends is derived based on the modified beam-column stability-function approach (Chen and Lui, 1991). The connections are modeled as rotational springs having the M–θ_r curves described by the three-parameter model given in Section 9.2. These springs are physically tied to the ends of the beam-column by enforcing equilibrium and compatibility at the connecting ends. Finally, the rotational degrees of freedom of the connections can be condensed out from the tangent-stiffness relationship of the beam-column element.

Consider a beam-column subjected to end moments M_A and M_B and axial force P with connections attached at both ends, as shown in Fig. 9.23. The presence of connections introduces relative incremental rotations of $\dot{\theta}_{rA}$ and $\dot{\theta}_{rB}$ at the Ath and Bth end of the member, respectively. By denoting R_{ktA} and R_{ktB} as the tangent stiffness of connections A and B, respectively, the relative incremental rotations between the joints and beam ends (i.e., the incremental rotational deformations of the connections) can be expressed as

$$\dot{\theta}_{rA} = \frac{\dot{M}_A}{R_{ktA}}, \qquad \dot{\theta}_{rB} = \frac{\dot{M}_B}{R_{ktB}} \tag{9.5.6}$$

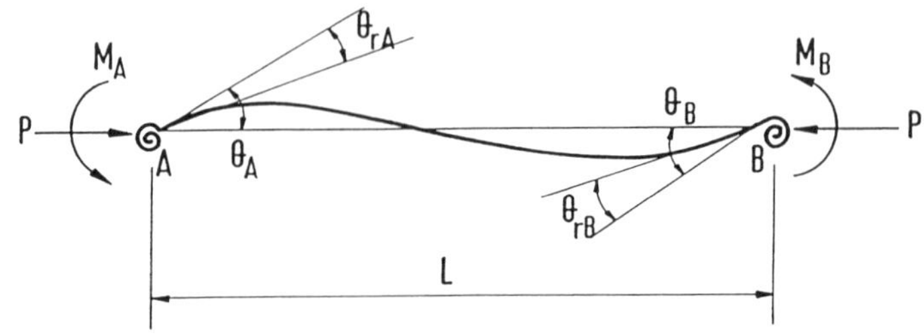

Fig. 9.23 End-spring model for a beam-column element with end connections.

The incremental force-displacement equations for the beam-column element with the presence of end connections can be expressed as (Chen and Lui, 1991):

$$\dot{M}_A = \frac{EI}{L}\left[S_{ii}\left(\dot{\theta}_A - \frac{\dot{M}_A}{R_{ktA}}\right) + S_{ij}\left(\dot{\theta}_B - \frac{\dot{M}_B}{R_{ktB}}\right)\right] \tag{9.5.7}$$

$$\dot{M}_B = \frac{EI}{L}\left[S_{ij}\left(\dot{\theta}_A - \frac{\dot{M}_A}{R_{ktA}}\right) + S_{jj}\left(\dot{\theta}_B - \frac{\dot{M}_B}{R_{ktB}}\right)\right] \tag{9.5.8}$$

S_{ii}, S_{ij}, and S_{jj} are defined in the following manner.

1. If an elastic or elastic-plastic-hinge assumption is assumed, then

$$S_{ii} = S_{jj} = S_1 \qquad \text{and} \qquad S_{ij} = S_2 \tag{9.5.9}$$

   where S_1 and S_2 are stability functions as defined in Section 6.2.3. Note that Eqs. (9.5.7) and (9.5.8) reduce to Eq. (6.2.16) for rigid end connections with R_{ktA} and R_{ktB} approach infinity.
2. If gradual stiffness degradation due to spread-of-plasticity effects in the element is considered, then

$$S_{ii} = \left[S_1 - \frac{S_2^2}{S_1}(1 - \phi_B)\right]\phi_A \tag{9.5.10}$$

$$S_{jj} = \left[S_1 - \frac{S_2^2}{S_1}(1 - \phi_A)\right]\phi_B \tag{9.5.11}$$

$$S_{ij} = \phi_A \phi_B S_2 \tag{9.5.12}$$

   and the elastic modulus E in Eqs. (9.5.7) and (9.5.8) should be replaced by the tangent modulus E_t.

The internal degrees of freedom can be condensed by solving Eqs. (9.5.7) and (9.5.8) for $\dot{M}_A$ and $\dot{M}_B$, which give

$$\dot{M}_A = \frac{E_t I}{L}[S_{ii}^* \dot{\theta}_A + S_{ij}^* \dot{\theta}_B] \tag{9.5.13}$$

$$\dot{M}_B = \frac{E_t I}{L}[S_{ij}^* \dot{\theta}_A + S_{jj}^* \dot{\theta}_B] \tag{9.5.14}$$

where

$$S_{ii}^{*} = \frac{S_{ii} + \dfrac{E_t I S_{ii} S_{jj}}{L R_{ktB}} - \dfrac{E_t I S_{ij}^2}{L R_{ktB}}}{\left(1 + \dfrac{E_t I S_{ii}}{L R_{ktA}}\right)\left(1 + \dfrac{E_t I S_{jj}}{L R_{ktB}}\right) - \left(\dfrac{E_t I}{L}\right)^2 \left(\dfrac{S_{ij}^2}{R_{ktA} R_{ktB}}\right)} \tag{9.5.15}$$

$$S_{jj}^{*} = \frac{S_{jj} + \dfrac{E_t I S_{ii} S_{jj}}{L R_{ktA}} - \dfrac{E_t I S_{ij}^2}{L R_{ktA}}}{\left(1 + \dfrac{E_t I S_{ii}}{L R_{ktA}}\right)\left(1 + \dfrac{E_t I S_{jj}}{L R_{ktB}}\right) - \left(\dfrac{E_t I}{L}\right)^2 \left(\dfrac{S_{ij}^2}{R_{ktA} R_{ktB}}\right)} \tag{9.5.16}$$

$$S_{ij}^{*} = \frac{S_{ij}}{\left(1 + \dfrac{E_t I S_{ii}}{L R_{ktA}}\right)\left(1 + \dfrac{E_t I S_{jj}}{L R_{ktB}}\right) - \left(\dfrac{E_t I}{L}\right)^2 \left(\dfrac{S_{ij}^2}{R_{ktA} R_{ktB}}\right)} \tag{9.5.17}$$

The incremental axial-force-displacement equation is

$$\dot{P} = \frac{E_t A}{L} \dot{e} \tag{9.5.18}$$

Finally, the tangent-stiffness relationship may be written as (Liew et al., 1993a, 1993b):

$$\begin{Bmatrix} \dot{M}_A \\ \dot{M}_B \\ \dot{P} \end{Bmatrix} = \frac{EI}{L} \begin{bmatrix} S_{ii}^{*} & S_{ij}^{*} & 0 \\ S_{ij}^{*} & S_{jj}^{*} & 0 \\ 0 & 0 & A/I \end{bmatrix} \begin{Bmatrix} \dot{\theta}_A \\ \dot{\theta}_B \\ \dot{e} \end{Bmatrix} \tag{9.5.19}$$

Equation (9.5.19) is the modified element tangent-stiffness equations that account for both the effects of distributed plasticity and connection flexibility on the element. They can be transformed to global coordinates to obtain the global element tangent-stiffness relationships; the details are described in Liew (1992).

Equation (9.5.19) accounts for the fact that the connection stiffness at both ends of the element need not be equal. In other words, this formulation allows the consideration of cases where the axial force P in the element is appreciable and the connections at the ends of the element may behave differently.

9.5.3 Plastic Hinges

In plastic-hinge analysis of semi-rigid frames, a plastic hinge will develop if the cross-section forces in a member is equal to the cross-section plastic resistance. Once a plastic hinge is formed (neglecting strain hardening), the subsequent change in the plastic moment capacity due to the change in axial force will affect the force-displacement relationships of the beam-column element. If the axial force is increased, the force point should move on the plastic strength surface. Analysis of semi-rigid frames, taking into consideration of concentrated plasticity in regions of plastic hinges, can be carried out by modifying the basic stiffness relationship as described in Chapter 6. However, if the connection reaches its ultimate capacity before a plastic hinge is developed in the connected members, a connection hinge is formed. The effect of this connection hinge must be accounted for by modifying Eq. (9.5.19) as follows:

1. If a connection hinge forms in connection A, set $R_{ktA} = 0$.
2. If a connection hinge forms in connection B, set $R_{ktB} = 0$.
3. If connection hinges form in both connections, set $R_{ktA} = 0$ and $R_{ktB} = 0$.

9.6 RECOMMENDED DESIGN PROCEDURE

In general, the design of a semi-rigid frame must consider the following factors that will eventually affect the performance of the structural system:

1. *Connection Stiffness, Strength, Rotation Capacity, Ductility, and Energy-Absorption Capability.* These are the critical factors for many structures, including rigidly joined frames. They are especially important for structures located in seismic regions.
2. *Serviceability of the Structure and Its Connections.* The prime considerations here are the deformations and other stiffness-related characteristics of the connections, and their effects on the service-load deformations of the frame.
3. *Strength and Stability of the Frame and Its Component Members.* The effects of a partial connection restraint on the ultimate strengths of the members and frame should be considered. The structure is acceptable with respect to strength if the nominal resistance of the structure is greater than the factored-load effects.

A design procedure for semi-rigid frames based on advanced inelastic analysis is proposed as follows:

1. Perform the preliminary analysis/design steps. A wind-moment method may be used to proportion members and connections. For an unbraced frame, the wind-moment method utilizes the connection stiffness to provide resistance to wind; but such resistance is ignored for proportioning beams under gravity loads. In its usual form, the method assumes the following: (a) under gravity load, the connections act as pins (Fig. 9.24a), and (b) under wind-load combinations, the connections behave as rigid joints, with points of contraflexure at the midheight of columns and the midlength of beams (Fig. 9.24b). The advantage of the method

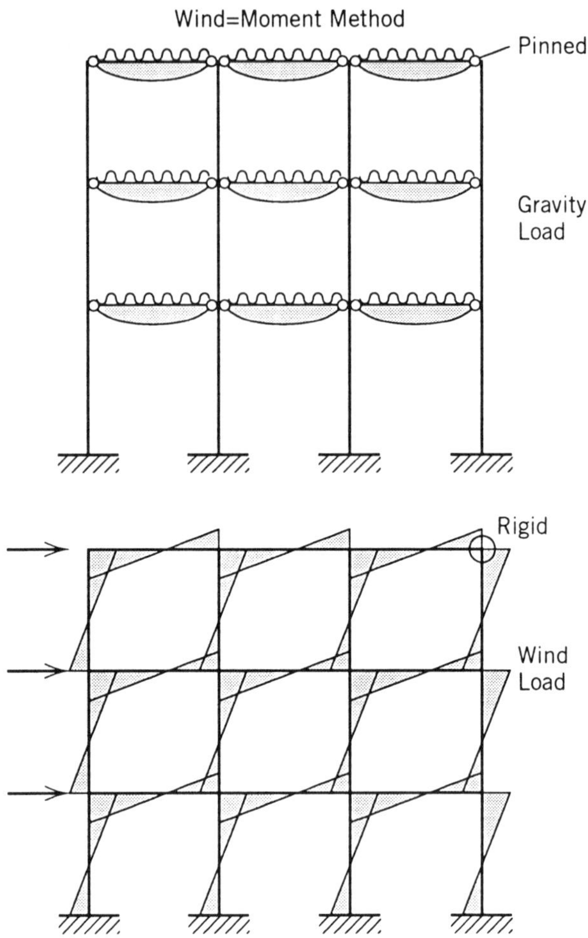

Fig. 9.24 Analysis model using the wind-moment method. (a) Gravity-load case. (b) Wind-load case.

is simplicity. As the frame is rendered statically determinate, internal moments and forces are not dependent on the relative stiffnesses of the members. Thus, member proportioning can be carried out conveniently without the use of computer analysis. The wind-moment method is also designated as "Type 2 Construction" by the AISC. The wind-moment method tends to overdesign beams and underdesign columns and connections. Beam design is usually governed by the sagging internal moment due to gravity load. The semi-rigid connections cause hogging moments at the supports, thus reducing the maximum sagging moment in the beam. The columns tend to be underdesigned because of the detrimental effect on the columns due to the hogging moments developed in the beams. These moments particularly affect external columns and other members subject to unbalanced loading, but the effect is not as significant as the overdesign of beams because the columns are also designed to support axial loads. The connections will generally be underdesigned using the wind-moment method, because of the second-order effects associated with semi-rigid sway frames, which are not considered in the analysis.

2. Select the type of semi-rigid connection to be used in the construction. The connection could be any type of standard connections, such as double web angles, single web angles, top and seat angles with or without double web angles, header plates, flushed or extended end plates, and so on, as shown in Fig. 9.25. Based on the information in step 1, make an initial guess of the various dimensions and thicknesses of the connecting material, and check the connections for bolt tension, bolt shear, and stresses in the connecting plates, angles, welds, and so on.
3. Determine the ultimate moment capacity and initial stiffness of the connection. The approximated shape of the moment–rotation curve of the connection should be known in advance. Because the currently available data do not include all conceivable types of connections, it is advisable to restrict connection designs to those for which M–θ_r curves have been well established, such as the connection types shown in Fig. 9.25. The AISC LRFD specification (AISC, 1993), however, allows use of any connection as long as its behavior can be demonstrated by either tests or by rational means.
4. Check connection ductility. This in essence requires that all connection materials such as bolts, welds, and plates not to be loaded beyond the resisting strengths under the simultaneous actions of factored gravity and wind loads.
5. Perform second-order inelastic analysis incorporating the effects of semi-rigid connections in the factored-load analysis.
6. Check the strength limit states. The structure is said to be satisfactory if its nominal resistance is greater than the factored-load effects.
7. Check the serviceability limit states. This is to ensure adequate stiffness for functionality at service-load levels.

The need to repeat the analysis to correspond to changed section sizes is unavoidable if one proceeds with second-order semi-rigid frame analysis. The connection designs need to be modified if the design resistance of the structure is greater than the factored-load effects. Alternatively, the member sizes may be adjusted to provide a more efficient design. Iteration of steps 2 to 7 gives results that will converge to the most cost-effective design satisfying the various design constraints imposed on the analysis.

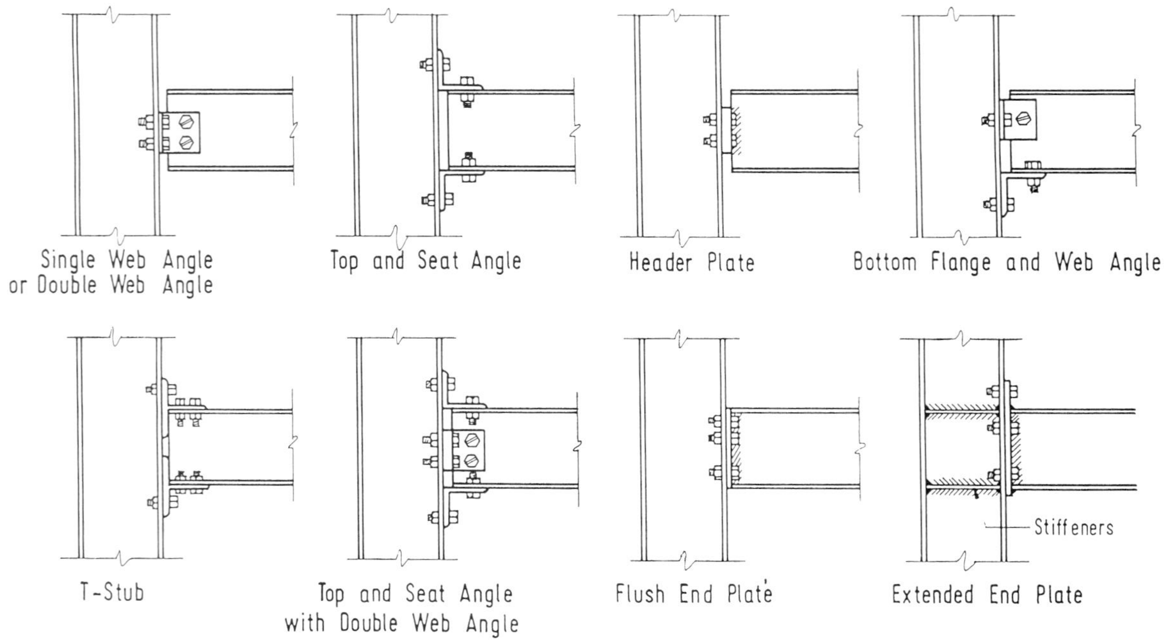

Fig. 9.25 Different types of beam-column connections.

9.7 A DESIGN EXAMPLE

9.7.1 Description of Frame and Loading

The geometry, dimensions, and nominal loads of the frame are shown in Fig. 9.26. The gravity loads are assumed to include both the contributions from dead and live loads. It is further assumed that the dead and live loads are of equal proportion, and the column sizes are kept the same for the top and bottom stories of the frame. For this two-story frame example, the preliminary beam sizes were chosen based on first-order plastic-hinge analysis assuming rigid frame action. The preliminary column sizes were selected according to the tributary gravity load supported by each column. After few cycles of design iterations using advanced inelastic analysis, the least-weight members satisfying the strength limit states are obtained. The member sizes are shown in Fig. 9.26.

The following assumptions are made for the analysis and design of the frame members and connections:

1. All members are made of ASTM A36 steel.
2. All girders are connected to the columns with their webs in the plane of the frame by top- and seat-angle connections with double web angles,

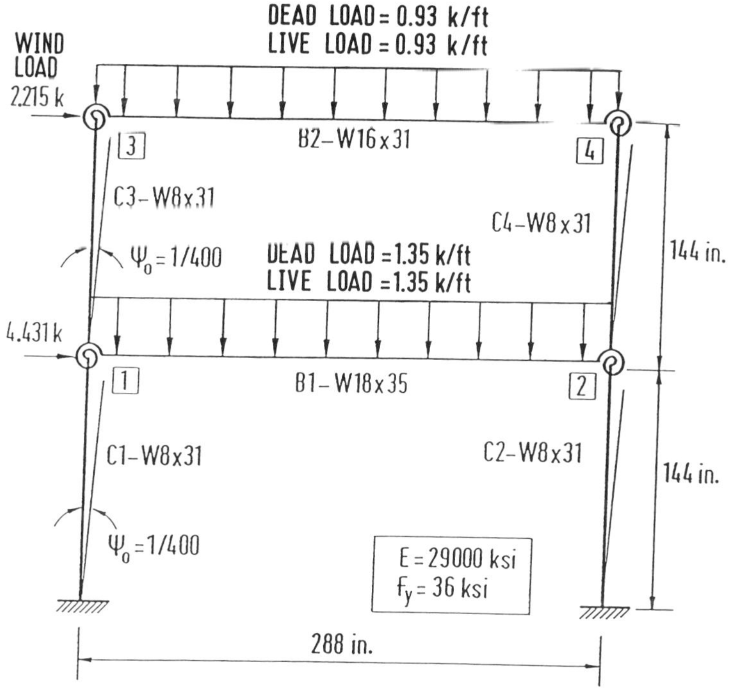

Fig. 9.26 Dimensions and loading of a two-story planar frame.

as shown in Fig. 9.27. The angles are fully bolted to the connection members. All connections should satsify the design requirements given in Section 9.3 in accordance with the AISC LRFD specifications (AISC, 1993).

3. All beams are adequately braced in the out-of-plane direction by the floor slabs.
4. Gravity loads are applied as concentrated loads at the beam quarter points.
5. Wind loads are applied as concentrated loads at the exterior joints.
6. Member lengths are defined by centerline dimensions.
7. Only bare steel frame behavior is assumed.

9.7.2 Connection Details

The three-parameter power model described in Section 9.2 is used to approximate the M–θ_r curves of the connections. The connection initial stiffness and ultimate moment capacity are computed based on the steel properties and the geometrical variables of the connection (Fig. 9.27) as follows:

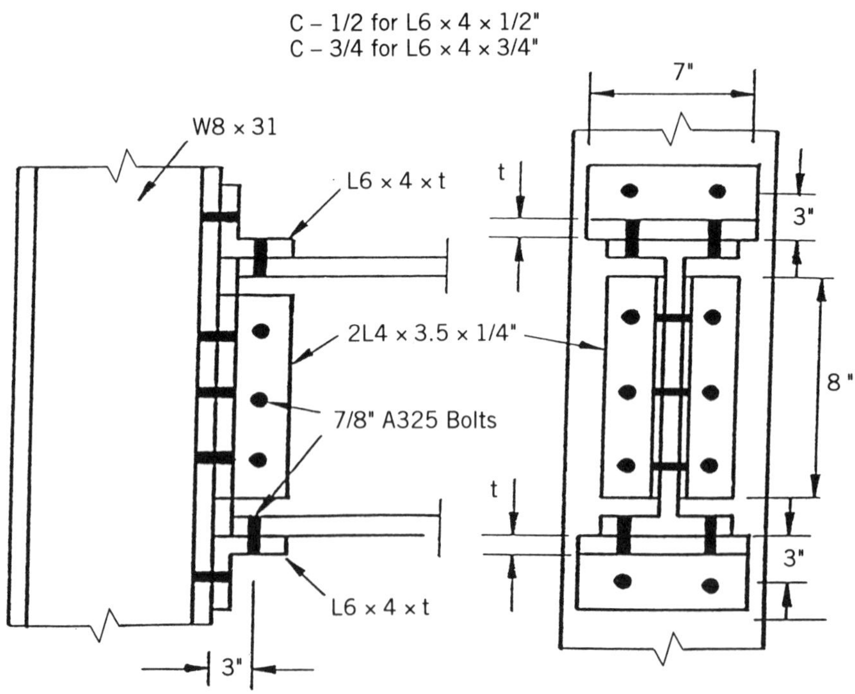

Fig. 9.27 Details of beam-to-column connections.

1. The thickness t_t and t_s of the top and seat angle, respectively. In the present example, $t_s = t_t$.
2. The thickness t_w of the web angle.
3. The angle lengths l_t, l_s, and l_w, which may be selected based on the column width, beam depth, and bolt gauges g_{ct} and g_{cs} (see Fig. 9.27).
4. The beam depth d.
5. The nut width W, which is dependent on the bolt diameter used in the connection. In the present example, A325 high-strength bolts of 7/8-inch diameter are used for all connections.

The details of all the connection asemblage are shown in Fig. 9.27. The resulting geometrical dimensions of a set of possible semi-rigid connections are summarized in Table 9.1. The connection type is denoted in the table by the thickness of the top and seat angle as C-3/8, C-1/2, and so on. The initial stiffnesses and ultimate moment capacities of these connections are shown in Table 9.2. The nondimensional moment–rotation curves of these connections are shown in Figs. 9.28a and 9.28b. The following illustrates how the initial stiffness and ultimate moment capacity of the C-3/8 connection are computed based on the step-by-step procedures described in Section 9.6. The computer program given in Chapter 11 may also be used to generate the required parameters for the analysis of semi-rigid frames.

9.7.2.1 Calculation of Initial Stiffness R_{ki} for Connection C-3/8 For $g_{ct} = 3$ in., $l_t = 7$ in., $t_t = 3/8$ in., $k_t = 7/16$ in., and $W_t = 1\text{-}7/16$ in., determine

$$\beta_t = \frac{g_{ct}}{l_t} = 0.42857, \qquad \gamma_t = \frac{l_t}{t_t} = 18.667, \qquad \kappa_t = \frac{k_t}{t_t} = 1.1667,$$

$$\omega_t = \frac{W_t}{t_t} = 3.8333$$

Similarly, for $g_{cw} = 3$ in., $l_w = 8$ in., $t_w = 1/4$ in., $k_w = 11/16$ in., and $W_w =$ 1-7/16 in., determine

$$\beta_w = \frac{g_{cw}}{l_w} = 0.375, \qquad \gamma_w = \frac{l_w}{t_w} = 32.0, \qquad \kappa_w = \frac{k_w}{t_w} = 2.75,$$

$$\omega_w = \frac{W_w}{t_w} = 5.75$$

Evaluate

TABLE 9.1 Connection Details

Connection Type	Bolt Diameter d (in.)	Nut Width W (in.)	Top and Seat Angle $2L6 \times 4 \times t_t$				Web Angles $L4 \times 3.5 \times 1/4$		
			t_t (in.)	l_t (in.)	g_{ct} (in.)	k_t (in.)	g_{cw} (in.)	l_w (in.)	k_w (in.)
C-3/8	7/8	1-7/16	3/8	7	3	7/8	3	8	11/16
C-1/2	7/8	1-7/16	1/2	7	3	1	3	8	11/16
C-9/16	7/8	1-7/16	9/16	7	3	1-1/16	3	8	11/16
C-5/8	7/8	1-7/16	5/8	7	3	1-1/8	3	8	11/16
C-3/4	7/8	1-7/16	3/4	7	3	1-1/4	3	8	11/16

TABLE 9.2 Connection Parameters

Connection Type	Beam W18 × 35 M_u (kips-in.)	Beam W18 × 35 R_{ki} (kips-in./rad)	Beam W18 × 35 n	Beam W16 × 31 M_u (kips-in.)	Beam W16 × 31 R_{ki} (kips-in./rad)	Beam W16 × 31 n
C-3/8	606	107,548	1.49	543	86,980	1.55
C-1/2	903	254,227	1.20	814	205,924	1.27
C-9/16	1,097	369,595	1.10	990	299,599	1.16
C-5/8	1,313	520,528	1.00	1,186	422,267	1.06
C-3/4	1,773	954,014	0.81	1,606	775,080	0.88

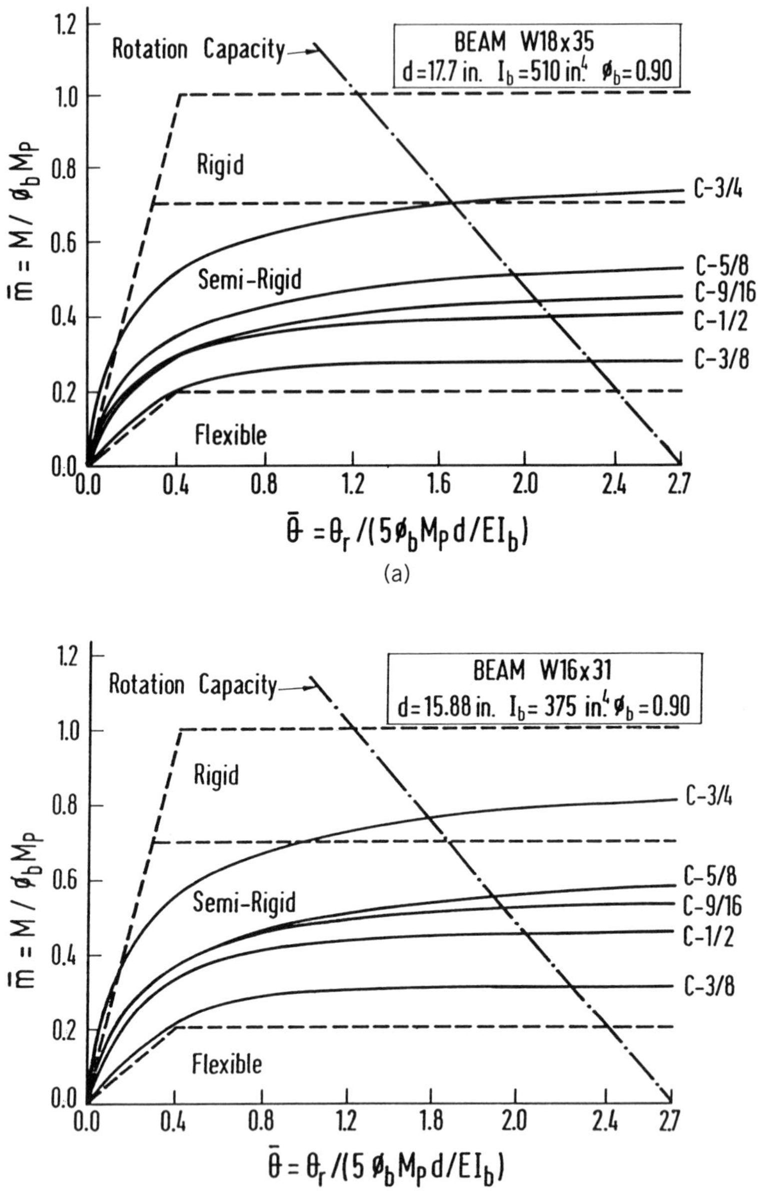

Fig. 9.28 Classification of connections: (a) At the lower floor beam (W18 $\times$ 35). (b) At the upper floor beam (W16 $\times$ 31).

$$\beta'_t = \beta_t - \frac{1}{2\gamma_t}(1 + \omega_t) = 0.2991$$

$$\beta'_w = \beta_w - \frac{1}{2\gamma_w}(1 + \omega_w) = 0.2695$$

For beam B1 (W18 × 35) with $d = 17.7$ in., evaluate $\delta_t = d/t_t = 47.2$. Determine

$$D_{ts} = \frac{3}{\beta'_t[\gamma_t^2(\beta'_t)^2 + 0.78]} = 0.313$$

and

$$D_w = \frac{3}{\beta'_w[\gamma_w^2(\beta'_w)^2 + 0.78]} = 0.074$$

Evaluate $\rho = t_w/t_t = 0.6667$, $I_{0t} = t_t^3/12 = 4.3945 \times 10^{-3}$ in.^{3}, and finally compute

$$R_{ki} = EI_{0t}(D_{ts} + \rho D_w)(1 + \delta_t)^2 = 107{,}548 \text{ kips-in.}$$

9.7.2.2 Calculation of Ultimate Moment M_u for Connection C-3/8 For $\beta_w = 0.375$, $\gamma_w = 32.00$, and $\kappa_w = 2.75$, determine

$$\beta_w^* = \beta_w\gamma_w - \kappa_w = 9.25$$

Evaluate ξ_w iteratively from Eq. (9.3.4) of which $\xi_w = 0.1081$.

For $\rho = 0.6667$, solve for

$$\frac{M_{uw}}{M_{0t}t_t} = \gamma_w(1 + \xi_w)\rho^3\left[\frac{\gamma_w(\xi_w - 1)}{3(\xi_w + 1)} + \delta_w + \frac{1}{\rho}\right] = 670$$

Similarly, for $\beta'_t = 0.2991$, $\gamma_t = 18.667$, and $\kappa_w = 1.16667$, determine

$$\beta_t^* = \beta'_t\gamma_t - \kappa_t = 4.4167$$

Evaluate ξ_t from Eq. (9.3.10) of which $\xi_t = 0.226$.

For $\delta_t = 47.2$, solve for

$$\frac{M_{uts}}{M_{0t}t_t} = \gamma_t(1 + \xi_t[1 + \beta_t^* + 2(\kappa_t + \delta_t)]) = 450$$

Evaluate $M_{0t}t_t = 0.4746$, and finally solve for

$$M_u = M_{0t}t_t\left(\frac{M_{uts}}{M_{0t}t_t} + \frac{M_{uw}}{M_{0t}t_t}\right) = 532 \text{ kips-in.}$$

Following these procedures, the R_{ki} and M_u parameters for the other connection types are obtained as shown in Table 9.2.

9.7.2.3 Connection Classifications All angle connections given in Table 9.1 are classified using the Bjorhovde et al. (1990) system of classification, and the results are shown in Figs. 9.28a and 9.28b. These connections can all be classified as semi-rigid both in terms of strength and stiffness. The rotational responses of the connections are highly nonlinear, and their resistances vary significantly, covering practically the entire semi-rigid range of connection behavior. The nondimensional $\overline{m}$–$\overline{\theta}$ curves for the upper- and lower-story beam connections are different because of the difference in beam depths.

9.7.3 Design Requirements

The frame shown in Fig. 9.26 is designed to satisfy the following requirements:

9.7.3.1 Ultimate Strength The frame is designed using the following load combinations (AISC, 1993):

$$1.2 \text{ Dead Load} + 1.6 \text{ Live Load} \tag{9.7.1}$$

$$1.3 \text{ Wind Load} + 1.2 \text{ Dead Load} + 0.5 \text{ Live Load} \tag{9.7.2}$$

The refined plastic-hinge analysis is carried out based on the factored cross-section strengths ϕP_y and ϕM_p with the LRFD. The LRFD tangent-modulus model is used in the refined plastic-hinge analysis when the compressive axial load in the column exceeds $0.39P_y$ (see Chapter 8). The resistance factor ϕ used for the cross-section design strengths is 0.90, which is same as that recommended by the Australian limit-states code (Standards Australia, 1990) for use in the plastic-zone analysis of structural steel frames. Because the LRFD tangent-modulus model is used for advanced inelastic analysis, the modeling of member initial out of straightness is not required. This approach simplifies the analysis/design procedures considerably, because explicit modeling of column initial out of straightness can be a cumbersome task for large structural systems. However, it should be noted that all the columns in this particular frame are not affected by the column tangent-modulus model. This is because column axial forces at the factored-load level are always less than $0.39P_y$. Similarly, if one wishes to use the notional-load plastic-hinge analysis, it is not necessary to apply member notional load as the design axial forces in these columns are small.

All columns in the frame are assumed to have an initial out of plumbness of $h/400$, where h is the story height. This column out-of-plumbness magnitude is obtained from ECCS provisions (ECCS, 1991) for use with second-order plastic-zone analysis of steel frames.

To ensure that the connection deformations are not unacceptably large, the relative joint rotations at the beam ends, which are computed from the inelastic analysis, are checked against the rotation capacities of the connections. The simplified expression given by Eq. (9.4.17) is used to calculate the connection rotational capacity.

The M_u values for the connections shown in Table 9.2 are not reduced by a design ϕ factor. Naturally, suitable resistance factors have to be developed based on the variability in the ultimate response of connections among other factors, such as connection strain-hardening effects, to achieve a more uniform reliability for the structure as a whole.

9.7.3.2 Serviceability The following requirements are imposed for determining the serviceability of the structure:

1. Service beam live-load deflections are limited to $L/360$. The service gravity-load combination used for computing the beam deflection is given as

$$1.0 \text{ Dead Load} + 0.8 \text{ Live Load} \qquad (9.7.3)$$

2. Lateral roof deflection of the frame under service wind load is limited to $H/400$, where H is the frame height. The service wind-load combination is given as

$$1.0 \text{ Dead Load} + 0.5 \text{ Live Load} + 1.0 \text{ Wind Load} \qquad (9.7.4)$$

   The gravity-load components are included in the drift analysis because of their contribution to the increase in connection deformation and the decrease in connection stiffness during the loading process.
3. Maximum interstory drift due to the service-load combination in Eq. (9.7.4) is limited to $h/250$, where h is the story height.
4. Plastic-hinge formation under service loads is prohibited to prevent the possibility of excessive localized damage before the full service load is reached.

9.7.3.3 Other Requirements The following additional requirements are imposed to satisfy typical conditions of practice:

1. For members with significant inelastic rotation upon yielding, their cross-section compactness is checked based on the ASCE plastic design guide (ASCE-WRC, 1971) to ensure that local buckling does not limit the member's rotation capacity. For I-sections, wide-flange beams, and welded H-shapes, the limitation on flange and web proportions for plas-

tic design sections (or Class 1 section) are as follows (ASCE-WRC, 1971):

Flanges in compression due to bending or axial force, or both (A36 steel):

$$\frac{b}{T} = 17 \tag{9.7.5}$$

Web under compression:

$$\frac{d}{t} = 43\sqrt{\frac{36}{F_y}} \tag{9.7.6}$$

Web under bending and compression:

$$\frac{d}{t} = 43\sqrt{\frac{36}{F_y}} \qquad \text{for } \frac{P}{P_y} > 0.27 \tag{9.7.9}$$

$$\frac{d}{t} = \left(70 - 100\frac{P}{P_y}\right)\sqrt{\frac{36}{F_y}}, \qquad \text{for } \frac{P}{P_y} \leq 0.27 \tag{9.7.10}$$

where

b = half the width of flange plate
d = section depth
F_y = design strength in ksi
t = thickness of web
T = thickness of flange

The provision given by ASCE WRC (1971) is more restrictive than those given in the AISC LRFD specification (AISC, 1993).

2. Beam-columns containing a plastic hinge are checked for adequate rotation capacity by limiting the lateral-brace spacing according to the LRFD plastic design provision, which is given as

$$L_b \leq \left(0.124 + 0.0759\frac{M}{M_p}\right)\frac{Er_y}{F_y} \tag{9.7.7}$$

where L_b is the maximum distance between the points of restraint. M/M_p is the ratio of end moments in which one end reaches a full plastic capacity M_p. The ratio is positive for double-curvature and negative for single-curvature bending.

3. Beam-column elements are checked by the following analytical expression to ensure that peak second-order moments occur at the member

end (Liew et al., 1991):

$$\frac{x}{L} = -\frac{1}{kL}\tan^{-1}\left[\frac{(M_1/M_2)\cos(kL) + 1}{(M_1/M_2)\sin(kL)}\right] > 1.0 \text{ or } < 0 \quad (9.7.8)$$

where x is the location of the maximum moment along the length of the member, L is the member length, $k = \sqrt{P/E_tI}$, P is the member axial load, and M_1 and M_2 are the smaller and larger of the beam-column end moments obtained from advanced analysis. M_1/M_2 is positive when the member is bent in double curvature, and negative when it is bent in single curvature. If the peak moment occurs in between the member ends, the frame is analyzed again by using two elements per member with a node inserted at the peak moment point.

4. All column sizes are kept the same in the design.
5. In the preliminary design, only one connection type available in Table 9.2 is used for all joints in the frame. The least-weight connection type that satisfies all the design requirements is selected. The suitability of the final design is then confirmed by studying the detailed response of the structure obtained by advanced analysis.

9.7.4 Summary of Results

Results showing the characteristics of the two-story frame with different connection configurations are summarized in Tables 9.3–9.6. For comparison, the pinned- and rigid-connection cases are also included in the studies. The results obtained from the inelastic analyses are compared with respect to (1) the applied load ratio at the formation of the first plastic hinge and at the limit of resistance of the frame; (2) the maximum beam deflection under the service gravity load, (3) the maximum lateral frame and interstory drift under the service wind-load combination, and (4) the rotation demands of connections under the factored gravity load.

TABLE 9.3 Applied Load Ratios for Frames with Semi-Rigid Connections

	1.2 Dead Load + 1.6 Live Load	
Connection Type	First Plastic Hinge	Limit Load
Pin	—	0.659
C-3/8	0.831	0.960
C-1/2	0.890	1.057
C-9/16	0.900	1.111
C-5/8	0.929	1.134
C-3/4	1.021	1.152
Rigid	1.143	1.157

TABLE 9.4 Maximum Beam Deflections under the Service Gravity Load

Connection Type	Beam Type	Section	Deflection (in.)
Pin	—	—	—
C-3/8	B1	W18 × 36	0.99
C-1/2	B1	W18 × 36	0.88
C-9/16	B1	W18 × 36	0.87
C-5/8	B1	W18 × 36	0.83
C-3/4	B1	W18 × 36	0.73
Rigid	B2	W16 × 31	0.66

Deflection at the service load is limited to $L/180 = 1.6$ in.

9.7.4.1 Applied Load Ratios The applied load ratios correspond to the formation of first plastic hinge and to the limit of resistance of the structure are shown in Table 9.3. The applied load ratio is the value of the applied load divided by the respective factored design load. An applied load ratio of 1.0 corresponds to the application of the full factored load. For this particular frame structure, the factored wind-load combination does not control the strength limit of the frame; therefore, only the results obtained from the gravity-load analysis are presented.

As expected, the limit load of the fully rigid frame is greater than that of the semi-rigid frames. For frames subjected to the gravity-load combination, only the semi-rigid frames with connection rigidity greater than that implied by C-3/8 reach their maximum strengths at an applied load ratio greater than 1.0. The maximum strength for the fully rigid frame is approximately 75% greater than that of the pin-connected frame, and the load ratios for the semi-rigid frames range from 80.6 to 99.6% of that for the fully rigid frame.

The load at the formation of the first plastic hinge is expressed as a ratio of the applied load to the full factored load. The plastic-hinge formation is determined using the design cross-section strength surface with resistance factor $\phi = 0.90$. For all semi-rigid frames (denoted by the connection types),

TABLE 9.5 Maximum Lateral Drifts under the Service Wind Load

Connection Type	Frame Drift (in.)	Interstory Drift (in.)
Pin	1.742	1.478
C-3/8	1.111	0.566
C-1/2	0.896	0.490
C-9/16	0.885	0.488
C-5/8	0.824	0.460
C-3/4	0.669	0.399
Rigid	0.531	0.342

The maximum frame drift is limited to $H/400 = 0.72$ in. The maximum interstory drift is limited to $h/250 = 0.576$ in.

TABLE 9.6 Connection Rotational Demands and Capacities at the Factored Gravity Load

Connection No.	Member Type	Section	Demand, $\bar{\theta}_d$	Capacity, $\bar{\theta}_c$	$\bar{\theta}_d / \bar{\theta}_c$
		Connection Type C-1/2			
1	B1	W18 × 36	8.791	2.102	4.182
2	B1	W18 × 36	8.955	2.102	4.260
3	B2	W16 × 31	0.780	2.024	0.385
4	B2	W16 × 31	1.037	2.024	0.512
		Connection Type C-9/16			
1	B1	W18 × 36	5.578	2.043	2.730
2	B1	W18 × 36	5.705	2.043	2.792
3	B2	W16 × 31	0.515	1.920	0.268
4	B2	W16 × 31	0.582	1.920	0.303
		Connection Type C-5/8			
1	B1	W18 × 36	2.101	1.941	1.082
2	B1	W18 × 36	2.202	1.941	1.134
3	B2	W16 × 31	0.419	1.873	0.224
4	B2	W16 × 31	0.465	1.873	0.248
		Connection Type C-3/4			
1	B1	W18 × 36	0.405	1.654	0.245
2	B1	W18 × 36	0.438	1.654	0.265
3	B2	W16 × 31	0.202	1.561	0.129
4	B2	W16 × 31	0.212	1.561	0.136

the first plastic hinge forms after the application of the service load. More specifically, for gravity-loaded frames, the first plastic hinge forms shortly after the application of the service load. For wind-load combinations, all semi-rigid frames achieve load factors greater than 1.0 prior to the first occurrence of a plastic hinge in any member.

9.7.4.2 Maximum Beam Deflections The maximum beam deflections under the service gravity load are shown in Table 9.4. For all cases, the maximum beam deflection is significantly lower than the deflection limit criteria of $L/180$. Therefore, the beam serviceability limit is not the controlling factor for the frame.

9.7.4.3 Maximum Frame Drifts In the refined plastic-hinge analyses, proportional loading is used where the loads are applied incrementally. The maximum frame and interstory drifts under service-load combinations, 1.0 Dead

Load + 0.5 Live Load + 1.0 Wind Load, are evaluated in Table 9.5. The results indicate that the frames with semi-rigid connections have maximum frame drifts in the order of 1.26 to 2.31 times that for the frame with rigid connections. Also, the interstory drift increases as the connection thickness of the top and seat angles reduces from 3/4 to 3/8 inch. Finally, except for the frame with pinned connections, all other semi-rigid frames satisfy the interstory drift criterion of $h/250$ imposed by the design requirements. Only the frames with a connection rigidity greater than or equal to that implied by C-3/4 satisfy the drift limit of $H/400 = 0.72$ inch.

9.7.4.4 Connection Rotation Demands Table 9.6 compares the normalized connection rotation demand, $\bar{\theta}_d$, at the beam ends with the required normalized rotation capacity $\bar{\theta}_c$, which is evaluated based on Eq. (9.4.17) as follows:

$$\bar{\theta}_c = \frac{1}{2}\left(5.4 - \frac{3M_u}{M_p}\right) \tag{9.7.6}$$

where M_u is the connection ultimate moment capacity, and M_p is the full plastic moment capacity of the beam.

The results in Table 9.6 show that for beam B_2, the normalized rotation demand of the connection, $\bar{\theta}_d$, is well below the rotation capacity angles, $\bar{\theta}_c$ for both semi-rigid frames. But for the lower-story beam B_1, the former is greater than the latter for connection type C-1/2. By inspection, the lightest connections in Table 9.1 that satisfy the rotation capacity requirements for the connections in beams B_1 and B_2 are C-3/4 and C-1/2, respectively. Thus, it is decided to redesign the frame by adopting connections C-3/4 for the lower-story beam and C-1/2 for the upper-story beam.

9.7.4.5 Final Analysis and Design The final selection of connections for the two-story frame is as follows:

1. Connections for the lower-story girder: the top- and seat-angle thickness of this connection type is 3/4 inch.
2. Connections for the upper-story girder: the top- and seat-angle thickness of this connection is 1/2 inch.

Load-displacement curves for the top and bottom floors are shown in Fig. 9.29. As in the previous case studies, the final design is controlled by the factored gravity-load combination. Therefore, only the detailed results for this load combination are presented.

The highly nonlinear response of the frame as the factored gravity load is applied is mainly due to the nonlinear response of the connections subjected to increasing end moments from the beams. The sudden increase in deflections at an applied load ratio slightly above 1.0 indicates that the frame failed by sidesway instability soon after the application of the factored gravity load.

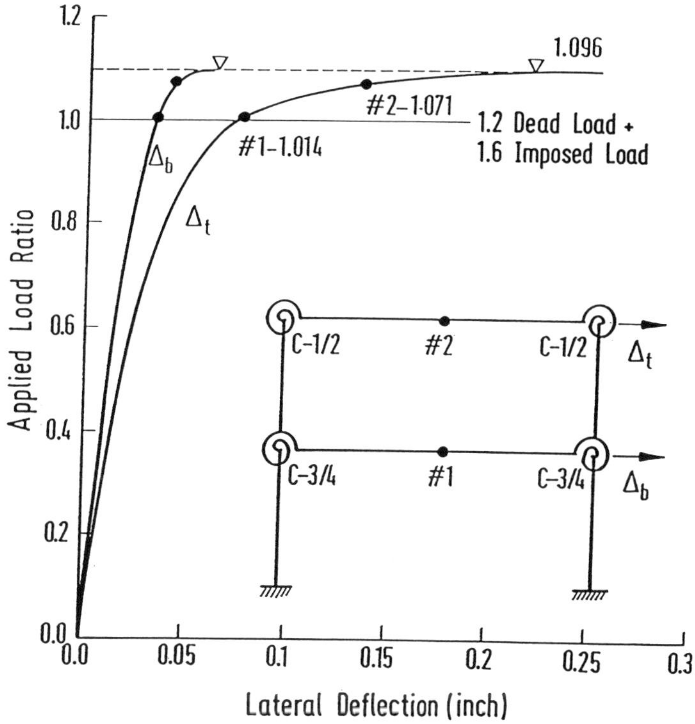

Fig. 9.29 Lateral load-displacement traces for the top and bottom floors.

The plastic-hinge-formation sequence is also shown in the load-displacement diagram. Only two plastic hinges are detected by the refined plastic-hinge analysis, which accounts for gradual stiffness degradation due to distributed plasticity in the members. The maximum load of this frame occurs at an applied load ratio of 1.096, which is about 10% higher than the full factored gravity load. The frame satisfies all the serviceability requirements. The first plastic hinge forms at an applied load ratio of 1.014, which is slightly above the full factored gravity load. The maximum frame and interstory drifts at service-load combinations are 0.702 and 0.401 inch, respectively. These lateral drifts are slightly below the frame and interstory drift limits of $H/400 = 0.72$ inch and $h/250 = 0.576$ inch, respectively. The ductility demands of the connections at the factored gravity loads are far below the connection capacities predicted by Eq. (9.4.17).

Moment diagrams and axial forces at the factored gravity load are shown in Fig. 9.30. The axial forces at the left and right columns of the frame are approximately equal. The maximum value of axial force in the column is $0.26P_y$; therefore, distributed yielding due to axial-force actions in the columns is small for this frame. The hogging moments at the beam ends reduce

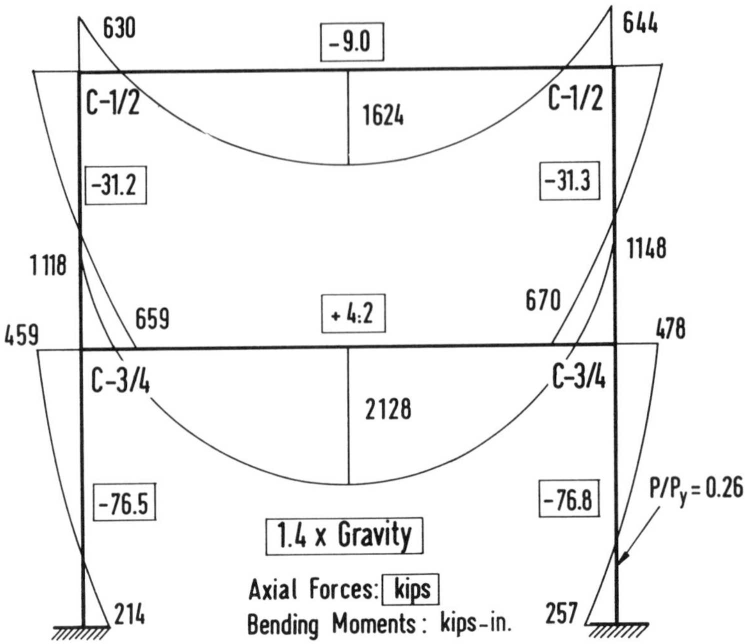

Fig. 9.30 Internal force distribution at the factored gravity load (1.4 × Gravity).

the beam deflection considerably, thus, permitting a better exploitation of the beam-section plastic strength. It is observed that the inelastic actions in the frame are concentrated primarily at the midspan of the beams.

9.8 EFFECTS OF SUPPORT FIXITY

The frame in Section 9.7 is designed based on the assumption that the column bases are fully fixed. In practice, the column bases are not fully rigid unless a massive foundation is constructed. To investigate the effect of support conditions on the frame behavior, the semi-rigid frame in Fig. 9.26 is reanalyzed by assuming the supports are pinned or with elastic rotational spring. For the elastic spring-support case, the supports are modeled by two linear-elastic springs with a spring constant R_{ki} of $2.5(EI/5d)_c$, where subscript c refers to the column.

Figure 9.31 shows the load-displacement curves of the moment resistant frame with different support conditions. An applied-load ratio of 1.0 in Fig. 9.31 corresponds to the fully factored wind-load combination of 1.2 Dead Load + 1.3 Wind Load + 0.5 Live Load. For the pinned-supported frame, the factored wind-load combination controls the strength limit of this frame. As expected, the maximum load of the pinned-supported frame is smaller

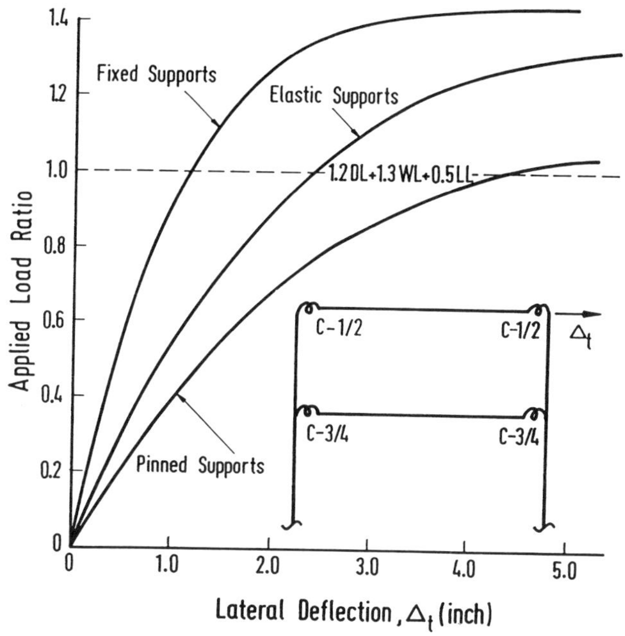

Fig. 9.31 Lateral-load-displacement traces for unbraced frames with different support fixities.

than the fixed-supported frame. The reduction in maximum strength is approximately 35%.

The lateral roof deflections of the frame at the service-load combinations, as obtained from the inelastic analysis, are summarized as follows:

- Fixed-base frame: 0.702 inch (less than $H/400 = 0.72$ inch)
- Elastic-supported frame: 1.452 inches (greater than 0.72 inch)
- Pinned-base frame: 2.256 inches (greater than 0.72 inch)

These results suggest that the frames with base-support conditions other than fully rigid may not satisfy the drift limit state. The lateral roof deflection of the frame with elastic supports is more than twice that of the frame with fixed supports at the service-load level. For the frame with pinned supports, the lateral frame drift increases dramatically as the loads are applied.

Figure 9.31 shows that the lateral drift of a semi-rigid frame structure can be controlled to a large extent by using heavy base connections. However, the additional cost of "rigid" base connections is likely to be expensive relative to the improvement in performance of the entire structure. Figure 9.31 gives an indication on how lateral frame drifts can be reduced by increasing the rotation stiffness at the base supports.

9.9 EFFECTS OF BRACING

One alternative way to reduce frame drift is to provide diagonal bracings in the frame. The inset of Fig. 9.32 shows the semi-rigid frame with diagonal braces provided at both stories. The diagonal braces are made of angles L3 × 2 × 1/4 and are pinned at their ends by simple shear connections made of an A325 high-strength bolt of 7/8 inch diameter. These diagonal members are strong in tension and weak in compression. When the wind force is applied from the left of the frame shown in Fig. 9.32, diagonal members 1 and 3 are in compression and their buckling strengths are small; therefore, these members are assumed to be inactive when the analysis due to wind is performed. The diagonal braces under tension are modeled as truss elements in which the effective net area of an angle member having a single hole is used in the analysis.

Figure 9.32 shows the load versus roof-displacement curves of the braced frames with different support conditions. Based on this study, the following observations can be made:

1. The use of bracing greatly increases the maximum loads of semi-rigid frames subjected to wind-load combinations. The gain is more pronounced for the frames with more flexible support connections. This

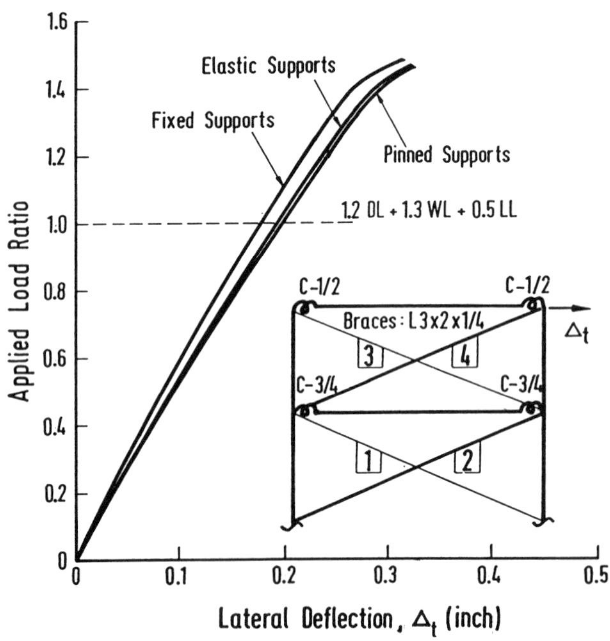

Fig. 9.32 Lateral-load-displacement traces for unbraced frames with diagonal braces.

fact is illustrated by comparing the maximum loads of the braced frames with different support conditions in Fig. 9.32 with those of the unbraced frames in Fig. 9.31. The comparisons show that the increase in the maximum strength is larger for the frame with pinned bases. This observation is in agreement with earlier research (Lui and Chen, 1988) that bracing is more effective in providing resistance to the more flexible semi-rigid frames.

2. The sensitivity of the frames to differences in support conditions is greatly reduced when bracings are provided. It is observed from Fig. 9.32 that the "spread" of the load-displacement curves and the maximum loads for the braced frames are much less than that of the unbraced frames in Fig. 9.31. For example, for unbraced frames, the maximum load ratio varies from 1.004 to 1.407, whereas for braced frames, the variation of the load ratio is almost negligible. It is further observed that the increase in load capacity as a result of the increase in the frame's support fixity is greatly diminished for the braced frames. Clearly, the effects of support conditions on the maximum load of semi-rigid frames are less important for braced frames than for unbraced frames.

The preceding observations have some important implications in the design of semi-rigid frames. The provision of bracing not only reduces frame drift, but it significantly increases the maximum strength of semi-rigid frames—the increase is more pronounced for the more flexible frames. Because lateral drift is often a problem for multistory semi-rigid frames, the use of bracing provides an effective means to satisfy both the serviceability and strength limit states in the design of semi-rigid frame structures. An added advantage of using advanced analysis for frame design is that a distinction between braced and unbraced or sway and nonsway frame systems is not required, because this distinction may be subject to question when a tall building framework is concerned. In advanced analysis, different framing systems, braced or unbraced, can be integrated to form a single structural system, and the system can be checked as a whole for overall stability and strength in the final design stage.

Another point to note is that the consideration of inelastic frame actions may not be as important in frames with very flexible connections because the connecting joints are not capable of developing the full capacity of the joining members. In this case, the elastic-analysis method, which takes into account the nonlinearity of connections and the geometrical nonlinearity of frames, can be used with sufficient accuracy to evaluate the strength and behavior of typical PR frames. However, when the connections are rather stiff and rigid, the inelastic-frame actions need to be considered in order to take advantage of the ductility of steel for inelastic-force redistribution due to the formation of plastic hinges.

9.10 ADVANTAGES OF SEMI-RIGID CONSTRUCTION

Based on the studies reported in Sections 9.7–9.9, the advantages of using semi-rigid connections for frame design may be summarized as follows:

1. In comparison with rigid-frame design, semi-rigid design leads to smaller column sizes, whereas in comparison with a simple framing system, it leads to smaller beam sizes. Thus, the use of semi-rigid connections for frame construction normally leads to an overall saving in steel weight.
2. Semi-rigid construction is comparable to other "fast-track" methods for speed of construction. Site assembly is easier with semi-rigid connections than with rigid connections. Thus, easy erection and speed of construction will lead to early completion of project. Labor cost saving and early return of investment can be realized eventually.
3. Semi-rigid steel frames need not be braced if they are designed for serviceability and ultimate strength limit states. Semi-rigid construction is preferred in low-rise multibay frames. For high-rise building design, service wind drift is always the main concern. In order to reduce the number of moment connections in high-rise building construction, the use of semi-rigid frames and braced frames may provide optimum balance between the dual objectives of buildability and functionality. Semi-rigid frames may be combined with rigid- or braced-frame systems to limit lateral drifts due to wind. A schematic diagram showing the ideas of mixed frames is shown in Fig. 9.33.

9.11 CONCLUSIONS

In this chapter, the three-parameter connection model proposed by Kishi and Chen (1990) is used with an advanced analysis/design procedure for proportioning semi-rigid frames whose beam-to-column connections are composed of angles. The three-parameter power model is based on connection parameters that can be determined readily from the connection geometry and material properties, thus making this model suitable for practical design use. A part of the chapter is devoted to the automation of calculating initial stiffness and ultimate moment capacity of connections composed of angles. Step-by-step procedures for the calculation of the connection parameters are presented, and design aids are generated to explain the behavior and characteristics of the three most commonly used angle connections. The mathematical expressions for M_u and R_{ki} are presented in nondimensionalized forms; they are functions of only the connection size parameters. By expressing the connection moment–rotation relationships in a form dependent only on nondimensional size parameters, the model is more general and can be applied to similar connections of different sizes. A computer program has

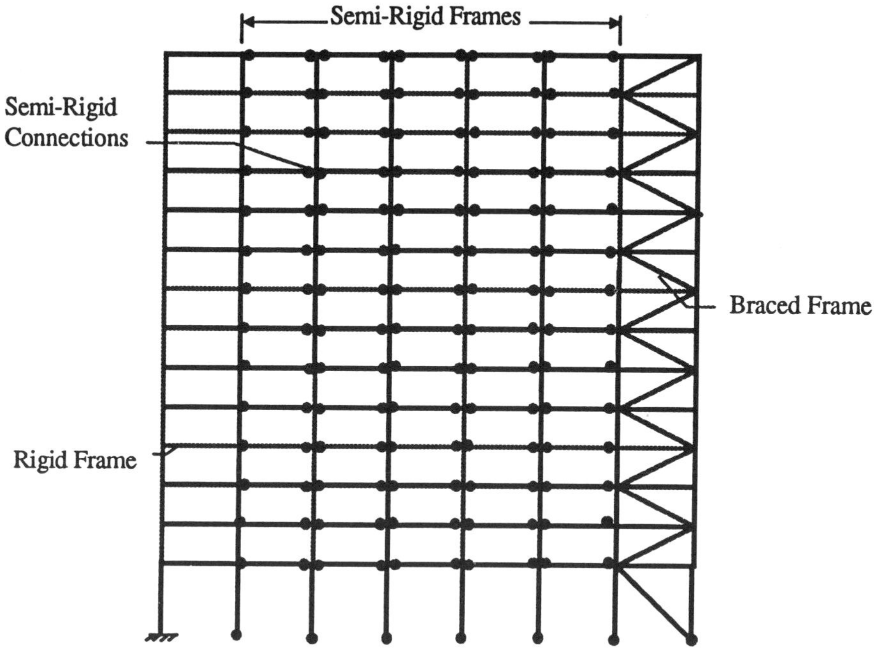

Fig. 9.33 Schematic illustration of semi-rigid frames combined with rigid or braced frame system.

been developed based on the procedure discussed in this chapter, and the software is provided in Chapter 11.

An inelastic-analysis method capable of analyzing the behavior of semi-rigid frames is presented. In this rigorous analysis, the assumption of material behavior being elastic-perfectly plastic at concentrated plastic-hinge locations in the member is removed. Yielding can be treated as a progressive process, and transition from elastic to plastic behavior considering the influence of residual stresses in a cross-section is modeled approximately by a column tangent modulus and a two-surface stiffness-degradation model. Thus, connection flexibility, frame imperfections, geometric nonlinearity, and gradual yielding effects can all be accounted for by the inelastic-analysis model. The analysis method can be applied to braced and unbraced frames. The validity of this analysis method has been confirmed in Chapter 8 against extensive benchmark solutions of rigidly connected braced and unbraced frames, and many other isolated beam-column cases. This method is found suitable for adoption in computer-based analysis and can be applied directly for frame proportioning without member-capacity checks.

A general approach for applying the inelastic analysis for proportioning of semi-rigid frames has been described. It is recognized that the design of semi-rigid frames is intrinsically more complex than the design of rigid frames because of the additional design variables associated with the connection parameters and moment–rotation behavior. Fortunately, studies reported by Bjorhovde et al. (1990), as summarized in Section 9.4, show that for practical purposes, connections can be classified into three main types, namely, rigid, semi-rigid, and flexible. Only the connections that are classified as semi-rigid have to be considered in an advanced analysis program that models the semi-rigid joint actions explicitly. Otherwise, the joints may be modeled as rigid or pin-ended following the conventional methods of analyses. The connection classification system also allows the rotation capacity of a connection to be determined once the moment–rotation curve is estimated from either tests or by analytical means.

The final section of the chapter demonstrates how the inelastic analysis can be used in the load and resistance factor design of semi-rigid frame structures. A two-dimensional two-story frame is investigated, and the design is checked for adequacy in resisting the effects of factored loads. The frame is also designed to meet serviceability conditions. The connections are proportioned such that they have sufficient rotation capacity to meet the deformation requirements at the strength limit state. The main objective of this chapter is to demonstrate how the advanced inelastic analysis can be used to design braced and unbraced semi-rigid frame structures. Chapter 11 provides some educational types of computer software for readers to perform planar frame analysis, so that a better understanding of the behavior and a more realistic prediction of system's strength and stability can be realized.

The influence of panel-zone shear deformations and the like are excluded in the studies because connection tests and models available so far have focused strictly on the characteristics of the connections. Panel-zone shear-deformation effects are usually reflected in the overall structural analysis using simple rules of thumb, or they can be neglected in the overall frame analysis by proper detailing of the joint panels to minimize shear deformation. However, several methods are now available for use in computing the panel-zone shear deformation, and thus allow it to be considered in the design process. The effects of panel-zone shear deformation on frame design are discussed separately in Chapter 10.

REFERENCES

Ackroyd, M. H. (1985). *Design of Flexibly Connected Steel Frames,* Final Research Report to the American Iron and Steel Institute for Project No. 333, Troy, NY: Rensselaer Polytechnic Institute.

AISC (1993). *Load and Resistance Factor Design Specifications for Structural Steel Buildings,* Chicago: American Institute of Steel Construction.

Al-Bermani, F. G. A., and S. Kitipornchai (1992). "Elastoplastic Nonlinear Analysis of Flexibly Jointed Space Frames," *Journal of Structural Engineering,* ASCE, Vol. 118, No. 1, pp. 108–127.

Altman, W. G., Jr., A. Azizinamini, J. H. Bradburn, and J. B. Radziminski (1982). *Moment-Rotation Characteristics of Semi-Rigid Steel Beam-to-Column Connections,* Columbia: Department of Civil Engineering, University of South Carolina.

Ammerman, D. J., and R. T. Leon (1987). "Unbraced Frames with Semi-Rigid Composite Connections," *Engineering Journal,* AISC, Vol. 24, No. 2, pp. 12–21.

ASCE-WRC (1971). *Plastic Design in Steel—A Guide and Commentary,* 2nd Ed., ASCE Manuals and Reports on Engineering Practice No. 41, New York: Joint Committee of the Welding Research Council and ASCE.

Azizinamini, A., J. H. Bradburn, and J. B. Radziminski (1985). *Static and Cyclic Behavior of Semi-Rigid Steel Beam-Column Connections,* Columbia: Structural Research Studies, Department of Civil Engineering, University of South Carolina.

Barakat, M., and W. F. Chen (1990). "Practical Analysis of Semi-Rigid Frames," *Engineering Journal,* AISC, Vol. 27, No. 2, pp. 54–68.

Barakat, M., and W. F. Chen (1991). "Design Analysis of Semi-Rigid Frames: Evaluation and Implementation," *Engineering Journal,* AISC, Vol. 28, No. 2, pp. 55–64.

Bijlaard, F. S. K., and P. Zoetmeijer (1986). "Influence of Joint Characteristics on the Structural Response of Frames," in *Steel Structures: Recent Research Advances and Their Applications to Design,* M. N. Pavlovic, ed., London: Elsevier.

Bijlaard, F. S. K., and C. M. Steenhuis (1991). "Prediction of the Influence of the Connection Behavior on the Strength, Deformations and Stability of Frames by Classification of Connections," in *Workshop on Connections in Steel Structures,* R. Bjorhovde, A. Colson, G. Haaijer, and J. W. B. Stark, eds., Chicago: AISC.

Bjorhovde, R., J. Brozzetti, and A. Colson (1990). "A Classification System for Beam to Column Connections," *Journal of Structural Engineering,* ASCE, Vol. 116, No. 11, pp. 3059–3076.

BSI (1990). *BS5950: Structural Use of Steelwork in Building. Part 1: Code of Practice for Design in Simple and Continuous Construction: Hot Rolled Sections,* London: British Standards Institution.

Chen, W. F., ed. (1987). *Joint Flexibility in Steel Frames,* London: Elsevier.

Chen, W. F., ed. (1988). *Steel Beam-to-Column Building Connections,* London: Elsevier.

Chen, W. F., ed. (1993). *Semi-Rigid Connections in Steel Frames,* New York: Council of Tall Buildings and Urban Habitat and McGraw-Hill.

Chen, W. F., and J. Y. R. Liew (1992). "Seismic Resistance Design of Steel Moment Resisting Frames Considering Panel-Zone Deformation," in *Stability and Ductility of Steel Structures under Cyclic Loading,* Y. Fukumoto and C. G. Lee, eds., Boca Raton, FL: CRC Press, pp. 323–334.

Chen, W.F., and E. M. Lui (1991). *Stability Design of Steel Frames,* Boca Raton, FL: CRC Press.

Chen, W.F., and I. S. Sohal (1995). *Plastic Design and Second-Order Analysis of Steel Frames,* New York: Springer-Verlag.

Chen, W. F., and S. Toma (1994). *Advanced Analysis of Steel Frames: Theory, Software, and Applications,* Boca Raton, FL: CRC Press.

CSA (1989). *Limit States Design of Steel Structures,* CAN/CSA-S16.1-M89, Toronto: Canadian Standards Association.

Dhillon, B. S., and S. Abdel-Majid (1990). "Interactive Analysis and Design of Flexibly Connected Frames," *Computer and Structures,* Vol. 36, No. 2, pp. 189–202.

ECCS (1991). *Essentials of Eurocode 3 Design Manual for Steel Structures in Building,* ECCS Advisory Committee 5, No. 65, Brussels: European Convention for Constructional Steelwork.

ECCS (1992). *Analysis and Design of Steel Frames with Semi-Rigid Joints,* ECCS Technical Committee 8, Structural Stability Technical Working Group 8.1/8.2 Skeletal Structures, No. 67.

Eurocode 3 (1992). *Design of Steel Structures: Part 1.1—General Rules and Rules for Buildings,* National Application Document for Use in the UK with ENV1993-1-1: 1991, Draft for Development.

Gibbons, C., P. A. Kirby, and D. A. Nethercot (1991). "Expeirmental Behaviour of 3-D Column Subassemblages with Semi-Rigid Joints," *Journal of Constructional Steel Research,* Vol. 19, No. 4, pp. 235–246.

Goto, Y., and W. F. Chen (1987). "Second-Order Elastic Analysis for Frame Design," *Journal of Structural Engineering,* ASCE, Vol. 113, No. 7, pp. 1505–1519.

Goto, Y., S. Satsuki, and W. F. Chen (1991). "Analysis of Critical Behavior of Semi-Rigid Frames with or without Load History in Connections," *International Journal of Solids and Structures,* Vol. 27, No. 4, pp. 467–483.

Hechtman, R. A., and B. G. Johnson (1947). *Riveted Semi-Rigid Beam-to-Column Building Connections,* Progress Report No. 1, Bethlehem, PA: AISC Research, Lehigh University.

Heish, S. H. (1990). *Analysis of Three-Dimensional Steel Frames with Semi-Rigid Connections,* Structural Engineering Report 90-1, Ithaca, NY: School of Civil and Environmental Engineering, Cornell University.

Jaspart, J. P., J. C. Gerardy, and R. Maquoi (1991). "Parametric Study of the Numerical Modelling for Semi-Rigid Joints," *Proceedings of 1991 Annual Technical Session,* Bethlehem, PA: Structural Stability Research Council.

Kishi, N., and W. F. Chen (1990). "Moment-Rotation Relations of Semi-Rigid Connections with Angles," *Journal of Structural Engineering,* ASE, Vol. 116, No. 7, pp. 1813–1834.

Kishi, N., W. F. Chen, K. G. Matsuoka, and S. G. Nomachi (1988a). "Moment-Rotation Relation of Single/Double Web-Angle Connections," in *Proceedings of the Workshop on Connections and the Behavior, Strength and Design of Steel Structures,* Cachan, France, May 25–27, London: Elsevier, pp. 135–149.

Kishi, N., W. F. Chen, K. G. Matsuoka, and S. G. Nomachi (1988b). "Moment-Rotation Relation of Top- and Seat-Angle Connections," in *Proceedings of the Workshop on Connections and the Behavior, Strength and Design of Steel Structures,* Cachan, France, May 25–27, London: Elsevier, pp. 121–134.

Kishi, N., W. F. Chen, K. G. Matsuoka, and K. G. Matsuoka (1993). "Design Aid of Semi-Rigid Connections for Frame Analysis," *Engineering Journal,* AISC, Vol. 30, No. 4, pp. 90–107.

Liew, J. Y. R. (1992). "Advanced Analysis for Frame Design," Ph.D. dissertation, School of Civil Engineering, Purdue University, West Lafayette, IN.

Liew, J. Y. R., D. W. White, and W. F. Chen (1991). "Beam-Column Design in Steel Frameworks—Insight on Current Methods and Trends," *Journal of Constructional Steel Research,* Vol. 18, No. 4, pp. 269–308.

Liew, J. Y. R., D. W. White, and W. F. Chen (1993a). "Limit-States Design of Semi-Rigid Frames Using Advanced Analysis: Part 1: Connection Modelling and Classification," *Journal of Constructional Steel Research,* Vol. 26, No. 1, pp. 1–27.

Liew, J. Y. R., D. W. White, and W. F. Chen (1993b). "Limit-States Design of Semi-Rigid Frames Using Advanced Analysis: Part 2: Analysis and Design," *Journal of Constructional Steel Research,* Vol. 26, No. 1, pp. 29–57.

Liew, J. Y. R., D. W. White, and W. F. Chen (1993c). "Second-Order Refined Plastic Hinge Analysis for Frame Design: Part 1," *Journal of Structural Engineering,* ASCE, Vol. 119, No. 11, pp. 3196–3216.

Liew, J. Y. R., D. W. White, and W. F. Chen (1993d). "Second-Order Refined Plastic Hinge Analysis for Frame Design: Part 2," *Journal of Structural Engineering,* ASCE, Vol. 119, no. 11, pp. 3217–3237.

Lui, E. M. (1985). "Effect of Connection Flexibility and Panel Zone Deformation on the Behavior of Plane Steel Frames," Ph.D. dissertation, School of Civil Engineering, Purdue University, West Lafayette, IN.

Lui, E. M., and W. F. Chen (1988). "Behavior of Braced and Unbraced Semi-Rigid Frames," *International Journal of Solids and Structures,* Vol. 24, No. 9, pp. 893–913.

Narayanan, R., ed. (1989). *Structural Connections—Stability and Strength,* London and New York: Elsevier.

Nethercot, D. A. (1989). "Frame Analysis and the Link between Connection Behavior and Frame Performance," in *Frame and Steel Structures,* G. G. T. Armer and D. B. Moore, eds., London: Butterworth, pp. 57–74.

Nethercot, D. A., and R. Zandonini (1989). "Methods of Prediction of Joint Behavior: Beam-to-Column Connections," in *Structural Connections, Stability and Strength,* R. Narayanan, ed., London: Elsevier, pp. 23–62.

Standards Australia (1990). *Steel Structures,* AS4100-1990, Sydney.

Stelmack, T. W., M. J. Marley, and K. H. Gerstle. (1986). "Analysis and Tests of Flexibly Connected Steel Frames," *Journal of the Structural Division,* ASCE, Vol. 112, no.7, pp. 1573–1588.

White, D. W., and W. F. Chen (1993). *Plastic Hinge Based Methods for Advanced Analysis and Design of Steel Frames,* Bethlehem, PA: Structural Stability Research Council.

Wu, F. H. (1988). "Semi-Rigid Connections in Steel Frames," Ph.D. dissertation, School of Civil Engineering, Purdue University, West Lafayette, IN.

Yu, C. H., and N. E. Shanmugam (1986). "Stability of Frames with Semi-Rigid Joints," *Computers and Structures,* Vol. 23, No. 5, pp. 639–648.

10

STEEL FRAMES WITH PANEL-JOINT DEFORMATIONS

10.1 INTRODUCTION

Experimental studies on frame subassemblies carried out by Fielding and Huang (1971); Bertero, Popov, and Krawinkler (1972); and Krawinkler (1978) have shown that beam-column joint panels can develop high shear resistance beyond initial yielding of the column web when the panels are subjected to repeated loads. A well-detailed panel zone usually possesses stable energy-dissipation characteristics and can sustain large cyclic inelastic deformations before weld fracture occurs. Recent seismic codes in the United States (SEAOC, 1988; AISC, 1990) specifically allow the panel zones to yield in shear prior to the full development of the beam capacity. In these seismic provisions, the nominal shear resistance of beam-column panel zones may be increased by considering the postyield shear strength of the boundary elements in the panels. These provisions not only allow plastic hinges to occur in the beams, but also allow panel zones to yield in shear.

By allowing the panel zones to yield in shear, some cost saving for the design of rigid joints is possible because the requirement of a doubler plate may be waived. However, excessive panel-zone deformations may influence the overall behavior of the frame subjected to sidesway. If the applied axial forces and moments are high, the panel zone may deform inelastically to the extent that the limit states of resistance and deflection of the frame will be affected. An increase in frame drift due to panel-zone shear deformations may render the frame unserviceable. Hence, proper measures must be taken to limit the frame drift or to prevent yielding of panel zones under the applied factored load. Otherwise, an advanced analysis method is required for accurate assessment of the overall system performance considering the second-order effects associated with panel-zone deformations (Chen and Liew, 1992).

This chapter provides insight to the background of the development and requirements for panel-zone design in the recent AISC LRFD (AISC, 1993) provisions. In particular, two criteria for the design of beam-column panel zones are discussed. The first method is based on the von Mises first-yield criterion for the panel-zone web, and the second method utilizes the postyield shear resistance of the panel derived from the flexural resistance of boundary elements surrounding the panel-zone joint. The nonlinear behavior of a rigid sway frame with panel-zone joints designed according to the previous two requirements is studied. A second-order elastic-plastic-hinge analysis is used to assess the strength and serviceability limit states of the frame. The initial imperfections effects (such as the column's initial out of straightness, the frame's initial out of plumbness, and initial residual stresses in members) are included in the second-order analysis through the use of notional lateral forces. From these studies, the effects of the panel-zone deformations on the overall frame resistance can be discussed.

10.2 RIGID CONNECTIONS

Full-strength rigid connections are commonly used in the construction of high-rise steel building frames. The use of rigid connections has contributed to the development of more compact and economical structural joints as well as providing greater stiffness and resistance against wind loads. A particular type of rigid connection that has gained wide acceptance in the United States and in Japan is shown in Fig. 10.1a. In this type of connection, the beam web is first field bolted to a shear tab, which initially is shop welded to the column flange. Full penetration bevel welds are then applied to join the beam flanges to the column flange. In order to prevent panel-zone deformation, a doubler plate or diagonal stiffener may be added, as shown in Figs. 10.1a and 10.1b respectively. However, if weak-axis framing is required, the diagonal stiffener will obstruct the beam from connecting to the column web. Therefore, a doubler plate is the preferred option if it is required to stiffen the joint.

For areas where field welding is not desired, beam-to-column flange connections can be made by shop welding of a beam to an end plate or directly to a column. The moment frame can then be erected by field bolting the end plate to the column flanges or splicing beams at inflection points in the beam (Figs. 10.1c and 10.1d).

10.3 BEHAVIOR OF PANEL ZONES

10.3.1 General

Figure 10.2 shows a possible system of forces that act on a panel zone of an interior beam-to-column joint. Under the action of these forces, the panel-

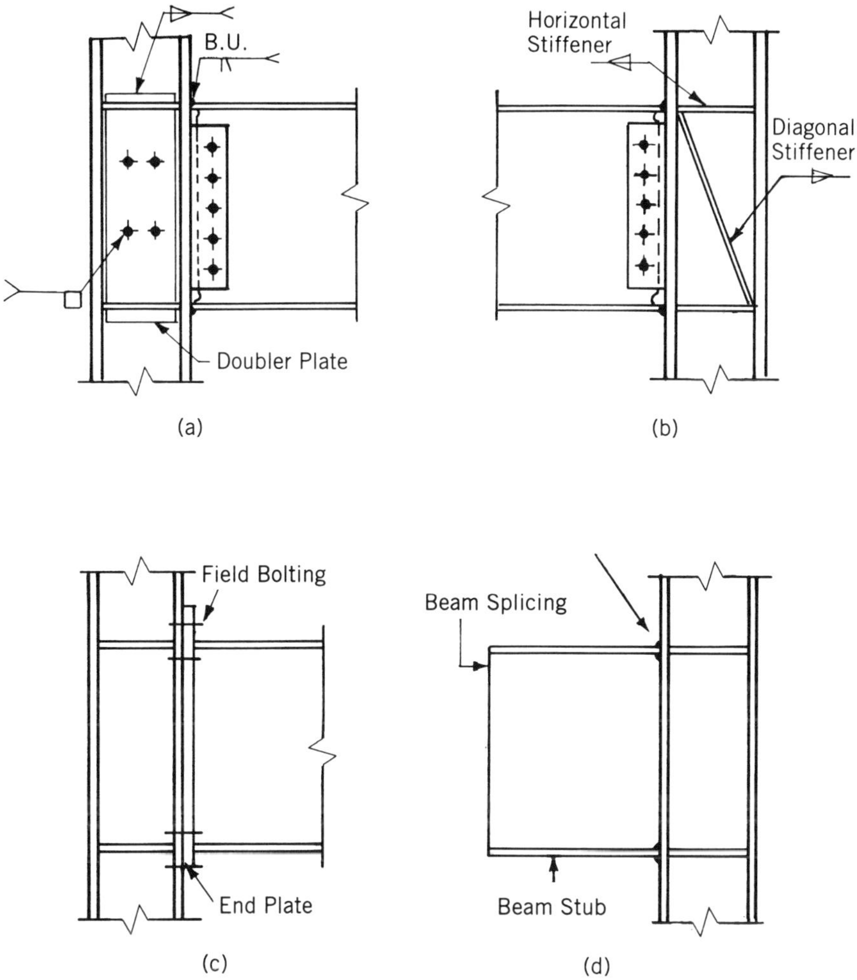

Fig. 10.1 (a) Bolted and welded connection with a doubler plate. (b) Bolted and welded connection with a diagonal stiffener. (c) Bolted end-plate connection. (d) Beam stub welded to a column.

zone joint will deform. In addition, the force actions may cause premature yielding of the panel-zone web, resulting in a reduction of resistance and stiffness of the joint.

A finite-element analysis has been performed by Tsai and Popov (1988) on a square panel zone consisting of its boundary elements and portions of the framing members. The deformed mesh of the specimen subjected to shear distortion is shown in Fig. 10.3. It is noted that the elements near the center of the panel specimen are highly distorted by shear, whereas those near the corners of the panel are less deformed. It can be observed that the shear

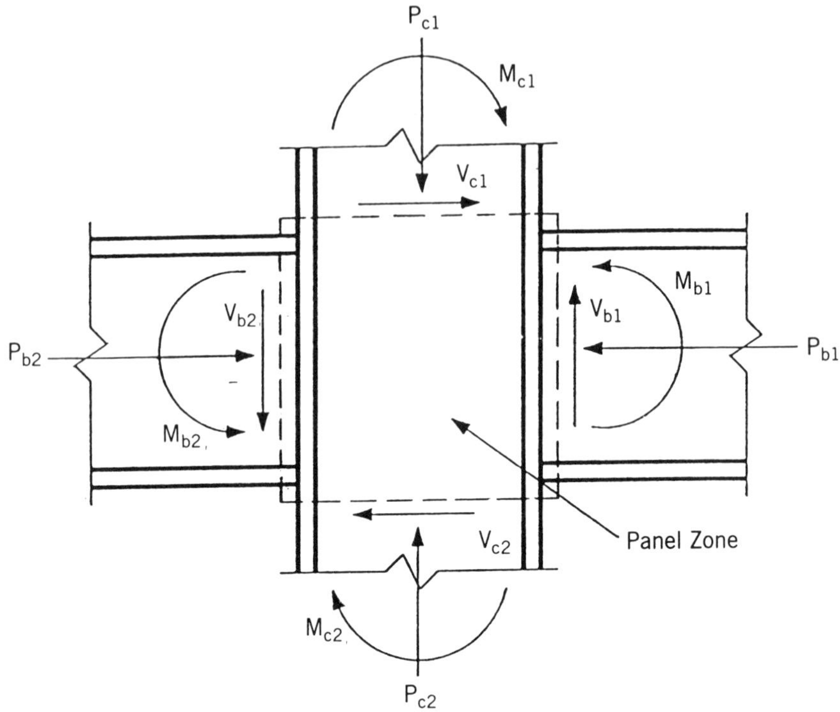

Fig. 10.2 Forces acting on a panel zone.

stresses in the panel zone are not uniformly distributed; instead, they are highest at the center of the panel, with a moderate drop toward the corners. The average shear stress in the panel zone is about 20% lower than the maximum shear stress at the center of the panel.

Most of the frame-analysis programs that model explicitly for panel-zone deformations are formulated based on the assumption that the shear distribution in the panel zone is uniformly distributed in the entire panel zone. It has been shown that design of a panel zone based on this assumption has the merit of simplicity and is accurate enough in estimating the resistance of steel beam-column joints (Fielding and Huang, 1971; Fielding and Chen, 1973; Krawinkler, 1978; Slutter, 1981; Chen and Liew, 1992).

10.3.2 Static Behavior

A number of large-scale tests on frame subassemblies have been performed to investigate the load-deformation behavior of the panel-zone joints (Fielding and Huang, 1971; Bertero et al., 1972; Krawinkler, 1978; Kato, Chen, and Nakao, 1988). Particular attention was focused on the shear resistance of the

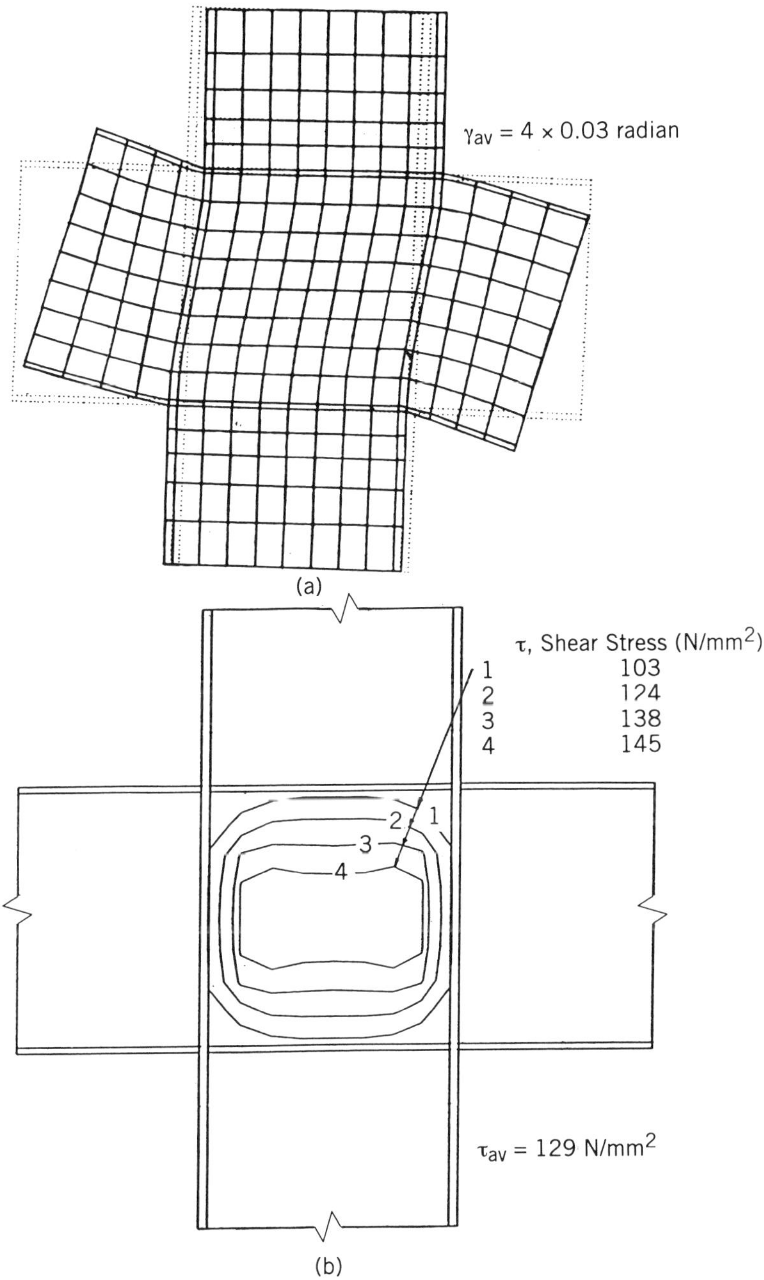

Fig. 10.3 (a) Deformed mesh of an interior panel zone. (b) Shear-stress contour at $M = M_{pb}$.

panel zone and the effects of panel-zone deformations on the strength and deformation of the frame subassemblages. Some of the significant behavior observed in these tests are summarized in what follows (Chen and Lui, 1991):

1. There are two distinct stiffnesses in the panel-zone shear-force-deformation response: an elastic stiffness, followed by a small range of gradually decreasing stiffness (may be ignored for practical reasons), and then stabilizing to a small and almost constant stiffness for a long range of deformation, as shown in Fig. 10.4. The postyield stiffness of a panel zone normally ranges from about 3 to 8% of the elastic stiffness.
2. The shear deformation of the panel zone is ductile and stable in nature. The failure of the panel zone is usually caused by a weld fracture occurring at very large plastic deformations.

The existence of the postyield stiffness in the shear-force-deformation response is attributed to the following:

1. The resistance of the boundary plate elements such as the column flanges and the stiffeners of the panel-zone joint. The column flanges bend in a frame-type manner after the "softening" of panel-zone web due to shear yielding.

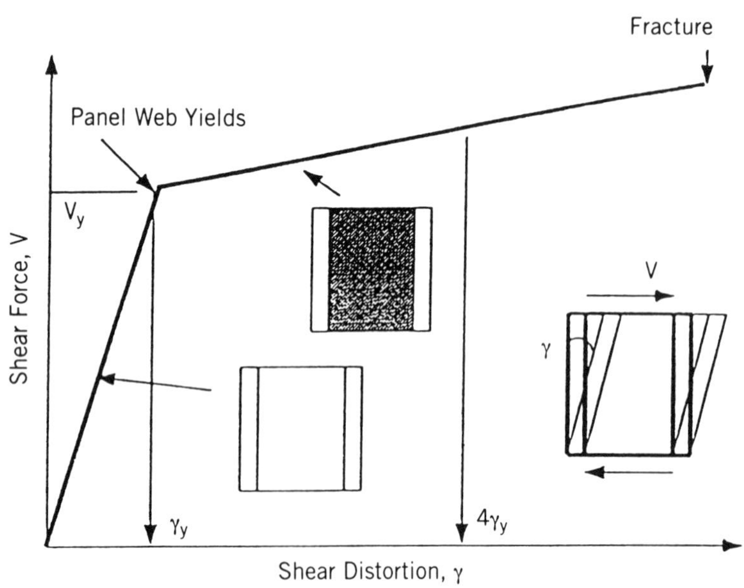

Fig. 10.4 Typical shear-force-distortion behavior of a joint panel.

2. The onset of strain hardening of the panel-zone web before complete yielding of the boundary elements.
3. The restraint from the beam flanges connecting the joint panel.

In the postyield range, the column web does not contribute to the additional shear capacity until strain hardening of the web begins. Although the web has yielded, the panel-zone boundaries (column flanges and horizontal web stiffeners) may still be elastic. Both experimental and analytical studies have confirmed that the postyield strength and stiffness of a panel zone are dependent on the stiffness of the surrounding elements, particularly, the flexural stiffness of the column flanges and the aspect ratio of the panel.

10.3.3 Cyclic Behavior

All tested panel zones reported in the past exhibited stable hysteresis loops and remain ductile under repeated and reversed loading for a range of large plastic deformations. For well-detailed panel joints, no reduction in capacity was observed even at extremely large inelastic deformations. However, for some specimens with thin webs, shear buckling in the panels was observed.

It is important to note that large panel-zone deformation may create local distortions and high strain concentrations at the regions where the beam flanges are welded to the column flanges, thus causing cracking of the flange welds (Krawinkler, 1978). However, weld fracture occurs only in a large number of cyclic loading and when the panel-zone distortion is large. Therefore, if the panel-zone joints are carefully detailed and all welds are properly done, the joints possess superior quality for energy dissipation without causing premature failure.

10.3.4 Simplified Behavior

The mathematical modeling of a panel-zone joint is complicated because the panel is subjected to the combined action of axial and shear forces from the column and bending moments from the adjoining beams (Fig. 10.5). The panel zone is also subjected to shear forces that vary according to the distribution of bending stresses in the beams. Hence, design criteria for panel zone require use of simplified expressions, which, nevertheless, should incorporate the most significant behavioral effects that contribute to the force transfer within the panel joint. A simple way to design panel zone is to assume that the average shear stress is distributed uniformly throughout the panel. A design procedure based on this assumption has been recommended by the AISC LRFD (AISC, 1993) for designing beam-to-column panel zones. Research has also shown that this assumption is adequate in evaluating the overall structural performance (Chen and Liew, 1992).

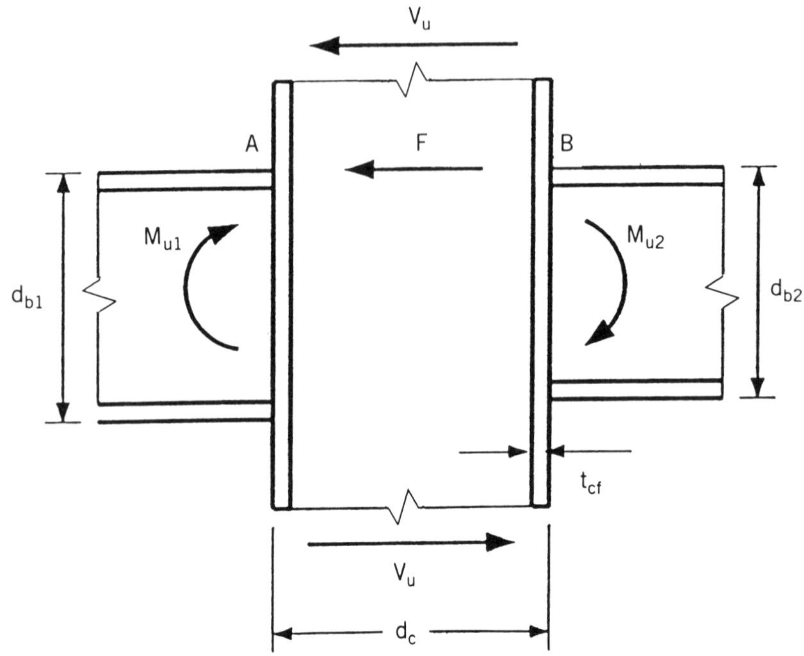

Fig. 10.5 Equilibrium of forces acting on the panel-zone web.

10.4 DESIGN OF BEAM-TO-COLUMN PANEL JOINTS

In the design of a rigid frame with rectangular panel joints, the column web shear forces are often quite high in the joint panel consisting two or more members whose webs lie in a common plane. Such webs should be reinforced when the calculated force, ΣF, along plane A–B in Fig. 10.5 exceeds column web resistance ϕR_v, where

$$\Sigma F = \frac{M_{u1}}{0.95d_{b1}} + \frac{M_{u2}}{0.95d_{b2}} - V_u \tag{10.4.1}$$

and

M_{u1}, M_{u2} = factored beam moments
d_{b1}, d_{b2} = beam depths
V_u = factored column shear force

The beam moment is assumed to act through the flanges as a couple with a moment arm of $0.95d_b$. For the LRFD (AISC, 1993), the design shear resistance of the panel zone, ϕR_v, must be greater than the factored shear force, ΣF. In such cases, no reinforcement in the web panel is required. If ϕR_v is smaller than ΣF, the web needs to be reinforced by either a doubler plate or by diagonal stiffeners.

In evaluating the shear resistance of a panel zone, two approaches are recommended by the LRFD (AISC, 1993). The first approach is based on first-yield criterion for the panel web considering the reduction of shear resistance due to the presence of an axial force. The second approach is based on the assumption that the panel joint can develop significant postyield shear strength beyond initial yielding. The second approach recommends that the panel-zone shear strength be increased by a factor that is related to the stiffness of the flange elements. These two approaches are discussed in what follows.

10.4.1 First-Yield Criterion

According to this approach, the shear resistance, R_v, of a panel zone is determined as

$$R_v = 0.6F_y d_c t_w, \qquad \text{for } P_u \leq 0.4P_y \tag{10.4.2a}$$

$$R_v = 0.6F_y d_c t_w(1.4 - P_u/P_y) \qquad \text{for } P_u > 0.4P_y \tag{10.4.2b}$$

where

d_c = column depth
P_u = design axial force
P_y = squash load
t_w = total thickness of the panel zone

Equations (10.4.2a) and (10.4.2b) represent a linearized form of the von Mises yield criterion for the panel web shear resistance, as shown in Fig. 10.6. For webs subjected to high shear in combination with an axial load exceeding $0.4P_y$, Eq. (10.4.2b) was developed to give a straight-line transition from the maximum shear force at $P = 0.4P_y$ to a maximum axial load at 40% of the maximum shear resistance, as shown in Fig. 10.6.

Test results reported by Johnson (1959); Naka, Kato, and Watabe (1966); Peters and Driscoll (1968); Tanaka et al. (1970); Becker (1971); and Fielding and Huang (1971) for panel zones under combined axial and shear forces at first yield ($\gamma = \gamma_y$) are plotted in Fig. 10.6 for comparison with the design equations. Equations (10.4.2a) and (10.4.2b), which are developed based on the first-yield criterion, compare well with the test results because all test results are shown for the panel zones at first yield.

Equations (10.4.2a) and (10.4.2b) should be used when the effect of panel-zone deformation on frame stability is not considered in the analysis. Most general structural analysis software cannot account for panel joint flexibility, so Eqs. (10.4.2a) and (10.4.2b) thus limit the panel-zone behavior to the

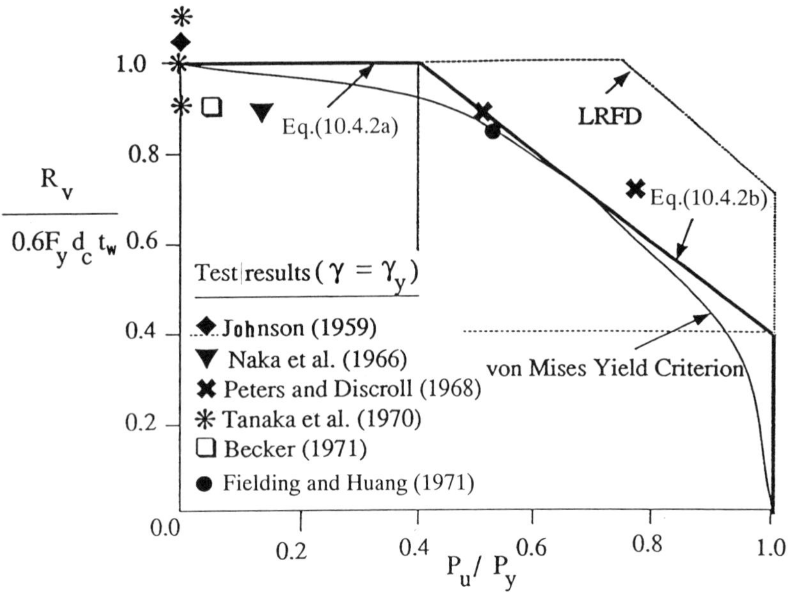

Fig. 10.6 Elastic shear- and axial-force interaction.

elastic range. Therefore, comparable results can be obtained by using a conventional method of analysis that does not account for panel-zone deformations.

10.4.2 Postyield Criterion

If adequate joint ductility is provided and the frame analysis considers inelastic panel-zone deformations, then an additional inelastic shear resistance of the panel zone may be recognized by applying the flange contribution factor (FCF) to Eqs. (10.4.2a) and (10.4.2b). The modified form of Eqs. (10.4.2a) and (10.4.2b) may be written as

$$R_v = 0.6F_y d_c t_w(\text{FCF}), \qquad \text{for } P_u \leq 0.75P_y \qquad (10.4.3a)$$

$$R_v = 0.6F_y d_c t_w\left(1.9 - \frac{1.2P_u}{P_y}\right)(\text{FCF}), \qquad \text{for } P_u > 0.75P_y \qquad (10.4.3b)$$

where FCF is the flange contribution factor, which is defined as

$$\text{FCF} = \frac{1 + 3b_{cf}t_{cf}^2}{d_b d_c t_w} \qquad (10.4.4)$$

b_{cf} = width of column flange
d_b = beam depth
d_c = column depth
F_y = yield strength of column web
t_{cf} = thickness of column flange
t_w = column web thickness (including doubler plate thickness when required)
P_y = F_yA, axial yield strength of the column
A = column cross-sectional area

Equations (10.4.3a) and (10.4.3b) are plotted in Fig. 10.7 for comparison with a test result reported by Fielding and Huang (1971). The axial- and shear-force-interaction expression in Eq. (10.4.3b) recognizes that when the panel-zone web has completely yielded in shear, the column axial load is carried by the flanges. In deriving Eqs. (10.4.3a) and (10.4.3b), it is assumed that the postyield stiffness of the joint panel is valid for the range $\gamma_y < \gamma < 4\gamma_y$, where $\gamma_y = F_y/\sqrt{3}G$ is the shear strain at first yield, G is the shear modulus defined as $0.5E/(1 + \nu)$, E is Young's modulus, and ν is Poisson's ratio.

The test results are shown in Fig. 10.7 in which the panel zone is subjected to a constant axial force $P = 0.5P_y$ and an increasing shear strain equal to two, three, and five times the shear strain at first yield. The test results show that the shear resistance defined by Eqs. (10.4.3a) is not attained even when shear strain γ is larger than $5\gamma_y$. However, the actual capacity of the tested

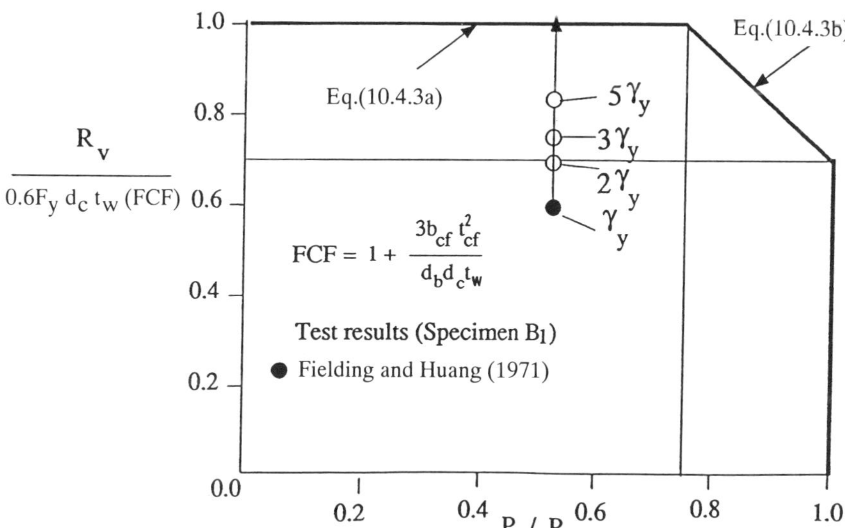

Fig. 10.7 Inelastic shear- and axial-force interaction.

panel zone can increase beyond $R_v = 0.6F_y d_c t_w$, provided that the distortion angle is large enough and a weld fracture in the panel-zone joint does not occur.

If the beam-to-column panel zones are ductile and the global frame analysis considers the inelastic panel-zone deformations, then Eqs. (10.4.3a) and (10.4.3b) may be used to account for the additional inelastic shear resistance given by the flange contribution factor defined in Eq. (10.4.4). The concept of permitting panel-zone shear yielding was recognized and adopted in the SEAOC (1988) provisions for the design of rigid steel frames subjected to reversal loading. This new provision enables the design of panel-zone joints with substantially reduced doubler-plate requirements. Consequently, the frame is more flexible and the inelastic panel-zone deformations must be considered in the drift calculations for rigid frames.

The nonlinear behavioral responses of a beam-column model with joint elements included can be quite different from a beam-column model without them. It is essential that structural analysis models explicitly the behavior of panel-zone joints for a better prediction of the postyield behavior of moment resisting sway frames.

10.5 ANALYTICAL MODELING OF PANEL ZONES

Over the years, several researchers have proposed behavioral models to describe the load-deformation relationship of panel zones (Nakao, 1975; Krawinkler, 1978; Lui and Chen, 1986; Kato et al., 1988). These models are either a semiempirical or an analytically based procedure. The Lui and Chen (1986) model, which is an analytically based model, is adopted herein to study the effect of panel zones on frame response.

The Lui and Chen model is based on finite-element formulation. The panel-zone joint is modeled using one web element and two flange elements. The flange elements used for the panel-zone model are the ordinary six-degrees-of-freedom beam-column element, and the web element is a 12-degree-of-freedom two-dimensional plane stress element, as shown in Fig. 10.8. The tangent stiffness relationships of the flange elements and the web element are formulated separately, and they are assembled to form a panel-joint element. The state of plasticity of the panel web is monitored using the von Mises yield criterion (or J_2 theory). Material strain hardening may be considered assuming an isotropic hardening rule. The detailed kinematics formulation relating the degrees of freedom of beam-columns and panel elements for different categories of connections were reported in Lui (1985).

The panel-zone model proposed by Lui and Chen (1986) accounts for the shear mode of deformation and the extension and bending deformations in the panel zone. In addition, the model accounts for yielding and strain hardening of the web panel and the postyield resistance from the surrounding

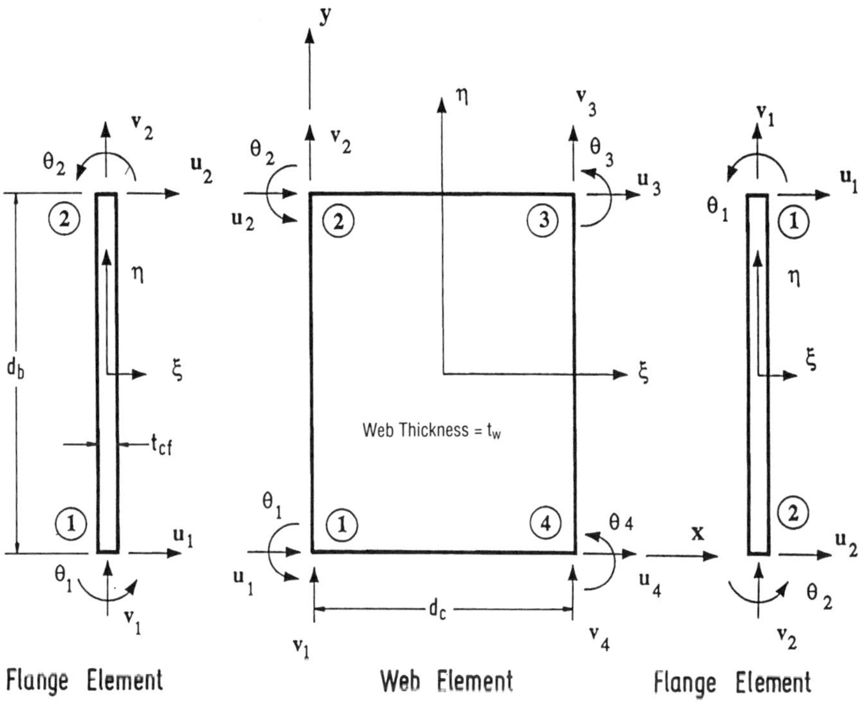

Fig. 10.8 Finite-element models of the panel-zone joint.

elements of the joint. The basic assumptions used by the analysis for modeling the panel zone are summarized as follows:

1. The stress–strain behavior of the web panel is assumed to be elastic-perfectly plastic with strain hardening.
2. Large rigid-body displacement of the joint panel is allowed, but the deformation or distortion of the panel remains small.
3. Local and lateral torsional buckling of the joint panel are not allowed. Only the maximum resistance of the joint panel is considered.
4. Full yielding of the panel web occurs when the average stress reaches the yield surface prescribed by the von Mises yield criterion.
5. The isotropic hardening rule is used to describe the subsequent yield surfaces.
6. Material and weld fractures are not considered.

The validity of this panel-zone model has been verified by comparison of results with experiments (Lui, 1985).

10.6 ANALYSIS OF A FRAME SUBASSEMBLAGE

The frame subassemblage shown in Fig. 10.9 was first studied by Fielding and Chen (1973). The frame is analyzed using a second-order elastic-plastic-hinge analysis, which models panel-zone deformation. The structure represents a one-story segment of a tall steel building frame with column inflection points assumed to exist midway between the floor level being analyzed and the floors above and below. The loading consists of factored uniform floor loads w_i, column loads P_i, and story shear force H. Loads w_i and P_i are applied initially as factored loads. Then the story drift angle ψ is computed based on the applied story shear force H.

Figure 10.9 shows the relationship between the total story shear H and the story drift angle ψ for the story subassemblage with different joint stiffnesses. For curve a, the total panel thickness is twice that of the column web. Curve b is for joint panels with no doubler plate, that is, the panel-zone thickness t is equal to the column web thickness t_w. Curve c represents the frame with rigid joint panels.

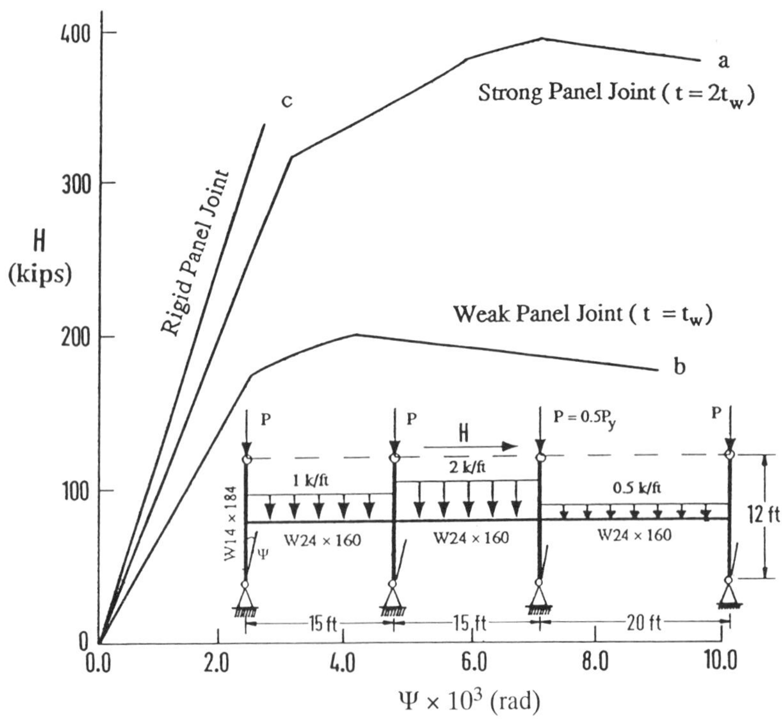

Fig. 10.9 Shear-force-deformation relationship of a story subassemblage.

The effect of increasing the column thickness by t_w is an increase in lateral stiffness by approximately 27% and an increase in story shear resistance of about 100%. A further increase in panel-zone stiffness does not significantly affect the maximum load of the subassemblage frame because the maximum load would be restricted by the formation of plastic hinges rather than by the formation of "panel hinges." A panel hinge is said to have formed when the shear capacity of the panel zone defined by the von Mises shear yield criterion is reached. The elastic stiffness of the frame with rigid panels is about 75% greater than that of the unstiffened frame.

The results of bending moment versus shear deformation in the column panels are shown in Fig. 10.10. The interior joint panels are deformed more than the exterior joint panels. It is important to note that all the joint panels do not deform more than twice the shear strain at first yield. This means that the additional shear resistance from the joint-panel boundaries is not available before reaching the limit load of the subassemblage frame. To meet the shear-capacity requirements indicated by Eqs. (10.4.3a) and (10.4.3b), the panel zone is required to deform at least four times or more times the shear strain at first yield (i.e., $\gamma \geq 4\gamma_y$).

10.7 BEHAVIORAL STUDY OF A SIX-STORY SWAY FRAME

The six-story two-bay moment resisting frame shown in Fig. 10.11 is analyzed using a second-order elastic-plastic-hinge analysis, which can model explicitly for the presence of panel-zone joints. To account for the effects of initial residual stresses and other initial imperfections, the elastic-plastic-hinge analysis is carried out using the notional-load method, as discussed in Chapter 7. The notional lateral load is computed based on 0.5% of the factored gravity load applied at each floor level. The notional load must be applied together with wind load W_i and gravity load ΣP_u that are acting on the structure (Liew, White, and Chen, 1994). The objectives of this study are (1) to evaluate the two different panel-zone design criteria, (2) to discuss the qualities and limitations associated with these design provisions, and (3) to examine the effects of panel-zone deformations on overall frame stability. Based on these studies, proper panel-zone requirements for the design of steel moment resisting frames are proposed.

10.7.1 Analysis Assumptions

1. All members are oriented with their webs in the plane of the frame. The detailed cross-section dimensions of the members are given in Table 10.1.

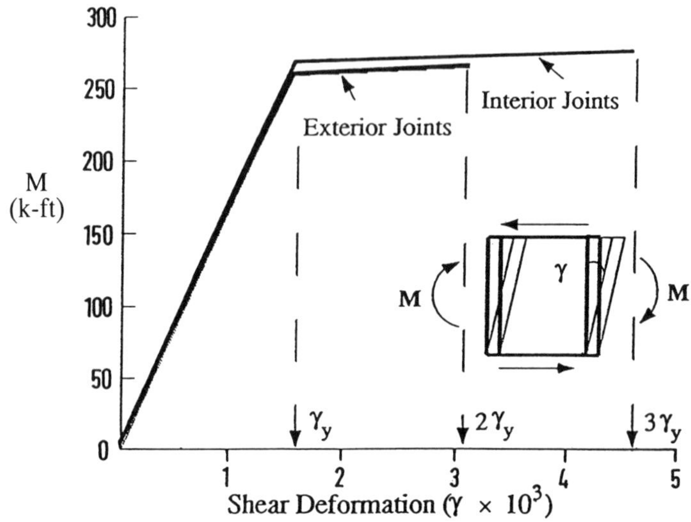

Fig. 10.10 Moment-shear-deformation relationship of the column panel-zone (for curve *b* in Fig. 10.9).

2. All beams are modeled by four discrete elements and all columns by two elements.
3. All panel zones are modeled using the finite-element technique discussed in Section 10.5. The material properties needed for modeling the panel-zone elements follow:
   (a) Young's modulus, $E = 200{,}000\ \text{N/mm}^2$
   (b) Design yield stress, $\phi F_y = 235\ \text{N/mm}^2$
   (c) Poisson's ratio, $\nu = 0.3$
   (d) Strain-hardening modulus, $E_{st} = 0.03E$
4. All members are assumed to be fully restrained against out-of-plane deformations.
5. Gravity loads are applied as concentrated loads at the beam quarter points. Wind loads combined with notional lateral loads equal to 0.5% of the story factored loads are applied as concentrated loads at the exterior joints.
6. The wind and lateral loads are applied proportionally in 5% increments with respect to the full factored loads until the full collapse load is attained.

10.7.2 Design of Panel Zones

The frame is first analyzed using the elastic-plastic-hinge analysis assuming rigid joints. The panel zones are then designed to resist the factored shear

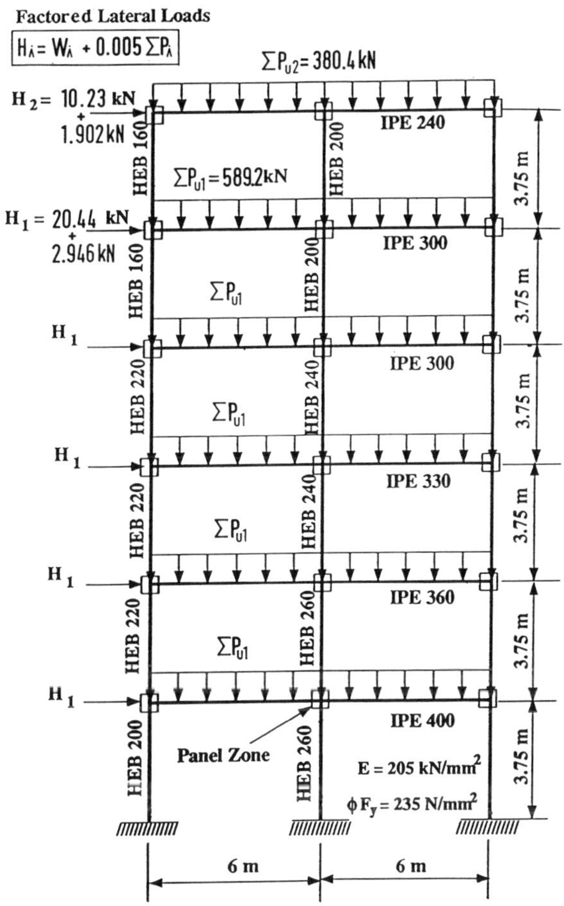

Fig. 10.11 Dimensions, loading, and member sizes of a six-story frame.

forces computed directly from the frame analysis. The design axial force P_u in Eqs. (10.4.2) and (10.4.3) and the design shear force V_u in Eq. (10.4.1) are calculated based on the average of the respective forces above and below the joint panel. The panel-zone dimensions and design forces are summarized in Tables 10.2 and 10.3, respectively.

The panel webs are reinforced by doubler plates when the calculated factored shear force F in the panel plane exceeds the column web shear resistance ϕR_v. Based on the two different shear-resistance criteria discussed, the following panel-zone joint designs are provided:

1. *Type 1 Joint.* A strong panel-zone design based on the first-yield criterion, Eqs. (10.4.2a) and (10.4.2b). This design ensures that the joint panel remains elastic under the application of full factored loads.

TABLE 10.1 Cross-Section Properties of Members

Member Attribute	A (cm^2)	I_x (cm^4)	I_y (cm^4)
IPE240	39.1	3,892	284
IPE300	53.8	8,356	604
IPE330	62.6	11,770	788
IPE360	72.7	16,270	1,043
IPE400	84.5	23,130	1,318
HEB160	54.3	2,492	889
HEB200	78.1	5,696	2,003
HEB220	91.0	8,091	2,843
HEB240	106.0	11,260	3,923
HEB260	118.0	14,920	5,135
HEB300	149.0	25,170	8,563
HEA340	133.0	27,690	7,436

2. *Type 2 Joint.* A ductile panel-zone design that utilizes the postyield shear resistance of the panel, Eqs. (10.4.3a) and (10.4.3b). Shear yielding of the panel zone is allowed prior to reaching the full factored loads.

The required thickness of doubler plates for the Type 1 and Type 2 joints are shown in Table 10.3. The doubler plate thickness is rounded to the nearest

TABLE 10.2 Detailed Dimensions of Panel Zones

Story No.	Panel Location	Beam Depth, d_b (cm)	Column Depth, d_c (cm)	Web Thickness, t_w (mm)	Flange Thickness, t_{cf} (mm)
1	Exterior	40	22	9.5	16.0
	Interior	40	26	10.0	16.0
2	Exterior	36	22	9.5	16.0
	Interior	36	26	10.0	17.5
3	Exterior	33	22	9.5	16.0
	Interior	33	24	10.0	17.0
4	Exterior	30	22	9.5	16.0
	Interior	30	24	10.0	17.0
5	Exterior	30	16	8.0	13.0
	Interior	30	20	9.0	15.0
Roof	Exterior	34	16	8.0	13.0
	Interior	24	20	9.0	15.0

TABLE 10.3 Double Plate Thickness for Panel-Zone Design

Story No.	Panel Location	Sum of Moments, $M_{u1} + M_{u2}$ (kN-m)	Average Shear Force, V_u (kN)*	Plate Thickness (mm)	
				Type 1 Joint	Type 2 Joint
1	Exterior	205	50.7	8	5
	Interior	323	56.0	19	10
2	Exterior	216	52.5	10	8
	Interior	278	45.4	14	10
3	Exterior	189	50.3	10	6
	Interior	234	32.0	13	10
4	Exterior	148	39.6	6	5
	Interior	216	21.4	13	10
5	Exterior	134	15.6	13	11
	Interior	247	8.9	22	20
Roof	Exterior	73	0.9	6	5
	Interior	167	1.8	18	14

*M_{u1}, M_{u2}, and V_u are the design forces at the factored load level obtained from the notional-load plastic-hinge analysis.

1/16 inch. As a result, the joint resistance is always greater than that required by Eq. (10.4.1).

In order to examine the effects of joint flexibility, two additional rigid-frame models, denoted as rigid joint and zero-size joint, are investigated. The use of a rigid-joint panel is the same as using 50% rigid end offsets in the structural analysis. The analysis model with zero-size joint is based on the member center-line dimensions instead of the member clear-span distance for member-stiffness formulation. As a result, the lateral stiffness of the frame model with a zero-size joint is often more flexible than that of the frame with rigid joints.

The load-displacement behaviors of sway steel frames modeled with rigid joints and zero-size joints are compared and contrasted with that of the frames with Type 1 and Type 2 joints. The results are summarized and described in what follows.

10.7.3 Results and Discussions

The load-displacement behaviors of rigid sway frames modeled with rigid joints and zero-size joints are compared and contrasted with those of the frames with Type 1 and Type 2 joints, as shown in Figs. 10.12 and 10.13. The lateral load-displacement plots are generated for the roof and the fourth story of the frame. It can be observed that the entire range of the load-

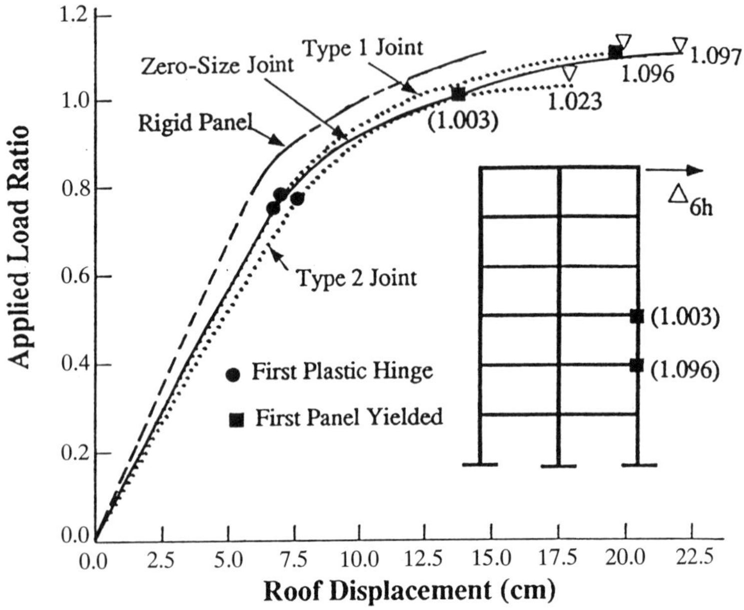

Fig. 10.12 Comparison of load-displacement curves at the roof level.

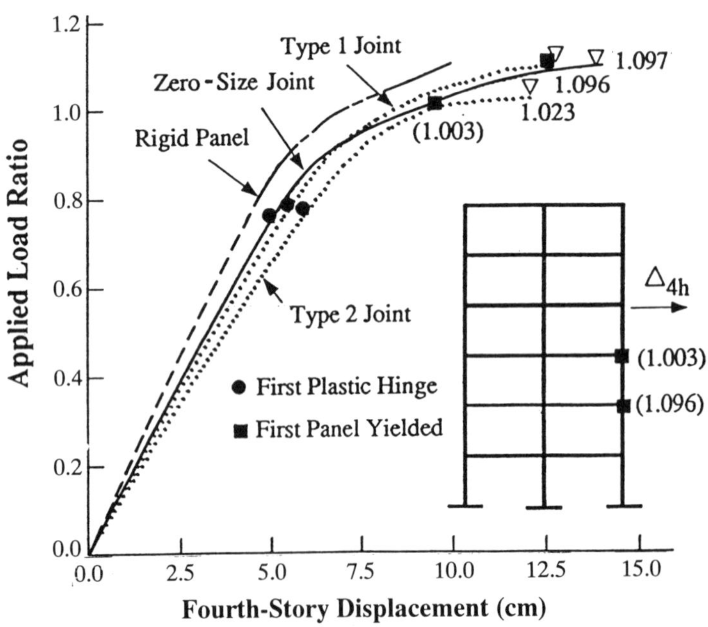

Fig. 10.13 Comparison of load-displacement curves at the fourth-story level.

displacement curves for the frame with a zero-size joint closely matches that of the frame model with a Type 1 joint design (strong panel). The maximum frame resistance is reached at a load parameter of 1.097 for the frame with zero-size joint and 1.096 for the frame with Type 1 joints. The limit loads predicted by these two frame models are very close.

The use of Type 2 joint design leads to a reduction in panel-zone stiffness. Also there is a noticeable decrease in the elastic frame stiffness at the onset of loading followed by a sharp decrease in frame stiffness in the postyield range. The loss of stiffness in the postyield range is attributed to the inelastic shear deformation of the panel zones and the subsequent formation of plastic hinges in the structure. Finally, the frame fails by sidesway instability due to excessive lateral displacement. Thus, allowing the panel zone to yield in shear at an early stage of loading leads to a reduction in the load-carrying capacity of the frame. The limit load predicted by the frame model with Type 2 joints is about 9% lower than that predicted by the frame model with Type 1 joints. The postyield stiffness of the Type 2 joint panel in the frame model is not significant enough to resist the additional P–Δ forces caused by the inelastic shear deformation of the yielded panel zones. Postyield resistance of the panel zone requires a large shear deformation, which is not achieved before reaching the limit load of the frame.

The lateral stiffness of a rigid sway frame is dependent on the rigidity of the panel-zone joints. The frame model with rigid joints overestimates the lateral stiffness of the frame model with Type 1 and Type 2 joints by about 20% and 30%, respectively. Therefore, it is more appropriate to analyze the comparable frame allowing for panel-zone flexibility. In general, not all the panel joints in a structural framework can undergo shear deformations that are large enough to quantify the assumed capacity increase defined by Eqs. (10.4.3a) and (10.4.3b). The postyield stiffness of the panel joints usually has little influence on the frame's limit of resistance; therefore, they may be ignored for stability analysis of rigid sway frames (Liew, 1992).

10.8 DESIGN IMPLICATIONS AND RECOMMENDATIONS

Under the prescribed lateral forces, the displacements of the frame model under factored loads with joint elements designed by the first-yield criterion, Eqs. (10.4.3a) and (10.4.3b), is larger than that calculated based on the member center-line distances. The difference is about 12% at the factored-load level. The advantage of this model is that the beam-column joints often require lighter doubler plates than that required by the Type 1 joint design.

The frame model with Type 2 joint elements shows an earlier yielding in the panel zones than the model using Type 1 joint elements. As discussed earlier, the panel-zone joint in this model is designed to deform beyond initial yielding prior to reaching full factored loads. However, Figs. 10.12 and 10.13 show that the first-yield criterion in the panel zones does not occur until the full factored load is reached (at an applied load ratio of 1.003). This is because the actual doubler plates as specified for the frame model with Type 2 joints are slightly thicker than the minimum required by Eq. (10.4.1). If the minimum required doubler-plate thickness is used instead, the frame analysis will predict an earlier yielding in the panel zone, and the load-displacement responses of the frame will be softer.

For tall building construction using deep beams and closely spaced columns, one would expect that an analysis based on member center-line dimensions may overestimate the lateral drift of such structural systems that would have otherwise worked had the rigidity of the joint been taken into consideration. This statement, however, is true only when the panel zones are designed based on the first-yield criterion. If the panel zone is flexible (such as with Type 2 joints), the deflection calculations based on a rigid panel may be unconservative. The calculated deflection for a frame with rigid joints at a load factor of 1.0 is about 18% smaller than the frame drift obtained for the frame with Type 1 joints, as shown in Fig. 10.13.

Experimental and analytical research (Fielding and Chen, 1973) suggested that panel-zone deformations should be accounted for in the analysis/design when the ratios of girder depth to story height (d_b/h) and column depth to girder span (d_c/L_b) are greater than the following limits:

$$\frac{d_b}{h} = \frac{1}{8}; \qquad \frac{d_c}{L_b} = \frac{1}{20} \tag{10.8.1}$$

For rigid frames in which the panel zones are unstiffened and the joint sizes are larger than that indicated by Eq. (10.8.1), the load-deflection behavior computed by second-order analysis based on member center-line distances and the analysis considering panel-zone deformations will be different (Liew, 1992).

Finally, it can be concluded that if the structural analysis program cannot account for panel-zone deformation, the design of the panel zone should be based on Eq. (10.4.1) with the shear-resistance term evaluated from Eqs. (10.4.2a) and (10.4.2b). For this type of frame model, the conventional structural analysis based on the member center-line to center-line distance is sufficiently accurate to approximate the effects of elastic panel-zone deformations on the frame under the prescribed factored loading. The analysis does not involve extra degrees of freedom or additional panel-zone elements in the stiffness formulation, so this simplified approach is favored in practical analysis procedures.

Otherwise, if the analysis program can account for the effects of panel-joint flexibility, Eqs. (10.4.3a) and (10.4.3b) may be used with advantage over Eqs. (10.4.2a) and (10.4.2b) for a more economical design of panel zones. Although some cost saving on the design of the beam-column joints is often possible, the analysis becomes more involved because of the extra degrees of freedom required for modeling the inelastic panel-joint behavior.

10.9 CONCLUSIONS

The use of doubler plates increases the local stiffness of panel zones and enhances the overall structural resistance and ductility against lateral-force action. Panel zones that are inadequately designed for resistance against full factored loads can undergo significant shear that will affect the strength and serviceability performances of the frame. Therefore, the effect of panel-zone deformations should be considered in the response analyses of such frames.

In this chapter, two methods for designing panel zones against high axial- and shear-force interaction have been identified. The first design method adopts the linearized form of von Mises general yield criterion for panel-web yielding. This design approach is adopted when the effect of panel-zone deformation on frame stability is not considered in the analysis. Conventional structural analysis based on the member center-line to center-line distance is sufficiently accurate to capture the frame response under the applied factored loading. The second design method utilizes the postyield inelastic shear resistance of the panel derived from the flexural resistance of flange elements surrounding the panel zone. In this approach, the additional inelastic shear resistance in the panel zone is recognized, and hence the overall frame stability, considering the effects of panel-zone deformation, should be considered in the analysis.

The studies presented in this chapter conclude that panel zones undergoing inelastic deformations will increase frame drift and reduce both the frame stiffness and resistance to sidesway instability. Also, the postyield stiffness of the panel joints is not significant enough to counteract the P–Δ effects caused by shear deformations of the yielded panel zones in the rigid frame. The static load analyses show that additional capacity available from the panel-zone boundaries is not significant enough to achieve the required postyield strength before reaching the frame's limit of resistance. Therefore, in the design of rigid frames against sidesway instability, the postyield shear resistance in the panel zones should not be relied upon in the design of a panel zone unless the analysis considers the direct effects of panel-zone deformations. The studies reported here are restricted to frames subjected to static loading only. Further studies are still required for structures with flexible panel joints under cyclic or earthquake types of loading.

REFERENCES

AISC (1990). *Load and Resistance Factor Design—Specification of Seismic Provisions for Structural Steel Buildings,* Chicago: American Institute of Steel Construction.

AISC (1993). *Load and Resistance Factor Design Specifications for Structural Steel Buildings,* 2nd ed., Chicago: American Institute of Steel Construction.

Becker, E. R. (1971). *Panel Zone Effect on the Strength and Stiffness of Rigid Steel Frames,* Structural Mechanics Laboratory Report USCCE001, Los Angeles: University of Southern California.

Bertero, V. V., E. P. Popov, and H. Krawinkler (1972). "Beam-Column Subassemblages under Repeated Loading," *Journal of the Structural Division,* ASCE, Vol. 98, No. ST5, pp. 1137–1159.

Chen, W. F., and J. Y. R. Liew (1992). "Seismic Resistance Design of Steel Moment Resisting Frames Considering Panel-Zone Deformation," in *Stability and Ductility of Steel Structures under Cyclic Loading,* Y. Fukumoto and C. G. Lee, eds., Boca Raton, FL: CRC Press, pp. 323–334.

Chen, W. F., and E. M. Lui (1991). *Stability Design of Steel Frames,* Boca Raton, FL: CRC Press.

Fielding, D. J., and W. F. Chen (1973). "Steel Frame Analysis and Connection Shear Deformation," *Journal of the Structural Division,* ASCE, Vol. 99, No. ST1, pp. 1–18.

Fielding, D. J., and J. S. Huang (1971). "Shear on Beam-to-Column Connections," *Welding Journal,* Vol. 50, no. 7, pp. 313s–326s.

Johnson, B. G. (1959). "Test on Welded Connections between I-Section Beams and Stanchions," *British Welding Journal,* Vol. 6, No. 1.

Kato, B., W. F. Chen, and M. Nakao (1988). "Effects of Joint-Panel Shear Deformation on Frames," in *Steel Beam-to-Column Building Connections,* W. F. Chen, ed., London and New York: Elsevier.

Krawinkler, H. (1978). "Shear in Beam-Column Joints in Seismic Design of Steel Frames," *Engineering Journal,* AISC, Vol. 5, No. 3, pp. 82–90.

Liew, J. Y. R. (1992). "Advanced Analysis for Frame Design," Ph.D. dissertation, School of Civil Engineering, Purdue University, West Lafayette, IN.

Liew, J. Y. R., D. W. White, and W. F. Chen (1994). "Notional Load Plastic Hinge Method for Frame Design," *Journal of Structural Engineering,* ASCE, Vol. 120, No. 5, pp. 1434–1454.

Lui, E. M. (1985). "Effect of Connection Flexibility and Panel Zone Deformation on the Behavior of Plane Steel Frames," Ph.D. dissertation, School of Civil Engineering, Purdue University, West Lafayette, IN.

Lui, E. M., and W. F. Chen (1986). "Analysis and Behavior of Flexibly-Jointed Frames," *Engineering Structures,* Vol. 8, No. 2, pp. 107–118.

Naka, T., B. Kato, and M. Watabe (1966). *Research on the Behavior of Steel Beam-to-Column Connections,* Laboratory for Steel Structures, Department of Architecture, University of Tokyo.

Nakao, M. (1975). "Research on the Behavior of Steel Beam-to-Column Rigid Connections," Doctoral dissertation, University of Tokyo, Tokyo (in Japanese).

Peters, J. W., and G. C. Driscoll, Jr. (1968). *A Study of the Behavior of Beam-to-Column Connections,* Fritz Laboratory Report 333.2, Bethlehem, PA: Lehigh University.

SEAOC (1988). *Recommended Lateral Force Requirements,* Sacramento, San Francisco, and Los Angeles: Seismology Committee, Structural Engineers Association of California.

Slutter, R. G. (1981). *Test of Panel Zone Behavior in Beam-Column Connection,* Report No. 200.81.403.1, Bethlehem, PA: Fritz Engineering Laboratory, Lehigh University.

Tanaka, K., et al. (1970). "Limit Analysis of Beam Column Connections, Experimental Verification," *Transactions of the Architectural Institute of Japan,* Vol. 170, No. IX.

Tsai, K. C., and E. P. Popov (1988). *Steel Beam-Column Joints in Seismic Moment Resisting Frames,* Report No. UCB/EERC, 88/19, Berkeley: Earthquake Engineering Research Center, University of California.

11

COMPUTER PROGRAMS FOR FRAME ANALYSIS AND CONNECTION MODELING

PART 1: A COMPUTER PROGRAM FOR ADVANCED INELASTIC ANALYSIS

11.1 INTRODUCTION

Part I describes a FORTRAN-based computer program, PHINGE, for advanced inelastic analysis of two-dimensional steel frames. The main purpose of developing this program is to provide an educational type software for engineers and graduate students to perform frame analysis for a more realistic prediction of system's strength and stability.

PHINGE has been tested in an IBM or equivalent personal computer system using Microsoft FORTRAN compiler Version 5.1. It also can be run on UNIX workstations using a standard FORTRAN 77 compiler. The memory required to run the program depends on the size of the problem. A computer with a minimum 640K of memory and a 30-MB hard disk is generally required. For PC application, the array sizes in the program have been reduced as follows:

Maximum total degrees of freedom:	MAXDOF=150
Maximum translational degrees of freedom:	MAXTOF=100
Maximum rotational degrees of freedom:	MAXROF=50
Maximum number of truss elements:	MAXTRS=20
Maximum number of connections:	MAXCNT=20

It is possible to run a larger job in a workstation by modifying the foregoing values in the parameter statements in the source code. The reader should read the file README, which is provided in the diskette, for the latest information about the programs.

PHINGE is developed based on a second-order elastic-plastic or refined plastic-hinge formulation using the stability function approach, which is discussed in Chapters 6 and 8. It is capable of performing first- and second-order elastic analysis subjected to proportional and nonproportional static loads. There are three basic elements in the analysis library: plane frame, plane truss, and connection elements. The connection element is represented by a zero-length rotational spring with a user-specified nonlinear moment–rotation curve in accordance with the three-parameter power model discussed in Chapter 9, Section 9.2.

Loading is allowed only at the nodal points with specified forces. Material nonlinearities are accounted for by using zero-length plastic hinges lumped at the element ends. The solution procedure is based on an incremental-load technique. The program can trace the force-displacement relationship up to the limit load.

The computer program consists of two main modules: INPUT and PHINGE. The INPUT module reads in the input data file, INFILE, and generates three working data files, DATA0, DATA1, and DATA2. The PHINGE module reads the working data files and enters into a run-time mode. Results are generated at each and every load step and they are written to three formatted text files. After the output files are generated, the user can view these files on the screen or print them for detailed information on element force distribution and nodal load versus displacement results. The schematic diagram shown in Fig. 11.1 explains the operation procedure of the computer program. Two executable program files named INPUT.EXE and PHINGE.EXE are supplied. These programs can be executed by issuing the commands "INPUT" and "PHINGE".

The complete source code listing for the computer program is provided in the enclosed diskette. The user is encouraged to modify or improve the program for his or her use. There are 19 source files for the complete program of PHINGE.

The necessary input instructions for the user to execute PHINGE are described in Section 11.5. The new user is advised to read all the instructions, paying particular attention to the appended notes, to gain an initial overall view of the data required to specify an analysis task to PHINGE. Readers without prior knowledge or experience in nonlinear analysis will benefit from careful study of the example problems given in Sections 11.6 and 11.7. These examples are intended to help clarify the data-generation process. Upon examining the input instructions, the reader should recognize that no system of units is assumed by the program. It is the user's responsibility to specify data in consistent units. Overlooking this requirement is a common source of error.

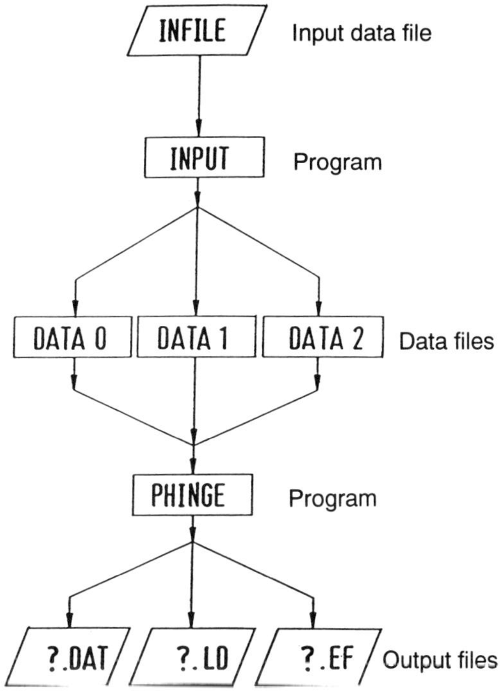

Fig. 11.1 Operating procedure of the PHINGE program.

11.2 INSTALLATION AND EXECUTION OF "PHINGE"

This section deals with the installation and execution of PHINGE on an MS DOS-based computer system, such as an IBM-compatible PC.

A computer diskette is provided, which contains the following:

1. PHINGE program executables (INPUT.EXE and PHINGE.EXE)
2. PHINGE batch file (RUN.BAT)
3. Sample examples and their output files (EX?.*)
4. Source codes written in FORTRAN 77 (in directory PHINGE)

The programs provided first must be copied onto the hard disk, that is, copying PHINGE.EXE, INPUT.EXE, INFILE, and RUN.BAT from the diskette to the hard disk. Before putting the program to work, the user should test the system by running the sample example provided on the same diskette.

Several input data files have been prepared and the details are explained in Sections 11.10 and 11.11. The user may test run the program by running the PHINGE program from the directory where the executable files are resident. After the execution of the PHINGE program, three working data files and three output files will be generated. The user can specify a file name for output, and the program will create three files with extensions of *.DAT, *.LD, and *.EF.

A batch file RUN.BAT has been created to facilitate the operation sequence. Assume that the data associated with the problem the user wishes to analyze have been created in a data file called INFILE. Entering the command RUN from the directory where the PHINGE.EXE and INPUT.EXE files are resident will activate a series of PHINGE programs in the required sequence for the analysis of the structure defined by the data file INFILE. The following sequence of actions is generated by the batch file, RUN.BAT, when it is activated:

1. Delete old data files, DATA0, DATA1, and DATA2.
2. Execute INPUT.EXE, which reads the input data file INFILE and generates three new data files: DATA0, DATA1, and DATA2.
3. Execute PHINGE.EXE, which reads the data files DATA0, DATA1, and DATA2, and generates three output files: *.DAT, *.LD, and *.EF. The PHINGE program will prompt the user to input the output file name, and the final output files consist of three files with a user-specified file name and the extensions DAT, LD and EF, which correspond to results for Data file, Load-Displacement and Element End Forces. The schematic diagram shown in Fig. 11.1 explains the sequential operation procedures generated by the batch file RUN.BAT.

After *.LD and *.EF files are generated, the user can view these files on the screen or print them for detailed information on force-distribution and load-deformation characteristics of the structure. To print an output file, the MS-DOS "PRINT" command may be used.

It should be noted that RUN is a batch command that facilitates the execution of PHINGE and clearing of old working files and output files. Entering the command RUN will erase the old files, DATA0, DATA1, and DATA2, before creating the new ones.

11.3 THEORY

Program PHINGE is developed based on the theory described in Chapters 6–9. The program can perform first-order elastic, second-order elastic-plastic, and second-order refined plastic-hinge analyses. Planar steel frames with rigid or semi-rigid connections can be analyzed. The second-order elastic-plastic-analysis procedure described in Chapters 6 and 7 can be reduced to first- and second-order elastic analysis by specifying appropriate load step increments and material properties in the analysis input data file. For first-order analysis, the total factored load should be applied in one load increment to suppress numerical iteration in the nonlinear analysis algorithm. For elastic analysis, section plastic moduli or material yield strength of arbitrary large values should be assumed for all beam-column elements to prevent the formation of plastic hinges.

Because second-order plastic-hinge-based analysis involves both geometric and material nonlinearities, recourse to the tangent-stiffness matrix, which is a linearized form of the nonlinear force-displacement relationships, is used to obtain solutions. In other words, instead of solving a set of nonlinear equilibrium equations, a set of linearized equations is solved at each cycle of calculation.

In the PHINGE program, the simple incremental method has been implemented for second-order elastic-plastic-hinge analysis. The refined plastic-hinge model is implemented using an automatic load-increment procedure. To prevent plastic hinges from forming within a constant-stiffness load increment, load step sizes less than or equal to the specified increment magnitude are internally computed such that plastic hinges form only after the load increment. Thus, the subsequent element stiffness formulation will account for the stiffness reduction due to the presence of plastic hinges.

For elements that are partially yielded at their ends, a limit is placed on the magnitude of the increment in the element end forces (Liew, 1992). Increments of these force components that are relatively large, compared to their respective plastic strengths, may indicate that the path taken from the initial yield surface to the plastic strength surface may have deviated excessively from the "exact" path, upon which the calculation of element effective stiffness may lead to some error. The analysis program automatically scales an attempted load increment when the change in the element stiffness parameter exceeds a predefined tolerance. A tolerance value has been specified in the solution scheme to suppress the error that may be incurred during the computation (Liew, 1992).

The nonlinear analysis routines in PHINGE are based on the matrix stiffness method and involved the solving of the following incremental force-displacement equation:

$$\mathbf{K}_t\dot{\mathbf{d}} = \dot{\mathbf{f}} \tag{11.3.1}$$

where $\dot{\mathbf{d}}$ is the vector of unknown incremental nodal displacements, $\dot{\mathbf{f}}$ is the vector of known equivalent incremental nodal loads, and $\mathbf{K}_t$ is the global tangent-stiffness matrix, which reflects the current state of the deformed structure with the corresponding effective stiffness. Matrix $\mathbf{K}_t$ is the assembly of the transformed element tangent-stiffness matrices, as discussed in Section 6.5.

11.4 CONVENTION AND TERMINOLOGY

This section deals with specific conventions and terminology used by PHINGE in the input preparation and the output interpretation phases of the analysis. Data preparation involves (1) describing the analysis method, size of the problems, and output format; (2) defining element properties and structural geometry; and (3) defining the loading conditions for which the structure needs to be analyzed.

11.4.1 Joints and Elements

The geometric dimensions of the structures are established by placing joints (or nodal points) on the structures. Each joint is given an identification number and is located in a plane associated with a global two-dimensional coordinate system.

The structural geometry is completed by connecting the predefined joints with structural elements that are of a specific type, namely, frame, truss, and connection. Each element has an identification number, too.

The following are some of the factors that need to be considered in locating joints on a structure:

1. The number of joints should be sufficient to describe the geometry and the response behavior of the structure.
2. Joints need to be located at points and lines of discontinuity, for example, at a change in material properties, section properties, and so on.
3. Joints should be located at points on the structure where forces and displacements are to be evaluated.
4. Joints should be located at points on the structure where concentrated loads are to be applied. The applied loading should be represented as concentrated loads acting on specific joints in the members.
5. Joints should be located at all support points. Support conditions are represented in the structural model by restricting the movement of the specific joints in specific directions.
6. Element mesh should be refined enough to capture the inelastic behavior and second-order effects in the regions of interest.

11.4.2 Coordinate Systems

For the generation of all the input and output data associated with the joints, a two-dimensional (X, Y) global coordinate system is used. The following input data are prepared with respect to the global coordinate system:

1. Joint coordinates
2. Joint restraints

3. Joint connection spring
4. Joint loading
5. Joint constraints

The following output is also referred in the global coordinate system:

1. Joint displacements
2. Joint reactions

11.4.3 Degrees of Freedom

A two-joint frame element has six displacement components, as shown in Fig. 11.2. Each joint can translate in the global X- and Y-directions, and it can rotate about the global Z-axis. The directions associated with these displacement components are known as *degrees of freedom* of the joint.

A two-joint truss element has four degrees of freedom, as shown in Fig. 11.3. Each joint consists of two translational degrees of freedom and with no rotational components.

If the displacement of a joint corresponding to any one of its joint's degrees of freedom is known to be zero, such as at a support point, then the degree of freedom is known as an inactive degree of freedom. Degrees of freedom at which the displacements are not known are called active degrees of freedom. In general, the displacement of an inactive degree of freedom is usually known, and the purpose of the analysis is to find the reaction in that direction. For an active degree of freedom, the applied load is usually known (could be

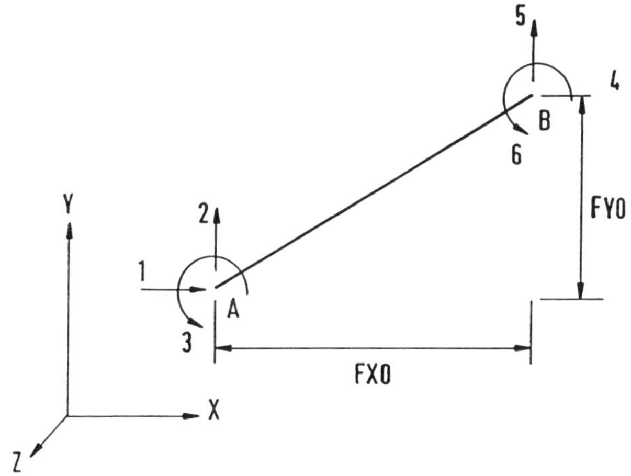

Fig. 11.2 Degrees-of-freedom numbering for the beam-column element.

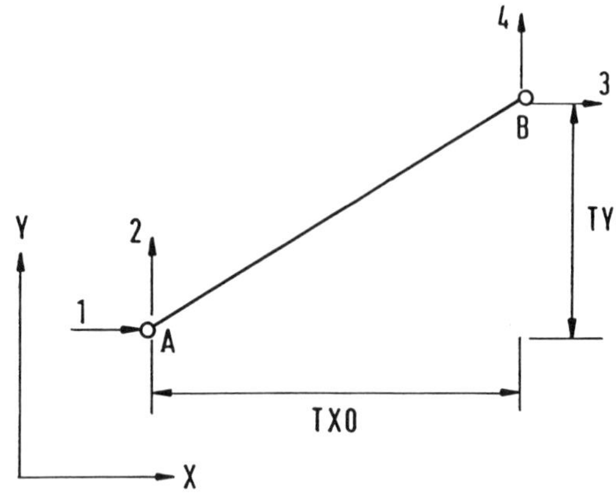

Fig. 11.3 Degrees-of-freedom numbering for the truss element.

zero) and the purpose of the analysis is to find the corresponding displacement.

11.4.4 Units

There are no built-in units in the PHINGE computer program. The user must prepare the input in a consistent set of units. The output produced by the program will then conform to the same set of units. Therefore, if the user chooses to use kips and inches as the input units, all the dimensions of the structure must be entered in inches and all the loads in kips. The material properties should also conform to these units. The output units will then be in kips and inches, so that the frame member axial force will be in kips, bending moments will be in kips-inch units, and displacements will be in inches. However, joint rotations are in radians, irrespective of units.

11.4.5 Accumulated and Incremental Loads

The initial tangent-stiffness matrix is formulated based on the state of zero axial force in the structure. The structure must be stable in this state. Successive load stages are *accumulated* by adding the load increment to the load from the previous level. For the case where a linearly elastic analysis is required, a one-step load increment should be used, and the design yield strength F_y should be set to a very large value so that a plastic hinge will not form anywhere in the structure.

11.4.6 Numerical Accuracy and Tolerance

To control the accuracy of the displacements and limit load, the load increment should be small enough so that accumulated errors can be minimized.

From experience, load step size of 0.05 times the factored load or less may be required. When the structure is approaching a state of equilibrium or when a plastic hinge is formed within the load increment, the analysis program automatically scales an attempted load increment in order to achieve the required accuracy.

11.5 DATA PREPARATION

Described in what follows is the input sequence and data structure to create an input file called INFILE. The analysis program, PHINGE, can analyze any frame structures with maximum degrees of freedom not more than 150. However, it is possible to recompile the source code to accommodate more degrees of freedom by changing the size of the arrays in the PARAMETER statements.

The input data file INFILE is prepared in a specific format. The following input sequence must be followed:

1. Title Line
2. Analysis Method
3. Analysis Attributes
4. Job Control
5. Element Types
6. Element Groups
7. Connection Properties
8. Frame Element Properties
9. Truss Element Properties
10. Connection Element Data
11. Frame Element Data and Geometry
12. Truss Element Data and Geometry
13. Nodal load

Some of the data input are mandatory; however, the existence of some of the elements in the input data depends on the problem being analyzed. The order in which the data lines occur in the input file must be followed.

The various functions and uses of these data are summarized in the following.

Line 1: Title Line

Note	Columns	Variable	Description
(1)	1–8	—	A comment card

Notes:

1. The comment line is skipped, without interpretation, at the execution of INPUT. This line is usually required to identify one input file from the other.

Line 2 (I5): Analysis Method

Note	Columns	Variable	Description
(1)	1–5	ISOLVE	Specify the analysis method. Specify ISOLVE = 0 for elastic-plastic-hinge analysis and ISOLVE = 1 for refined plastic-hinge analysis.

Notes:

1. IF ISOLVE = 0 is specified, the second-order elastic-plastic hinge-analysis is activated. If ISOLVE = 1 is specified, the second-order refined plastic-hinge method described in Section 8.7 is initiated.

Line 3 (2I5): Analysis Attributes

This line is skipped if ISOLVE = 0.

Note	Columns	Variable	Description
(1)	1–5	IET	Tangent-modulus model IET = 1, CRC tangent modulus IET = 2, LRFD tangent modulus
(2)	6–10	IPHI	Element stiffness-degradation model (IPHI = 1) IPHI = 1, parabolic stiffness reduction

Notes:

1. If IET = 1 is selected, the CRC tangent-modulus formula, as given in what follows, is adopted in the refined plastic-hinge analysis for members subjected to compression or tension:

$$\frac{E_t}{E} = \begin{cases} 1.0, & \text{for } P \le 0.5P_y \\ \dfrac{4P}{P_y}\left(1 - \dfrac{P}{P_y}\right), & \text{for } P > 0.5P_y \end{cases} \tag{11.5.1}$$

If IET = 2 is specified, the LRFD tangent-modulus formula, as shown in what follows, is effected for members subjected to axial compression only:

$$\frac{E_t}{E} = \begin{cases} 1.0, & \text{for } P \leq 0.39P_y \\ -2.7243\left(\dfrac{P}{P_y}\right) \ln\left(\dfrac{P}{P_y}\right), & \text{for } P > 0.39P_y \end{cases} \quad (11.5.2)$$

For members subjected to tension, the CRC tangent-modulus model is automatically selected, as discussed in Chapter 8, Section 8.3.1.

2. If IPHI = 1 is selected, the parabolic stiffness-reduction function given in what follows will be used in the refined plastic-hinge analysis:

$$\phi = 4\alpha(1 - \alpha), \qquad \text{for } \alpha > 0.5 \quad (11.5.3)$$

Equation (11.5.3) is a parabolic stiffness-degradation function based on an initial yield surface that is the same shape and one-half the size of the plastic strength surface. It has been shown in Chapter 8, Section 8.7, that the function produces sufficiently accurate results in comparison with the more "exact" plastic-zone solutions. Other possible stiffness-reduction schemes used in calibrating the refined plastic-hinge model to best fit the "exact" plastic-zone solutions are given in the subroutine ESTIFF. The reader is referred to Chapter 8, Section 8.5, or Liew et al. (1993a, 1993b) for detailed information pertinent to this aspect.

Line 4 (4I5): Job Control

Note	Columns	Variable	Description
(1)	1–5	NDOFS	Number of degrees of freedom of the structure (≤ 150)
(2)	6–10	NINCRE	Number of load increments
(3)	11–15	NSEQN	Number of load sequences (≤2)
(4)	16–20	IPRTFC	Flag for printing member end forces IPRTFC = 0, do not print element end forces IPRTFC = 1, print element end forces

Notes:

1. The number of degrees of freedom of the structure is required. This value should not be larger than the default limit of 150. If this value is exceeded, the user needs to recompile the source code by setting a larger value for "MAXDOF" in the PARAMETER statements.
2. The total load incremental steps are required. The analysis is terminated when the number of the total load incremental steps is reached. For first-order analysis, the total factored load should be applied in one load increment to suppress numerical iteration in the nonlinear analysis algorithm, that is, let NINCRE = 1.

3. Only two load sequences are allowed in the present version of the program.
4. Member end forces are not printed if IPRTFC = 0. These forces are printed at each step of load increment if IPRTFC = 1; this means a larger storage capacity is required for OUT.EF.

Line 5 (3I5): Element Types

Note	Columns	Variable	Description
(1)	1–5	NCTYPE	Number of connection types (≤ 10)
(1)	6–10	NFTYPE	Number of frame element types (≤ 20)
(1)	11–15	NTTYPE	Number of truss element types (≤ 10)

Note:

1. If the limits specified before are exceeded, the user needs to recompile the program by specifying suitable values for the variables in the PARAMETER statements of the source code.

Line 6 (3I5): Element Groups

Note	Columns	Variable	Description
(1)	1–5	NUMCNT	Number of connection elements (≤ 10)
(1)	6–10	NUMFRM	Number of frame elements (≤ 150)
(1)	11–15	NUMTRS	Number of truss elements (≤ 20)

Note:

1. If the limits specified before are exceeded, the user needs to recompile the program by specifying suitable values for the variables in the PARAMETER statements of the source code.

Line 7 (I5, 3D10.0): Connection Properties

This line should be omitted if NCTYPE = 0. Otherwise, this line must be supplied NCTYPE times.

Note	Columns	Variable	Description
(1)	1–5	ICTYPE	Connection-type identifier (an integer)
(2)	6–15	MU	Connection ultimate moment capacity
(2)	16–25	RKI	Connection initial stiffness
(2)	26–35	n	Connection shape parameter

Notes:

1. Connection property sets must be specified in the order as shown. Users should refer to Section 8.8 for the definition of the connection parameters.
2. The connection property set is stored in an array CTYPE(ICTYPE, J=1,3). Step-by-step procedures for calculating M_u and R_{ki} for semi-rigid connections composed of angles are given in Section 9.3.

Line 8 (I5, 5D10.0): Frame Element Properties

If NFTYPE ≠ 0, this line must be supplied NFTYPE times. Otherwise, this line should be omitted.

Note	Columns	Variable	Description
	1–5	IFTYPE	Frame-type identifier (an integer)
(1)	6–15	A	Cross-sectional area
(1)	16–25	I	Moment of inertia
(1)	26–35	Z	Plastic section modulus
(1)	36–45	E	Modulus of elasticity
(1) & (2)	46–55	FY	Material yield strength

Notes:

1. Material property sets for the frame elements must be specified in the order as shown. The material set is stored in an array FTYPE(IFTYPE, J=1,5).
2. For elastic analysis, material yield strength of arbitrary large values should be assumed for all elements to prevent the formation of plastic hinges due to bending actions, say, FY = 1E+05.

Line 9 (I5, 4D10.0): Truss Element Properties

If NTTYPE ≠ 0, this line must be supplied NTTYPE times. Otherwise, this line should be omitted.

Note	Columns	Variable	Description
	1–5	ITTYPE	Truss-type identifier (an integer)
(1)	6–15	A	Cross-sectional area
(1)	16–25	I	Moment of inertia
(1)	26–35	E	Modulus of elasticity
(1)	36–45	FY	Yield stress

Note:

1. Material property sets for the truss elements must be specified in the order as shown. For a truss element under axial compression, the anal-

ysis is terminated once the axial force in the truss element exceeds the column strength implied by the LRFD column-strength equations 8.3.1 and 8.3.2. Therefore, the moment of inertia for weak-axis bending should be specified for a pinned-end truss element. For an element subjected to tensile force, the cross-section effective area (with allowance for bolt holes) may be specified. The material set is stored in an array TTYPE(ITTYPE,J=1,4).

Line 10 (4I5): Connection Element Data

This line should be omitted if NUMCNT = 0. Otherwise, this line must be supplied NUMCNT times.

Note	Columns	Variable	Description
	1–5	LCNT	Connection element number
	6–10	IFMCNT(LCNT)	Beam-column element number to which the connection is attached
(1)	11–15	IEND(LCNT)	Beam-column element end to which the connection is attached IEND = 1 for connection at end A IEND = 2 for connection at end B
	16–20	JDCNT(LCNT)	Connection-type number

Note:

1. See Fig. 11.2 for the respective end of the beam-column element.

Line 11 (I5, 2D10.0, 7I5): Frame Element Data and Geometry

This line should be omitted if NUMFRM = 0. Otherwise, this line must be supplied NUMFRM times.

Note	Columns	Variable	Description
	1–5	LFRM	Frame element number
(1)	6–15	FXO(LFRM)	Horizontal projected length
(1)	16–25	FYO(LFRM)	Vertical projected length
	26–30	JDFRM(LFRM)	Frame-type number
(2)	31–35	NFRMCO(JDFRM,1)	Node number for degree of freedom 1
(2)	36–40	NFRMCO(JDFRM,2)	Node number for degree of freedom 2
(2)	41–45	NFRMCO(JDFRM,3)	Node number for degree of freedom 3
(2)	46–50	NFRMCO(JDFRM,4)	Node number for degree of freedom 4
(2)	51–55	NFRMCO(JDFRM,5)	Node number for degree of freedom 5
(2)	56–60	NFRMCO(JDFRM,6)	Node number for degree of freedom 6

Note:

1. The projected lengths for a frame element are shown in Fig. 11.2. The nodal numbering sequence must follow the order of the degrees of freedom for the frame element shown in Fig. 11.2.

Line 12 (I5, 2D10.0, 5I5): Truss Element Data and Geometry

This line should be omitted if NUMTRS = 0. Otherwise, this line must be supplied NUMTRS times.

Note	Columns	Variable	Description
	1–5	LTRS	Truss element number
(1)	6–15	TXO(LTRS)	Horizontal projected length (in inches)
(1)	16–25	TYO(LTRS)	Vertical projected length (in inches)
	26–30	JDTRS(LTRS)	Truss-type number
(2)	31–35	NTRSCO(JTRS,1)	Node number for degree of freedom 1
(2)	36–40	NTRSCO(JTRS,2)	Node number for degree of freedom 2
(2)	41–45	NTRSCO(JTRS,3)	Node number for degree of freedom 3
(2)	46–50	NTRSCO(JTRS,4)	Node number for degree of freedom 4

Note:

1. The projected lengths for a truss element are shown in Fig. 11.3. The nodal numbering sequence must follow the order of the degrees of freedom for the truss element shown in Fig. 11.3.

Line 13 (I5, 3D10.0): Nodal Load

Note	Columns	Variable	Description
(1)	1–5	NLDOF	Degree of freedom at which a load is applied
(2)	6–15	FAMG(I,1)	Magnitude of load increment of load sequence 1
(3)	16–35	FAMG(I,2)	Magnitude of load at which load sequence 1 ends
(3)	36–45	FAMG(I,3)	Magnitude of load increment of load sequence 2

Notes:

1. All concentrated forces must be applied at the nodal point in the X and Y directions of the global coordinates by specifying the translational degrees of freedom of the node. A concentrated moment may be applied at the nodal point by specifying the rotational degree of freedom of the node.
2. If the number of load increments NINCRE = 1, the values assigned to FAMG(I,2) and FAMG(I,3) will not affect the analysis results. In this case, the analysis is terminated when the number of load increment NINCRE is reached or when the structural stiffness matrix is nonposi-

tive definite after attempting several load-reduction steps near the maximum load point.

3. If NINCRE = 2, the user must specify values for FAMG(I,2) and FAMG(I,3) so that the analysis knows when to apply load sequence 2 once the total load parameter for load sequence 1 is reached.

11.6 EXAMPLE 1: ANALYSIS OF A ONE-STORY PORTAL FRAME

11.6.1 Rigid Plastic Analysis

Figure 11.4 shows a rectangular portal frame subjected to two concentrated vertical loads acting at the one-third points of the beam. This example is similar to the problem given in Section 5.5.1 in which the members are proportioned based on plastic analysis.

The collapse mechanism based on plastic analysis is a beam mechanism shown in Fig. 11.5. By equating the external work due to applied loads to the internal energy dissipation, the collapse load factor λ is obtained as

$$80\lambda(20\theta) + 60\lambda(10\theta) = M_p(2\theta) + 1.94M_p(3\theta) + M_p\theta$$

$$\lambda = \frac{8.82}{2200}M_p$$

$$\Rightarrow \lambda = \frac{8.82}{2200}\left(\frac{303}{12}\right) = 1.215 \qquad (11.6.1)$$

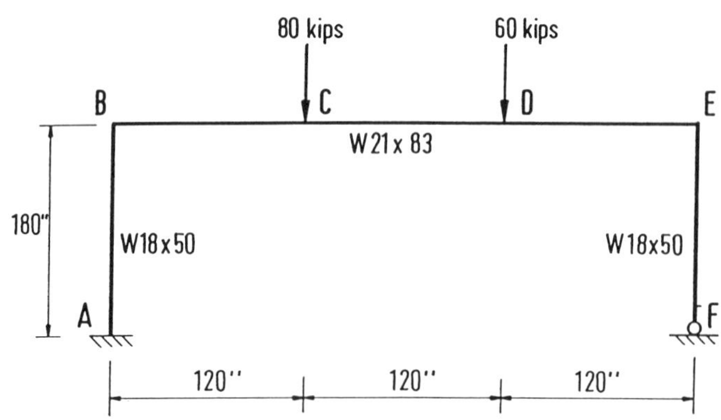

Fig. 11.4 Portal frame subjected to two concentrated loads.

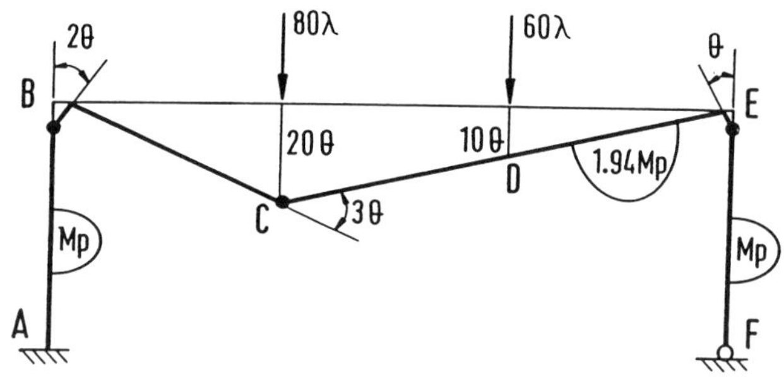

Fig. 11.5 Collapse mechanism based on plastic analysis.

From equilibrium, the bending moment diagram and reaction forces at collapse (i.e., $\lambda = 1.215$) are shown in Figure 11.6. One of the disadvantages of plastic analysis is that the state of the frame at the factored load level cannot be determined readily. One has to perform a hinge-by-hinge analysis in order to obtain the "correct" force distribution in the frame for the purpose of member proportioning.

In order to perform member-capacity checks, the bending moment diagram at the design load factor (i.e., at $\lambda = 1.0$) must be established. This can be obtained by factoring the bending moment diagram at collapse by a factor of 1.215 to achieve the bending moment diagram at the factored load level. The resulting bending moment diagram is shown in Fig. 11.7.

11.6.2 Second-Order Elastic-Plastic-Hinge Analysis

A second-order elastic-plastic-hinge analysis considers both the effects of geometric and material nonlinearities in a more exact manner. The effects of

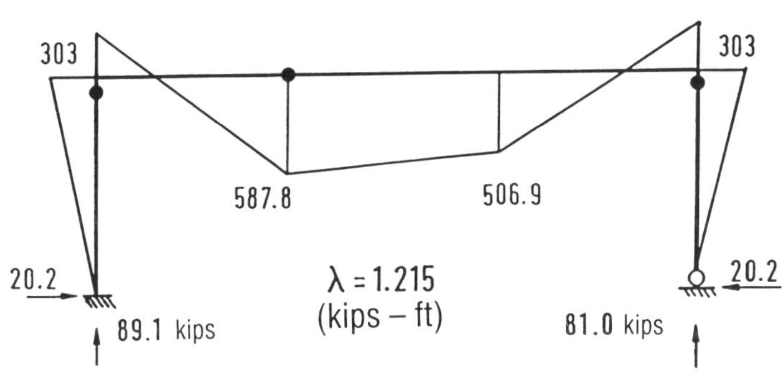

Fig. 11.6 Force distribution at the incipience of collapse based on plastic analysis.

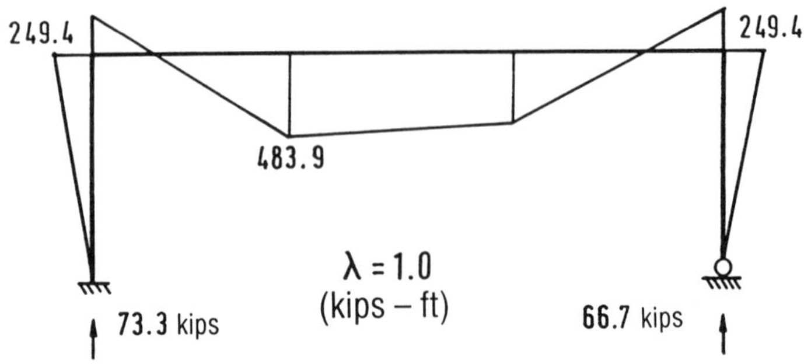

Fig. 11.7 Force distribution at the factored load level based on plastic analysis.

residual stress and geometric initial imperfections, however, are not modeled in the analysis. The problem is nonlinear, so the analysis has to be carried out using small load increments.

For the frame shown in Fig. 11.4, the beam is discretized into three elements, and the columns are represented by one element per member. The computer model is shown in Fig. 11.8. The input file, INFILE, to perform the second-order elastic-plastic analysis is shown in Fig. 11.9. Kips-inch units are used. All members are modeled using beam-column elements, and the loads are applied proportionally to collapse.

The resulting bending moment diagram and support reaction forces at the maximum load factor (at Load Step 31, $\lambda = 1.166$) and at the design load factor (at Load Step 21, $\lambda = 1.0$) are shown in Figs. 11.10 and 11.11, respectively. The first plastic hinge is formed at the top right-hand column at

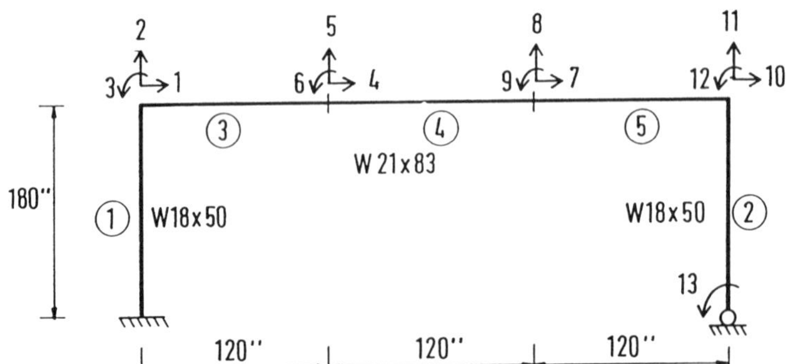

Fig. 11.8 Structural modeling of the portal frame shown in Fig. 11.4.

```
/*SECOND-ORDER ELASTIC-PLASTIC-HINGE ANALYSIS
   0
  13    50       1      1
   0     2       0
   0     5       0
   1     14.70        800.0      101.00      29000.       36.00
   2     24.30       1830.0      196.00      29000.       36.00
   1       0.0        180.0      1     0     0     0     1     2     3
   2       0.0        180.0      1     0     0    13    10    11    12
   3     120.0          0.0      2     1     2     3     4     5     6
   4     120.0          0.0      2     4     5     6     7     8     9
   5     120.0          0.0      2     7     8     9    10    11    12
   5     -4.00        -80.0       0.000
   8     -3.00        -60.0       0.000
```

Fig. 11.9 Input data file, INFILE, for the second-order elastic-plastic-hinge analysis of a portal frame ("EX1" in the diskette).

$\lambda = 0.95$, and the second plastic hinge is at the larger concentrated load point at $\lambda = 1.02$. The last plastic hinge is formed at the top left-hand column at the incipience of collapse with the applied load factor $\lambda = 1.166$.

The complete output data files are stored in the diskette and are renamed as EX1, EX1.DAT, EX1.LD, and EX1.EF.

11.6.3 Second-Order Refined Plastic-Hinge Analysis

The frame model shown in Fig. 11.8 is now analyzed using the second-order refined plastic-hinge analysis. The analysis adopted the LRFD tangent modulus in which the explicit modeling of member initial imperfections is not

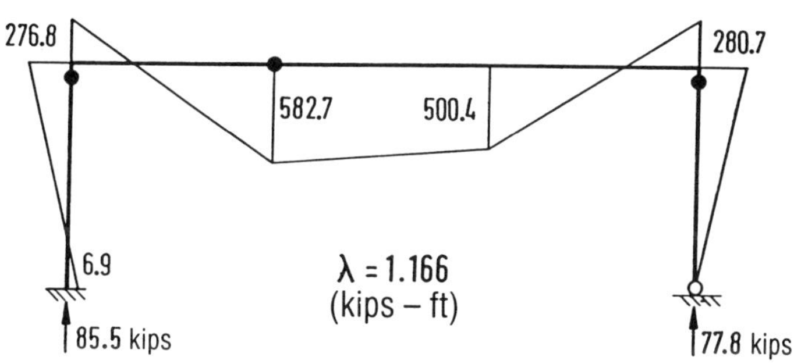

Fig. 11.10 Force distribution at the maximum load based on the second-order elastic-plastic-hinge analysis.

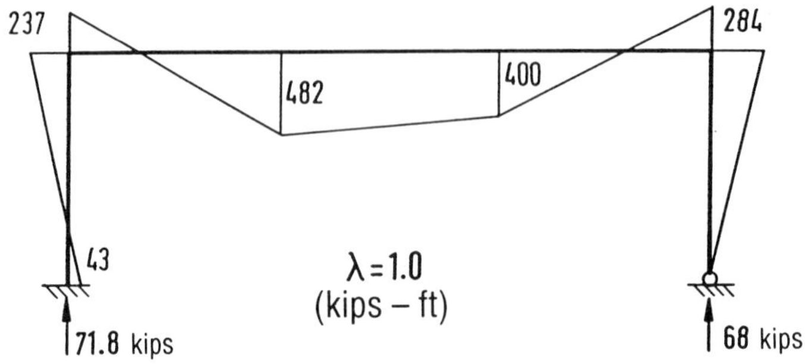

Fig. 11.11 Force distribution at the factored load level based on the second-order elastic-plastic-hinge analysis.

required. The input file, INFILE, for this problem is shown in Fig. 11.12. A large number of load increments (NINCRE = 40) have been specified for limit-load analysis. The bending moment diagrams at the design-load factor (at Load Step 21) and at the limit-load factor (at Load Step 30) are shown in Figs. 11.13 and 11.14, respectively.

The complete output data files are stored in the diskette, and they are renamed as EX1a (for INFILE), EX1a.DAT, EX1a.LD, and EX1a.EF.

```
/*SECOND-ORDER REFINED PLASTIC-HINGE ANALYSIS
  1
  2    1
 13   40     1     1
  0    2     0
  0    5     0
  1     14.70      800.0    101.00    29000.      36.00
  2     24.30     1830.0    196.00    29000.      36.00
  1       0.0      180.0    1    0    0    0    1    2    3
  2       0.0      180.0    1    0    0   13   10   11   12
  3     120.0        0.0    2    1    2    3    4    5    6
  4     120.0        0.0    2    4    5    6    7    8    9
  5     120.0        0.0    2    7    8    9   10   11   12
  5     -4.00      -80.0     0.000
  8     -3.00      -60.0     0.000
```

Fig. 11.12 Input data file, INFILE, for second-order refined plastic hinge analysis of a portal frame ("EX1a" in the diskette).

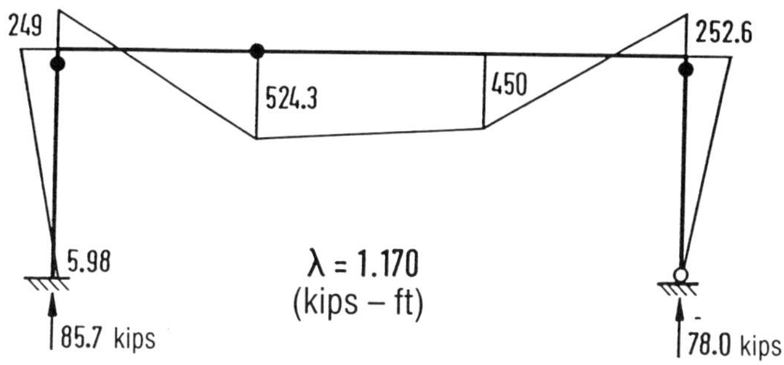

Fig. 11.13 Force distribution at the maximum load based on the second-order refined plastic-hinge analysis.

11.6.4 Comparison of Results

Load deflection results from the three analyses are compared in Fig. 11.15. The rigid plastic analysis predicts a larger collapse load at $\lambda = 1.215$, followed by the second-order elastic-plastic hinge at $\lambda = 1.166$ and the refined plastic-hinge analysis at $\lambda = 1.170$, in which the latter two results are very close to each other. In the elastic-plastic-hinge analysis, the first plastic hinge is formed at the beam's right end, denoted as point E in the frame. The applied load factor corresponding to the formation of the first plastic hinge is $\lambda = 0.95$. Because the elastic-plastic-hinge analysis neglects the effects of residual stresses and gradual plastification in the beam, the distribution of internal forces during the loading process is somewhat different from that predicted by the refined plastic-hinge method. However, for this particular frame, the

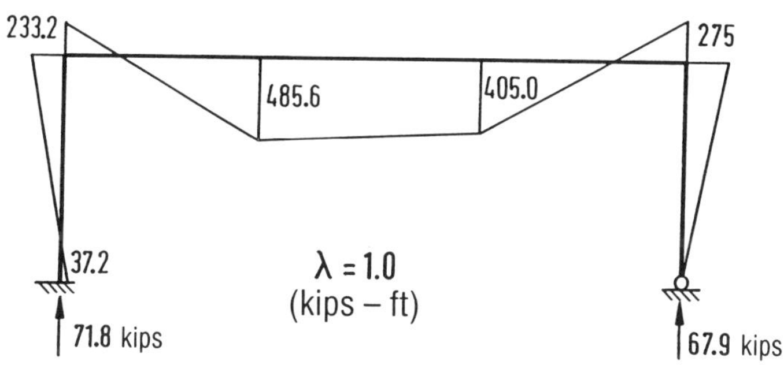

Fig. 11.14 Force distribution at the factored load level based on the second-order refined plastic-hinge analysis.

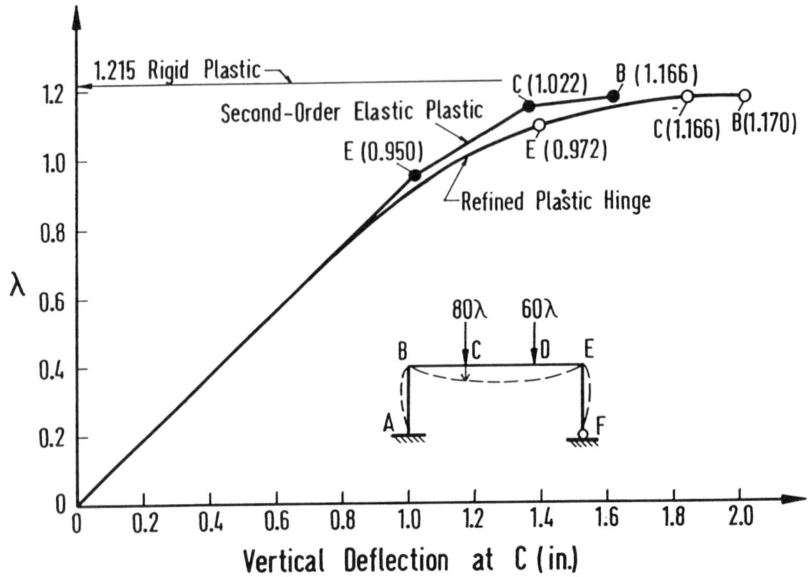

Fig. 11.15 Comparison of load-displacement curves for the portal frame.

inelastic-force-redistribution process does not have any influence on the prediction of the ultimate strength. This is because the second-order effect is virtually absent, and frame failure is due to the formation of a beam mechanism.

In the refined plastic-hinge analysis, the first plastic hinge is formed at point E in the frame. The load factor corresponding to the formation of the first plastic hinge is $\lambda = 0.972$, which is slightly higher than the load at the first hinge predicted by the plastic-hinge analysis. The formation of the first plastic hinge is delayed in the refined plastic-hinge analysis because the inelastic-force redistribution occurs before the attainment of full plastic section capacity at E. The gradual plastification at E tends to shed load to other less critical locations and therefore leads to a smooth transition of stiffness from the elastic to the inelastic range.

Analyses based on rigid plastic analysis, second-order elastic-plastic, and second-order refined plastic-hinge methods predict different load-displacement behavior because of the different degree of accuracy in predicting the second-order and inelasticity effects in the frame. The plastic analysis does not provide any information concerning the load-deflection behavior of the frame. The elastic-plastic-hinge approach predicts a piecewise linear curve, and the refined plastic-hinge approach predicts a smooth curve with gradual stiffness degradation when the limit load is approached.

11.7 EXAMPLE 2: ANALYSIS OF A SEMI-RIGID BRACED FRAME

Figure 11.16 shows a semi-rigid frame braced by truss diagonals and subjected to the following load combination:

$$1.0 \text{ Gravity} + 1.3 \text{ Wind} \tag{11.7.1}$$

The computer model for the braced semi-rigid frame is shown in Fig. 11.17. For illustration purposes, only the refined plastic-hinge analysis is used. All beams are modeled by four discrete elements and all columns by one element. The loads are applied at the nodal points in 5% increments with respect to the full factored loads. Two load sequences are prescribed. The first load sequence applies incremental forces up to the factored load level, and is then followed by the second load increment applied up to the maximum load point. Frames with column bases assumed to be fully pinned or fixed are analyzed.

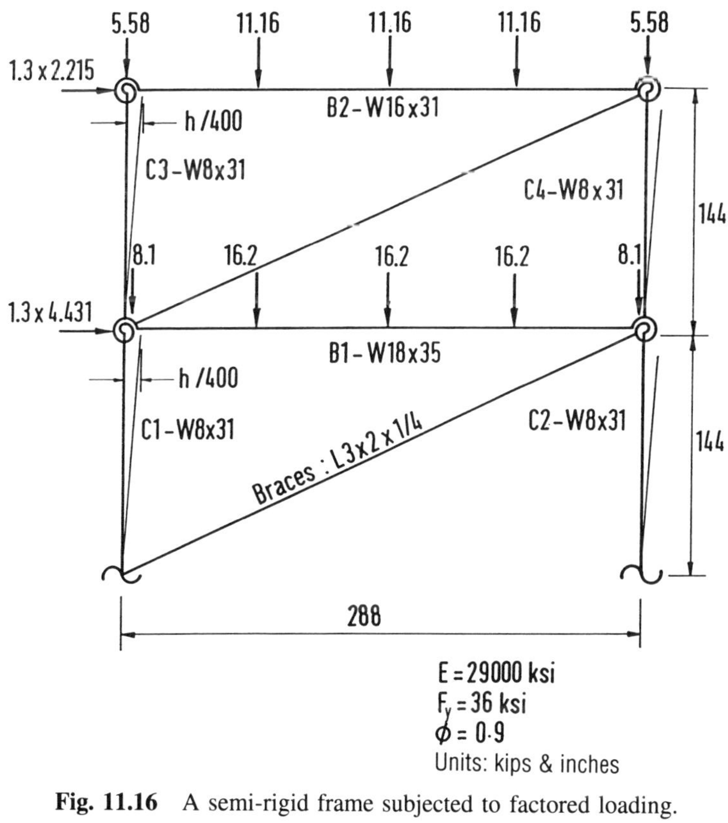

Fig. 11.16 A semi-rigid frame subjected to factored loading.

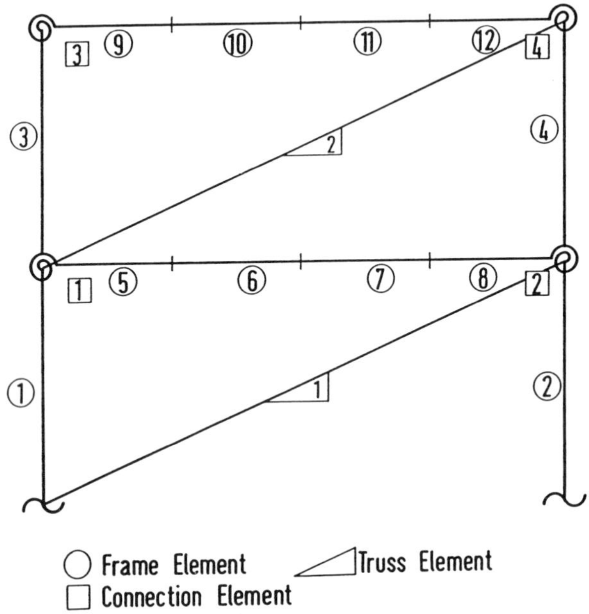

Fig. 11.17 Structural modeling of the semi-rigid frame.

The input data files for the semi-rigid frame with pinned and fixed bases are shown in Figs. 11.18 and 11.19, respectively. In the refined plastic-hinge analysis, the LRFD tangent modulus is activated by specifying IET = 2, as given in Figs. 11.18 and 11.19. However, the tangent-modulus model will not affect the ultimate strength results of this frame because the axial forces in the columns at factored loads are small (less than $0.3P_y$). The design yield strength is taken as 0.9 times the nominal yield strength of 36 ksi. The resistance factor of 0.9 for material yield strength is recommended by the Australian limit states code (Standards Australia, 1990) for use in advanced analysis of structural steel frames.

All columns are assumed to have initial out-of-plumbness of $h/400$, where h is the story height. This column out-of-plumbness magnitude may be obtained from ECCS provisions for use with second-order plastic-zone analysis (ECCS, 1991). The column initial out of straightness is not modeled explicitly in the refined plastic-hinge analysis because the axial forces in all the columns are smaller than $0.26P_e$, where P_e is the Euler buckling load.

The diagonal braces are made of angles L3 × 2 × 1/4 and are pinned at their ends by using simple bolt connections made of A325 bolts of 7/8 inch diameter. The diagonal braces are modeled using truss elements in which the

```
/* REFINED PLASTIC-HINGE ANALYSIS OF FRAME WITH SIMPLY SUPPORTED BASES
  1
  2    1
 32   80     1     2
  2    3     1
  4   12     2
  1    1773.0  954013.0        0.80
  2     814.0  205924.0        1.57
  1      9.13      110.        30.4     29000.       32.4
  2      10.3      510.        66.5     29000.       32.4
  3      9.12      375.        54.0     29000.       32.4
  1      1.19     0.392     29000.0         32.4
  1    5     1     1
  2    8     2     1
  3    9     1     2
  4   12     2     2
  1       0.0      144.    1    0    0    1    3    4    5
  2       0.0      144.    1    0    0    2   15   16   17
  3       0.0      144.    1    3    4    5   18   19   20
  4       0.0      144.    1   15   16   17   30   31   32
  5      72.0        0.    2    3    4    5    6    7    8
  6      72.0        0.    2    6    7    8    9   10   11
  7      72.0        0.    2    9   10   11   12   13   14
  8      72.0        0.    2   12   13   14   15   16   17
  9      72.0        0.    3   18   19   20   21   22   23
 10      72.0        0.    3   21   22   23   24   25   26
 11      72.0        0.    3   24   25   26   27   28   29
 12      72.0        0.    3   27   28   29   30   31   32
  1     288.0     144.0    1    0    0   15   16
  2     288.0     144.0    1    3    4   30   31
  3    0.2880    5.7603     0.2880
 18  0.143975    2.8795   0.143975
  4    -0.405      -8.1     -0.405
  7    -0.810     -16.2     -0.810
 10    -0.810     -16.2     -0.810
 13    -0.810     -16.2     -0.810
 16    -0.405      -8.1     -0.405
 19   -0.2790     --5.58   -0.2790
 22   -0.5580    -11.16    -0.5580
 25   -0.5580    -11.16    -0.5580
 28   -0.5580    -11.16    -0.5580
 31   -0.2790     -5.58    -0.2790
```

Fig. 11.18 Input data file, INFILE, for the second-order refined plastic-hinge analysis of a semi-rigid braced frame with a pinned base ("EX2" in the diskette).

net effective area of the angle section, after reduction for a single hole, is used in the analysis.

Figure 11.20 shows the load-displacement curves of the frame with two different support conditions. An applied load ratio of 1.0 corresponds to the full factored wind-load combination of 1.0 Gravity + 1.3 Wind. The analysis shows that diagonal bracing provides an effective means to control the frame drift and enhance the lateral stiffness of the frame. The increase in load-carrying capacity is more pronounced for more flexible frames.

```
/*REFINED PHINGE ANALYSIS OF A FRAME WITH FIXED BASES
 1
 2    1
30   80      2      1
 2    3      1
 4   12      2
 1    1773.0  954013.0        0.80
 2     814.0  205924.0        1.57
 1      9.13      110.        30.4      29000.         32.4
 2      10.3      510.        66.5      29000.         32.4
 3      9.12      375.        54.0      29000.         32.4
 1   0.95563      0.703    29000.0         32.4
 1    5      1      1
 2    8      2      1
 3    9      1      2
 4   12      2      2
 1     0.360        144.     1     0     0     0     1     2     3
 2     0.360        144.     1     0     0     0    13    14    15
 3     0.360        144.     1     1     2     3    16    17    18
 4     0.360        144.     1    13    14    15    28    29    30
 5      72.0          0.     2     1     2     3     4     5     6
 6      72.0          0.     2     4     5     6     7     8     9
 7      72.0          0.     2     7     8     9    10    11    12
 8      72.0          0.     2    10    11    12    13    14    15
 9      72.0          0.     3    16    17    18    19    20    21
10      72.0          0.     3    19    20    21    22    23    24
11      72.0          0.     3    22    23    24    25    26    27
12      72.0          0.     3    25    26    27    28    29    30
 1    288.36       144.0     1     0     0    13    14
 2    288.36       144.0     1     1     2    28    29
 1    0.2880        5.76    0.2880
16  0.143975      2.8795  0.143975
 2    -0.405        -8.1    -0.405
 5    -0.810       -16.2    -0.810
 8    -0.810       -16.2    -0.810
11    -0.810       -16.2    -0.810
14    -0.405        -8.1    -0.405
17   -0.2790       -5.58   -0.2790
20   -0.5580      -11.16   -0.5580
23   -0.5580      -11.16   -0.5580
26   -0.5580      -11.16   -0.5580
29   -0.2790       -5.58   -0.2790
```

Fig. 11.19 Input data file, INFILE, for the second-order refined plastic-hinge analysis of a semi-rigid braced frame with fixed bases ("EX2a" in the diskette).

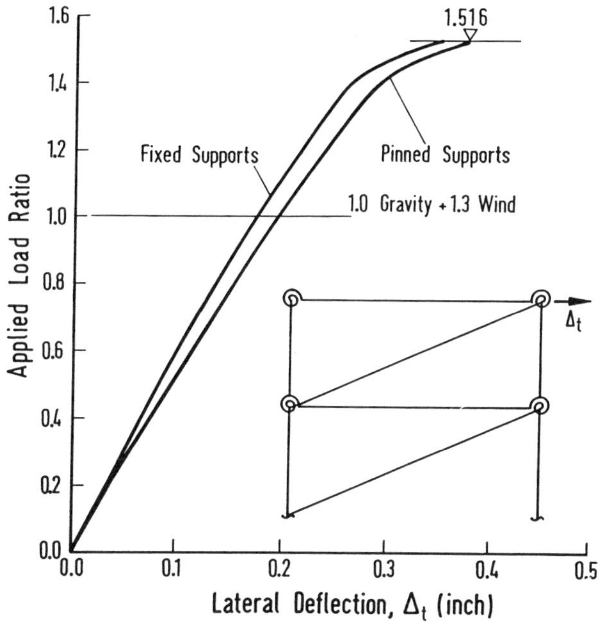

Fig. 11.20 Load-displacement traces of semi-rigid braced frames with pinned and fixed supports.

PART II: COMPUTER PROGRAM FOR CONNECTION MODELING AND CLASSIFICATION

11.8 PROGRAM "CON_ANA"

Part II describes the *connection modeling and classification* software, "CON_ANA," which can be used to generate the moment–rotation characteristics of selected types of connections for use in advanced analysis of semi-rigid frames.

The process of modeling a large number of connections in a typical steel framework is tedious, prone to error, and time-consuming. The automation of connection modeling and classification helps to reduce the time span from data preparation to final analysis. It also helps the engineer to select the appropriate type of connections and hence a suitable frame design and analysis method.

11.9 SCOPES

The program is written to calculate the ultimate moment, initial stiffness, and shape parameters of angle types of connections. It can also perform connection classification. The Kishi–Chen model is used to describe the moment–rotation curves of the connection. Connection classification is presented in a graphical form in which the moment–rotation curve of the connection is plotted against the classification system specified by the user. The program deals with the following four types of angle connections:

1. Single web-angle connection
2. Double web-angle connection
3. Top- and seat-angle connection
4. Top- and seat angle with double web cleat connection

11.10 INSTALLATION

The program "CON_ANA" can be activated from either a floppy disk or from a hard disk. The monitor can be monochrome or color. The program has been written using C language under the QuickC environment. Data entry can be made interactively via screen menus. The results can be stored in an output file, which then can be viewed on the screen or output to a printer.

To install the CON_ANA program on the hard disk or another disk, simply copy the file CON_ANA.EXE from the A drive to the C drive using the DOS command

```
C:\> copy a:\CON_ANA.EXE
```

The executable file CON_ANA.EXE will appear in the current directory of the C drive. The connection program can be executed by typing

```
C:\> CON_ANA
```

11.11 MAIN MENU

The computer program has three main options that appear in the main manual:

1. Angle-connection analysis
2. Connection classification
3. Quit

11.11.1 Angle-Connection Analysis

Under this option, there are four types of angle connections that can be analyzed:

1. Single web-angle connection
2. Double web-angle connection
3. Top- and seat-angle connection
4. Top- and seat-angle with double web-angle connection

A data input screen appears when the option number of 1 to 4 is selected. Information needed by the program to perform analysis calculation is conveniently grouped in the screen. The information requested in the screen input depends on the connection type chosen.

11.11.2 Connection Classification

This option in the main manual allows the user to classify a connection either by the Bjorhovde et al. (1990) or Eurocode 3 (1992) system. The following information is required for connection classification:

1. Plastic moment capacity of the beam about the major axis, M_{pb}
2. Beam depth, d_b
3. Modulus of elasticity of the beam and connection material, E
4. Second moment of area of the beam, I_b
5. Connection ultimate moment capacity, M_u, which is referred to as "MP" in the program.
6. Connection initial stiffness, R_{ki}, which is referred to as "Ki" in the program
7. Shape parameter (the n parameter in Chapter 9), which is referred to as "Beta" in the program
8. Length of the beam (if the Eurocode 3 system is chosen).

After the user has entered all the required data, the computer plots the nondimensional moment–rotation curve of the connection together with the three regions of classification (i.e., rigid, semi-rigid, and flexible) according to Bjorhovde et al. (1990) or Eurocode 3 (1990).

11.11.3 Quit

This final option of the main menu terminates the program and returns control to the DOS environment.

11.12 GRAPHICS SCREEN DUMP

Graphical display screens can be "dumped" to standard 9-pin printers. To dump to Laserjet and Deskjet printers it is necessary to load the GRAPHICS command. This can be done, for example, by typing from the DOS prompt:

```
c:\> Graphics Deskjet
```

This would enable screen dumps to a Deskjet printer.

Alternatively, the same command can be placed in the AUTOEXEC.BAT file such that this command is automatically loaded each time the computer is switched on. For the types of printers supported by the GRAPHICS command, consult the DOS manual.

11.13 SAVE RESULTS

The results from the analysis can be saved by using the Save the results option, which is available after the connection is analyzed. Different output files will be created for different types of angle connections. The results will be stored in the files SINGLEW.DAT, DOUBLEW.DAT, TOPSEAT.DAT, and TSDW.DAT for the single web-angle connection, the double web-angle connection, the top- and seat-angle connection, and the top- and seat-angle with double web-cleat connection, respectively. All the results in these files will be overwritten by the new results if the save option is chosen.

11.14 RESTART OR ANALYZE AGAIN

These two options are available after the analysis. They return control from the output menu to the main menu. From the main menu, you can run the program again using new data or run a different program.

11.15 EXAMPLE 3: ANALYSIS OF A TOP- AND SEAT-ANGLE DOUBLE WEB-ANGLE CONNECTION

Choose option 1 in the main menu (see Fig. 11.21) by pressing "1", and a screen as shown in Fig. 11.22 appears.

Choose option 4 by pressing "4" to activate the input screen for "top and seat with double web angles," which is shown in Fig. 11.23. The top window is the message window. Any instruction given by the computer appears here. The two middle windows are data windows. The user has to enter the appropriate values listed in items 1 to 8 in these windows. The upper middle window is for the properties of top and seat angle and the lower middle

1. Angle Connection Analysis .

2. Connection Classification .

3. Quit !

Fig. 11.21 Main menu of "CON_ANA."

window is for the properties of the web angle. The lower left-hand window displays the three parameters of the connection, namely, M_u, R_{ki}, and n. The lower right-hand window displays the plot of the nondimensional moment–rotation curve together with the connection classification boundaries (i.e., rigid, semi-rigid, and flexible). It also displays the rotation capacity value if the Bjorhovde et al. classification is chosen.

For this example, enter the values for the top and seat angles in the upper window, as shown in Fig. 11.23 (note that the units shown are in inches and kips). Press ⟨enter⟩. Press "n" to confirm that no change in data is required. The input screen for web angles appears as shown in Fig. 11.24. Enter the values as shown in this figure and press ⟨enter⟩. Follow the instruction in the upper window by pressing "n" to confirm that no change in data is required for the web-angle connections. The user is prompted to enter the values for the plastic section modulus and the second moment of area of the beam. In this case, enter the values shown in Fig. 11.25. The user will be prompted to select the option for the connection classification, as shown in Fig. 11.26. Select "B" for the Bjorhovde et al. or "E" for the Eurocode 3 approach. In this example, option "B" is selected, and the results for the three connection parameters appear in the bottom left window.

The nondimensional moment–rotation curve of the connection is shown in the bottom right window. The graph is normally too small to be visualized. The user may choose to view the complete graph in enlarged form, as shown in Fig. 11.27, by choosing option "v," which is the "(v)iew detail graph," as shown in Fig. 11.28. The example connection shown in Fig. 11.27 is classified

CONNECTION TYPES

1. SINGLE WEB ANGLE

2. DOUBLE WEB ANGLE

3. TOP AND SEAT ANGLE CONNECTION

4. TOP AND SEAT WITH DOUBLE WEB ANGLES

0. Exit

Fig. 11.22 Select connection types.

```
ANGLE PROPERTIES :   Top/Seat
1.MODULUS OF ELASTICITY             =    29000___
2.THICKNESS OF THE ANGLE              = 0.4375__
3.DIST. OF BOLT-LINE FROM EDGE        = 3_______
4.LENGTH OF THE ANGLE                 = 7_______
5.DIST. OF THE FILLET END FROM EDGE = 0.875___
6.WIDTH OF THE NUT                    = 1.4375__
7.STEEL STRENGTH                      = 36______
8.DEPTH OF THE BEAM                   = 17.7____
```

Fig. 11.23 Dimensions and material properties of top- and seat-angle connections.

```
ANGLE PROPERTIES :   Top/Seat
1.MODULUS OF ELASTICITY             =     29000.00
2.THICKNESS OF THE ANGLE              =        0.44
3.DIST. OF BOLT-LINE FROM EDGE        =        3.00
4.LENGTH OF THE ANGLE                 =        7.00
5.DIST. OF THE FILLET END FROM EDGE =        0.88
6.WIDTH OF THE NUT                    =        1.44
7.STEEL STRENGTH                      =      36
8.DEPTH OF THE BEAM                   =       17.70

ANGLE PROPERTIES :  Web
1.MODULUS OF ELASTICITY             =    29000___
2.THICKNESS OF THE ANGLE              = 0.4375__
3.DIST. OF BOLT-LINE FROM EDGE        = 3_______
4.LENGTH OF THE ANGLE                 = 7_______
5.DIST. OF THE FILLET END FROM EDGE = 0.875___
6.WIDTH OF THE NUT                    = 1.4375__
7.STEEL STRENGTH                      = 36______
```

Fig. 11.24 Dimensions and material properties of web-angle connections.

```
 Please Provide The Following Informations :
1.The Plastic Section Modulus of Beam (about x-x) =66.5____
2.The Second Moment of Area of Beam (about x-x)   =510_____

          ANGLE PROPERTIES :   Top/Seat
     1.MODULUS OF ELASTICITY             =     29000.00
     2.THICKNESS OF THE ANGLE               =       0.44
     3.DIST. OF BOLT-LINE FROM EDGE         =       3.00
     4.LENGTH OF THE ANGLE                  =       7.00
     5.DIST. OF THE FILLET END FROM EDGE =       0.88
     6.WIDTH OF THE NUT                     =       1.44
     7.STEEL STRENGTH                       =     36
     8.DEPTH OF THE BEAM                    =     17.70

          ANGLE PROPERTIES :  Web
     1.MODULUS OF ELASTICITY             =     29000.00
     2.THICKNESS OF THE ANGLE               =       0.44
     3.DIST. OF BOLT-LINE FROM EDGE         =       3.00
     4.LENGTH OF THE ANGLE                  =       7.00
     5.DIST. OF THE FILLET END FROM EDGE =       0.88
     6.WIDTH OF THE NUT                     =       1.44
     7.STEEL STRENGTH                       =     36
```

Fig. 11.25 Plastic section modulus and second moment of area of the connecting beam.

```
     CONNECTION CLASSIFICATION: (B)jorhovde etal or (E)urocode3 ?

          ANGLE PROPERTIES :   Top/Seat
     1.MODULUS OF ELASTICITY             =     29000.00
     2.THICKNESS OF THE ANGLE               =       0.44
     3.DIST. OF BOLT-LINE FROM EDGE         =       3.00
     4.LENGTH OF THE ANGLE                  =       7.00
     5.DIST. OF THE FILLET END FROM EDGE =       0.88
     6.WIDTH OF THE NUT                     =       1.44
     7.STEEL STRENGTH                       =     36
     8.DEPTH OF THE BEAM                    =     17.70

          ANGLE PROPERTIES :  Web
     1.MODULUS OF ELASTICITY             =     29000.00
     2.THICKNESS OF THE ANGLE               =       0.44
     3.DIST. OF BOLT-LINE FROM EDGE         =       3.00
     4.LENGTH OF THE ANGLE                  =       7.00
     5.DIST. OF THE FILLET END FROM EDGE =       0.88
     6.WIDTH OF THE NUT                     =       1.44
     7.STEEL STRENGTH                       =     36
```

Fig. 11.26 Choice of connection classifications.

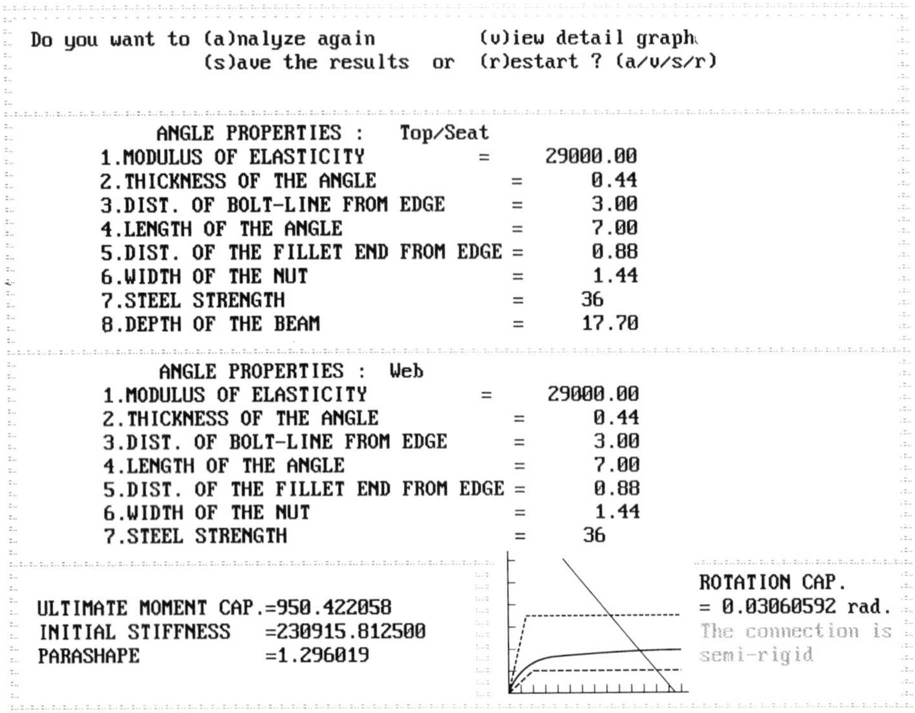

Fig. 11.27 Final analysis and results.

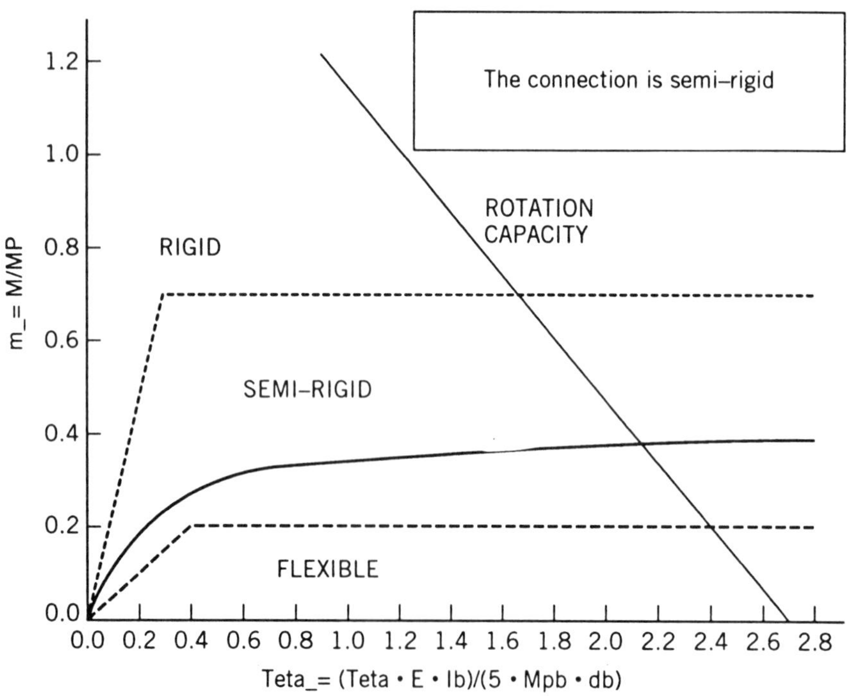

Fig. 11.28 View detail graph.

according to Bjorhovde et al. (1990). The connection is classified as semi-rigid both in terms of initial stiffness and ultimate strength.

REFERENCES

Bjorhovde, R., J. Brozzetti, and A. Colson (1990). "A Classification System for Beam to Column Connections," *Journal of Structural Engineering,* ASCE, Vol. 116, No. 11, pp. 3059–3076.

ECCS (1991). *Essentials of Eurocode 3: Design Manual for Steel Structures in Buildings,* ECCS Advisory Committee 5, No. 65, Brussels: European Convention for Constructional Steelwork.

Eurocode 3 (1992). *Design of Steel Structures: Part 1.1—General Rules and Rules for Buildings,* National Application Document for use in the UK with ENV1993-1-1: 1991, Draft for Development.

Liew, J. Y. R. (1992). "Advanced Analysis for Frame Design," Ph.D. dissertation, School of Civil Engineering, Purdue University, West Lafayette, IN.

Liew, J. Y. R., D. W. White, and W. F. Chen (1993a). "Limit-States Design of Semi-rigid Frames Using Advanced Analysis: Part 1: Connection Modelling and Classification," *Journal of Constructional Steel Research,* Vol. 26, No. 1, pp. 1–27.

Liew, J. Y. R., D. W. White, and W. F. Chen (1993b). "Limit-States Design of Semi-rigid Frames Using Advanced Analysis: Part 2: Analysis and Design," *Journal of Constructional Steel Research,* Vol. 26, No. 1, pp. 29–57.

Standards Australia (1990). *Steel Structures,* AS4100-1990. Sydney.

DISCLAIMER

The authors shall not accept any liability whatsoever for loss or damage arising from any errors that might be contained in the software, documentation, or operation of the program supplied.

It shall be the responsibility of the user (1) to check the documentation and the operation of the programs supplied, (2) to ensure that the person operating the programs is suitably qualified and experienced, and (3) to ensure that program operation is carried out in accordance with the instructions given in this chapter.

INDEX